BIOMASS
AND
BIOFUELS

BIOMASS
AND
BIOFUELS

Advanced Biorefineries for Sustainable
Production and Distribution

Shibu Jose • Thallada Bhaskar

CRC Press
Taylor & Francis Group
Boca Raton   London   New York

CRC Press is an imprint of the
Taylor & Francis Group, an **informa** business

CRC Press
Taylor & Francis Group
6000 Broken Sound Parkway NW, Suite 300
Boca Raton, FL 33487-2742

First issued in paperback 2018

© 2015 by Taylor & Francis Group, LLC
CRC Press is an imprint of Taylor & Francis Group, an Informa business

No claim to original U.S. Government works

ISBN 13: 978-1-138-89415-0 (pbk)
ISBN 13: 978-1-4665-9531-6 (hbk)

Library of Congress Cataloging-in-Publication Data

Biomass and biofuels : advanced biorefineries for sustainable production and distribution / editors, Shibu Jose and Thallada Bhaskar.
  pages cm
Includes bibliographical references and index.
ISBN 978-1-4665-9531-6 (alk. paper)
1. Biomass energy. I. Jose, Shibu. II. Bhaskar, Thallada.

TP339.B5473 2015
662'.88--dc23   2014043738

Visit the Taylor & Francis Web site at
http://www.taylorandfrancis.com

and the CRC Press Web site at
http://www.crcpress.com

Contents

Part I
Biomass Feedstock and Logistics

Part II
Conversion Processes

Thallada Bhaskar, Bhavya Balagurumurthy, Rawel Singh, and Azad Kumar

**Otavio Cavalett, Tassia L. Junqueira, Mateus F. Chagas, Lucas G. Pereira,
and Antonio Bonomi**

Muhammad A. Sokoto

Preface

The long-held tenets of the energy sector are being rewritten in the twenty-first century. Major importers are now becoming exporters, and several countries that have been long-defined as major energy exporters are on the road to becoming leading centers of global demand growth. The right combination of policies and technologies can prove that the links among economic growth, energy demand, and energy-related CO_2 emissions can be weakened. The rise of unconventional oil and gas and of renewables is transforming our economies and improving our understanding of the distribution of the world's energy resources and their impacts. A complete knowledge of the dynamics underpinning energy markets is necessary for decision-makers who are attempting to reconcile economic, energy, and environmental objectives. Those who anticipate global energy developments successfully can derive an advantage, while those who fail to do so risk making poor policy and investment decisions. The center of gravity for energy demand is moving toward emerging economies such as those of China, India, and the Middle East, as they drive global energy use one-third higher.

Contributing around two-thirds of global greenhouse gas emissions, the energy sector will determine if climate change goals are achieved or not. There are several carbon abatement schemes that have been developed in recent times—such as the President's Climate Action Plan in the United States, the Chinese plan to limit the share of coal in the domestic energy mix, the European debate on 2030 energy and climate targets, and Japan's discussions on a new energy plan—that all have the potential to limit growth in energy-related CO_2 emissions. It has been proposed that primary energy demand will increase by 41% between 2012 and 2035, with growth averaging 1.5% per year. Among nonfossil fuels, renewables (including biofuels) are seen to gain shares rapidly from around 2% currently to 7% by 2035. The level of carbon emissions would continue to grow (1.1% per annum)—slightly slower than energy consumption but faster than that recommended by the scientific community. The biggest challenge though is in terms of sustainability of biomass–biofuel production, processing, and distribution systems.

In this scenario of increasing efforts on behalf of biomass conversion to value-added hydrocarbons and energy, the editors felt the need for a book that provides a holistic view of the entire biomass–biofuel supply chain—from feedstock to the end product. Production of biofuels from first-generation feedstocks has been documented to have a net negative impact on the environment and on climate. Research findings show that production of biofuel from food crops is unsustainable in the long run because artificial shortages in food supply and subsequent impacts will destabilize the global economy. So long as there is ethanol from food crops, the markets for corn and oil will be linked and the food versus fuel debate will continue. The question we need to ask is: How can we balance our food and fuel needs without compromising one or the other? In other words, there is no question of food versus fuel; we need both. The second-generation biofuels, specifically cellulosic biofuels, have helped balance this debate somewhat and have reduced some of the negative impacts of first-generation biofuels. For example, most second-generation feedstock can be grown on marginal lands, thereby reducing pressure on prime agricultural lands. They also have smaller greenhouse gas footprints than corn-based or first-generation feedstocks. These lignocellulosic feedstocks, which could promote large-scale energy production, include crop residues, perennial grasses, biomass sorghum, and short rotation woody crops, among others. When managed properly, the high productivity of these plants and their relatively high tolerance to soil constraints make them ideal feedstocks for biofuel production.

The advanced biorefinery concept has been garnering a lot of attention in recent years as a model of decentralized production of advanced biofuels, particularly in rural areas. Such decentralized small- to medium-scale biorefineries seem to be having the greatest potential for increasing biofuel production and accelerating economic revitalization of rural communities.

These decentralized biorefineries help in effective utilization of agricultural/forest residues or energy crops in a particular area in addition to offering a number of local and regional job opportunities. This book aims to discuss the various feedstocks that can be used as raw material in biorefineries, the methods that can be used for biomass conversion, and the effective integration of biomass to make a biorefinery more sustainable—economically, environmentally, and socially.

Acquiring new scientific information and rapidly incorporating new knowledge and experiences into planning and actions are of utmost importance in the dynamics of the renewable energy sector. It is critical to provide relevant and timely information to professionals, policy makers, and the general public so that they can make informed decisions. We recognize that one book alone cannot fill this niche. However, we hope that the current volume will serve as a reference book for students, scientists, professionals, and policy makers who are involved in the biomass and biofuel sector the world over.

We are grateful to a large number of individuals for assistance in accomplishing this task, particularly the authors for their commitment to the project and their original research or synthesis of the current knowledge. Also, the invaluable comments and suggestions made by the referees significantly improved the clarity and content of the chapters. We also wish to extend our sincere thanks to John Sulzycki and Jill Jurgensen of CRC Press for their timely efforts in publishing this book.

Editors

Dr. Shibu Jose is the Harold E. Garrett Endowed Chair Professor and director of the Center for Agroforestry at the University of Missouri, Columbia, Missouri. Prior to his current appointment, he was professor of forest ecology at the School of Forest Resources and Conservation at the University of Florida, Gainesville, Florida. He earned his BS from Kerala Agricultural University, India, and MS and PhD from Purdue University, West Lafayette, Indiana. His current research efforts focus on ecosystem services of agroforestry systems and ecological sustainability of biomass and biofuel production systems. Dr. Jose leads a regional consortium focused on commercializing integrated biomass and biofuels production systems, the Mississippi/Missouri River Advanced Biomass/Biofuel Consortium (MRABC). Dr. Jose has authored over 150 refereed publications, edited 8 books, and secured nearly $39 million in funding (~$10 million as PI). He has also served as major professor for 14 PhD students and 24 MS students during the past 17 years. He serves as editor-in-chief of *Agroforestry Systems*, academic editor of *PLOS ONE*, and editorial board member of the *International Journal of Ecology*. His awards and honors include a Fulbright Fellowship (U.S. Department of State), Aga Khan International Fellowship (Switzerland), Nehru Memorial Award for Scholastic Excellence (India), Award of Excellence in Research by the Southeastern Society of American Foresters (SAF), Stephen Spurr Award by the Florida Division SAF, the Young Forester Leadership Award by the National SAF, Barrington Moore Award by SAF, and the Scientific Achievement Award by the International Union of Forest Research Organizations (IUFRO).

Dr. Thallada Bhaskar, a senior scientist, is currently heading the Thermo-Catalytic Processes Area, Bio-Fuels Division (BFD), at CSIR–Indian Institute of Petroleum, Dehradun, India. He earned a PhD for his work at CSIR–Indian Institute of Chemical Technology (IICT), Hyderabad, India. He carried out postdoctoral research at Okayama University, Okayama, Japan, after which he joined there as faculty and has worked at the level of research assistant professor for ~7 years. Dr. Bhaskar has about 100 publications in journals of international repute with an *h-index* of 27 and around 2000 citations, contributed 16 book chapters to renowned publishers (Elsevier, ACS, John Wiley, Woodhead Publishing, CRC Press, Asiatech, etc.), and 13 patents to his name in his field of expertise, in addition to 250 national and international symposia presentations. His 20 years of research experience cover various fields of science revolving around his expertise in heterogeneous catalysis and thermochemical conversion of biomass, waste plastics, and e-waste (WEEE) plastics into value-added hydrocarbons. He has prepared several catalysts and thrown light on the structure–activity relationships of novel catalytic materials for hydrotreatment of fossil-based crudes. In view of his expertise, he is on the editorial board of three international peer-reviewed journals and editor for two books from Elsevier and CRC. Dr. Bhaskar received the Distinguished Researcher award from The National Institute of Advanced Industrial Science and Technology (AIST) (2013), Japan, and the Most Progressive Researcher award from Research Association for Feedstock Recycling of Plastics (FSRJ), Japan (2008). He is also the Fellow of the Biotech Research Society of India (FBRS)

and a member of the Board of Directors (BRSI). He received the Raman Research Fellowship for the year 2013–2014. He was also a JSPS Visiting Scientist to the Tokyo Institute of Technology, Japan, during 2009. He has organized several international symposia in India and abroad in this area of expertise and has visited several countries to deliver invited/plenary lectures, for collaborative research and also as a subject expert.

Contributors

Janaki Alavalapati
Department of Forest Resources
  and Environmental Conservation
Virginia Polytechnic Institute and State
  University
Blacksburg, Virginia

Kent G. Apostol
Department of Forest, Rangeland, and
  Fire Sciences
College of Natural Resources
University of Idaho
Moscow, Idaho

Bhavya Balagurumurthy
Bio-Fuels Division
CSIR–Indian Institute of Petroleum
Dehradun, India

Sougata Bardhan
The Center for Agroforestry
University of Missouri
Columbia, Missouri

William E. Berguson
Natural Resources Research Institute
University of Minnesota Duluth
Duluth, Minnesota

Robert F. Berry
European Bioenergy Research Institute
School of Engineering and Applied Science
Aston University
Birmingham, UK

Thallada Bhaskar
Bio-Fuels Division
CSIR–Indian Institute of Petroleum
Dehradun, India

Antonio Bonomi
Brazilian Bioethanol Science and Technology
  Laboratory (CTBE)
Brazilian Center of Research in Energy and
  Materials (CNPEM)
Campinas, São Paulo, Brazil

Pralhad Burli
Department of Earth and Environmental
  Studies
Montclair State University
Montclair, New Jersey

Otavio Cavalett
Brazilian Bioethanol Science and Technology
  Laboratory (CTBE)
Brazilian Center of Research in Energy and
  Materials (CNPEM)
Campinas, São Paulo, Brazil

Mateus F. Chagas
Brazilian Bioethanol Science and Technology
  Laboratory (CTBE)
Brazilian Center of Research in Energy and
  Materials (CNPEM)
Campinas, São Paulo, Brazil

Rashmi Chandra
Bioengineering and Environmental Centre
CSIR–Indian Institute of Chemical Technology
Hyderabad, India

Surinder Chopra
Plant Science Department
Pennsylvania State University
University Park, Pennsylvania

Patrick Dube
Department of Agricultural and Biological
  Engineering
University of Florida
Gainesville, Florida

Puneet Dwivedi
Daniel B. Warnell School of Forestry
  and Natural Resources
University of Georgia
Athens, Georgia

Sandra D. Ekşioğlu
Department of Industrial Engineering
Clemson University
Clemson, South Carolina

Larry Godsey
The Center for Agroforestry
University of Missouri
Columbia, Missouri

Kannaiah R. Goud
Bioengineering and Environmental Centre
CSIR–Indian Institute of Chemical Technology
Hyderabad, India

Ignacio E. Grossmann
Department of Chemical Engineering
Carnegie Mellon University
Pittsburgh, Pennsylvania

Shibu Jose
The Center for Agroforestry
University of Missouri
Columbia, Missouri

Tassia L. Junqueira
Brazilian Bioethanol Science and Technology
  Laboratory (CTBE)
Brazilian Center of Research in Energy and
  Materials (CNPEM)
Campinas, São Paulo, Brazil

Cerry M. Klein
Department of Industrial and Manufacturing
  Systems Engineering
University of Missouri
Columbia, Missouri

Arvind Kumar
AcSIR
and
Salt and Marine Chemicals Division
CSIR–Central Salt and Marine Chemicals
  Research Institute
Bhavnagar, India

Azad Kumar
Bio-Fuels Division
CSIR–Indian Institute of Petroleum
Dehradun, India

Pankaj Lal
Department of Earth and Environmental
  Studies
Montclair State University
Montclair, New Jersey

Rasika B. Mane
Chemical Engineering and Process
  Development Division
CSIR–National Chemical Laboratory
Pune, India

Mariano Martín
Departamento de Ingeniería Química
Universidad de Salamanca
Salamanca, Spain

S. Venkata Mohan
Bioengineering and Environmental Centre
CSIR–Indian Institute of Chemical
  Technology
Hyderabad, India

Eric J. Ogdahl
Natural Resource Science and Management
University of Minnesota
St. Paul, Minnesota

Priyanka Ohri
Bio-Fuels Division
CSIR–Indian Institute of Petroleum
Dehradun, India

Amit R. Patel
School of Mechanical, Materials and
  Energy Engineering
Indian Institute of Technology Ropar
Rupnagar, India

Lucas G. Pereira
Brazilian Bioethanol Science and
  Technology Laboratory (CTBE)
Brazilian Center of Research in Energy and
  Materials (CNPEM)
Campinas, São Paulo, Brazil

Reddy Shetty Prakasham
CSIR–Indian Institute of Chemical
  Technology
Hyderabad, India

Pratap Pullammanappallil
Department of Agricultural and Biological
  Engineering
University of Florida
Gainesville, Florida

P. Parthasarathy Rao
International Crops Research Institute for
  the Semi-Arid Tropics (ICRISAT)
Hyderabad, India

P. Srinivasa Rao
International Crops Research Institute for
  the Semi-Arid Tropics (ICRISAT)
Hyderabad, India

Chandrashekhar V. Rode
Chemical Engineering and Process
  Development Division
CSIR–National Chemical Laboratory
Pune, India

M. V. Rohit
Bioengineering and Environmental Centre
CSIR–Indian Institute of Chemical Technology
Hyderabad, India

Sudhakar Sagi
European Bioenergy Research Institute
School of Engineering and Applied Science
Aston University
Birmingham, UK

Osamu Sato
Research Center for Compact
  Chemical System
National Institute of Advanced Industrial
  Science and Technology (AIST)
Sendai, Japan

Masayuki Shirai
Department of Chemistry and Bioengineering
Faculty of Engineering
IWATE University
Morioka, Japan

Harpreet Singh
School of Mechanical, Materials and Energy
  Engineering
Indian Institute of Technology Ropar
Rupnagar, India

Rawel Singh
Bio-Fuels Division
CSIR–Indian Institute of Petroleum
Dehradun, India

Muhammad A. Sokoto
Department of Pure and Applied Chemistry
Usmanu Danfodiyo University
Sokoto, Nigeria

Tushar J. Trivedi
AcSIR, CSIR–Central Salt and Marine
  Chemicals Research Institute
Bhavnagar, India

Timothy A. Volk
College of Environmental Science and Forestry
State University of New York
Syracuse, New York

Carol L. Williams
Department of Agronomy
University of Wisconsin–Madison
Madison, Wisconsin

Jeff Wright
Hardwood Development
ArborGen Inc.
Ridgeville, South Carolina

Aritomo Yamaguchi
Research Center for Compact Chemical
  System
National Institute of Advanced Industrial
  Science and Technology (AIST)
and
Japan Science and Technology Agency (JST),
  PRESTO
Sendai, Japan

Diomides S. Zamora
Department of Forest Resources
University of Minnesota Extension
University of Minnesota

PART I

Biomass Feedstock and Logistics

CHAPTER 1

Biomass Feedstocks
Types, Sources, Availability, Production, and Sustainability

Carol L. Williams
Department of Agronomy, University of Wisconsin–Madison, WI, USA

CONTENTS

1.1 PURPOSE

This chapter provides a comprehensive overview of biomass feedstocks used in production of bioenergy. Particular emphasis is given to biomass feedstocks associated with advanced biofuels derived through biological conversion technologies. A broad geographical scope is presented. The discussion is limited to feedstocks currently considered feasible, or plausibly so, in the near future, resulting from current momentum in technology, policy, and economic change. The reader will gain from this chapter an appreciation of the scope of biomass supply and related issues and will be able to apply the lens of biomass supply while reading other chapters of this book.

1.2 INTRODUCTION

As world demand for energy increases, energy providers are turning to new technologies and new sources of renewable energy. *Biomass*—recently living biological material and animal wastes—has been used since early history to cook and heat spaces where humans live and labor. Since the eighteenth century, biomass has been used to provide heat, steam, and power for work processes. Today, biomass has an expanded role in the global demand for energy. *Bioenergy*—energy produced from biomass—is a promising solution to environmental challenges and a driver of economic development from local to global levels (Coleman and Stanturf 2006; Kleinschmidt 2007).

To meet bioenergy demand, energy providers must continuously secure a sufficient and reliable supply of biomass at prices allowing them to operate profitably. As global attention increasingly focuses on sustainability of resource use, biomass producers must balance market pressures for an ever-increasing supply at low prices with demands for nonmarket benefits of sustainable production systems, such as soil conservation, water quality protection, and biodiversity enhancement. Hence, policymakers and researchers are seeking to innovate solutions that will reduce the potential trade-offs between economic development and resource conservation, and competition for land resources.

Supply and sustainability needs are driving scientific and business interests in new and improved sources of biomass feedstocks. The quest for ideal biomass feedstocks includes exploration for new biomass types, sources, and production systems, as well as improvements to existing sources and production systems. This quest generally follows in tandem with innovation in production and harvest equipment and with breakthroughs in biomass pretreatments and conversion technologies (Ragauskas et al. 2006). Universities, government agencies, and public–private partnerships provide centers of innovation and information for these discoveries (e.g., Brazilian Centers for Excellence in Bioenergy Research and Development, the European Bioenergy Research Institute, and the Great Lakes Bioenergy Research Center at the University of Wisconsin, USA).

To understand the challenges of bioenergy production, it is first necessary to know and understand the various forms of bioenergy and the types of biomass materials and sources from which bioenergy is produced. Although the primary focus of this chapter is plant sources of biomass, animal waste is also briefly considered. At the conclusion of this chapter, readers should be able to answer the following questions: What is the difference between "biomass" and "feedstock"? What are the key differences between woody and nonwoody biomass? What are the primary sources for woody biomass and nonwoody biomass? Which agricultural crops are currently feedstock sources? What are the emerging feedstock sources for the future? What are the advantages and disadvantages associated with different types of biomass? What sorts of trade-offs do decision-makers face as they attempt to optimize biomass production?

1.3 FORMS OF BIOENERGY

Bioenergy is a form of renewable energy. Forms of bioenergy include power, heat, and solid, liquid, and gas fuels. Uses of these various forms of bioenergy include industrial, residential, and commercial applications. *Biofuel* refers to liquid and gas fuels used for transportation and industrial processes. Liquid and gas biofuels are produced through fermentation, gasification, pyrolysis, and torrefaction. (For more information about conversion technologies, see Faaij 2006.) Bioethanol and biodiesel are major forms of biofuel. Biofuels produced from oils, sugars, and starches originating in food crops are known as *first-generation biofuels*. First-generation biofuels are produced through relatively simple and established technologies. Conversion technologies still under development allow the creation of *second-generation biofuels*, also known as *advanced biofuels*, which are produced from nonfood crops such as perennial grasses and woody materials and from nonfood portions of food crops. *Third-generation biofuels* are produced from algae (Goh and Lee 2010; Lee and Lavoie 2013). The form of bioenergy and conversion technology determines the quantity and quality of biomass needed.

1.4 FEEDSTOCK MATERIALS

Bioenergy *feedstocks* are biomass-derived materials that are converted to energy through the application of microbial activity, heat, chemicals, or through a combination of these processes. Feedstocks are biomass materials that have been at least minimally processed to be ready for conversion into bioenergy. That is, biomass does not usually exist in a form that can be converted directly into energy without some alteration. Combustion of fuelwood for household use is an exception. It is frequently necessary to process biomass into a form that is more economical to transport from where it is grown to where it is converted into energy. Specifically, bulk density of biomass is relatively low (McKendry 2002). Bulk density is the weight of biomass per volume of biomass. Low bulk density of biomass means it takes up space in transport vehicles that are otherwise equipped to handle heavier loads, and that means more hauling trips are necessary compared to materials with greater bulk density. A first step then is *aggregation*—the process of gathering up harvested biomass into easily handled units such as bales (Figure 1.1). Low bulk density also translates into low energy density, particularly compared to other sources of energy such as coal. Processing of biomass to reduce moisture and increase bulk density; for example, *densification*—the application of pressure and other processes to create solid fuel (Tumuluru et al. 2010)—increases energy density. Increased energy density improves conversion efficiency and therefore reduces costs associated with conversion (Stephen et al. 2010). *Pelletization*—densification into pellets—is a common method of increasing bulk density that improves storage, logistical, and transport characteristics of biomass (Figure 1.2).

As biomass moves from where it was grown to where it is converted, it passes through various stages of processing and handling. Each stage adds value to what started as a relatively low-value material. The sequence of processes is known as a supply chain, sometimes also called a value chain. Because of low bulk density and subsequent low energy density, optimal conversion facility size is frequently dependent on biomass haulage costs, and feedstock supply/value chains trend toward smaller, more distributed, and more localized facilities (Jack 2009; Searcy et al. 2007).

There are three main types of biomass materials from which bioenergy feedstocks are derived: lipids, sugars/starches, and cellulose/lignocellulose. Lipids are energy-rich water-insoluble molecules such as fats, oils, and waxes. Lipids are a feedstock source derived from nonwoody plants and algae. Soyabean (*Glycine max*), oil palm (*Elaeis guineensis* and *Elaeis oleifera*), and various seed crops such as sunflower (*Helianthus annuus*) are common agricultural sources of oils for biodiesel.

Figure 1.1 Baling of biomass is a form of aggregation that makes handling more efficient. (Courtesy of C. L. Williams, 2012.)

Figure 1.2 Pelletized biomass. Cattail (*Typha* spp.) biomass has been milled and pressed into a relatively high density solid fuel. (Courtesy of C. L. Williams, 2013.)

Sugars and starches are carbohydrates typically found in the edible portions of food crops, such as corn (*Zea mays*) grain. Cellulosic/lignocellulosic biomass is composed of complex carbohydrates and noncarbohydrate molecules typically found in the leaves and stems of plants. Cellulose/ lignocellulose is chemically accessible by only a narrow range of organisms and is therefore of little or no food value to humans. Advanced biofuels offer an opportunity to take these relatively low

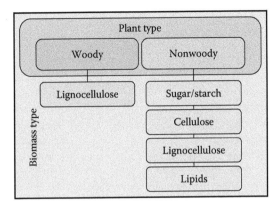

Figure 1.3 Woody and nonwoody plants are the sources of four different types of plant materials: cellulose and lignocellulose (both from plant cell walls), noncellulosic carbohydrates sugar and starch, and fatty acids (oils, fats, and waxes).

value materials and use them in the production of high value energy products (Clark et al. 2006). Thus, the remainder of this section further details the nature of cellulosic/lignocellulosic biomass.

There are two broad categories of plants from which cellulosic/lignocellulosic feedstocks are derived: woody and nonwoody (Figure 1.3). Cellulose is a fibrous glucose polymer found in plant cell walls. Cellulose provides physical strength to plant cells. Cellulose can be broken down into simple sugars, which can then be converted into ethanol and other fuels, typically through biological conversion (i.e., fermentation). In addition to cellulose, many plants also contain hemicellulose and lignin. Hemicellulose is a heteropolymer (i.e., a very large, complex carbohydrate molecule) that helps cross-link cellulose fibers in plant cell walls. Lignin is a noncarbohydrate polymer that fills the spaces between cellulose and hemicellulose. When cellulose, hemicellulose, and lignin are present together they are referred to as *lignocellulose*. Trees, for example, contain high amounts of lignocellulose; nonwoody plants, such as grasses, typically contain more cellulose than hemicellulose and lignin. Hemicellulose can be broken down into fermentable sugar and then converted into ethanol and other fuels. Lignin is difficult to convert into other usable forms and is therefore considered a by-product (i.e., waste) that is sometimes burned for heat energy (Hahn-Hagerdal et al. 2006). As technologies for transforming lignin improve, new markets for its use may emerge. In which case, lignin could become a higher-value co-product of biorefining (Hahn-Hagerdal et al. 2006).

In biological conversion of lignocellulosic feedstocks, *pretreatment* is required. Pretreatments break down cellulose and hemicellulose into sugars and separate lignin and other plant constituents from fermentable materials. The form of pretreatment will depend on the nature of the feedstock. Pretreatment technologies are physical, biological, and combinatorial. Physical pretreatment includes gamma ray exposure; chemical pretreatment methods include the use of acids, alkali, and ionic liquids; and biological methods include the use of microorganisms to degrade lignin and hemicellulose (Zheng et al. 2009). For more information on pretreatment technologies and interrelated developments in agronomic qualities of bioenergy crops, see Coulman et al. (2013) and Sticklen (2006).

1.5 BIOMASS SOURCES AND TYPES

Biomass for bioenergy comes from a variety of sources. Forests, agriculture, and wastes are currently the world's major sources of biomass (Figure 1.4). However, alternative sources such as agroforestry, conservation lands, and algae may grow in importance as demand for bioenergy grows. Forests are the primary source of woody biomass (Figure 1.4). Agriculture and

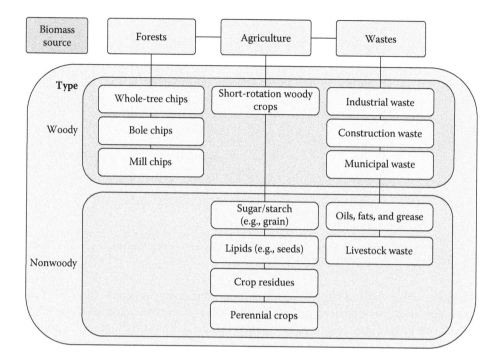

Figure 1.4 Biomass sources and types. Most biomass for biofuels comes from three sources: forests, agriculture, and wastes. Each source provides different types of woody and nonwoody biomass.

waste sources provide both woody and nonwoody biomass (Figure 1.4). Each of these sources has limitations of biomass availability and quality and has issues of accessibility. Additionally, these sources typically have competing uses for their biomass, which may affect price as well as availability (Suntana et al. 2009).

1.5.1 Forest-Based Feedstocks

Woody biomass from forests is the original source of bioenergy (Demirbas 2004). It remains the most important source of biofuel for cooking and space heating throughout the world, particularly among subsistence cultures (Cooke et al. 2008). Few extended rotation forests (i.e., growth harvest cycles of decades), whether public or private, are or will likely be managed specifically to provide biomass for bioenergy (Hedenus and Azar 2009). Instead, biomass for bioenergy is typically a coproduct of forest management activities (e.g., fuel hazard removal) or commercial activities emphasizing higher value materials such as merchantable wood. However, fast-growing tree species are sometimes purpose grown for bioenergy (White 2010). Short-rotation plantations typically receive more intensive management than timber plantations, and rotations can range from 3 to 12 years depending on management and on the species cultivated. Because of these characteristics, and the limited acreage on which they usually occur, short-rotation plantations are frequently considered as agroforestry rather than as forestry production.

In general, only wood that is not merchantable as lumber or pulp is used in bioenergy production (Figure 1.4). There are two main ways low grade wood is removed from forests for bioenergy use: as bark and as wood chips. Bark is typically burned to fire wood kilns at mills, or it is sold in higher value markets such as for landscaping materials. Although bark has a high energy density (more than wood chips), it has high silica and potassium content that affect its quality as a feedstock (Lehtikangas 2001). Woodchips, however, can be used directly as a solid fuel (for combustion) or they can be refined and densified into pellets. There are three main types of woodchips: mill chips, whole-tree chips, and bole chips (Figure 1.4). Mill chips are produced from waste wood (off-cuts and slabs from sawing logs into lumber). Because logs are debarked before sawing, mill chips are usually very clean. Whole-tree chips originate from managed forests with little commercial value for lumber and where removal of trees could improve future commercial timber value. Whole-tree chips are produced by either chipping the entire low-grade tree or from only the tops and limbs severed from logs. Although a majority of whole-tree chips are generated from forest management activity, they are also produced from land clearing and land-use conversion projects making way for roads, parking lots, buildings, and open spaces, for example. The felled trees are typically chipped on site. Bole chips are produced from low-grade or pulp logs usually from managed forests. The difference between whole-tree chips and bole chips is that bole chips do not include branches or foliage.

1.5.2 Agriculture-Based Feedstocks

Agriculture is a source of sugars, starches, lipids, nonwoody cellulosic materials, and woody materials (i.e., lignocellulosic biomass; Figure 1.4). Agriculture-based biomass comes from crops grown specifically for bioenergy production, or *dedicated bioenergy crops*, as well as agricultural *residues*. Agricultural residues are nonedible cellulosic materials that remain after harvest of edible portions of crops. Dedicated bioenergy crops include annual crops grown for their sugars, starches, or oils, and perennial herbaceous nonfood plants grown for their cellulose. Agricultural residues include plant leaves and stems. Some annual crops, such as corn, can be dedicated bioenergy crops for both their grain and their cellulosic residues.

Most of the world's first-generation bioethanol is made from feedstocks derived from annual food crops. Annual row crops are grown and harvested in a single year and must be planted every year. Sugarcane (*Saccharum officinarum*) and corn are the primary feedstock sources for first-generation bioethanol. However, bioethanol is also produced from cereal crops, sugar beets (*Beta vulgaris*), potatoes (*Solanum tuberosum*), sorghum (*Sorghum bicolor*), and cassava (*Manihot esculenta*) as well. Sugarcane is the primary feedstock in Brazil, and corn grain is the primary feedstock in the United States. These two feedstock sources are converted into approximately 62% of the world's bioethanol (Kim and Dale 2004).

The primary agricultural sources of lipids for first-generation biodiesel are annual row crops—soybean, palm, and oilseed rape (or rapeseed, *Brassica napus*). Soybean is the primary feedstock source for biodiesel produced in the United States, Europe, Brazil, and Argentina—the world leaders in biodiesel production (Bergmann et al. 2013). Palm, a tropical plant, is the primary feedstock source in Southeast Asia (e.g., Malaysia and Indonesia), while oilseed rape is grown in Europe, Canada, the United States, Australia, China, and India (Rosillo-Calle et al. 2009). Inedible oil crops are being examined for commercial potential in second-generation biodiesel production, including castor (*Ricinus communis*) and Camelina (*Camelina sativa*; Atabani et al. 2012). For more information on biodiesel feedstocks and production technologies, see Salvi and Panwar (2012).

Perennial crops are the primary sources of lignocellulosic biomass for second-generation biofuels. They have received considerable attention because they are not food crops, and they provide long-term yield potential and environmental benefits not usually achieved in annual row crop agriculture (Sanderson and Adler 2008, and references therein). These potential environmental benefits include wildlife habitat, soil erosion prevention, and water quality improvement. Perennial crops live for more than one growing season and do not have to be planted every year. Perennial crops include *herbaceous plants* (plants lacking permanent woody stems) and woody plants. Perennial grasses in particular are of considerable value in advanced biofuels as are fast-growing trees such as hybrid poplars (*Populus* spp.) and willows (*Salix* spp.). Whether they are herbaceous or woody, perennial dedicated bioenergy crops are typically grown with some amount of agronomic intensity (e.g., inputs of fertilizer and pesticides), which is why they are considered as crops.

Agricultural residues also are an important source of cellulosic feedstocks (Figure 1.4). Their use potentially limits the impacts of biofuels on food security (Kim and Dale 2004). Global residue biomass is estimated as 3.8 billion Mg yr^{-1} (Lal 2005). Availability of residues differs by region, country, and within countries according to climate and soil variations affecting the growth suitability of particular crops. For example, rice straw is readily available in Asia, and stover (corn residue) is available in the United States, Mexico, and Europe (Kim and Dale 2004). The amount of residue available differs widely among crops (Lal 2005; see further discussion in Section 1.6). Use of agricultural residues must be carefully planned and managed due to their important role in soil erosion control and maintenance of soil quality, and their use as forage, fodder, and bedding for livestock (Lal 2005).

1.5.3 Waste-Based Feedstocks

Waste-based biomass includes organic materials left over from industrial processes, agricultural liquid and solid wastes (e.g., manure), municipal solid wastes, and construction wastes (Figure 1.4). Many industrial processes and manufacturing operations produce residues, wastes, or coproducts that can be potentially used for bioenergy. Major sources of nonwoody wastes include waste paper, liquid left over from paper production (called black liquor), and textile manufacturing. Major sources of woody waste materials include used pallets, sawmill by-products such as sawdust and shavings, cut-offs from furniture manufacturing, and composite wood products containing non-wood resins, adhesives, and/or fillers. Conversion technologies for these wastes are potentially the same as for virgin wood (Antizar-Ladislao and Turrion-Gomez 2008).

Agricultural wastes include by-products of agro-industrial processes and manure from livestock. Agro-industrial processes such as animal processing, grain milling, starch production, and sugar production result in by-products that may be used as bioenergy feedstocks. Bagasse, the fibrous material left over from sugarcane and sorghum crushing in sugar production, for example, is sometimes used as a fuel source for heat in sugar mills but it can also be converted to bioethanol (Botha and Blottnitz 2006). Animal processing generates large quantities of feathers, bones, and other materials. These animal by-products are a potential source of diseases that have public and/or animal health risks (e.g., bovine spongiform encephalopathy), and rigorous protocols must be followed to eliminate the possibility of spread of disease. Accordingly, animal by-products are used as feedstocks in anaerobic digesters that kill potential pathogens and that produce biogas (i.e., methane). Biogas is a substitute for propane, kerosene, and firewood, and it is used to produce heat and electric power. It can also be compressed and liquefied for use as a transportation fuel.

Manure can be used as a fertilizer on agricultural fields, and land application is often an important component of on-farm nutrient management (Binford 2005). However, manure application can be a highly regulated agricultural activity (e.g., Concentrated Animal Feeding Operations), and its disposal can present challenges to farm profitability (Centner and Newton 2008; Keplinger and Hauck 2006). In some circumstances, manure cannot be applied directly to fields because the ground is frozen, or the amount of manure available exceeds the amount that can be put onto fields without endangering nearby water resources with contamination (Funk et al. 2014). Use of manure as a bioenergy feedstock, then, is an opportunity for turning a potentially expensive liability into a benefit. Livestock manure is converted into biogas via anaerobic digestion.

Municipal solid waste is a major source of biomass. Also called trash and urban solid waste, municipal solid waste is predominantly household or domestic waste. Municipal solid waste includes biodegradable waste such as kitchen food waste and food packaging; clothing and toys; recyclable materials such as paper, plastics, and metals; appliances and furniture; and debris. Most municipal solid waste is diverted to landfills, but in some locations, it is incinerated to make electricity. Portions that are not incinerated can be converted to syngas through gasification. Syngas can be cofired in boilers with coal, for example, to produce electricity.

Construction waste consists of wood, plastic, and metal debris. Although plastic and metal may be used in combustion for production of power, for example, only woody construction wastes are feedstocks for *bio*energy. Construction waste varies greatly in composition and by location. Currently, the primary conversion technology for construction waste is combustion for heat, steam, and biopower; although as lignocellulosic material, it can potentially be used in biological and other conversion technologies for biofuels (Antizar-Ladislao and Turrion-Gomez 2008).

1.5.4 Agroforestry Feedstocks

Agroforestry is the intentional integration of perennial nonfood crops with annual food crops on a farm. This may occur as *alley cropping*—the planting of trees or shrubs in rows of wide spacing that allow for the planting of crops in between rows of woody crops (Holzmueller and Jose 2012). Alternatively, fast-growing, intensively managed woody crops may be grown in monoculture as part of a diverse farm enterprise (Dickmann 2006). Regardless of the production system, agroforestry is an emerging source of lignocellulosic feedstocks for second-generation biofuels. Short-rotation woody crops (SRWC) typically grow to harvestable size in less than 15 years; depending on species and management this could be as soon as three years (Volk et al. 2004). Globally, *Eucalyptus* is the most extensively planted species, although other hardwoods predominate in temperate regions—such as Europe (Rockwood et al. 2008). In temperate regions, SRWC include hybrid poplars (*Populus* spp.), willows (*Salix* spp.), and maples (*Acer* spp.). Most SRWC are shade-intolerant, which makes them suitable to the openness of farm fields. Many temperate SRWC have the ability to coppice (sprout new growth from stumps) when harvested. Hence, coppicing will produce harvestable biomass on shorter rotations.

A potentially important source of whole plant biomass for biorefineries is jatropha (*Jatropha curcas*), an oil-bearing tree that can be grown in agroforestry systems (Achten et al. 2007). Jatropha is native to Mexico, Central America, and parts of South America. It is drought-resistant, easily propagated, and performs well in a wide variety of soils, including degraded lands. Jatropha contains inedible oil and is toxic to humans and animals. The biodiesel production with Jatropha results in valuable by-products such as seed cake and husks. These characteristics make it an attractive biorefinery feedstock candidate. India, in particular, has set ambitious goals for establishing Jatropha on degraded lands in rural areas to replace diesel used in transportation (Achten et al. 2010). Challenges remain in the commercial-scale use of Jatropha, however. Among them are the issue of relatively low yields on wastelands and agriculturally marginal lands (i.e., lower revenues), relatively high costs of establishment, efficiency of harvesting, and logistics (Achten et al. 2010; Francis et al. 2005).

1.5.5 Biomass from Conservation Lands

To avoid potential competition with production of food and forage on prime agricultural lands, government authorities and researchers are considering the potential benefits and risks associated with periodic harvest of biomass from conservation lands, such as those set aside in agriculturally dominated landscapes for purposes of soil conservation, water quality improvement, wildlife habitat, hunting access, or other nonagricultural purposes (Adler et al. 2009; Fargione et al. 2009; Rosch et al. 2009). Conservation lands, whether privately or publically owned, typically require management for maintaining cover types and various conservation goals. Harvest may be a viable form of management (Figure 1.5). Biomass resulting from habitat management actions has the potential to be used in a variety of conversion technologies. Land managers, therefore, may be able to offset management costs with the sale of biomass resulting from periodic management actions. However, the long-term impacts of removal of biomass from conservation lands, such as nutrient loss and soil compaction, are currently unknown. Hence, greater scrutiny is necessary

Figure 1.5 **(See color insert.)** Harvest of grassland biomass for habitat management on public conservation land in Wisconsin. (Courtesy of C. L. Williams, 2012.)

to understand the impacts of biomass harvest on lands set aside for wildlife and other resource management goals, and ultimately on their contribution to world energy needs.

1.5.6 Algae

Challenges in meeting the demands for bioenergy include competition for land, water, and other resources needed to produce plant-based feedstocks (Dale et al. 2011). Algae are promising sources of feedstocks for advanced biofuels because they do not compete for additional land use, and they have minimal water requirements compared to land crops (Dismukes et al. 2008). Micro- and macroalgae are thus being explored as commercially viable feedstocks for *third-generation biofuels*. Microalgae are unicellular and simple multi-cellular organisms, including prokaryotic microalgae and eukaryotic microalgae. Macroalgae are macroscopic, multicellular marine algae. Algae are sources of lipids and carbohydrates for biofuels. However, they are most frequently used as lipid sources for second-generation biodiesel. Carbohydrates can be recovered after oil extraction and fermented into bioethanol. Algae are capable of year-round production; therefore, their yield can exceed that of oilseed crops (Brennan and Owende 2010). For a comprehensive review of microalgae cultivation for biofuels, see Brennan and Owende (2010), John et al. (2011), and Mata et al. (2010).

1.6 BIOMASS SUPPLY AND AVAILABILITY

A chief question regarding the potential for bioenergy to provision a growing world population is the size of the global biomass supply and the amount of energy available within biomass. Many studies have been conducted to answer this question at the global level—with widely varying results (Beringer et al. 2011; Berndes et al. 2003; Hoogwijk et al. 2003; Tilman et al. 2006). At issue are the various assumptions necessary to model and compare potential yields of different plants in different natural and human systems of biomass production and the various factors influencing them including climate, soils, and topography. This task is made all the more challenging by rapid climate change and by the different approaches to modeling impacts on agriculture (see, e.g., Lobell and Fields 2007).

There is additional uncertainty about availability of land for biomass production (i.e., competition for other land uses such as food production), improvements in plant yields due to human innovation over time, differing capabilities among countries to address the gap between potential yields and actual yields, and differences in energy yields among feedstock types and conversion technologies (Johnson et al. 2009). The issue of land availability is particularly acute as the world population grows and diets drift toward more calories and the increased proportion of calories provided by meat (McMichael et al. 2007). Many studies of potential biomass supply consider biodiversity protection and other conservation measures, at least to some minimal extent (e.g., Beringer et al. 2011). So in effect, the question of potential biomass supply must be answered by (1) considering intrinsic productivity of lands, (2) deciding which plant species to include in the analysis, (3) estimating potential biomass yield of plant species, varieties, and cultivars, (4) identifying the differences between potential yields and actual yields, (5) identifying how much land will be allotted to biomass production over time, and (6) determining the degree to which environmental impacts of biomass production will curb future biomass supply.

Biomass supply is likely to be sufficient to play a significant role in global energy consumption, estimated as 285 EJ in 2005 (IEA 2008). Use of residues only could produce about 100 EJ yr^{-1}, although use of all biomass sources could potentially be converted into 1500 EJ yr^{-1} (Dornburg et al. 2008). A cautious range of 200–500 EJ yr^{-1} from all biomass sources has been found in a survey of global studies (Dornburg et al. 2008). Potential biomass supply has also been modeled at regional and national levels under a variety of assumptions, including economic drivers. For example, the U.S. Department of Energy (2011) estimated U.S. bioenergy production in 2012 at about 450 Mg DM yr^{-1}, or about enough to displace approximately 30% of current petroleum consumption. The European supply is estimated at up to 11.7 EJ yr^{-1} (Ericsson and Nilsson 2005). Studies also show that potential biomass supply is unevenly distributed such that it is/will be abundant at some locations and less available in others (e.g., Milbrandt 2005). These differences will have profound effects on economies and on trade (Milbrandt 2005) and are likely to impact poorer rural areas in particular (Phalan 2009). Therefore, it is recommended that the entire value/supply chain of advanced biofuels be carefully planned and managed to limit negative effects on human livelihoods (Bailey et al. 2011).

1.7 AGRICULTURAL CELLULOSIC BIOMASS PRODUCTION

Forests and waste sources of biomass alone do not meet current demand for advanced biofuels and are unlikely to expand to the degree necessary to meet anticipated future needs (Simmons et al. 2008). Agriculture, therefore, has a vital role in bioenergy and is the focus of much innovation particularly in crop improvements, cropping systems, and related technologies. This section highlights some important crops, production systems, and related issues in agricultural production of cellulosic biomass.

1.7.1 Perennial Grass Crops

Perennial grasses are an important feedstock source for second-generation biofuels. Traditionally used as forages, perennial grasses have recently been the focus of breeding research to improve yields and other traits important in conversion (e.g., lignin content; Coulman et al. 2013). Cool season (C3 photosynthetic pathway) grass species are not generally recommended for bioenergy use because of their poor feedstock quality (Lewandowski et al. 2003). Warm season grasses (C4 photosynthetic pathway), however, demonstrate great promise for yield and feedstock quality, as well as water use efficiency (McLaughlin et al. 2006, Boehmel et al. 2008, and Carroll and Somerville 2009). Warm season grasses, however, are slow to establish. Depending on species, peak yields are not usually achieved until three to five years after planting. Even with slow establishment though, overall operating costs of perennial grass crops may be lower than conventionally managed annual row crops (Sanderson and Adler 2008). When agronomically managed, perennial grasslands can be maintained in long-term rotations (10-plus years; McLaughlin and Kszos 2005).

Governments and academic researchers have embarked on rigorous evaluations of a variety of perennial herbaceous plants as candidates for advanced biofuels (Lewandowski et al. 2003). Switchgrass has been extensively studied for second-generation biofuels particularly in North America (Lewandowski et al. 2003; McLaughlin and Kszos 2005; Wright and Turhollow 2010); and miscanthus has been widely evaluated in Europe (e.g., Christian et al. 2008). Hence, these two bioenergy crops are further detailed here.

Switchgrass is a C4 grass that has evolved as a component of diverse tallgrass prairie ecosystems in the eastern two-thirds of the United States (Parrish and Fike 2005). It has been used there since the arrival of Europeans to graze ruminant livestock; over time, it has been intentionally managed and improved for forage. In the last 20 years, switchgrass has been scrutinized for bioenergy purposes (Wright and Turhollow 2010). Switchgrass as managed forage and in bioenergy research is typically grown as a pure grass sward (i.e., monoculture), although interest is high in its use in *polycultures*—diverse plant mixtures that may include different plant functional groups (e.g., grasses, forbs, and legumes; Tilman et al. 2006). Polycultures are receiving research and development attention for their biomass yields as well as wildlife habitat and environmental benefits compared to monocultures (Sanderson and Adler 2008; Tilman et al. 2006, 2009). However, there is conflicting evidence regarding the effects of plant diversity on biomass yields. Tilman et al. (2006) reported increases in biomass yields with increases in the number of species in a polyculture. Others report lower yields in polycultures containing switchgrass compared to switchgrass monocultures (Wang et al. 2010). Therefore, further study is required to more fully understand the effect of species richness and plant functional types on biomass yields.

There are numerous switchgrass varieties and cultivars, each having different responses to characteristics of location (e.g., soil, day length) and fertilization (Casler et al. 2007; Fike et al. 2006; Virgilio et al. 2007). Choice of variety or cultivar will depend on characteristics of the location where it will be grown (e.g., growing season) and on the management that will be used. Management of switchgrass in monocultures can be quite different than polycultures in which it is a component. Whether in monoculture or polyculture, switchgrass is grown from seed. Seedbed preparation ranges from "conventional" well-tilled soil to no-till seed drilling and "frost-seeding" during soil freezing and thawing activity (Lewandowski et al. 2003; Teel and Barnhart 2003). Weed control during establishment is critical. Weed control strategy will be affected by weed species present; however, chemical control is common (Parrish and Fike 2005). Pest control may be necessary during sward maturity depending on the cultivar (Coulman et al. 2013).

Reports of yield responses of switchgrass to fertilizer, particularly nitrogen, vary greatly (Heggenstaller et al. 2009; Vogel et al. 2002), and consensus for any nutrient has not emerged (Parrish and Fike 2005). However, yield declines are reported over time without nitrogen fertilization (Mitchell et al. 2008). Harvesting of switchgrass involves cutting, swathing, and aggregating (e.g., baling). It is usually not harvested in its first growing season (Lewandowski et al. 2003). Depending on location and cultivar, switchgrass may be harvested once or twice annually (Parrish and Fike 2005). However, one annual harvest after senescence is recommended for plant nutrient management and wildlife considerations (Hull et al. 2011). Reported yields for switchgrass vary according to cultivar, location, fertilizer use, and other factors, and generally range from 5.3 to 21.3 Mg DM ha^{-1} yr^{-1} (Fike et al. 2006; Lemus et al. 2002; Lewandowski et al. 2003).

Perennial C4 grasses of the genus *Miscanthus* originate in the tropics and subtropics of East Asia. Due to their high yields and wide climatic adaptability, they have received much attention as potential bioenergy crops (Lewandowski et al. 2000). *Miscanthus* × *giganteus* (hereafter, miscanthus) is a hybrid of *M. sinensis* and *M. sacchariflorus* and is the frontrunner in bioenergy crop research, development, and production in Europe (Lewandowski et al. 2003), although it has recently begun receiving attention in the United States (e.g., Heaton et al. 2004, 2008).

Miscanthus is a sterile hybrid that does not produce seeds and instead reproduces vegetatively from rhizomes. It is grown in monocultures that are established through manual or mechanical planting of rhizomes or rhizome pieces (Lewandowski et al. 2000). Miscanthus plantations, therefore, are monocultures of clones (i.e., genetically identical plants). The rhizomes are grown in nursery fields where they are mechanically collected and divided just before planting in fields for biomass production (Lewandowski et al. 2000). Mechanization of miscanthus culture and management has been a source of rapid innovation (Anderson et al. 2011). Nonetheless, relatively high costs of propagation and planting are barriers to adoption of miscanthus as a bioenergy crop (Atkinson 2009; Coulman et al. 2013).

Plowing is the recommended soil preparation method for miscanthus planting (Lewandowski et al. 2000). Winter kill can be a problem in miscanthus cultivation (Heaton et al. 2010; Lewandowski et al. 2000). Miscanthus has low fertilizer demand (Lewandowski et al. 2000). In soils with sufficient nitrogen mineralization from soil organic matter, there is no effect of nitrogen fertilization on miscanthus yield (Lewandowski et al. 2000). Weed control is necessary during establishment of miscanthus (Anderson et al. 2011). Mechanical and chemical controls are used in Europe, but in the United States no herbicides are registered for biofuel plantings of miscanthus (Anderson et al. 2011). There is currently no evidence of pest or pathogen issues affecting yields of miscanthus; hence, pesticide and other interventions are not yet developed (Anderson et al. 2011).

Miscanthus is harvested only once per year, usually after senescence (Lewandowski et al. 2000). In Europe, miscanthus is typically harvested in early spring because stems dry during winter and chemical constituents are leached consequently improving feedstock quality, although yield losses may be as much as 25% (Lewandowski et al. 2000, 2003). Harvest consists of mowing, swathing, and aggregating (e.g., baling). Standard mowing machines for grain and grass do not work well with miscanthus because it is taller and stiffer than the crops for which these machines are designed. Equipment modifications have therefore been necessary (Lewandowski et al. 2000). Biomass yields of miscanthus vary widely depending on location, use of irrigation, and harvest timing. Yield reports range from 7 to 40 Mg DM ha^{-1} yr^{-1} (Lewandowski et al. 2000, 2003; Price et al. 2004). For a comprehensive overview of miscanthus improvements, agronomy, and biomass characteristics, see Lewandowski et al. (2000) and Jones and Walsh (2001).

1.7.2 Short-Rotation Woody Crops and Agroforestry

The oil embargo of the Organization of Petroleum Exporting Countries (OPEC) in 1973 was a boon to research and development of SRWC (Dickmann 2006). The embargo forced governments and international agencies to investigate alternative sources of energy, and as a result of the upheaval, SRWC gained a new status as candidate domestic bioenergy crops (Wright 2006). Generous government funding facilitated creation of new hybrids and genetic transformations of practical advantage, improvement of propagation methods, invention of high-density cropping systems, and innovations in stand management were enabled (Dickmann 2006).

SRWC are genetically improved tree species purpose-grown in short cycles, usually 1–15 years, and using intensive cultural techniques of fertilization, irrigation, and weed control—often relying on coppice regeneration (Dickmann 2006; Drew et al. 1987; Hinchee et al. 2009). Essentially, SRWC are grown more like annual commodity crops than traditional pine, oak, or spruce forests. In temperate regions, SRWC tend to be grown in plantations (i.e., extensive monocultures), although in the tropics, very fast growing trees tend to be alley-cropped (Dickmann 2006; Holzmueller and Jose 2012).

Site preparation typically involves soil tillage and removal of plant debris (Tubby and Armstrong 2002). Unrooted cuttings (clones) are planted using mechanical equipment, and weed control is required. Cover crops can be established on erosion-prone sites (Volk et al. 2004). When planted with other crops in agroforestry systems, SRWC species produce multiple benefits such as increased yields and improved water quality (Holzmueller and Jose 2012). Some SRWC species tolerate alley cropping, a form of intercropping where trees are placed in rows of wide spacing creating alleys for growing of agricultural or horticultural crops, including grains and forages (Headlee et al. 2013). Other agroforestry practices include placement of trees and shrubs in shelterbelts and along riparian areas (Holzmueller and Jose 2012).

There are several important pests of SRWC including grazing mammals, boring and defoliating insects, and disease (Mitchell et al. 1999). Tending of SRWC will therefore require interventions as necessary to avert biomass loss. Harvest is conducted with agricultural equipment that cuts and chips the biomass in a single operation (Berhongaray et al. 2013; Mitchell et al. 1999). Yields of SRWC vary widely depending on location, species (or hybrid), water availability, pests, management, and harvest timing (Dickmann 2006). Holzmueller and Jose (2012) summarize woody biomass crop yields in annual and short-rotation systems of the U.S. Upper Midwest and report a range of 5.4–30.0 Mg DM ha^{-1} yr^{-1}. Labrecque and Teodorescu (2005) report on 12 willow and poplar clones in southeastern Canada, finding a biomass yield range of 5.6–16.4 Mg DM ha^{-1} yr^{-1}. Yield ranges of 6 to more than 20 Mg ha^{-1} yr^{-1} are reported for poplars and willows in Europe (Hoffmann and Weih 2005).

1.7.3 Annual Row Crop Residues

Around the world, human populations depend on the production of commercial crops to satisfy daily calorie needs. As a result, annual food crop agriculture occupies almost half of the Earth's land surface (Ramankutty et al. 2008). There exists, therefore, great potential for biofuel use of the cellulosic residues remaining after harvest of the food portions of these crops. Corn stover, rice, wheat straw, and bagasse have been considered for bioenergy production. Conventionally managed, these annual crops require external inputs including mineral fertilizer, pesticides, and herbicides, and in some places, irrigation. The energy value of global crop residues is estimated at 69.9 EJ yr^{-1} (Lal 2005). Graham et al. (2007) conclude that 30% of corn stover produced in the United States could be harvested with existing equipment—enough to produce more than the current volume of corn grain ethanol.

Harvesting crop residues has been associated with declining soil quality and productivity (Lal 2005; Moebius-Clune et al. 2008). Trade-offs exist among beneficial effects of residue harvest—such as faster warming of soils in spring, better seed germination, and less favorable habitat for plant pathogens, and the potential adverse effects—such as organic matter declines, greater soil temperature fluctuations, and faster losses of stored soil moisture (Mann et al. 2002; Wilhelm et al. 2004). Crop residues are typically important reservoirs of elements necessary for crop growth (e.g., C, K, Ca, N, and P), thus their return to the soil after harvest is essential for sustaining grain and biomass yields (Blanco-Canqui and Lal 2004). The economic benefit of harvesting crop residues must therefore be weighed against the potentially negative effects that such management may have on soil quality (Moebius-Clune et al. 2008). Trade-offs also exist with livestock agriculture. As previously mentioned, some residues are used as fodder and bedding for animals.

1.8 SUSTAINABILITY ISSUES

It is beyond the scope of this chapter to discuss the complexity of specific challenges to sustainability of biofuels. Hence, a broad, integrated definition of sustainability as well as a statement about its importance is offered here. A brief survey of major sustainability issues of biomass supply is then given. These issues include: food insecurity, climate change, invasive and transgenic plants, marginal lands, water supply and quality, and rural development and social justice.

1.8.1 Sustainability Defined

There are many definitions of sustainability, each supporting various principles and concepts. Essentially, however, sustainability can be described as a set of goals and the practices and behavior that support such goals. As a set of goals, sustainability describes desired conditions of the environment and human well-being as a result of interaction with the environment, now and in the future. As practices and behaviors, sustainability describes human actions that support and enhance the environment and human benefits. Sustainability is important because the choices and actions of today affect everything in the future. Sound decisions at present may prevent undesirable outcomes in the future.

Bioenergy is frequently evoked as an important tool in improving environmental conditions, as well human lives and livelihoods (Domac et al. 2005; Faaij and Domac 2006; Tilman et al. 2009). However, much remains to be understood about the impacts of bioenergy on the environment and human society. Ultimately, however, sustainability of bioenergy will depend on the goals defined, and when, where, and by whom those goals are defined; what actions and behaviors people are willing and able to adopt to support those goals; and the ability of science to assist human knowledge of connections among the many aspects of bioenergy and sustainability goals. In the meantime, governments, international agencies, and nongovernment organizations at different levels have produced white papers and various guidelines in an effort to encourage sustainable practices in biomass production (Hull et al. 2012; RSB 2011; UNEP 2009).

1.8.2 Food Insecurity

Chief among the concerns over the impacts of biomass production is food insecurity. This concern is related to abrupt rises in short-term food commodity prices (e.g., corn and grain) that lead to hunger or starvation in some areas (Economic Research Service 2013). The food insecurity impact of biofuels stems largely from a mid-2008 crisis when an unexpected rise in grain prices created supply shortages in some countries (Nonhebel 2012). Critics were quick to label biofuels as the leading cause of the crisis, but years later the cause is still being debated. Many analysts have concluded that tight interlinkages of global commodity supply and trade, price speculation, and other factors were as much if not more to blame as biomass production (Godfray et al. 2010; Mittal 2009; Mueller et al. 2011). This is not to say, however, that in a globalized world economy biomass and biofuels are not at all connected to price and supply fluctuations and the effects thereof. Indeed, sustainability analyses appear to converge on a multitactic approach for resolving the multiple challenges of providing food, energy, and environmental protection for the world population (Groom et al. 2008; Reijnders 2006; Tilman et al. 2009). These concerns have motivated a focus on inedible feedstocks for bioenergy and on the use of nonagricultural lands for biomass production. For more consideration of food insecurity challenges and connections to biomass production, see Bryngelsson and Lindgren (2013), Foley et al. (2011), and Tilman et al. (2009).

1.8.3 Climate Change

Climate change and climate change mitigation are major concerns in the production of biofuels. The concerns center on emissions of greenhouse gases associated with land use and land-use change in production of biomass for biofuel (e.g., carbon capture in, or release from, soil due to agricultural practices; Delucchi 2011). There is contradictory evidence as to whether advanced biofuels and cellulosic biomass production are solutions or problems (Georgescu et al. 2009; Searchinger et al. 2008). This suggests more study is needed and that much depends on the ability of science—in coordination with policy, to deliver solutions. For more on climate change and sustainability of biomass/biofuels, see Fargione et al. (2008), Robertson et al. (2008), and Menten et al. (2013).

1.8.4 Invasive and Transgenic Plants

There is growing concern about the invasiveness of plants used for biomass production because the traits of ideal biomass crops are also commonly found among invasive plant species (Raghu et al. 2006). Invasive species are of concern due to adverse environmental and ecological impacts and the economic costs associated with lost productivity of natural ecosystems and the services they provide, as well as costs associated with invasive species control (Pimentel et al. 2005). The invasiveness issue is particularly acute for wildlife and biodiversity managers in public agencies and in nongovernment organizations (Smith et al. 2013). Warm season grass monocultures, for example, are seen as providing very little value as wildlife habitat (Fargione et al. 2009; Hartman et al. 2011). Switchgrass is another example of concern. Although it is native to North America, many switchgrass ecotypes and improved germplasm are being introduced to new locations and subsequent outcross with local ecotypes could erode native biodiversity at local and regional levels (Kwit and Stewart 2012). Use of transgenic plants, particularly SRWC in agroforestry, is also of grave concern to managers of ecological systems and wildlife species (Hinchee et al. 2009). There are no easy solutions for these challenges, and many decision-makers must seek to balance the benefits of biofuels and biomass production with known and potential risks. For a further overview of invasive plants and biofuels, see Gordon et al. (2011) and Smith et al. (2013).

1.8.5 Marginal Lands

Definitions of marginality with regard to productivity of arable lands vary greatly but, in general, marginal lands are those that have one or more characteristics not conducive to annual crop production. Characteristics such as steep slopes, shallow soils, excessive wetness, or drought-proneness generally have negative effects on profitability of agricultural use; hence, marginal lands usually are of fairly low value (i.e., comparatively low price per acre for rent or taxation purposes). Row crop production on marginal lands is associated with land degradation and decreased productivity over time as a result (Pimentel 1991). Hence, production of perennial crops is seen as a potential source of resource protection and income for farmers.

It has been suggested that marginal lands be targeted for production of biomass for bioenergy not only for meeting renewable energy goals but also as a potential means for avoiding land-use conflicts contributing to food insecurity (Achten et al. 2013; Campbell et al. 2008). The 2007 U.S. Census of Agriculture, for example, identifies approximately 12 million ha of idle lands, land in cover crops for soil improvement, and fallow rotations as potentially available for biomass production (COA 2009). Conversion of steep or wet land currently in food crop production (i.e., row crop) to less intensive bioenergy crops such as high diversity, low-input perennial mixes is thought to have the potential to generate more ecosystem services (Tilman et al. 2009). However, some researchers caution that conversion of marginal lands, particularly those in set-aside programs (i.e., currently idle or planted in perennial cover), to more intensively managed bioenergy cropping systems could lead to permanent land degradation and a net increase in greenhouse gases as well as food insecurity (Bryngelsson et al. 2013; Zenone et al. 2013). Government, academic, and private sector research is needed to assess whether and to what degree marginal lands can be relied upon for meeting future bioenergy demands, while policymakers and other decision-makers address questions of whether and to what degree marginal lands should be relied upon for bioenergy needs.

1.8.6 Water Supply and Quality

Cultivation of crops for biomass, food, feed, and fiber requires vast amounts of water. In many temperate areas, there is sufficient water from precipitation to meet crop needs during the growing season. In other areas, crop production requires irrigation. There is growing concern, therefore, over expansion of bioenergy crop production into drier areas, and hence requiring more irrigation, and the associated impacts on food crops and therefore food insecurity (De Fraiture and Berndes 2009). Floodplains represent a potential opportunity for growing biomass crops without threats to water supply and without displacing food crops. Food crops planted in floodplains are prone to failure because of flooding and soil erosion, but perennial biomass crops such as SRWC and perennial grasses are less vulnerable (Bardhan and Jose 2012). Such biomass crops could help reduce floodplain soil erosion while providing a source of income for farmers and a source of renewable energy (Bardhan and Jose 2012).

Also of concern is the use of agrochemicals in agricultural production of biomass. Water quality and aquatic habitats can be affected by agricultural drainage of fertilizers, pesticides, and sediments. Thus, expansion of bioenergy crop production must be carefully managed to avoid water pollution (Gopalakrishnan et al. 2009). For more information on the opportunities to limit and mitigate the *water footprint* of biofuels, see Dominguez-Faus et al. (2009).

1.8.7 Rural Development and Social Justice

In rural areas, community leaders are reconsidering traditional drivers of economic activity in search of sustainable, diversified, and environmentally friendly options. Bioenergy may be a viable economic development option for communities that can grow dedicated energy crops and that can develop energy industries to process those crops into power or fuel. The development of a bioenergy industry may be particularly well suited for local economies—given adequate investors—in that the costs of transporting bioenergy crops makes local processing necessary. Thus, economic activity and economic benefit may stay local; although local net benefits are not always guaranteed when balanced against negative impacts to community life and well-being, such as food insecurity, increased truck and/or train traffic (i.e., noise, air quality, traffic safety), and odors and noise from the biomass conversion facility (Selfa et al. 2011). Economic benefits must also be weighed against impacts to water supply. Each community and situation is different, and local decisions around the choice of energy crops, processing systems, and markets will define the economic benefits, while state and federal policy can provide incentives and influence outcomes. An additional issue, however, is rural self-determination and empowerment. Government policies tend to overlook social considerations in biofuel development strategies (Mol 2007; Rossi and Hinrichs 2011), leading to macrolevel goals that adversely affect local-level realities. For a comprehensive review of rural development and social justice issues of biofuels, see Dale et al. (2013) and van der Horst and Vermeylen (2011).

1.9 SUMMARY

Biomass feedstocks for bioenergy, particularly advanced biofuels, have an important role in global, regional, and local energy consumption and economic development. Expansion of biomass production, processing into feedstocks, handling, transportation, and storage, if done sustainably, may provide supply/value chains that support renewable energy goals while enhancing rural livelihoods. Although second-generation lipid feedstocks are important in the production of biodiesel, it is cellulosic/lignocellulosic feedstocks that hold the greatest potential for transforming fuel energy portfolios while simultaneously transforming agriculture and resource management. Crop residues have the potential to contribute substantially to second-generation biofuels, but their collection and use must be carefully planned and managed so as not to degrade soils and water, nor to generate shortages in fodder and bedding for livestock. Perennial grasses are chief among cellulosic/lignocellulosic feedstock sources at the center of research and development efforts—in part because of their yield potentials and environmental benefits, but due in greater part to their lack of food value. Innovations in biomass crops and cropping systems must occur to increase their overall contribution to global renewable energy consumption. These innovations must run apace with improvements in agricultural equipment and conversion technology. Sustainability of biomass production and the feedstocks derived therefrom will depend on the ability of science and public policy to limit competition for land use that leads to food insecurity; to curtail land-use practices that contribute to climate change, environmental degradation, and reduction in water supply and quality; to enhance rural self-determination and empowerment; and to prevent creation of energy poverty. Ultimately, decision-makers at all levels must consider advantages and disadvantages among specific biomass types and production systems and make informed decisions with regard to desired goals for the present and the future.

ACKNOWLEDGMENTS

David Duncan and Mary Julia Laycock are gratefully acknowledged for their suggestions in improving this chapter. Special thanks to David Williams for his loving support.

REFERENCES

Achten WMJ, J Almeida, V Fobelets, E Bolle, E Mathijs, VP Singh, DN Tewari, LV Verchot, and B Muys. 2010. Life cycle assessment of *Jatropha* as transportation fuel in rural India. *Applied Energy* 87: 3652–3660.

Achten WMJ, E Mathijs, L Verchot, VP Singh, R Aerts, and B Muys. 2007. *Jatropha* biodiesel fueling sustainability? *Biofuels, Bioproducts and Biorefining* 1: 283–291.

Achten WMJ, A Trabucco, WH Maes, LV Verchot, R Aerts, E Mathijs, P Vantaomme, VP Singh, and B Muys. 2012. Global greenhouse gas implications of land conversion to biofuel crop cultivation in arid and semiarid lands—Lessons learned from *Jatropha*. *Journal of Arid Environments* 98: 135–145.

Adler PR, MA Sanderson, PJ Weimer, and KP Vogel. 2009. Plant species composition and biofuel yields of conservation grasslands. *Ecological Applications* 19: 2202–2209.

Anderson E, R Arundale, M Maughan, A Oladeinde, A Wycislo, and T Voigt. 2011. Growth and agronomy of *Miscanthus x giganteus* for biomass production. *Biofuels* 2: 167–183.

Antizar-Ladislao B and JL Turrion-Gomez. 2008. Second-generation biofuels and local bioenergy systems. *Biofuels, Bioproducts and Biorefining* 2: 455–469.

Atabani AE, AS Silitonga, IA Badruddin, TMI Mahlia, HH Masjuki, and S Mekhilef. 2012. A comprehensive review on biodiesel as an alternative energy resource and its characteristics. *Renewable and Sustainable Energy Review* 16: 2070–2093.

Atkinson CJ. 2009. Establishing perennial grass energy crops in the UK: A review of current propagation options for *Miscanthus*. *Biomass and Bioenergy* 33: 752–759.

Bailey C, JF Dyer, and L Teeter. 2011. Assessing the rural development potential of lignocellulosic biofuels in Alabama. *Biomass and Bioenergy* 35: 1408–1417.

Bergmann JC, DD Tupinamba, OYA Costa, JRM Aleida, CC Barreto, and BF Quirino. 2013. Biodiesel production in Brazil and alternative biomass feedstocks. *Renewable and Sustainable Energy Reviews* 21: 411–420.

Berhongaray G, O El Kamioui, and R Ceulemans. 2013. Comparative analysis of harvesting machines on an operational high-density short rotation wood crop (SRWC) culture: One-process verses two-process harvest operation. *Biomass and Bioenergy* 58: 333–342.

Beringer T, W Lucht, and S Schaphoff. 2011. Bioenergy production potential of global biomass plantations under environmental and agricultural constraints. *Global Change Biology—Bioenergy* 3: 299–312.

Binford CD. 2005. Nutrient management. In *Encyclopedia of Soils in the Environment*, eds JL Hatfield, DS Powlson, C Rosenzweig, KM Scow, MJ Singer, and DL Sparks. Waltham, MA: Elsevier, pp. 71–75.

Blanco-Canqui H and R Lal. 2004. Mechanisms of carbon sequestration in soil aggregates. *Critical Reviews in Plant Sciences* 23: 481–504.

Boehmel C, I Lewandowski, and W Claupein. 2008. Comparing annual and perennial energy cropping systems with different management intensities. *Agricultural Systems* 96: 224–236.

Botha T and HV Blottnitz. 2006. A comparison of the environmental benefits of bagasse-derived electricity and fuel ethanol on life-cycle basis. *Energy Policy* 34: 2654–2661.

Bardhan S and S Jose. 2012. The potential for floodplains to sustain biomass feedstock production systems. *Biofuels* 3: 575–588.

Berndes G, M Hoogwijk, and R van den Broek. 2003. The contribution of biomass in the future global energy supply: A review of 17 studies. *Biomass and Bioenergy* 25: 1–28.

Brennan L and P Owende. 2010. Biofuels from microalgae—A review of technologies for production, processing, and extractions for biofuels and co-products. *Renewable and Sustainable Energy Reviews* 14: 557–577.

Bryngelsson DK and K Lindgren. 2013. Why large-scale bioenergy production on marginal is unfeasible: A conceptual partial equilibrium analysis. *Energy Policy* 55: 454–466.

Campbell JE, DB Lobell, RC Genova, and CB Field. 2008. The global potential of bioenergy on abandoned agricultural lands. *Environmental Science & Technology* 42: 5791–5794.

Casler MD, KP Vogel, CM Taliaferro, NJ Ehlke, JD Berdahl, EC Brummer, RL Kallenbach, CP West, and RB Mitchell. 2007. Latitudinal and longitudinal adaptation of switchgrass populations. *Crop Science* 47: 2249–2260.

Carroll A and C Somerville. 2009. Cellulosic biofuels. *Annual Review of Plant Biology* 60: 165–182.

Centner TJ and GL Newton. 2008. Meeting environmental requirements for the land application of manure. *Journal of Animal Science* 86: 3228–3234.

Christian DG, AB Riche, and NE Yates. 2008. Growth, yield and mineral content of *Miscanthus x giganteus* grown as a biofuel for 14 successive harvests. *Industrial Crops and Products* 28: 320–327.

Clark JH, V Budarin, FEI Deswaarte, JJE Hardy, FM Kerton, AJ Hunt, R Luque, et al. 2006. Green chemistry and the biorefinery: A partnership for a sustainable future. *Green Chemistry* 8: 853–860.

COA. 2009. *2007 Census of Agriculture*. AC-07-A-51. Washington, DC: U.S. Department of Agriculture.

Coleman MD and JA Stanturf. 2006. Biomass feedstock production systems: Economic and environmental benefits. *Biomass and Bioenergy* 30: 693–695.

Cooke P, G Kohlin, and WF Hyde. 2008. Fuelwood, forests and community management—Evidence from household studies. *Environment and Development Economics* 13: 103–135.

Coulman B, A Dalai, E Heaton, CP Lee, M Lefsrud, D Levin, PG Lemaux, et al. 2013. Developments in crops and management systems to improve lignocellulosic feedstock production. *Biofuels, Bioproducts, & Biorefining* 7: 582–601.

Dale VH, RA Efroymson, KL Kline, MH Langholtz, PN Leiby, GA Olaldosu, MR Davis, ME Downing, and MR Hilliard. 2013. Indicators for assessing socioeconomic sustainability of bioenergy systems: A short list of practical measures. *Ecological Indicators* 26: 87–102.

Dale VH, KL Kline, LL Wright, RD Perlack, M Downing, and RL Graham. 2011. Interactions among bioenergy feedstock choice, landscape dynamics, and land use. *Ecological Applications* 21: 1039–1054.

De Fraiture C and G Berndes. 2009. Biofuels and water. In *Biofuels: Environmental consequences and interactions with changing land use*. Proceedings of the Scientific Committee on Problems of the Environment (SCOPE) International Biofuels Project Rapid Assessment, eds RW Howarth and S Bringezu. Ithaca, NY: Cornell University, pp. 139–153.

Delucchi M. 2011. A conceptual framework for estimating the climate impacts of land-use change due to energy crop programs. *Biomass and Bioenergy* 35: 2337–2360.

Demirbas A. 2004. Combustion characteristics of different biomass fuels. *Progress in Combustion and Energy Science* 30: 219–230.

Dickmann DI. 2006. Silviculture and biology of short-rotation woody crops in temperate regions: Then and now. *Biomass and Bioenergy* 30: 696–705.

Dismukes GC, D Carrieri, N Bennett, GM Ananyev, and MC Posewitz. 2008. Aquatic phototrophs: Efficient alternatives to land-based crops for biofuels. *Current Opinion in Biotechnology* 19: 235–240.

Domac J, K Richards, and S Risovic. 2005. Socio-economic drivers in implementing bioenergy projects. *Biomass and Bioenergy* 28: 97–106.

Dominguez-Faus R, SE Powers, JG Burken, and PJ Alvarez. 2009. The water footprint of biofuels: A drink or drive issue? *Environmental Science and Technology* 43: 3005–3010.

Dornburg V, A Faaij, P Verweij, H Langeveld, G van de Ven, F Wester, H van Keulen, et al. 2008. *Assessment of global biomass potentials and their links to food, water, biodiversity, energy demand and economy*. The Netherlands Research Programme on Scientific Assessment and Policy Analysis for Climate Change. Netherlands Environmental Assessment Agency MNP, WAB Secretariat, Bilthoven, The Netherlands.

Drew AP, L Zsuffa, and CP Mitchell. 1987. Terminology relating to wood plant biomass and its production. *Biomass* 12: 79–82.

Economic Research Service. 2013. *Food Security in the U.S.* U.S. Department of Agriculture. Available at: http://www.ers.usda.gov/topics/food-nutrition-assistance/food-security-in-the-us.aspx (Accessed September 8, 2013).

Ericsson K and LJ Nilsson. 2005. Assessment of the potential biomass supply in Europe using a resource-focused approach. *Biomass and Bioenergy* 30: 1–15.

Faaij A. 2006. Modern biomass conversion technologies. *Mitigation and Adaption Strategies for Global Change* 11: 343–375.

Faaij APC and J Domac. 2006. Emerging international bio-energy markets and opportunities for socio-economic development. *Energy for Sustainable Development* X: 7–19.

Fargione J, J Hill, D Tilman, S Polasky, and P Hawthorne. 2008. Land clearing and the biofuel carbon debt. *Science* 319: 1235–1238.

Fargione JE, TR Cooper, DJ Flashpohler, J Hill, C Lehman, T McCoy, S McLeod, EJ Nelson, KS Oberhauser, and D Tilman. 2009. Bioenergy and wildlife: Threats and opportunities for grassland conservation. *BioScience* 59: 767–777.

Fike JH, DJ Parrish, DD Wolf, JA Blasko, JT Green, Jr, M Rasnake, and JH Reynolds. 2006. Switchgrass production for the upper southeastern USA: Influence of cultivar and cutting frequency on biomass yields. *Biomass and Bioenergy* 30: 207–213.

Foley JA, N Ramankutty, KA Brauman, ES Cassidy, JS Gerber, M Johnston, ND Mueller, et al. 2011. Solutions for a cultivated planet. *Nature* 478: 337–342.

Francis G, R Edinger, and K Becker. 2005. A concept for simultaneous wasteland reclamation, fuel production, and socio-economic development in degraded areas in India: Need, potential and perspectives of *Jatropha* plantations. *Natural Resources Forum* 29: 12–24.

Funk T, B Paulsrud, R Fonner, and S Bretthauer. 2014. *EZRegs: Illinois NPDES general permit—Livestock.* University of Illinois Extension. Available at: http://web.extension.illinois.edu/ezregs/ezregs.cfm?section=viewregs_byq&QuestionID=54&searchTerm=&ProfileID=5 (Accessed June 23, 2014).

Georgescu M, DB Lobell, and CB Field. 2009. Potential impact of U.S. biofuels on regional climate. *Geophysical Research Letters* 36: L21806.

Godfray HCJ, JR Beddington, IR Crute, L Haddad, D Lawrence, JF Muir, J Pretty, S Robison, SM Thomas, and C Toulmin. 2010. Food insecurity: The challenge of feeding 9 billion people. *Science* 327: 812–818.

Goh CS and KT Lee. 2010. A visionary and conceptual macroalgae-based third-generation bioethanol (TGB) biorefinery in Sahab, Malaysia as an underlay for renewable and sustainable development. *Renewable and Sustainable Energy Reviews* 14: 842–848.

Gopalakrishnan G, MC Negri, M Wang, M Wu, SW Snyder, and L LaFreniere. 2009. Biofuels, land, and water: A systems approach to sustainability. *Environmental Science and Technology* 43: 6094–6100.

Gordon DR, KJ Tancig, DA Onderdonk, and CA Gantz. 2011. Assessing the invasive potential of biofuel species proposed for Florida and the United States using the Australian Weed Risk Assessment. *Biomass and Bioenergy* 35: 74–79.

Graham RL, R Nelson, J Sheehan, RD Perlack, and LL Wright. 2007. Current and potential U.S. corn stover supplies. *Agronomy Journal* 99: 1–11.

Groom MJ, EM Gray, and PA Townsend. 2008. Biofuels and biodiversity: Principles for creating better policies for biofuel production. *Conservation Biology* 22: 602–609.

Hahn-Hagerdal B, M Galbe, MF Gorwa-Grauslund, G Liden, and G Zacchi. 2006. Bio-ethanol—The fuel of tomorrow from the residues of today. *Trends in Biotechnology* 24: 549–556.

Hartman JC, JB Nippert, RA Orozco, and CJ Springer. 2011. Potential ecological impacts of switchgrass (*Panicum virgatum* L.) biofuel cultivation in the Central Great Plains, USA. *Biomass and Bioenergy* 35: 3415–3421.

Headlee WL, RB Hall, and RS Zalensy, Jr. 2013. Establishment of alleycropped hybrid aspen "Crandon" in Central Iowa, USA: Effects of topographic position and fertilizer rate on aboveground biomass production and allocation. *Sustainability* 5: 2874–2886.

Heaton E, T Voigt, and SP Long. 2004. A quantitative review comparing the yields of two candidate C4 perennial biomass crops in relation to nitrogen, temperature and water. *Biomass and Bioenergy* 27: 21–30.

Heaton EA, FG Dohleman, and SP Long. 2008. Meeting US biofuel goals with less land: The potential of *Miscanthus*. *Global Change Biology* 14: 2000–2014.

Heaton EA, FG Dohleman, AF Miguez, JA Juvik, V Lozavaya, J Widholm, OA Zabotina, et al. 2010. *Miscanthus*: A promising biomass crop. *Advances in Botanical Research* 56: 75–137.

Hedenus F and C Azar. 2009. Bioenergy plantations or long-term carbon sinks?—A model based analysis. *Biomass and Bioenergy* 33: 1693–1702.

Heggenstaller AH, KJ Moore, M Liebman, and RP Anex. 2009. Nitrogen influences biomass and nutrient partitioning by perennial, warm-season grasses. *Agronomy Journal* 101: 1363–1371.

Hinchee M, W Rottmann, L Mullinax, C Zhang, S Chang, M Cunningham, L Pearson, and N Nehra. 2009. Short-rotation woody crops for bioenergy and biofuels applications. *In vitro Cellular Development Biology—Plant* 45: 619–629.

Hoffmann D and M Weih. 2005. Limitations and improvement of the potential utilisation of woody biomass for energy derived from short rotation woody crops in Sweden and Germany. *Biomass and Bioenergy* 28: 267–279.

Hoogwijk M, A Faaij, R van den Broek, G Berndes, D Gielen, and W Turkenberg. 2003. Exploration of the ranges of the global potential for biomass energy. *Biomass and Bioenergy* 25: 119–133.

Holzmueller EJ and Jose S. 2012. Bioenergy crops in agroforestry systems: Potential for the U.S. North Central Region. *Agroforestry Systems* 85: 305–314.

Hull S, J Arntzen, C Bleser, A Crossley, R Jackson, E Lobner, L Paine, et al. 2011. *Wisconsin sustainable planting and harvest guidelines for nonforest biomass.* Madison, WI: Wisconsin Bioenergy Council.

Hull S, J Arntzen, C Bleser, A Crossley, R Jackson, E Lobner, L Paine, et al. 2012. *Wisconsin sustainable planting and harvest guidelines for nonforest biomass.* Wisconsin Department of Agriculture and Consumer Protection. Available at: http://datcp.wi.gov/uploads/About/pdf/WI-NFBGuidelinesFinalOct2011.pdf Verified January 20, 2015).

IEA (International Energy Agency). 2008. *Worldwide trends in energy use and efficiency: Key insights from IEA indicator analysis.* Paris, France: International Energy Agency.

Jack MW. 2009. Scaling laws and technology development strategies for biorefineries and bioenergy plants. *Bioresource Technology* 100: 6324–6330.

John RP, GS Anisha, KM Nampoothiri, and A Pandey. 2011. Micro and macroalgal biomass: A renewable resource for bioethanol. *Bioresource Technology* 102: 186–193.

Johnson M, JA Foley, T Holloway, C Kucharik, and C Monfreda. 2009. Resetting global expectations from agricultural biofuels. *Environmental Research Letters* 4: 014004.

Jones MB and M Walsh, eds. 2001. *Miscanthus for energy and fibre.* London: James and James.

Keplinger KO and LM Hauck. 2006. The economics of manure utilization: Model and application. *Journal of Agricultural and Resource Economics* 31: 414–440.

Kim S and BE Dale. 2004. Global potential bioethanol production from wasted crops and crop residues. *Biomass and Bioenergy* 26: 361–375.

Kleinschmidt J. 2007. *Biofueling rural development: Making the case for linking biofuel production to rural revitalization.* Policy Brief no. 5. Durham: Carsey Institute, University of New Hampshire.

Kwit C and CN Stewart. 2012. Gene flow matters in switchgrass (*Panicum virgatum* L.), a potential widespread biofuel feedstock. *Ecological Applications* 22: 3–7.

Labrecque M and TI Teodorescu. 2005. Field performance and biomass production of 12 willow and polar clones in short-rotation coppice in southern Quebec (Canada). *Biomass and Bioenergy* 29: 1–9.

Lal R. 2005. World crop residues production and implications for its use as a biofuel. *Environment International* 31: 575–584.

Lee RA and J-M Lavoie. 2013. From first- to third-generation biofuels: Challenges of producing a commodity from a biomass of increasing complexity. *Animal Frontiers* 3: 6–11.

Lemus R, EC Brummer, KJ Moore, NE Molstad, CL Burras, and MF Barker. 2002. Biomass yield and quality of 20 switchgrass populations in southern Iowa, USA. *Biomass and Bioenergy* 23: 433–442.

Lehtikangas P. 2001. Quality properties of pelletised sawdust, logging residues and bark. *Biomass and Bioenergy* 20: 351–360.

Lewandowski I, JC Clifton-Brown, JMO Scurlock, and W Huisman. 2000. *Miscanthus*: European experience with a novel energy crop. *Biomass and Bioenergy* 19: 209–227.

Lewandowski I, JMO Scurlock, E Lindvall, and M Christou. 2003. The development and current status of perennial rhizomatous grasses as energy crops in the US and Europe. *Biomass and Bioenergy* 25: 335–361.

Lobell DB and CB Fields. 2007. Global scale climate-crop yield relationships and the impacts of recent warming. *Environmental Research Letters* 2: 014002, p. 7.

Mann L, V Tolbert, and J Cushman. 2002. Potential environmental effects of corn stover removal with emphasis on soil organic matter and erosion: A review. *Agriculture, Ecosystems and Environment* 89: 149–166.

Mata TM, AA Martins, and NS Caestano. 2010. Microalgae for biodiesel production and other applications: A review. *Renewable and Sustainable Energy Reviews* 14: 217–232.

McKendry P. 2002. Energy production from biomass (part 1): Overview of biomass. *Bioresource Technology* 83: 37–46.

McLaughlin SB, JR Kiniry, CM Taliaferro, and D De La Torre Ugarte. 2006. Projecting yield and utilization potential of switchgrass as an energy crop. *Advances in Agronomy* 90: 267–297.

McLaughlin SB and LA Kszos. 2005. Development of switchgrass (*Panicum virgatum*) as a bioenergy feedstock in the United States. *Biomass and Bioenergy* 28: 515–535.

McMichael AJ, JW Powles, CD Butler, and R Uauy. 2007. Food, livestock production, energy, climate change, and health. *The Lancette* 370: 1253–1263.

Menten F, B Cheze, L Patouillard, and F Bouvart. 2013. A review of LCA greenhouse gas emissions results for advanced biofuels: The use of meta-regression analysis. *Renewable and Sustainable Energy Reviews* 26: 108–143.

Milbrandt A. 2005. *A geographical perspective on the current biomass resource availability in the United States.* Technical Report NREL/TP-560-39181. Washington, DC: National Renewable Energy Laboratory, U.S. Department of Energy, Office of Energy Efficiency & Renewable Energy.

Mitchell CP, EA Stevens, and MP Watters. 1999. Short-rotation forestry—Operations, productivity and costs based on experience gained in the UK. *Forest Ecology and Management* 121: 123–136.

Mitchell R, KP Vogel, and G Sarath. 2008. Managing and enhancing switchgrass as a bioenergy feedstock. *Biofuels, Bioproducts and Biorefining* 2: 530–539.

Mittal A. 2009. *The 2008 food price crisis: Rethinking food security policies.* G-24 Discussion Paper Series, Research papers for the Intergovernmental Group of Twenty-Four on International Monetary Affairs and Development. New York: United Nations.

Moebius-Clune BN, HM van Es, OJ Idowu, RR Schindelbeck, DJ Boebius-Clune, DW Wolfe, GS Abawi, JE Thies, BK Gungino, and R Lucey. 2008. Long-term effects of harvesting maize stover and tillage on soil quality. *Soil Science Society of America Journal* 72: 960–969.

Mol APJ. 2007. Boundless biofuels? Between environmental sustainability and vulnerability. *Sociologia Ruralis* 47: 297–315.

Mueller SA, JE Anderson, and TJ Wallington. 2011. Impact of biofuel production and other supply and demand factors on food price increases in 2008. *Biomass and Bioenergy* 35: 1623–1632.

Nonhebel S. 2012. Global food supply and the impacts of increased use of biofuels. *Energy* 37: 115–121.

Parrish DJ and JH Fike. 2005. The biology and agronomy of switchgrass for biofuels. *Critical Reviews in Plant Sciences* 24: 423–459.

Phalan B. 2009. The social and environmental impacts of biofuels in Asia: A review. *Applied Energy* 86: S21–S29.

Pimentel D. 1991. Ethanol fuels: Energy security, economics, and the environment. *Agricultural and Environmental Ethics* 4: 1–13.

Pimentel D, R Zuniga, and D Morrison. 2005. Update on the environmental and economic costs associated with alien-invasive species in the United States. *Ecological Economics* 52: 273–288.

Price L, M Bullard, H Lyons, S Anthony, and P Nixon. 2004. Identifying the yield potential of *Miscanthus x giganteus*: An assessment of the spatial and temporal variability of *M x giganteus* biomass productivity across England and Whales. *Biomass and Bioenergy* 26: 3–13.

Ragauskas AJ, CK Williams, BH Davison, G Britovsek, J Cairney, CA Eckert, WJ Fredrick, Jr., et al. 2006. The path forward for biofuels and biomaterials. *Science* 311: 484–489.

Raghu S, RC Anderson, CC Daehler, AS Davis, RN Wiedenmann, D Sumberloff, and RN Mack. 2006. Adding biofuels to the invasive species fire? *Science* 313: 1742.

Ramankutty N, AT Evan, C Monfreda, and JA Foley. 2008. Farming the planet: 1. Geographic distribution of global agricultural lands in the year 2000. *Global Biogeochemical Cycles* 22: GB1003.

Reijnders L. 2006. Conditions for the sustainability of biomass based fuel use. *Energy Policy* 34: 863–876.

Robertson GP, VH Dale, OC Doering, SP Hamburg, JM Melillo, MM Wander, WJ Parton, et al. 2008. Sustainable biofuels redux. *Science* 322: 49–50.

Rockwood DL, AW Rudie, SA Ralph, JY Zhu, and JE Winandy. 2008. Energy product options for *Eucalyptus* species grown as short rotation woody crops. *International Journal of Molecular Science* 9: 1361–1378.

Rosch C, J Skarka, K Raab, and V Stelzer. 2009. Energy production from grassland—Assessing the sustainability of different process chains under German conditions. *Biomass and Bioenergy* 33: 689–700.

Rosillo-Calle F, L Pelkmans, and A Walter. 2009. *A global overview of vegetable oils, with reference to biodiesel*. Report for the IEA Bioenergy Task Force 40. Paris, France: International Energy Agency.

Rossi AM and CC Hinrichs. 2011. Hope and skepticism: Farmer and local community views on the socio-economic benefits of agricultural bioenergy. *Biomass and Bioenergy* 35: 1418–1428.

RSB (The Round Table on Sustainable Biofuels). 2011. *Consolidated RSB EU RED principles & criteria for sustainable biofuel production*. RSB-STD-11-001-01-001 (Ver 2.0). Lausanne, Switzerland: The Round Table on Sustainable Biofuels.

Salvi BL and NL Panwar. 2012. Biodiesel resources and production technologies—A review. *Renewable and Sustainable Energy Reviews* 16: 3680–3689.

Sanderson MA and PR Adler. 2008. Perennial forages as second generation bioenergy crops. *International Journal of Molecular Sciences* 9: 768–788.

Searchinger T, R Heimlich, RA Houghton, F Dong, A Elobeid, J Fabiosa, S Tokgoz, D Hayes, and T-H Yu. 2008. Use of U.S. croplands for biofuels increases greenhouse gases through emissions from land use. *Science* 319: 1238–1240.

Searcy E, P Flynn, E Ghafoori, and A Kumar. 2007. The relative cots of biomass energy transport. *Applied Biochemistry and Biotechnology* 136–140: 639–652.

Selfa T, L Kulcsar, C Bain, R Goe, and G Middendorf. 2011. Biofuels bonanza? Exploring community perception of the promises and perils of biofuels production. *Biomass and Bioenergy* 35: 1379–1389.

Simmons BA, D Loque, and HW Blanch. 2008. Next-generation biomass feedstocks for biofuel production. *Genome Biology* 9: 242–248.

Smith AL, N Klenk, S Wood, N Hewitt, I Henriques, N Yan, and DR Bazley. 2013. Second generation biofuels and bioinvasions: An evaluation of invasive risks and policy responses in the United States and Canada. *Renewable and Sustainable Energy Reviews* 27: 30–42.

Stephen JD, WE Mabee, and JN Saddler. 2010. Biomass logistics as a determinant of second-generation biofuel facility scale, location and technology selection. *Biofuels, Bioproducts and Biorefining* 4: 503–518.

Sticklen M. 2006. Plant genetic engineering to improve biomass characteristics for biofuels. *Current Opinion in Biotechnology* 17: 315–319.

Suntana AS, KA Vogt, EC Turnblom, and R Upadhye. 2009. Bio-methanol potential in Indonesia: Forest biomass as a source of bio-energy that reduces carbon emissions. *Applied Energy* 86: 5215–5221.

Teel A and S Barnhart. 2003. *Switchgrass seeding recommendations for the production of biomass fuel in southern Iowa*. Ames, IA: Iowa State University Cooperative Extension.

Tilman D, J Hill, and C Lehman. 2006. Carbon-negative biofuels from low-input high-diversity grassland biomass. *Science* 314: 1598–1600.

Tilman D, R Socolow, JA Foley, J Hill, E Larson, L Lynd, S Pacala, J Reilly, T Searchinger, C Somerville, and R Williams. 2009. Beneficial biofuels—The food, energy and environment trilemma. *Science* 325: 270–271.

Tubby I and A Armstrong. 2002. *Establishment and management of short rotation coppice*. Edinburgh, UK: Silviculture and Seed research Branch, Forest Research, Forestry Commission.

Tumuluru JS, CT Wright, KL Kenny, and JR Hess. 2010. *A technical review on biomass processing: Densification, preprocessing, modeling and optimization*. ASABE Paper No. 1009401. St. Joseph, MI: American Society of Agricultural and Biological Engineers.

UNEP (United Nations Environment Programme). 2009. *Towards sustainable production and use of resources: Assessing biofuels*. United Nations Environment Programme. Available at: http://www.unep.org/pdf/biofuels/Assessing_Biofuels_Full_Report.pdf (Accessed September 8, 2013).

U.S. Department of Energy. 2011. *U.S. Billion-ton update: Biomass supply for a bioenergy and bioproducts industry*. RD Perlack and BJ Stokes (leads). ORNL/TM-2011/224. Oak Ridge, TN: Oak Ridge National Laboratory.

Van der Horst D and S Vermeylen. 2011. Spatial scale and social impacts of biofuel production. *Biomass and Bioenergy* 35: 2435–2443.

Virgilio ND, A Monti, and G Venturi. 2007. Spatial variability of switchgrass (*Panicum virgatum* L.) yield as related to soil parameters in a small field. *Field Crops Research* 101: 232–239.

Volk TA, T Verwijst, PJ Tharakan, LP Abrahamson, and EH White. 2004. Growing fuel: A sustainability assessment of willow biomass crops. *Frontiers in Ecology and the Environment* 2: 411–418.

Vogel KP, JJ Brejda, DT Walters, and DR Buxton. 2002. Switchgrass biomass production in the Midwest USA: Harvest and nitrogen management. *Agronomy Journal* 94: 413–420.

Wang D, DS Lebauer, and MC Dietze. 2010. A quantitative review comparing the yield of switchgrass in monocultures and mixtures in relation to climate change and management factors. *Global Change Biology—Bioenergy* 2: 16–25.

White EM. 2010. *Woody biomass for bioenergy and biofuels in the United States—A briefing paper.* General Technical Report PNW-GTR-825. United State Department of Agriculture, Forest Service, Pacific Northwest Research Station, Portland, Oregon (USA).

Wilhelm WW, JMF Johnson, JL Hatfield, WB Voorhees, and DR Linden. 2004. Crop and soil productivity response to corn residue removal. *Agronomy Journal* 96: 1–17.

Wright L. 2006. Worldwide commercial development of bioenergy with a focus on energy crop-based projects. *Biomass and Bioenergy* 30: 706–714.

Wright L and A Turhollow. 2010. Switchgrass selection as a "model" bioenergy crop: A history of the process. *Biomass and Bioenergy* 34: 851–868.

Zenone T, I Gelfand, J Chen, SK Hamilton, and GP Robertson. 2013. From set-aside grassland to annual and perennial cellulosic biofuel crops: Effects of and use change on carbon balance. *Agricultural and Forest Meteorology* 182–183: 1–12.

Zheng Y, Z Pan, and R Zhang. 2009. Overview of biomass pretreatment for cellulosic ethanol production. *International Journal of Agricultural and Biological Engineering* 2: 51–68.

CHAPTER 2

Sorghum as a Sustainable Feedstock for Biofuels

P. Srinivasa Rao,[1] **Reddy Shetty Prakasham,**[2] **P. Parthasarathy Rao,**[1]
Surinder Chopra,[3] **and Shibu Jose**[4]

[1]International Crops Research Institute for the Semi-Arid Tropics (ICRISAT), Hyderabad, India
[2]CSIR–Indian Institute of Chemical Technology, Hyderabad, India
[3]Plant Science Department, Pennsylvania State University, University Park, PA, USA
[4]The Center for Agroforestry, University of Missouri, Columbia, MO, USA

CONTENTS

2.1 INTRODUCTION

Biofuels have the potential to contribute significantly in meeting the challenges of the global energy crisis as they replace fossil fuels and provide a number of environmental and economic benefits. The most widely used biofuel for transportation worldwide so far is bioethanol produced from corn grain, sugarcane, and biomass. Reducing dependence on fossil fuel is a key element of the energy policy adapted by many nations (Srinivasarao et al. 2013). At present, global production of ethanol exceeds 85 billion liters per year using sugarcane (60%), maize grain, sorghum grain, wheat grain, and sugar beet (40%) as feedstocks. The U.S. Energy Independence and Security Act RFS2 (EISA 2007) has set an ambitious target to triple the current 12 billion gallons a year biofuel use by 2022, with 21 billion gallons coming from advanced biofuels. In 2011, the United States used 40% of its corn production for ethanol, generating serious concerns about the future of food versus fuel. Generating biofuels from food crops is unsustainable because the high demand for biofuels will create a shortage in supplies because more and more of these crops will be diverted from food supply to biofuel production. Although crops such as corn and sugarcane are expected to remain important sources of biofuels, greater emphasis is being placed on other potential energy crops such as grasses and nonfood crops.

According to the International Energy Agency (IEA), there was a 27 million ha increase in harvested areas globally during 2005–2010 for 13 major crops such as wheat (*Triticum aestivum* L.), rice (*Oriza sativa* L.), maize (*Zea mays* L.), soybeans (*Glycine max* (L.) Merr.), pulses (legumes), barley (*Hordeum vulgare* L.), sorghum (*Sorghum bicolor* (L.) Moench), pearl millet (*Pennisetum glaucum* (L.) R. Br.), cotton (*Gossypium hirsutum* L.), rape seed (*Brassica napus* L.), groundnut (*Arachis hypogaea* L.), sunflowers (*Helianthus annuus* L.), and sugarcane (*Saccharum officinarum* L.). This rapid expansion of crops, many of them currently used also for biofuel production, indicates that there is a scope to expand their area globally to meet the rising demand for energy.

Sorghum is the fifth most important cereal crop and is the staple food of more than 500 million people in more than 90 countries, primarily in the developing world. It belongs to the family Poaceae and is a highly efficient C_4 photosynthetic crop that is well adapted to tropical and arid climates. Due to its distinct physiological characteristics, the crop is extremely efficient in using water, carbon dioxide, nutrients, and solar radiation (Kundiyana 1996; Serna-Saldivar 2010). It is also able to tolerate high temperatures and drought, and it is considered as one of the most successful crops in the semiarid tropics, such as in Africa and Asia (Woods 2000). Its ability to tolerate high temperatures and drought stress is due to its C_4 photosynthetic metabolism, which contributes to sugar accumulation during the night, ensuring low photorespiration rates during the daytime that consequently reduce water loss (Keeley and Rundel 2003). Sorghum has an advantage over other biofuel crops such as maize because a lower seed rate of 10–15 kg ha^{-1} is required (Kundiyana 1996). In some parts of the world, it is possible to produce multiple crops per year, either from seed or from ratoons (Saballos 2008; Turhollow et al. 2010). All the features mentioned here make sorghum an environmentally sustainable crop.

Sweet sorghum produces food (grain) and fuel (ethanol from stem sap) and the stalks contain 10–15% sugars (Srinivasarao et al. 2009). The sweet stalks are traditionally used as livestock fodder due to their ability to form excellent silage, whereas the stalk juice is extracted, fermented, and distilled to produce ethanol. Therefore, the grain, juice, and bagasse (the fibrous residue that remains after juice extraction) can be used to produce food, fodder, ethanol, and cogeneration. This chapter will focus on the efficient features of sorghum that make it a sustainable source for biofuel production.

2.2 TYPES OF SORGHUM

Sorghum is a unique multipurpose crop used as food, feed, and fodder, as well as fuel. Sorghum is mainly used as food in various forms in different regions, specifically African and South Asian countries. In some countries, it is also used as feed for animals in the form of silage, owing to its sugar-rich stalks. It is capable of producing high biomass under low input production systems, and hence, is amenable for the production of biofuel and bioproducts (Rooney et al. 2007; Venuto and Kindiger 2008; Prakasham et al. 2014). Based on utility patterns, sorghum can be broadly classified as grain sorghum, sweet sorghum, forage sorghum, brown midrib sorghum, and energy sorghum; each class of sorghum has more than one usage in common (Figure 2.1).

2.2.1 Grain Sorghum

Grain sorghum is a major food crop in the semiarid regions of Africa and Asia, and the second most important feed grain in the United States. Grain sorghums grow 3–5 feet tall, depending upon variety and conditions. They are usually not considered for forage production because of low dry matter yield (Undersander 2003). People eat the grain as porridge; grind the grains into flour for unleavened bread, cookies, and cakes; and ferment it into a malted drink. Leading sorghum

Figure 2.1 **(See color insert.)** Types of sorghum: (a) grain sorghum, (b) brown midrib sorghum, (c) energy sorghum, (d) sweet sorghum, (e) forage sorghum, (f) broomcorn, and (g) sudangrass.

producers around the world are Nigeria (19.3%), the United States (16.3%), India (11.7%), and Mexico (10.5%) (Sorghum U.S. Grains Council 2010). Grain sorghums have a high panicle-to-green-biomass ratio, with dwarf, low-tillering hybrids suitable for combine harvest. Sorghum grain is rich in Ca, Mg, P, and K, and hence, is considered a good feed grain as long as it is properly supplemented. Sorghums without a pigmented testa have 95% or greater feeding value of yellow dent maize for all species of livestock. Ethanol production has become an important new market for grain sorghum due to the classification of grain sorghum as an advanced biofuel feedstock in the US 2008 Farm Bill. In the United States, about 30% of sorghum grain goes for ethanol production. In developing countries such as India, a significant portion of rainy season produce is converted to either ethanol or beverages based on market demand. The yield potential of sorghum stover after grain harvest is identical to that of corn with an estimated harvest of 4 tons dry stover per hectare (Saballos 2008).

2.2.2 Sweet Sorghum

Sweet sorghum is similar to grain sorghum, except that the stalk in the former accumulates sugar that can be used for various purposes. Cultivation practices are similar for grain and sweet sorghums. Sweet sorghum can be grown under a wide range of environmental conditions. The sugar-rich juice obtained from sweet sorghum stalks can be used for ethanol production (Prasad et al. 2007). It has reasonably good tolerance to drought, waterlogging, and saline-alkaline conditions (Srinivasarao et al. 2009; Zegada-Lizarazu and Monti 2012). Compared to sugarcane or maize, the two major feedstocks used globally for ethanol production, ethanol from sweet sorghum is cheaper, quicker to produce, and poses little risk to the environment. In addition, it is a seed-propagated, short-duration crop with hybrid technology in place. The presence of reducing sugars in sweet sorghum prevents crystallization and has 90% fermentation efficiency. These important traits, in addition to the crop's suitability for seed propagation and mechanized crop and ethanol production as compared to sug-arcane molasses, give it an edge as a viable alternative raw material for ethanol production. Sweet sorghum–based ethanol is sulphur-free and cleaner than molasses-based ethanol, when mixed with gasoline (Basavaraj et al. 2013b). The ethanol conversion process from sweet sorghum juice generates low or negligible effluents in the spent wash, and the same can be composted with press-mud to pro-duce organic fertilizer (Semra et al. 1997; Srinivasarao et al. 2009). The usage of bioethanol blended with gasoline fuel for automobiles will significantly reduce petroleum use and reduce greenhouse gas (GHG) emissions (Balat and Balat 2009). The ethanol production potential of sweet sorghum (5600 L ha^{-1} yr^{-1} from 140 t ha^{-1} with two crops annum^{-1} @ 40 L t^{-1}) is almost equivalent to ethanol production from sugarcane (6500 L ha^{-1} from 85 to 90 t ha^{-1} per crop @ 75 L t^{-1}) (Srinivasarao et al. 2009, 2010a, 2010b; Shoemaker and Bransby 2010). Although the production cost varies from region to region, the cost of ethanol production from sugarcane has been estimated to vary between US$0.39 and US$0.52; whereas for sweet sorghum, it is US$0.27–0.31 per liter (Raju et al. 2012), thus making sweet sorghum a viable feedstock. However, it is to be noted that production cost primarily varies with procurement price of feedstock, distance between feedstock production area and processing unit, taxes, and incentives because processing technology is standardized.

2.2.3 Forage Sorghum

Forage sorghum grows 6–12 feet tall, produces more dry matter than grain sorghum, and its coarse stem is used for silage. It displays abundant tillering ability, and some types are peren-nial or semiperennial in warmer regions. Forage types are traditionally referred to as sorghum-sudangrass or sorghum-sudangrass hybrids. Sorghum-sudangrass hybrids are a cross between the two forage types that have intermediate yield potential and can be used for pasture, hay, or silage (Undersander et al. 1989).

2.2.4 Brown Midrib Sorghum

The diethyl sulfate (DES)-induced mutation in two grain sorghum resulted in the generation of brown midrib (*bmr*) lines (Porter et al. 1978). These mutations are phenotypically characterized by the presence of brown vascular tissues in the leaf blade and leafsheath as well as in the stem. The *bmr* phenotype is very distinct until the four leaf stage, later it fades with maturity. Although the intensity of the brown coloration is not a measure of reduction in lignin, it is an indicator that the *bmr* gene(s) are present (Miller and Stroup 2003). The *bmr* mutants significantly reduce the level of enzyme-resistant lignin in plants and increase their palatability and digestibility (Cherney et al. 1991). The silage, with and without protein supplements, is known to significantly increase the milk yield of lactating cows (Cherney et al. 1991; Oba and Allen 1999). The reduced lignin content of *bmr* sorghum increases its energy conversion efficiency and nutritive value as a livestock feed (Gressell 2008). Glucose yields for the sorghum biomass improved by 27%, 23%, and 34% for bmr-6, bmr-12, and the double mutant, respectively, as compared to the wild type; thus, reducing lignin content increases conversion efficiency of lignocellulose to sugars and ethanol (Dien et al. 2009). It is believed that *bmr* sorghums can play a greater role in lignocellulosic biofuel production due to reduced pretreatment costs, besides its use in forage-fodder industry.

2.2.5 Energy Sorghum

Dedicated cellulosic energy crops will need to produce large quantities of biomass over a short period of time with low or limited inputs such as fertilizer and water. Because sorghum is a short day plant, photoperiod-sensitive (PS) varieties are available in which flowering does not initiate until day length falls below 12 hours, 20 minutes (in some locations it is less than 12 hours); the plant would continue to produce biomass until it flowers. The PS characteristic allows these sorghums to accumulate vegetative dry matter for longer periods throughout the growing season. These hybrids can be derived from the cross of two photoperiod-insensitive parental sorghum lines (Rooney et al. 2007). It has been estimated that corn stover and switchgrass can produce 3.14 and 15.6 t ha^{-1} harvestable dry mass, respectively (TAES 2006). On the contrary, sorghum can produce 29.12 t ha^{-1} harvestable dry mass, and it has the potential to produce 33.6–44.8 t ha^{-1} stover through genetic improvement, which is quite realistic considering that corn has already undergone much genetic improvement so far. McCollum et al. (2005) reported dry matter yields of PS sorghum to be 26–43% greater than non-*bmr* and *bmr* forage sorghums. The highest biohydrogen yield was observed in energy sorghum line IS 27206 followed by IS 22868 and ICSV 93046 (Nagaiah et al. 2012).

2.3 POTENTIAL USES OF SORGHUM

2.3.1 Resilient Dryland Crop

Sorghum is a fast-growing C$_4$ plant with a wide adaptability to different environmental conditions. Because sorghum can grow well on marginal lands and less fertile salinity prone lands, it does not compete with food crops. Sorghum has the capacity to survive dry spells and resume growth upon receipt of rain. Sorghum also withstands wet extremes better than many other cereal crops, especially maize. Sorghum continues to grow, though not well, in flooded conditions; maize by contrast will die. In another study, the results for the measured variables—carbon exchange rate (CER), transpiration, transpiration ratio (CER/transpiration), leaf diffusive resistance, leaf water potential, and osmotic adjustment—showed a general trend for greater drought resistance in sorghum than in millet, indicating that the commonly observed adaptation of the millets to dry environments may be due to other factors, such as drought escape or heat tolerance

(Blum and Sullivan 1985). Sugarcane requires 900 m³ of water for producing one ton of dry matter while sorghum requires only 200 m³ of water. Further, it is not possible to increase the area under sugarcane beyond a certain limit because increasing the area under sugarcane will be at the cost of diverting land from other staple food crops (Shinoj et al. 2011). Sorghum has relatively lower agronomic requirements compared to other sugar crops such as sugarcane or sugar beet and produces 30% more dry matter per unit of water (Shoemaker and Bransby 2010).

2.3.2 Food/Feed/Fodder Crop

Sorghum grain is used for preparing various food products such as porridges, flat bread, chips, *bhakri*, *suhali*, *khichri*, *dalia*, *shakkerpera*, *ugali*, *tortias*, and *to*, and is also used extensively in making baked products, extruded products, health products, and in weaning and supplementary foods (Sehgal et al. 2003). As sorghum is a gluten-free cereal with high fibre content it is more preferred for celiac patients. Sorghum has emerged as a potential feedstock because of its versatility, yield potential, and growth characteristics. Among the cultivated sorghums, *bicolor* sorghums have been selected not only for grain production but also for fiber, forage, and sugar production (Dogget 1998). The sugar type is characterized by genotypes that produce tall juicy stalks rich in sugars. In general, the content of nonstructural carbohydrates is higher in sweet sorghum types than in forage or fiber ones (Guiying et al. 2000). The fiber and forage types are predominantly composed of structural carbohydrates. Under arid and semiarid conditions in saline lands, the combination of drought tolerance and salt tolerance makes sorghum a very interesting feed resource (Al-Khalasi et al. 2010; Fahmy et al. 2010; Khanum et al. 2010). Sorghum grain can replace maize in poultry feed to a great extent in view of the similarity in chemical composition of the grain. The results on egg production and broiler weight were similar when sorghum or maize was fed as a source of energy. Mold-infected grain is increasingly used to produce beverages, rectified spirits, or ethanol. Most recently, various forage accessions have been assessed for their potential for renewable fuel production. Near infrared technologies have been developed to quickly and cost effectively screen large numbers of accessions for such compositional characteristics as ash, lignin, glucan, xylan, galactan, and arabinan, all of which have unique properties related to various bioconversion technologies. In the future, utilization of sorghum as a health food, feed, and bioenergy source will make it an attractive renewable resource (Dahlberg et al. 2011).

2.3.3 Biofuel Feedstock

Sorghum is a candidate in the search for efficient energy crops due to an increased interest in the conversion of biomass to energy. The bagasse can be utilized to generate electricity or steam as part of a cogeneration scheme or as a feedstock for cellulosic biofuel production. Several high biomass sorghum hybrids have been developed and improved for the production of sugar, starch, and lignocellulosic feedstocks (Rooney et al. 2007); further its development as an energy crop is also gaining importance. Sweet sorghum, which accumulates sugar in the stalks, can be used as the only feedstock for biofuel production through three different pathways, with negligible trade-offs on food security. These are starch-to-ethanol, sugar-to-ethanol from sweet sorghum, and cellulosic ethanol yield (Gnansounou et al. 2005; Srinivasarao et al. 2011, 2012; Damasceno et al. 2014). The sorghum-based ethanol pathways can achieve substantial fossil energy savings compared to gasoline. The grain sorghum–based ethanol production using fossil natural gas (FNG) as the process fuel can achieve moderately lower GHG emission reductions relative to baseline conventional gasoline than corn ethanol. Sweet sorghum ethanol achieves about 71–72% well to wheels (WTW) GHG emission reductions. Forage sorghum ethanol has WTW fossil energy use and GHG emission reductions similar to grain sorghum ethanol using renewable natural gas (Cai et al. 2013). Within the biorefinery concept, biomass can be converted to useful biomaterials and/or energy carriers in an integrated manner, thereby maximizing the economic value of the feedstock used while reducing the waste streams produced (Thomsen 2005). Biorefineries for the production of several products and by-products from bagasse—such as bioethanol, heat or electricity, and feed—have been highlighted in recent years (Luo et al. 2010). When compared to other feedstocks such as sugarcane, sugar beet, maize, and wheat, sorghum yields a better energy output/input ratio (Almodares and Hadi 2009).

2.3.4 Other Industrial Uses

Sorghum, like other cereals, is an excellent source of starch and protein and can be processed into starch flour, grits, and flakes that can be used to produce a wide range of industrial products (Palmer 1991). Sorghum grain is also used for the industrial manufacture of potable alcohol, malt, beer, liquid gruels, starch, adhesives, core binders for metal casting, ore refining, and grits as packaging material. Sorghum fibers are used in wallboard, fences, biodegradable packaging materials, and solvents. Dried stalks are used as cooking fuel, and dye extracted from the plant is used to color local leather. The classic example of industrial use of sorghum in Europe is broomcorn (*Sorghum vulgare* var. *technicum* [*Korn.*]). Trends favoring ecological and natural products of all kinds have led to renewed interest in old-fashioned, biodegradable, wooden handled brooms, which had a positive impact on broomcorn production (Dahlberg et al. 2011). Sorghum malt increases protein content, fat, fiber, minerals, in vitro protein digestibility, and in vitro starch digestibility (Sehgal et al. 2003). In cancer research and glycemic control, novel technologies are permitting sorghum germplasm to be screened for high levels of antioxidants that show potential.

2.4 BIOETHANOL FROM SORGHUM: STATUS

The leading nations in bioethanol production are the United States and Brazil, with the United States being the world's largest producer. Approximately one-third of the U.S. sorghum crop is used for biofuel production and contributes to 31% of ethanol's market share. Naturally drought tolerant, sorghum can be used for different types of alcohol production including starch-based, sugar-based, and cellulosic ethanol production. Sorghum and corn are interchangeable in the grain-based ethanol market. One kilogram of grain sorghum produces as much ethanol as a kilogram of corn. Sorghum dried distillers grains with solubles (DDGs), a coproduct in starch-to-ethanol production systems, tends to be lower in fat and higher in protein than corn DDGs. A few developing countries in Asia and many African nations use sorghum grain for making beverages and ethanol. In India, mold-infected rainy season sorghum produce is widely used for the production of ethanol and beverages.

Of late, sweet sorghum is increasingly used as a raw material for ethanol production. Approximately 50–85 t ha^{-1} of sweet sorghum stalks yields 39.7–42.5 t ha^{-1} of juice, which upon fermentation yields 3450–4132 L ha^{-1} ethanol (Serna-Saldivar et al. 2012). Other studies have shown similar ethanol production levels: 3296 L ha^{-1} (Kim and Day 2011) and 4750–5220 L ha^{-1} (Wu et al. 2010). The Government of India (GOI) approved the National Policy on Biofuels on December 24, 2009. The policy encourages use of renewable energy resources (sweet sorghum, cassava, sugarbeet) as alternate fuels to supplement transport fuels and had proposed an indicative target of replacing 20% of petroleum fuel consumption with biofuels (bioethanol and biodiesel) by the end of the 12th Five-Year Plan (2017). The Rusni distillery was the first sweet sorghum distillery established in 2007 near Sangareddy, Medak district of Andhra Pradesh, India, amenable to using multiple feedstocks for transport grade ethanol production. It generated 99.4% of fuel ethanol with a total capacity of 40 kiloliters per day (KLPD). It also produced 96% extra neutral alcohol (ENA) and 99.8% pharma alcohol from agro-based raw materials such as sweet sorghum juice, molded grains, broken rice, cassava, and rotten fruits. ICRISAT has incubated sweet sorghum ethanol production in partnership with Rusni Distilleries through its Agri-Business Incubator. Rusni is a 40 KLPD ethanol production unit located 25 km from ICRISAT headquarters. A pilot scale sweet sorghum distillery of 30 KLPD capacity was established in 2009 at Nanded, Maharashtra. It used commercially grown sweet sorghum cultivars such as CSH 22SS, ICSV 93046, sugargrace, JK Recova, and RSSV 9 in the 25-km radius of the distillery to produce transport grade ethanol and ENA from 2009 to 2010. However, it could not continue operations further due to the low mandated ethanol price. On November 22, 2012, the Cabinet Committee of Economic Affairs (CCEA) of the GOI recommended a 5% mandatory blending of ethanol with gasoline (GAIN report 2013). The government's current target of 5% blending of ethanol in gasoline has been partially successful in years of surplus sugar production and unfilled when sugar production declines. The interim price of US$0.44 per liter would no longer hold as the price would now be decided by market forces. It is expected this decision will have a positive effect on upcoming distilleries in India.

Nonfood crops and materials such as cassava and sweet sorghum are the priority choice for biofuel ethanol production in China. The Sorghum Research Institute (SRI) of the Liaoning Academy of Agricultural Sciences (LAAS) is the lead organization involved in sweet sorghum research in China since the 1980s. So far, 17 promising sweet sorghum hybrids have been released nationally. A few industries such as ZTE energy company limited (ZTE, Inner Mongolia), Fuxin Green BioEnergy Corporation (FGBE), Xinjiang Santai Distillery, Liaoning Guofu Bioenergy Development Company Ltd., Binzhou Guanghua Biology Energy Company Ltd., Jiangxi Qishengyuan Agri-Biology Science and Technology Company Ltd., Jilin Fuel Alcohol Company Ltd., and Heilongjiang Huachuan Siyi Bio-fuel Ethanol Company Ltd. either conducted large-scale sweet sorghum–processing trials or are in the commercialization stage.

In the Philippines, sweet sorghum has been proven to be a technically and economically viable alternative feedstock for bioethanol production. The plantation, agronomic performance, and actual bioethanol production of sweet sorghum was evaluated on different plantation sites nationwide. A hectare of a sweet sorghum plantation can potentially provide farmers with an annual net income of US$1860.47 at a stalk selling price of US$22 and a grain price of US$0.3. San Carlos Bioenergy Inc. (SCBI) became the first commercial distillery to process sweet sorghum bioethanol in Southeast Asia under the Philippine Department of Agriculture (DA) and produced 14,000 L of fuel grade ethanol in 2012. The Ecofuels 300 KLPD distillery at San Mariano, Isabela, is planning to use sugarcane and sweet sorghum as feedstocks for commercial ethanol production. The sweet sorghum growers are enthusiastic because the ratoon yields are about 20–25% higher than that of the plant crop. The Bapamin enterprises based in Batac have been successfully marketing vinegar and hand sanitizer made from sweet sorghum since 2009.

In the United States, a sweet sorghum distillery is under construction in South Florida by South Eastern Biofuels Ltd. In Brazil, large-scale sweet sorghum pilot trials are being conducted by Ceres, Inc., Chromatin, Inc., Advanta Inc., and Dow Agro Sciences besides Empresa Brasileira de Pesquisa Agropecuária (EMBRAPA) since last three years in areas of sugarcane renovation to commercialize sweet sorghum. The Government of Brazil has identified 1.8 m ha for sweet sorghum plantations to augment fuel grade ethanol production. In African countries such as Mozambique, Kenya, South Africa, and Ethiopia, sweet sorghum adaptation trials are being conducted on a pilot scale to assess feasibility of sweet sorghum for biofuel production.

2.5 BIOFUEL PRODUCTION MODELS

The biofuel production value chain model for sweet sorghum encompasses the production, transportation, crushing, ethanol production, and the blending with gasoline. In this section, two primary models for the production of bioethanol from sweet sorghum feedstock—a centralized unit (CU) and a decentralized crushing unit (DCU)—will be examined.

2.5.1 CU Model

In the CU model, a constellation of villages within a 50-km radius from the crushing units are targeted for the cultivation of the crop, along with the transportation of the sweet sorghum stalks to the crushing units within 24 hours of harvesting from the farmer's field (Srinivasarao et al. 2013). The centralized model of the sweet sorghum supply chain linking farmers to the distillery is illustrated in Figure 2.2. A CU of 40 KLPD capacity requires crops from about 8000 ha yr^{-1} for two seasons (e.g., Rusni Distilleries Ltd., Sangareddy, Medak District, Andhra Pradesh, India, and Tata Chemicals Ltd., Nanded, Maharashtra, India), although probably a lesser area is required for ZTE Ltd., Inner Mongolia, China, due to the higher productivity of the stalk (Srinivasarao et al. 2013). Supply of the sweet sorghum stalks directly from the farmer's field to the distillery should happen in quick time while keeping the eye on in-stalk fermentation, an issue in the tropics. During the post rainy season, it would be really difficult to find 4500 ha of land with irrigation facilities for cultivation of sweet sorghum. In turn, it would be difficult to convince farmers to cultivate sweet sorghum because they find other crops such as soybean, maize, rice, and wheat to be more profitable under irrigation. The major issues in the centralized model are (i) price fixing for the procurement of the stalks by the distilleries, (ii) harvesting and transportation of the stalks to the distillery in time, (iii) maintenance of the quality of the sugar content in the stalks, (iv) planning of different dates of sowing for continuous supply of the feedstock, and (v) mechanical harvesting. The DCU model overcomes some of these difficulties.

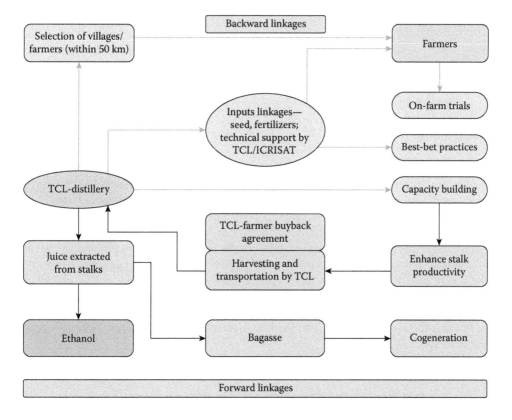

Figure 2.2 Centralized model of the sweet sorghum supply chain linking farmers to distillery. (Reddy, C. R., et al., Community seed system: Production and supply of sweet sorghum seeds, In: *Developing a sweet sorghum ethanol value chain*, eds. B. V. S. Reddy, et al., ICRISAT, Hyderabad, India, 2013, 38–44.)

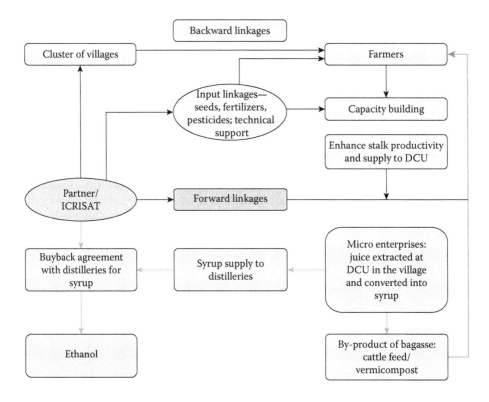

Figure 2.3 Decentralized model—a village enterprise to crush stalks and produce syrup linked with a centralized model to produce ethanol from syrup. (From Srinivasarao, P. et al., Sweet sorghum: Genetics, breeding and commercialization, In: *Biofuel crops: Production, physiology and genetics*, ed. B. P. Singh, CABI, Cambridge, MA, 2013, 172–198.)

2.5.2 DCU Model

The DCU model involves the crushing of sweet sorghum stalks, extracting and boiling the juice for syrup production in the area of crop production, when it is beyond 50 km from CU. The advantage of this model is localized production and processing, and it ensures bagasse usage to feed livestock (Figure 2.3). Hence, a farmer's cooperative can manage DCU as a microenterprise. This model also assures a link between DCU and CU with a buyback agreement for syrup supply at a preagreed-upon price. However, some of the issues that draw stakeholders' attention are (i) the establishment of a DCU requires a high initial investment, (ii) assurance is required from the CU to buy small quantities of syrup, (iii) price fixation criteria by CU, (iv) whether payments to farmers are based on either stalk weight or syrup quality, and so forth, and (v) use of the by-products or giving them back to the farmers. The DCUs were operated at Ibrahimbad, Andhra Pradesh (2008–2011), and at Parbhani, Maharashtra (2010–2012), by farmers' cooperatives and ICRISAT.

2.6 SWEET SORGHUM ETHANOL VALUE CHAIN: SWOT ANALYSIS

As described earlier, CU and DCU are the two pilot models tested for conversion of sweet sorghum into bioethanol. The following sections discuss the strengths, weaknesses, opportunities, and threats (SWOT) analysis of CU and DCU.

2.6.1 SWOT Analysis of the CU

Several attempts were made to use sweet sorghum for ethanol production in a centralized model though difficulties such as raw material availability, transporting the harvested stalks in a timely manner, and so on were major impediments to success (Srinivas et al. 2013). The following section elaborates the SWOT of CU.

2.6.1.1 Strengths of the CU for Ethanol Production

- In this model, groups of adjoining villages are targeted to cultivate the sweet sorghum crop within the 50-km radius from the distillery.
- The CU model minimizes the transportation cost of the stalks to the distillery (Srinivasarao et al. 2013).
- It offers the advantage of crushing the raw material in bulk quantities.
- Because it is highly mechanized, the efficiency of sugar recovery is high and the bagasse can be used either for power or for fertilizer.
- The CU model also creates jobs in this sector.

2.6.1.2 Weaknesses of the CU for Ethanol Production

- The probability of operating a processing unit at full capacity is minimal due to lack of raw material availability (Srinivas et al. 2013).
- Technologies for microdistilleries not yet mature.

2.6.1.3 Opportunities of the CU for Ethanol Production

The opportunities that could be explored for this model are as follows:

- To overcome land constraints, CU can enter into contractual agreements with the small farmers.
- Both the CU and farmers can profit if contract farming is adopted as a form of commercial agricultural production.
- A long-term relationship between the farmers and the stakeholders can be established by linking farmers directly to the markets.
- A case study of the ICRISAT–Rusni partnership in various activities of the ethanol value chain provides improved crop production technology to the farmers and the partners.
- Farmers are linked with the distillery via a buyback agreement (Srinivas et al. 2013).
- The private sector is expected to play a key role in fostering linkage with various farmers groups.
- Changes in policies and emergence of new market opportunities for ethanol fuel (e.g., aviation, heavy vehicles, households, rural electrification) may provide further opportunities.

2.6.1.4 Disadvantages of the CU for Ethanol Production

- Moving toward alternative transport, including electric vehicles, may affect biofuel markets worldwide.
- Fluctuations in global prices of agricultural commodities including ethanol as well as trade policies may negatively affect the prospects of CU.

2.6.2 SWOT Analysis of DCU Model

A SWOT analysis is carried out on the sweet sorghum to ethanol value chain in DCU to form a better picture of the conversion of the sweet sorghum to syrup and further to ethanol in the distillery unit established in 2008 at Ibrahimbad village, Medak district, Andhra Pradesh, India (Basavaraj et al. 2013a).

2.6.2.1 Strengths of the DCU for Syrup-Ethanol Production

- DCU model for sweet sorghum ethanol production reduces the transportation of feedstock over very long distances.
- The by-product (bagasse) obtained after crushing can be used as a livestock feed and as an organic matter by farmers (Srinivasarao et al. 2013).
- DCU generates additional income from syrup production for the farmers, providing economic security.

2.6.2.2 Weaknesses of the DCU for Ethanol Production

- In operating the model, there is a lack of coordination and implementation of strategies.
- There is a lack of management skills in operating and handling DCU by farmers (Basavaraj et al. 2013b).
- The costs involved in the establishment, processing, and operating of DCU are high.
- DCU are incapable of providing large quantities of syrup to the industry for ethanol production.
- Farmers' inability to bargain for the price of syrup is a drawback in this model.

2.6.2.3 Opportunities for the DCU

- To mechanize and standardize the processing activities, DCU provides future opportunities by the addition of other by-products to the markets.
- In rural areas, promotion of the DCU as a small-scale agro-enterprise could be beneficial for strengthening the production of ethanol.
- DCU provides opportunities for a lower income group of farmers to become microlevel entrepreneurs through establishment and management of the DCU.

2.6.2.4 Threats for the DCU

- DCU's huge dependency on distilleries is one major reason for its uncertain and uneconomical market price.
- Opting for large-scale syrup production can be counterproductive because most of the processing activities are labor intensive (Basavaraj et al. 2013b).

2.7 ENVIRONMENTAL AND ECONOMIC SUSTAINABILITY

2.7.1 Environmental Sustainability

Environmental sustainability is based on (i) the needs of the poor and (ii) flexibility in meeting present and future needs (i.e., long-term attention to the ecosystem so that a healthy and flourishing natural environment is sustained for future generations). This means that the environment needs to be the primary consideration at all stages of the development (i.e., planning, implementation, monitoring, and follow-up). The benefits of an energy value chain of any biofuel crop are primarily decided by net energy ratio and GHG balance. Sweet sorghum has a high net energy balance of 3.63 when compared to grain sorghum's (1.50) and corn's (1.53) energy balance (Wortmann et al. 2008). Depending on yield, production methods, and the land cover prior to sweet sorghum cultivation, GHGs between 1.4 and 22 kg CO_2 equivalents can be saved (Köppen et al. 2009). An estimation of the potential of sweet sorghum biofuels to reduce GHG in Mozambique when it is used for electricity energy generation showed a savings potential of 1515.9–1203.5 t CO_2 eq^{-1}. However, more research in this area is necessary on the type and efficiency of conversion technology, the use of by-products (e.g., bagasse), crop yield per cultivation area, land-use changes, and on the type of fossil energy carriers that are replaced because these vary with specific scenarios (FAO 2013).

Even though for sweet sorghum the ethanol yield per unit weight of feedstock is lower as compared to sugarcane, the much lower production costs and water requirement for sorghum crops compensate for the difference; hence, it still returns a competitive cost advantage in the production of ethanol in India (Farrell et al. 2006). Production of cellulosic biofuels and other bio-based products are expected to increase national energy independence, improve rural economies, and to reduce GHG as compared to conventional transportation fuels (Demirbas 2008). The food-versus-fuel debate triggered during the sharp rise in food prices in 2007–2008 cannot be attributed to sorghum as mold affected grain and stalk sugars form the raw material for biofuels in the case of sorghum. In case of sweet sorghum, the data on parameters that either determine or contribute to environmental sustainability is lacking from commercially operated plants in the public domain. In light of the ongoing debate surrounding the competition between bioenergy and food, the United Nation's Food and Agriculture Organization (FAO) study comes out in favor of sweet sorghum cultivation for its multiple uses as food, as a first- and second-generation bioethanol, and as a fertilizer (FAO 2013). Yet the expected large gap between future demand and potential domestic supply still requires expanding biofuel production in developing countries that have the land and the climate needed to produce feedstocks on a large scale.

2.7.2 Economic Sustainability

To use available resources to their best advantage, economic sustainability is important for identifying various strategies for production of biofuels. To provide future long-term benefits, economic sustainability is essential to promote the use of those resources in ways that are available and efficient.

2.7.2.1 Trade and Policies

The current ethanol trade represents just 20% of total ethanol demand, but the share has steadily risen from about 12% in 2002 (Licht 2008). The European Union (EU) imports ethanol and is mandated to meet 20% of total energy through renewable sources by 2020. In the EU, the largest importers are the Netherlands, Germany, and the United Kingdom. Some policies were adopted by a number of governments that act as barriers to the biofuel trade to safeguard their emerging biofuel industries. The U.S. government dropped the tax credit (i.e., subsidy) provided to its domestic ethanol sector at the end of 2011 in response to political pressure toward fiscal austerity and by domestic producers of livestock who perceived the tax credit as supporting higher feed prices. However, the rapid growth of the U.S. ethanol sector from 2005 to 2008 was related more to the high levels of profitability due to high oil prices (Babcock 2011). Brazilian exports of biofuels, including volume reexported from countries in the Caribbean Basin initiative, account for about 45% of their global trade. In Brazil, pure ethanol is used in approximately 40% of cars. The remaining vehicles use blends of 24% ethanol with 76% gasoline. Brazil's annual ethanol consumption is nearly 4 billion gallons. Apart from meeting ethanol consumption, Brazil also facilitates ethanol exports to other countries.

Over the past two decades, China's booming economic growth has driven rapid increases in its energy demand. Presently, China is the world's second-largest and fastest-growing energy market; as urbanization and incomes increase, China's energy demand will continue to be driven by heavy industrial, residential, and transport usage. According to the IEA, China's primary energy needs will grow from 1742 Mt of oil equivalent (Mtoe) in 2005 to 3819 Mtoe in 2030, implying an average annual increase rate of 3.2% (IEA 2007). The IEA also predicts that between 2005 and 2030 China will account for almost 30% of the increase in the world's oil consumption. China was able to meet much of its energy requirements from domestic sources until 2001, but since then it has become increasingly dependent on importing energy. Currently, in the world scenario, China ranks as the third-largest importer of oil. China has looked to biofuels to achieve some of these objectives, such as to increase its energy efficiency and its use of alternative, domestically produced (and cleaner) sources of energy. In 2002, China launched its ethanol promotion program (GSI, report). The program has steadily expanded over time, with the provision of support shifting from direct subsidies to tax breaks and low interest loans. With maize constituting about 80% of the feedstock in 2007, China produced around 1.6 Mt of fuel ethanol per year. China's energy sector is heavily regulated, with new ethanol plants requiring central government approval. All transport fuel prices in China are controlled by the government, and the ethanol price is set at 0.911 times the ex-factory price of gasoline (research octane 90) with a sales price within US\$584–US\$730 per ton. Following the government's decision to lift petrol prices in 2008, the ethanol prices rose considerably. Now China has mandated to use nonfood crops for biofuel production and is targeted to produce 10 Mt of ethanol and 2 Mt of biodiesel by 2020.

The Indian government's national policy on biofuels has approved an indicative target of 20% blending of biofuels, for biodiesel and bioethanol, to be reached by 2017 (FAO 2013). There is assured demand from beverage and pharmaceutical industries under the current tendering system for ethanol. Now, the ethanol price in India is fixed by market forces—unlike the earlier ad hoc price of US\$0.44 L^{-1}. Similarly, many nations have fixed targets to increase the contribution of renewable sources of energy in their total energy basket with some emphasis on the transport sector.

2.7.2.2 Economic Analysis of the Sweet Sorghum Value Chain

For analyzing the economics for ethanol production from sweet sorghum if commercially cultivated, the major areas on which to focus are the supply and demand for ethanol in India and the potential for sweet sorghum as an alternative feedstock and the economics of ethanol production from sweet sorghum under CU and DCU to meet a small proportion of mandated blending requirements set by the government for ethanol (Basavaraj et al. 2013a). Accordingly, the area under cultivation, input supply, production (grain and stalk), gains in productivity, cost, and returns were monitored and recorded for the agricultural economic analysis.

2.7.2.2.1 Economic Analysis of CU

The economics of processing sweet sorghum for ethanol production were analyzed based on discussions with Rusni Distilleries (CU) on the recovery of ethanol per ton of stalk and the costs incurred in processing. Table 2.1 shows the economics of ethanol production without accounting for capital costs. Based on an average recovery rate of ethanol at 4.5% (45 L t^{-1} of stalk), feedstock priced at US\$9.67 t^{-1} and ethanol priced at US\$0.43 L^{-1}, the benefit–cost ratio worked out to 0.01.

The data on various parameters used for economic viability assessment of ethanol production from sweet sorghum were collected from the distillery in India. The capacity of the plant is 40 KLPD operating for 180 days. The reference year chosen is 2010, and the economic life of the project is 20 years. All economic costs and benefits (including by-products) are valued at current prices. The prevailing administered price of US\$0.43 L^{-1} of ethanol announced by the GOI and the recovery rate of 4.5% per ton of sweet sorghum was considered for financial and economic viability assessment. The landed cost of feedstock during 2010 was US\$19.35 t^{-1} of stalk. The indicators of economic viability showed negative net present value (NPV) of the project at a discount rate of 10% (bank rate) and a benefit–cost ratio of 0.89 with a feedstock price at US\$19.35 t^{-1} and an ethanol price of US\$0.43 L^{-1} (Basavaraj et al. 2013b). It would thus be difficult for the industry to take off under the current scenario of ethanol price, feedstock price, and recovery rate. Sensitivity analysis was carried out on key parameters where the project NPV becomes zero. The key parameters were identified and include recovery rate, feedstock price, and ethanol price. Small improvement in the aforementioned parameters—particularly recovery rate—would contribute to making the NPV positive. One of the major shortcomings of the CU is extensive coordination and planning

Table 2.1 Costs and Returns of Sweet Sorghum Production

Particulars	Cost (US$)
Sweet Sorghum (Production)	
Average stalk yield (t ha^{-1})	0.23
Variable costs of production excluding family labor ($ ha^{-1})	124.45
Gross returns ($ ha^{-1})	172.87
Net returns excluding family labor ($ ha^{-1})	48.37
Sweet Sorghum (Ethanol Production)	
Cost of the raw material ($ t^{-1})	9.67
Cost of processing ($ t^{-1})	6.19
Recovery of ethanol (L t^{-1})	0.72
Cost of ethanol ($ L^{-1})	0.35
Price of ethanol received ($ L^{-1})	0.43
Benefit–cost ratio	0.01

Source: Basavaraj, G. et al. 2013b. In: *Developing a sweet sorghum ethanol value chain*, eds. B. V. S. Reddy, A. A. Kumar, C. H. Ravinder Reddy, P. Parthasarthy Rao, and J. V. Patil, pp. 110–132. Hyderabad, India: ICRISAT.

Note: The price of ethanol was US$0.34 L^{-1} when the centralized unit was established, and it increased to US$0.43 L^{-1} during 2010 (where US$1 = Rs 62). Medak, Andhra Pradesh, 2007, Centralized Unit.

requirements in supply chain management. Delay in crushing stalks beyond 24 hours of harvest causes low recovery of ethanol per ton of stalk. Additionally, the distillery faced some issues in terms of the functioning of crushers, boilers, and other equipment. Furthermore, mobilizing farmers to cultivate sweet sorghum and sourcing the raw material becomes difficult. However, the observations have shown that, under the centralized system, considerable opportunities exist for increasing the efficiency of the value chain at the crop production and processing stages.

2.7.2.2.2 Economic Analysis of DCU

A DCU was set up at Ibrahimabad village, Medak district, India, on a pilot basis to conquer some of the shortfalls of the centralized unit system due to its delay in transportation of the stalks to the industrial site, which resulted in a loss of juice quality and yield. In the DCU, the stalk will be crushed close to the villages where it is grown, and the juice is converted into syrup and stored in cans. The final product of the DCU (syrup) is transported to the ethanol industrial site for further processing into ethanol. The main advantage of this model is that the storage time for syrup is increased to six to eight months before it is converted into ethanol, thus permitting flexibility in transportation and in conversion into ethanol. Accordingly, the area under cultivation, input supply, production (grain and stalk), gains in productivity, cost, returns, and gains in productivity were monitored and recorded for the agricultural years 2008–2009, 2009–2010, and for 2010–2011 for economic analysis (Basavaraj et al. 2013b). On average, the cost incurred in processing 1 kg of syrup was US$0.41 in 2009 (Table 2.2).

The gross returns and total costs per hectare realized from sweet sorghum for syrup production worked out to US$155.96 and US$399.72, respectively, with a net deficit of US$243.75 (Basavaraj et al. 2013b). The loss is because the cost of sweet sorghum syrup production is US$0.41 kg^{-1} even though the distillery is willing to pay only US$0.16 kg^{-1}. To make the DCU viable or to break even, the syrup cost needs to be reduced by increasing the recovery of juice and brix percent content or by reducing the cost of production. One of the major limitations of the financial viability assessment studies is that they look at benefits only from the financial returns. If incorporated into

Table 2.2   Cost of Syrup Production in Decentralized Unit, Rainy Season, 2009

Item	Total Costs (US$[a])	Cost per Kilogram of Syrup (US$)	Percent to Total Loss
Cost of Raw Materials			
Cost of stalks	6773.06	0.23	57
Processing Costs			
Labor costs	3400.48	0.11	29
Chemical costs	336.29	0.01	3
Fire wood costs	174.59	0.00	1
Operating Expenses			
Fuel costs	763.85	0.00	6
Repair and maintenance	255.95	0.00	2
Miscellaneous	213.95	0.00	2
Total Costs	11,927.87	0.41	100

Source: Basavaraj, G. et al. 2013b. In: *Developing a sweet sorghum ethanol value chain*, eds. B. V. S. Reddy, A. A. Kumar, C. H. Ravinder Reddy, P. Parthasarthy Rao, and J. V. Patil, pp. 110–132. Hyderabad, India: ICRISAT.
Note: Ibrahimbad, Medak, Andhra Pradesh.
[a] US$1 = Rs 62.

the viability assessment, the environmental benefits in the cultivation of sweet sorghum/energy sorghum for ethanol production should be more attractive, and hence, make a case for justifying policy support and for an enabling environment.

2.7.2.3 Farmers' Insight into Sweet Sorghum Cultivation

Small- and medium-sized groups of farmers grow more sweet sorghum. The supply of input on a credit basis, the low cost of cultivation, the low risk, and the short duration of the sweet sorghum crop compared to maize were cited as the main reasons for growing sweet sorghum. When considering expanding the area under cultivation in the future, only 38% of households were interested in growing sweet sorghum. Of these households, 55% stated that they would replace maize with sweet sorghum, while 21% of households would expand the area under sweet sorghum in fallow lands (Parthasarathy et al. 2013). The availability of input on credit was the major reason for planting sweet sorghum in the near future. To promote sweet sorghum cultivation in the future, income from the sale of the grain and the stalk as well as the lower cost of cultivation and the lower risk were strong incentives that guided the interested group of farmers to assure planting of sweet sorghum. Nonavailability of additional dryland households that grew sweet sorghum would not further increase the area under cultivation of sweet sorghum (Parthasarathy et al. 2013).

2.7.3 Biorefinery Concept in Sorghum

Sweet sorghum has been considered as a viable energy crop for alcohol fuel production. Biorefining of sweet sorghum stems produces multiple valuable products, such as ethanol, butanol, and wood plastic composites (Jianliang et al. 2012). The biorefinery approach—the production of multiple bioproducts coupled with alcohol fuels—improves the potential profits and return on investment. Ethanol production from sweet sorghum is currently not commercialized. The major constraint is the high capital cost involved in building a processing plant. Hence, a distillery should have the option of using multiple feedstocks. Likewise, producing multiple products in one location but in different seasons may allow for year-round facility operation.

Technology development in ethanol production from these feedstocks should be encouraged to improve their energy and environmental performance, especially for sugarcane. A study conducted by Liang et al. (2012) on 11 types of bio-ethanol feedstock in China using a mixed unit input–output life cycle assessment model concluded that sugarcane based bioethanol had the potential to provide positive economic, energy, and environmental impact. Corn grain and wheat grain showed higher negative economic, energy, and environmental impact when compared to sweet sorghum, cassava, sugar beet, and sugarcane, which showed better economic performance but increasing negative energy and environmental impacts. However, cellulose-based feedstocks in general showed posi-tive economic, energy, and environmental performance but may lead to increasing negative impact on freshwater use, global warming, toxicity, and aquatic ecotoxicity.

Within the biorefinery concept, biomass can be converted to useful biomaterials and/or energy in an integrated manner, thereby maximizing the economic value of the feedstock used while reducing the waste streams produced (Thomsen 2005). Biorefineries for the production of several products (sugars, polyols, levulinic acid, and furan derivatives) and by-products from bagasse (such as bioethanol, heat or electricity, and feed) have been highlighted in recent years (Luo et al. 2010; Prakasham et al. 2014). With a low delivered cost per ton of carbohydrate and soluble sugar content, sweet sorghum is an advantageous feedstock choice for biochemical conversion pathways. Replacing fossil-based energy and fuel feedstock with cellulosic feedstock would greatly reduce cradle-to-gate GHG emissions.

2.8 PROPOSED MODEL FOR ENHANCED SUSTAINABILITY

A sustainable economic and environmental model based on sorghum needs to embrace an integrated biorefinery approach where not only biofuels (ethanol, butanol, and hydrogen) are produced but an array of by-products or bioproducts having higher value (sugars, polyols, acetic acid, ethylene glycol, levulinic acid and furan, renewable biopolymers, adhesives, resins, phenolic-based chemicals, vanillin, bio-oils, biochar, etc.) are produced by using all parts of the plant in a way similar to the sugar industry (Table 2.3). An integrated biorefinery can use a biochemical or thermochemical route or a combination of both to process the feedstocks into useful products. The whole life cycle of a biorefinery system must be considered—from the cultivation and harvesting of biomass, to its collection and transportation to the plant, its conversion to fuels and chemicals, and to their consumption by the end users. This is necessary to avoid shifting sustainability impacts from one part of the supply chain to another—for example, reducing GHG emissions from the refinery only to increase them through transportation of feedstock to the refinery. The GHG emissions from ethanol produced in an integrated biochemical refinery based on sorghum can then be negative.

Table 2.3 Possible Products from Integrated Sorghum Biorefinery

Plant Part	Primary Product	Secondary Products
Grain	A variety of food products for humans; feed for animals, poultry	Food for celiac patients, ethanol, beverages, tannins, polyphenols, pigments, natural colors, DDGs
Stalk juice	Ethanol, beverages, butanol, biodiesel	Fuel oil, vinasse, pressmud cake, CO_2, methane, anticancer drugs
Leaves/bagasse	Livestock feed, biopower, syngas, ethanol, butanol, hydrogen, methane	Chemicals: lactate, propionate, biopolymer, lignin nanotubes, compost, polyols, furan, LA, EG, adhesives, biochar, etc.
Roots	Organic matter/compost	Sorgoleone

2.9 CONCLUSIONS

Sorghum is a noninvasive species with adaptation to diverse crop ecologies; in particular, its resilience to dryland rainfed systems with a short duration of four months makes it an invaluable feedstock, not only for biofuel production but also for various bioproducts. More intensive efforts on R_4D in the areas of crop improvement and crop management, besides focused efforts in developing efficient processing technologies in an integrated biorefinary platform and necessary policy support, are called for in making this novel energy species an economically, environmentally, and ecologically sustainable feedstock.

ACKNOWLEDGMENT

We are thankful to the Department of Biotechnology (DBT), Government of India and US Department of Energy for their financial support through Indo-US Joint Clean Energy Consortium and DBT funded project on development of low-lignin high-biomass sorghums suitable for biofuel production.

REFERENCES

Al-Khalasi, S. S., Mahgoub, O., Kadim, I. T., Al-Marzouqi, W. and Al-Rawahi, S. 2010. Health and performance of Omani sheep fed salt-tolerant sorghum (*Sorghum bicolor*) forage or Rhodes grass (*Chloris gayana*). *Small Ruminant Research* 91(1): 93–102.

Almodares, A. and Hadi, M. R. 2009. Production of bioethanol from sweet sorghum: A review. *African Journal of Agricultural Research* 5(9): 772–780.

Babcock, B. A. 2011. The Impact of U.S. Biofuel Policies on Agricultural Price Levels and Volatility. Issue Paper 35, International Centre for Trade and Sustainable Development, Geneva.

Balat, M. and Balat, H. 2009. Recent trends in global production and utilization of bioethanol. *Applied Energy* 8: 2273–2282.

Basavaraj, G., Parthasarathy, R. and Reddy, C. R. 2013a. SWOT analysis of sweet sorghum ethanol value chain. In *Developing a sweet sorghum ethanol value chain*, eds. B. V. S. Reddy, A. A. Kumar, C. H. Ravinder Reddy, P. Parthasarthy Rao, and J. V. Patil, pp. 197–200. Hyderabad, India: ICRISAT.

Basavaraj, G., Rao, P. P., Basu, K., Reddy, C. R., Reddy, B. V. S. and Kumar, A. A. 2013b. Sweet sorghum ethanol production—An economic assessment. In *Developing a sweet sorghum ethanol value chain*, eds. B. V. S. Reddy, A. A. Kumar, C. H. Ravinder Reddy, P. Parthasarthy Rao, and J. V. Patil, pp. 110–132. Hyderabad, India: ICRISAT.

Blum, A. and Sullivan, C. Y. 1985. The comparative drought resistance of landraces of sorghum and millet from dry and humid regions. *Annals of Botany* 57(6): 835–846.

Cai, H., Dunn, J. B., Wang, Z., Han, J. and Wang, M. Q. 2013. Life-cycle energy use and greenhouse gas emissions of production of bioethanol from sorghum in the United States. *Biotechnology for Biofuels* 6: 141.

Cherney, J. H., Cherney, R. D. E., Akin, D. E. and Axtell, J. D. 1991. Potential of brown midrib, low lignin mutants for improving forage quality. *Advances in Agronomy* 46: 157–198.

Dahlberg, J., Berenji, J., Sikora, V. and Latković, D. 2011. Assessing sorghum [*Sorghum bicolor* (L.) Moench] germplasm for new traits: Food, fuels and unique uses. *Maydica* 56(1750): 85–92.

Damasceno, B. M. C., Schaffert, E. R. and Dweikat, I. 2014. Mining genetic diversity of sorghum as a bioenergy feedstock, plants and bioenergy. *Advances in Plant Biology* 4: 81–106.

Demirbas, A. 2008. Biofuels sources, biofuel policy, biofuel economy and global biofuel projections. *Energy Conversation and Management* 49: 2106–2116.

Dien, B. S., Sarath, G., Pedersen, J. F., Sattler, S. E., Chen, H., Harris, D. L. F., Nichols, N. N. and Cotta, M. A. 2009. Improved sugar conversion and ethanol yield for forage sorghum [*Sorghum bicolor* (L.) Moench] lines with reduced lignin contents. *BioEnergy Research* 2: 153–164.

Doggett, H. 1998. *Sorghum*. 2nd ed. Tropical Agriculture Series. Essex, UK: Longman Scientific and Technical Publishers, p. 512.

EISA (Energy Independence and Security Act). 2007. http://www.gpo.gov/fdsys/pkg/BILLS-110hr6enr/pdf/BILLS-110hr6enr.pdf

Fahmy, A. A., Youssef, K. M. and El Shaer, H. M. 2010. Intake and nutritive value of some salt-tolerant fodder grasses for sheep under saline conditions of South Sinai, Egypt. *Small Ruminant Research* 9: 110–115.

FAO (Food and Agriculture Organization of the United Nations). 2013. FAO Statistics Division. http://www.fao.org/economic/ess/en/ (accessed February 6, 2014).

Farrell, A. E., Plevin, R. J., Turner, B. T., Jones, A. D., O'Hare, M. and Kammen, D. M. 2006. Ethanol can contribute to energy and environmental goals. *Science* 27: 506–508.

GAIN report. 2013. India Biofuels Annual. pp. 1–18. http://gain.fas.usda.gov/Recent%20GAIN%20Publications/Biofuels%20Annual_New%20Delhi_India_8-13-2013.pdf

Gnansounou, E., Dauriat, A. and Wyman, C. E. 2005. Refining sweet sorghum to ethanol and sugar: Economic trade-offs in the context of North China. *Bioresource Technology* 96(9): 985–1002.

Gressell, J. 2008. Transgenics are imperative for biofuel crops. *Plant Science* 174: 246–263.

Guiying, L., Weibin, G., Hicks, A. and Chapman, K. R. (2000). *Training manual for sweet sorghum*, FAO, Bangkok, Thailand.

IEA (International Energy Agency). 2007. World energy outlook: China and India insights. http://www.worldenergyoutlook.org/media/weowebsite/2008-1994/weo_2007.pdf (accessed March 6, 2014).

Jianliang, Y., Zhang, T., Zhong, J., Zhang, X. and Tan, T. 2012. Biorefinery of sweet sorghum stem. *Biotechnology Advances* 30: 811–816.

Keeley, J. E. and Rundel, P. W. 2003. Evolution of CAM and C_4 carbon-concentrating mechanisms. *International Journal of Plant Science* 164(3): S55–S77.

Khanum, S. A., Hussain, H. N., Hussain, M. and Ishaq, M. 2010. Digestibility studies in sheep fed sorghum, *Sesbania* and various grasses grown on medium saline lands. *Small Ruminant Research* 91(1): 63–68.

Kim, M. and Day, D. 2011. Composition of sugar cane, energy cane, and sweet sorghum suitable for ethanol production at Louisiana sugar mills. *Journal of Industrial Microbiology and Biotechnology* 38(7): 803–807.

Köppen, S., Reinhardt, G. and Gärtner, S. (2009). *Assessment of energy and greenhouse gas inventories of sweet sorghum for first and second generation bioethanol*. FAO Environmental and Natural Resources Service Series, 30. http://foris.fao.org/static/data/nrc/SweetSorghumGHGIFEU2009.pdf (accessed February 8, 2013).

Kundiyana, D. K. 1996. Sorganol: In-field production of ethanol from sweet sorghum. http://digital.library.okstate.edu/etd/umi-okstate-1974.pdf (accessed February 6, 2014).

Liang, S., Xu, M. and Zhang, T. 2012. Unintended consequences of bioethanol feedstock choice in China. *Bioresource Technology* 125: 312–317.

Licht, F. O. 2008. *World fuel ethanol production*. Renewable Fuels Association. http://ethanolrfa.org/pages/World-Fuel-Ethanol-Production (accessed February 6, 2014).

Luo, L., van der Voet, E. and Huppesa, G. 2010. Biorefining of lignocellulosic feedstock technical, economic and environmental considerations. *Bioresource Technology* 101(13): 5023–5032.

McCollum, T., McCuistion, K. and Bean, B. 2005. *Brown midrib and photoperiod sensitive forage sorghums*. AREC 05-20. Amarillo, TX: Texas A&M University Agricultural Research and Extension Center.

Miller, F. R. and Stroup, J. A. 2003. Brown midrib forage sorghum, Sudangrass, and corn: what is the potential? http://alfalfa.ucdavis.edu/+symposium/proceedings/2003/03-143.pdf (accessed February 3, 2014).

Nagaiah, D., Srinivasarao, P., Prakasham, R. S., Uma, A., Radhika, K., Yoganand, B. and Umakanth, A. V. 2012. High biomass sorghum as a potential raw material for biohydrogen production: A preliminary evaluation. *Current Trends in Biotechnology and Pharmacy* 6(2): 183–189.

Oba, M. and Allen, M. S. 1999. Effects of brown midrib-3 mutation in corn silage on dry matter intake and productivity of high yielding dairy cows. *Journal of Dairy Science* 82: 135–142.

Palmer, G. H. 1991. Sorghum—Food beverage and brewing potentials. *Review Process Biochemistry* 27(3): 145–153.

Parthasarathy, R. P., Basavaraj, G., Basu, K., Reddy, C. R., Kumar, A. A. and Reddy, B. S. 2013. Economics of sweet sorghum feedstock production for bioethanol. In *Developing a sweet sorghum ethanol value chain*, eds. B. V. S. Reddy, A. A. Kumar, C. H. Ravinder Reddy, P. Parthasarathy Rao, and J. V. Patil, pp. 99–109. Hyderabad, India: ICRISAT.

Porter, K. S., Axtell, J. D., Lichtenberg, V. L. and Colenbrander, V. F. 1978. Phenotype, fiber composition and in vitro organic matter digestability of chemically induced brown midrib mutants of sorghum. *Crop Science* 18: 205.

Prakasham, R. S., Nagaiah, D., Vinutha, K. S., Uma, A., Chiranjeevi, T., Umakanth, A. V., SrinivasaRao, P. and Yan, N. 2014. Sorghum biomass: A novel renewable carbon source for industrial bioproducts. *Biofuels* 5(2): 159–174.

Prasad, S., Singh, A., Jain, N. and Joshi, H. C. 2007. Ethanol production from sweet sorghum syrup for utilization as automotive fuel in India. *Energy Fuels* 21: 2415–2420.

Raju, S. S., Parappurathu, S., Chand, R., Kumar, P. and Msangi, S. 2012. *Biofuels in India: Potential, policy and emerging paradigms*. New Delhi, India: National Centre for Agricultural Economics and Policy Research. Policy Paper 27, pp. 1–90.

Reddy, C. R., Kumar, A. A. and Reddy, B. V. S. 2013. Community seed system: Production and supply of sweet sorghum seeds. In *Developing a sweet sorghum ethanol value chain*, eds. B. V. S. Reddy, A. A. Kumar, C. H. Ravinder Reddy, P. Parthasarthy Rao, and J. V. Patil, pp. 38–44. Hyderabad, India: ICRISAT.

Rooney, W. L., Blumenthal, J., Bean, B. and Mullet, J. E. 2007. Designing sorghum as a dedicated bioenergy feedstock. *Biofuels Bioproducts and Biorefinery* 1: 147–157.

Saballos, A. 2008. Development and utilization of sorghum as a bioenergy crop. In *Genetic improvement of bioenergy crops*, ed. W. Vermerris, pp. 211–248. New York: Springer Science and Business Media, LLC.

Sehgal, S., Kawatra, A. and Singh, G. 2003. Recent technologies in pearl millet and sorghum processing and food product development. In *Proceedings of Expert Meeting on Alternative Uses of Sorghum and Pearl Millet in Asia*, July 1–4, pp. 60–92.

Semra, T., Uzun, D. and Türe, E. I. 1997. The potential use of sweet sorghum as a nonpolluting source of energy. *Energy* 22(1): 17–19.

Serna-Saldivar, S. O. 2010. *Cereal Grains: Properties, Processing, and Nutritional Attributes*, CRC Press, Florida, USA.

Serna-Saldivar, S. O., Cristina, C., Esther, P. and Erick, H. 2012. Sorghum as a multifunctional crop for the production of fuel ethanol: Current status and future trends. In *Bioethanol*, ed. M. A. P. Lima. http://cdn.intechopen.com/pdfs-wm/27350.pdf (accessed February 6, 2014).

Shinoj, P., Raju, S. S. and Joshi, P. K. 2011. India's biofuel production programme: Need for prioritizing alternative options. *Indian Journal of Agricultural Sciences* 81(5): 391–397.

Shoemaker, C. E. and Bransby, D. I. 2010. The role of sorghum as a bioenergy feedstock. In *Sustainable alternative fuel feedstock opportunities, challenges and roadmaps for six U.S. regions*, eds. R. Braun, D. Karlen, and D. Johnson, *Proceedings of the Sustainable Feedstocks for Advance Biofuels Workshop*, Atlanta, GA, September 28–30, pp. 149–159. Ankeny, IA: Soil and Water Conservation Society.

Srinivas, I., Adake, R. V., Reddy, B. V. S., Korwar, G. R., Reddy, C. R., Venkateswarlu, B., Dange, A. and Udaykumar, M. 2013. Mechanization of sweet sorghum production and processing. In *Developing a sweet sorghum ethanol value chain*, eds. B. V. S. Reddy, A. Ashok Kumar, C. H. Ravinder Reddy, P. Parthasarthy Rao, and J. V. Patil, pp. 63–75. Hyderabad, India: ICRISAT.

Srinivasarao, P., Prasad, J. V. N. S., Umakanth, A. V. and B. V. S. Reddy. 2011. Sweet sorghum (*Sorghum bicolor* (L.) Moench)—A new generation water use efficient bioenergy crop. *Indian Journal of Dryland Agriculture* 26(1): 65–71.

Srinivasarao, P., Rao, S. S., Seetharama, N., Umakanth, A. V., Sanjana Reddy, P., Reddy, B. V. S. and Gowda, C. L. L. 2009. *Sweet sorghum for biofuel and strategies for its improvement*. Information Bulletin No. 77. Hyderabad, India: ICRISAT, p. 80.

Srinivasarao, P., Ravikumar, S., Prakasham, R. S., Deshpande, S. and Reddy B. V. S. 2010a. *Bmr*—From efficient fodder trait to novel substrate for futuristic biofuel: Way forward. In *Brown midrib sorghum—current status and potential as novel ligno-cellulosic feedstock of bioenergy*, eds. P. Srinivasarao, R. S. Prakasham, and S. Deshpande, pp. 99–112. Saarbrücken, Germany: Lap Lambert Academic Press.

Srinivasarao, P., Reddy, B. V. S., Blümmel, M., Subbarao, G. V., Chandraraj, K., Sanjana Reddy, P. and Parthasarathy Rao, P. 2010b. *Sweet sorghum as a biofuel feedstock: Can there be food-feed-fuel tradeoffs?* http://www.corpoica.org.co/sitioweb/Documento/JatrophaContrataciones/SWEETSORGHUMASABIOFUELSFEEDSTOCK.pdf

Srinivasarao, P., Reddy, C. R., Ashok Kumar, A., Blümmel M., Rao P. P., Basavaraj, G. and Reddy, B. V. S. 2012. Opportunities and challenges in utilizing by-products of sweet sorghum-based biofuel industry as livestock feed in decentralized systems. In *Biofuel co-products as livestock feed—Opportunities and challenges*, ed. H. P. S. Makkar, pp. 229–242. FAO, Rome.

Srinivasarao, P., Umakanth, A. V., Reddy, B. V. S., Dweikat, I., Bhargava, S., Kumar, C. G., Braconnier, S., Patil, J. V. and Singh, B. P. 2013. Sweet sorghum: Genetics, breeding and commercialization. In *Biofuel crops: Production, physiology and genetics*, ed. B. P. Singh, pp. 172–198. Cambridge, MA: CABI.

TAES. (2006). *Sustainable bioenergy initiatives across Texas A&M Agriculture.* College Station, TX: Texas A&M AgriLife Research.

Thomsen, M. 2005. Complex media from processing of agricultural crops for microbial fermentation. *Applied Microbiology and Biotechnology* 68(5): 598–606.

Turhollow, A. F., Webb, E. G. and Downing, M. E. (2010). *Review of sorghum production practices: Applications for bioenergy.* Oak Ridge, TN: Oak Ridge National Laboratory.

Undersander, D. 2003. Sorghums, Sudan grasses and sorghum-Sudan hybrids focus on forage. http://www.uwex.edu/ces/crops/uwforage/SorghumsFOF.pdf (accessed February 4, 2014).

Undersander, D. J., Smith, L. H., Kaminski, A. R., Kelling, K. A. and Doll, J. D. 1989. Alternative field crops manual. http://www.hort.purdue.edu/newcrop/afcm/forage.html (accessed February 3, 2014).

USDA (United States Department of Agriculture). 2014. Data and Statistics. http://www.usda.gov/wps/portal/usda/usdahome?navid=DATA_STATISTICS (accessed February 4, 2014).

U.S. Grains Council. 2010. World sorghum production and trade. http://www.grains.org/index.php?option=com_contentandview=articleandid=74andItemid (accessed February 4, 2014).

Venuto, B. and Kindiger, B. 2008. Forage and biomass feedstock production from hybrid forage sorghum and sorghum-Sudan-grass hybrids. *Grassland Science* 54(4): 189–196.

Woods, J. 2000. Integrating Sweet sorghum and sugarcane for bioenergy: Modelling the potential for electricity and ethanol production in SE Zimbabwe, Ph.D. Thesis, King College, London.

Wortmann, C., Ferguson, R. and Lyon, D. 2008. Sweet sorghum as a biofuel crop in Nebraska. *Paper presented at the 2008 Joint Annual Meeting, Celebrating the International Year of Planet Earth*, October 5–9, 2008, Houston, TX. http://crops.confex.com/crops/2008am/techprogram/P44581.HTM (accessed February 4, 2014).

Wu, X., Staggenborg, S., Propheter, J. L., Rooney, W. L., Yu, J. and Wang, D. 2010. Features of sweet sorghum juice and their performance in ethanol fermentation. *Industrial Crops and Products* 31(1): 164–170.

Zegada-Lizarazu, W. and Monti, A. 2012. Are we ready to cultivate sweet sorghum as a bioenergy feedstock? A review on field management practices. *Biomass and Bioenergy* 40: 1–12.

Figure 1.5 Harvest of grassland biomass for habitat management on public conservation land in Wisconsin. (Courtesy of C. L. Williams, 2012.)

Figure 2.1 Types of sorghum: (a) grain sorghum, (b) brown midrib sorghum, (c) energy sorghum, (d) sweet sorghum, (e) forage sorghum, (f) broomcorn, and (g) sudangrass.

Figure 3.2 Twelve-year-old hybrid poplar plantation in a single-stem, single-harvest management system in Minnesota, USA, ready for harvest. (Courtesy of Bernard McMahon, University of Minnesota, Duluth.)

Figure 3.5 A New Holland (NH) forage harvester fitted with a specially designed NH FB130 coppice header harvesting four-year-old willow biomass crops in central New York. (Courtesy of D. Angel, SUNY-ESF, Syracuse, NY.)

Figure 5.2 Local applications of biomass.

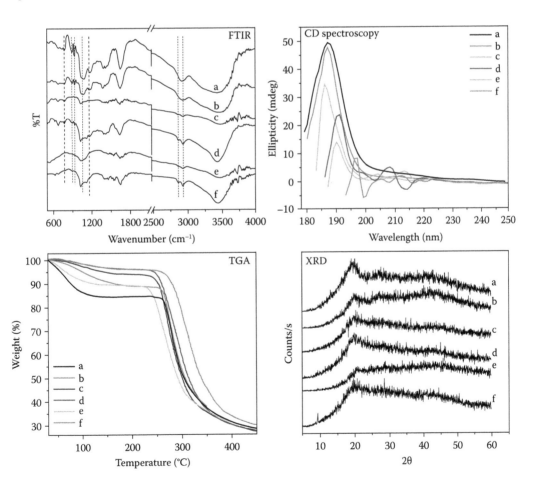

Figure 7.4 Spectral analysis of agarose regenerated from various solutions: (a) native agarose, (b) water, (c) [C$_4$mim][C$_1$OSO$_3$], (d) [C$_8$mim][Cl], (e) [C$_4$mim][Cl], and (f) [C$_4$mpy][Cl].

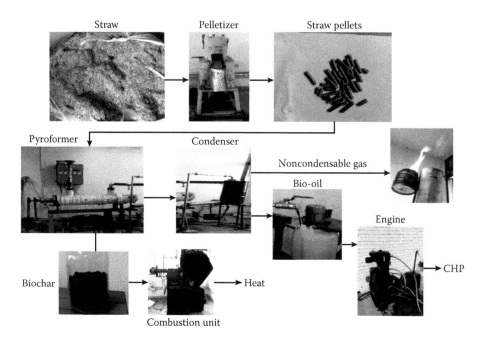

Figure 10.6 Conversion of straw to bio-oil and biochar.

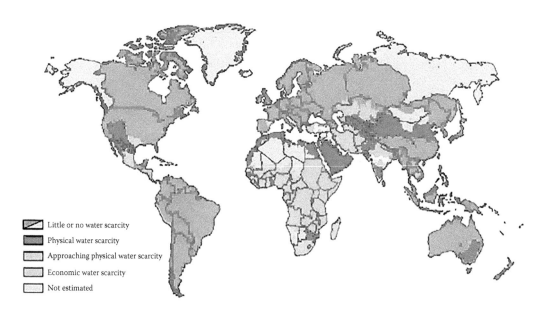

Little or no water scarcity

Physical water scarcity

Approaching physical water scarcity

Economic water scarcity

Not estimated

Figure 12.3 Water availability. (From IWMI, *Insights from the comprehensive assessment of water management in agriculture*, World Water Week, Stockholm, 2006.)

CHAPTER 3

Short Rotation Woody Crops Biomass Production

Diomides S. Zamora,[1] **Kent G. Apostol,**[2] **William E. Berguson,**[3] **Timothy A. Volk,**[4]
Jeff Wright,[5] **and Eric J. Ogdahl**[6]

[1]Department of Forest Resources, University of Minnesota Extension, MN, USA
[2]Department of Forest, Rangeland, and Fire Sciences, College of Natural Resources
 University of Idaho, Moscow, ID, USA
[3]Natural Resources Research Institute, University of Minnesota Duluth, Duluth, MN, USA
[4]College of Environmental Science and Forestry, State University of New York, Syracuse, NY, USA
[5]Hardwood Development, ArborGen Inc., Ridgeville, SC, USA
[6]Natural Resource Science and Management, University of Minnesota, St. Paul, MN, USA

CONTENTS

3.1 INTRODUCTION

The development and use of biomass resources for bioenergy has become a critical priority in North America, Europe, Australia, and elsewhere (US DOE 2011; Zalesny et al. 2011; Kiser and Fox 2013). Rising concerns about long-term energy security, human health, global climate change, and a variety of other social and environmental concerns have combined to make developing renewable and sustainable alternatives to fossil fuels a critical national priority (Abrahamson et al. 1998; NRC 2000; Verwijst 2001; Hoogwijk et al. 2005; Volk et al. 2011). The global energy supply of biomass has been predicted to range from 100 to 400 EJ yr^{-1} by 2050 (Berndes et al. 2003). Under the Renewable Fuels Standard (RFS) detailed in the Energy Independence and Security Act (EISA) of 2007, the U.S. economy is mandated to use 136 billion liters (bl) of renewable transportation fuel per year in its transportation fuel supply by 2022. Of that 136 bl yr^{-1}, 75 bl yr^{-1} is mandated to consist of cellulosic ethanol and other advanced biofuels (US EISA 2007). It is expected that, in some regions of the globe, short rotation woody crops (SRWCs) will be a critically important feedstock supply in meeting those goals.

The U.S. Department of Energy (US DOE) examined the potential for the sustainable production of one billion dry tons of biomass annually for bioenergy to displace approximately 30% of the U.S. petroleum supply. The U.S. Billion-Ton Update (US DOE 2011) indicates that SRWCs have the potential to provide between 126 and 315 million dry Mg of biomass annually across the United States by 2030, based on assumptions regarding annual yield and economics at $60 per dry ton in harvested form ready for hauling to a conversion facility (US DOE 2011). A range of SRWC systems are being developed to meet the growing need for woody biomass such as hybrid poplars (*Populus* spp.) and willow shrubs (*Salix* spp.) in northern temperate areas and eucalyptus (*Eucalyptus* spp.) in the subtropics (Volk et al. 2004; Langholtz et al. 2005; Dougherty and Wright 2012). Other species, such as pines (*Pinus* spp.), sycamore (*Plantanus occidentalis*), and sweetgum (*Liquidambar styraciflua*), have potential for planting in biomass-oriented plantations but these species either do not coppice (thus enabling repeated harvest from a single planting) or yields in SRWC systems have not been fully explored. We chose to focus on the three species listed here as these are the species of greatest global interest currently.

Biomass can come from a number of different sources, including forests, agricultural crops, grass monocultures, polycultures (mixture of grasses/forbs), and dedicated SRWCs (US DOE 2011). SRWCs are defined as highly productive, purpose-grown plantations in which the bolewood and much of the limbs and tops are used as feedstocks for energy (US DOE 2011). SRWCs are ideal management systems because they are renewable energy feedstocks for biofuels that can be strategically placed in the landscape to provide a plethora of ecological services including carbon sequestration and wildlife habitat (Volk et al. 2011). Also, these crops are ideal for genetic improvement because of their considerable genetic variability, ease of propagation, and relatively short generation time of 3–10 years (Berguson et al. 2010; Vance et al. 2010).

In this chapter, key aspects of SRWCs are summarized and synthesized, including biology, the production system, silviculture, breeding, and genetic improvements. Much more extensive information is presented in previous reviews by Stettler et al. (1996), Dickmann 2006, Zalesny et al. (2011), and Stanturf et al. (2013). Current efforts to maximize bioenergy production and efforts to address barriers to utilization of *Populus*, *Salix*, and *Eucalyptus* as SRWCs are described, and the opportunities afforded by SRWCs are highlighted. Considerable research effort is underway worldwide on these species due to their wide geographic range from warm southern climates to very cold mid-continental regions and their ability for repeated coppicing to achieve greater amount of biomass in a short period of time. A case study of the production of willow as influenced by various planting designs is presented.

3.2 BRIEF HISTORY AND BACKGROUND OF SRWCs PROGRAM

Although the discipline of SRWCs was highly refined in the twentieth century and is increasingly being researched in the twenty-first century, the program itself is rooted in Old World practices. Romans coppiced short-rotation willow osiers for baskets, ropes, and agriculture tools, while calculating yields and labor-requirements for coppice plantations. The management of poplar cultures for timber, fuel, and windbreaks also dates back several millennia in Rome and China. Hybrid tree plantations, namely poplar, were first documented in the eighteenth century, after the North American cottonwood (*Populus deltoides*) was introduced to Europe and found to spontaneously produce desirable hybrids with the native black poplar (*P. nigra*) (Dickmann 2006). The background information is well detailed in previous reviews (Dickmann 2006).

Currently, the use of native and hybrid SRWCs is a worldwide practice, with a particular emphasis recently on bioenergy due to increasing interest in the production of liquid fuels having a reduced carbon dioxide emission footprint as well as the potential for production of electricity from biomass. The first formal research initiative devoted to SRWCs was the *Istituto di Sperimentazione per la Pioppiocultura* (ISP), established in 1937 in the Po River Valley of Italy. ISP research initially focused on poplar culture but now encompasses both poplar and willow with the goal of creating fast-growing, pest-resistant trees producing high-quality wood (Dickmann 2006). In 1947, the founding of the International Poplar Commission (IPC), a statutory body of the United Nations Food and Agriculture Organization (UN FAO), marked the first initiative to coordinate SRWC projects worldwide. The IPC currently comprises 37 member countries with the aim of promoting "the cultivation, conservation, and utilization of poplars and willows of the Salicaceae family" (IPC 2013). Other SRWC umbrella organizations following the foundation of IPC include the short-rotation forestry division (Unit Division 1.03) of the International Union of Forest Research Organizations (IUFRO) and the SRWC Operations Working Group in the United States (Dickmann 2006). These organizations have helped to greatly advance the practice of SRWCs.

The contribution of SRWCs to global bioenergy is expanding. In 1978, the International Energy Agency (IEA) established a Bioenergy Agreement to facilitate cooperation and "information exchange between countries that have national programs on bioenergy research, development, and deployment" (Wright 2006). In January 2007, the IEA established Task Force 30 (known as Short-Rotation Crops for Bioenergy Systems) that completed its task on December 2012. The Task Force 30 was formed "to acquire, synthesize, and transfer theoretical and practical knowledge of sustainable short rotation biomass production systems and thereby to enhance market development and large-scale implementation in collaboration with the various sectors involved" (IEA 2013). The Task Force 30 comprised member countries such as Australia, Brazil, Canada, New Zealand, Sweden, the United Kingdom (UK), and the United States (Wright 2006). Within Task Force 30 countries, the amounts of land devoted specifically to growth of SRWCs range from approximately 25 km^2 in the UK to 30,000 km^2 in Brazil (Wright 2006). The total contribution of short-rotation crop biomass to Task Force 30 countries' total energy consumption ranges from 0.6% in the UK to 27.2% in Brazil; however, by the IEA's definition of short-rotation crops, these percentages may include SRWCs as well as lignocellulosic crops, such as reed canary grass (*Phalaris arundinacea*) and *Miscanthus*. Other countries that are not a part of Task Force 30 also are developing bioenergy from SRWCs. China, for example, has an estimated 7–10 million hectares of land under SRWC production for fuelwood and obtains 16.4% of its total energy consumption from biomass (Wright 2006). From this brief overview of the global status of SRWCs and bioenergy, it is apparent that SRWCs are a promising source of renewable energy.

3.3 SHORT ROTATION WOODY CROPS

Background on SRWCs such as hybrid poplars, willows, and eucalyptus is highlighted in this section. Included in this discussion are biology, adaptation, and production systems, as well as breeding and genetic improvements. Hybrid poplars, willows, and eucalyptus are of interest due to their demonstrated high growth rate as well as their ability to clonally propagate from cuttings and to resprout from stumps after harvest thus enabling a repeated coppice management and continual periodic harvest of biomass from recurring growth without replanting.

3.3.1 Hybrid Poplar

3.3.1.1 Introduction

Species of the genus *Populus* are often pioneers of disturbed forest ecosystems in temperate climates. They also have a wide genetic base and geographic range, occurring throughout North America—where *Populus tremuloides* is the most geographically widespread tree (Jelinski and Cheliak 1992), as well as in Europe and Asia. Throughout this chapter, the term "poplar" will be used to refer to genotypes from the *Aegiros* section (cottonwood and black poplars—*P. deltoides* and *P. nigra*, respectively) and the *Tacamahaca* section (balsam poplars—*P. maximowiczii* and *P. trichocarpa*). These species can be readily hybridized to produce clones having a wide range of adaptability, growth habit, wood characteristics, and growth rates. For this chapter, the term hybrid poplar refers to hybrids produced through controlled breeding among these species.

Due to its widespread geographic adaptation, rapid growth, ease of reproduction, and coppicing ability, *Populus* species are ideal candidates for SRWCs. In fact, *Populus* is the most widely researched genus among all SRWC taxa. It is also a widely researched tree genus in general, with the genome of *P. trichocarpa* the first to be sequenced among woody plants (Tuskan et al. 2006). *Populus* spp. and their hybrids (i.e., poplars) are a significant component of the total biofuels and bioenergy feedstock resource in the United States, Canada, and in Europe. They are capable of producing in excess of 7 Mg ha^{-1} yr^{-1} by age six even in the harsh climate of the North Central region of the United States (Riemenschneider et al. 2001; Zalesny et al. 2009). Although there are some interests in aspens (from the *Populus* and *Leucoides* sections) in some regions in the United States, they are generally not amenable to vegetative propagation and growth rates have not been shown to equal or exceed cottonwood or hybrid poplar. Further genetic tests are currently underway on aspen clones (notably, *P. grandidentata* × *alba* hybrids).

In order for poplars to be successfully produced commercially, research is needed to produce genetic material that exhibits all of the necessary attributes of high rate of establishment success from vegetative cuttings, high biomass productivity rates, disease resistance, and coppicing ability. Due to the long-term nature of these perennial systems, it is necessary that research be done in each geographic zone of commercial interest to produce genotypes possessing all of these characteristics. Although there are areas in North America, Europe, and Asia where commercial production has been demonstrated (Berguson et al. 2010), research such as that underway by the Sun Grant Poplar Woody Crops Program is attempting to increase yield and genetic diversity as well as increase the geographic range of new hybrids for commercial production.

3.3.1.2 Biology and Production System

Populus species are angiosperms with dioecious trees bearing male or female catkins. Worldwide, there are 29 recognized species, with 12 native to North America. They possess abundant genetic variation, occurring at the genus, sectional, species, and clonal levels (Berguson et al. 2010; Zalesny et al. 2011). Features that make them suitable for biomass production include their

Figure 3.1 Two-year-old hybrid poplar trees planted in a loam soil in central Minnesota, USA. (Courtesy of Bernie McMahon, University of Minnesota-Duluth.)

ease of rooting and vegetative propagation, quick establishment and fast growth, and their ability to resprout following repeated harvesting (Figure 3.1).

Poplar plantation establishment has relied upon the use of unrooted, hardwood stem cuttings that are approximately 20–30 cm in length (depending on the climatic region). The use of vegetative propagation allows for easy and inexpensive planting of stem cuttings from dormant material. Once harvested, cuttings are stored under refrigeration and then planted in the field when soil temperatures reach levels appropriate to specific regions and genotypes.

Plant spacing of poplars can vary widely depending on the intended end product. In those situations where debarked wood is required (e.g., paper, oriented strandboard, potentially biochemical liquid fuels), lower plant populations ranging from 750 to 1700 stems per hectare are used to produce large diameter trees that facilitate debarking—typically on a 10- to 12-year rotation. However, plantations oriented toward repeated coppice production of woody biomass for energy applications may be planted at higher densities ranging from 5000 to 10,000 stems per hectare (Al Afas et al. 2005). Plantations require intensive weed management during the first one to three years, until canopy closure is reached. Assuming effective weed control, poplars establish quickly and weed control activity can be eliminated or greatly reduced after two years.

Poplars can be managed in a number of ways, depending on the desired end product, management system, and target rotation age. Poplars are flexible in terms of rotation and management system. More closely spaced systems have the potential advantage of reaching canopy closure more quickly and of producing higher biomass yield. However, this advantage is partially offset by higher planting costs, thus affecting economics (Berguson et al. 2012). *Populus* fiber production systems are harvested every 7–15 years (Zalesny et al. 2011). *Populus* bioenergy plantations require shorter rotation periods, although more research on bioenergy rotation periods, as well as how yields from these compare to longer rotation periods, is needed.

The most important drivers of poplar SRWC system productivity have been suggested to be proper species or hybrid selection for site conditions, correct site preparation, pest/disease resistance (Hansen et al. 1994; Berguson et al. 2010; Zalesny et al. 2011), and weed control at establishment (Hansen 1994; Kauter et al. 2003). Typical operational procedures may include tillage, fertilization,

Figure 3.2 (See color insert.) Twelve-year-old hybrid poplar plantation in a single-stem, single-harvest management system in Minnesota, USA, ready for harvest. (Courtesy of Bernard McMahon, University of Minnesota, Duluth.)

and irrigation (Sartori et al. 2007). Planting densities range from 1500 cuttings per hectare (Sartori et al. 2007) in more traditional plantation designs to 10,000 cuttings per hectare (Al Afas et al. 2005) in double-row systems that more closely resemble willow SRWC systems. Thinning is sometimes conducted at higher planting densities to provide a supply of biomass at intermediate points in the rotation.

Populus species are typically harvested after several years, with a rotation period of five years or less for bioenergy crops. Rotation lengths for poplars are expected to range from three to five years in coppice regeneration as opposed to 5–15 years in more traditional single-rotation, single-stem plantations (Hansen et al. 1994; Berguson et al. 2010; Zalesny et al. 2011) (Figure 3.2). The means through which trees resprout after harvest, or coppicing, varies with species. In the case of the *Aeigeros* and *Tachamahaca* sections, resprouting is limited to shoot initiation from the cambium of the stump with little root sprouting occurrence. Stand vigor may diminish with age and frequent coppicing. Conversely, species such as *P. tremuloides* can produce both stump and root sprouts starting at one year of age with as many as 25,000 stems per hectare arising from the existing rootstock (Perala 1977). Sprouting characteristics affect ease of harvest and machinery requirements because species that are restricted to stump sprouting retain the original planted row orientation. The ability for many *Populus* species to resprout after harvests defrays the cost of replanting in successful plantations.

3.3.1.3 Breeding/Genetic Improvements

Populus tree improvement programs in the United States began in the 1920s (Northeast) and 1930s (Midwest) (Berguson et al. 2010; Zalesny et al. 2011). Also, significant work on genetic improvement of *Populus* has been done in the Pacific Northwest by Stettler et al. 1996 and continues in this region as part of GreenWood Resources commercial operations near Boardman, Oregon.

In the case of development of poplars for commercial production in the Upper Midwest region of the United States, most genetic improvement work in woody crops consisted of a series of tests in the region using clones that were developed previously in other programs. These clones were propagated and distributed to the various regions, and some were used commercially for shelterbelts and plantations. Also, clones were being produced in Europe in Belgium, Germany, and Italy and imported to Canada and the United States with clone tests at multiple locations throughout the Midwest (Hansen et al. 1994). Of more than 80 clones that were tested in Minnesota, only three clones, DN34, DN5, and NM6 (DN = *deltoides* × *nigra*, NM = *nigra* × *maximowiczii*), were found to be fast growing and disease resistant (*Septoria* canker, *Melampsora* white leaf rust, and *Marsonina* leaf spot) (Hansen et al. 1994); DN34 was eliminated due to slow growth early in the rotation and associated increased weed control costs. Between NM6 and DN5, NM6 dominates the current commercial acreage due to a perceived problem with DN5 in terms of growth form (bole sweep). However, current plantations of NM6 show infestation of severe canker problems. Obviously, genetics of woody crops are not diverse with only two clones currently of commercial interest. Genetic variation among pure-species collections and hybrids is very high in *Populus,* which presents opportunities to improve biomass yield and nutrient efficiency. Breeding efforts are underway at various institutions across the globe, particularly in North America (Table 3.1). In light of the lack of genetic diversity in the Midwest, since 1996, more than 100,000 *Populus* offspring have been produced there at the University of Minnesota, Duluth (Berguson et al. 2010), through the breeding program—with many of these genotypes

Table 3.1   Currently Active *Populus* spp. Breeding Programs in North America

Company	Location	Region of Concentration	Area	Focus
Greenwood Resource Institute	Boardman, Oregon, USA	USA	9300 ha	• Operating on a 15-year rotation with the output going into the saw log market. • Signed an agreement with ZeaChem, a cellulosic ethanol company, to supply them with feedstock from their hybrid poplar farm.
Alberta-Pacific Forest Products	Alberta, Canada	Canada	8000 ha	• Growing hybrid poplar trees to feed pulp mills in Alberta, Canada, and nearby regions. • Ensuring fiber supply availability in response to the deforestation activities in Alberta caused by tar sand development.
ArborGen	USA, New Zealand, Australia, and Brazil	Worldwide		• Largest global supplier of seedling products and a leading provider of improved technologies to the commercial forest industry. • Developing high value products through conventional breeding and genetic improvement. • Working to improve the sustainability of working forests while helping to meet the world's growing need for wood, fiber, and energy.
University of Minnesota, Duluth, Natural Resources Research Institute	USA	USA		• Working to increase yield and genetic diversity of poplars for use as an energy crop. • Collecting data from clonal trials, yield tests, spacing trials, and large-scale genetics tests with some locations containing more than 900 genotypes.

Table 3.2   Genetic Improvement of *Populus* spp. in the United States Based on Geographical Locations

Location	Genetic Improvement Focus	Authors
Upper Midwest	DN34, DN5, and NM6 were found to be disease resistant and fast growing under a range of conditions. Large breeding program has produced new hybrid to increase growth rate, genetic diversity, and disease resistance.	Hansen et al. (1994); Berguson et al. (2010)
Southern USA (Lower Mississippi Alluvial Valley)	*P. deltoides* performs best on those sites that are high in nutrient availability and that have consistent rainfall. Unlike pure *P. deltoides*, limited testing of hybrid poplars in the South has not determined the incidence of *Septoria* canker, a potentially serious disease of some poplars.	Berguson et al. (2010)
Pacific Northwest	Identify superior clones of *P. trichocarpa* and produce hybrids of *P. deltoides* and *P. trichocarpa* for commercial production.	Stanton et al. (2002)

being tested in long-term field tests in Minnesota. The experience in Minnesota and at other locations worldwide confirms the need to produce populations using parents putatively adapted to the region followed by field tests to identify those unique genotypes possessing all of the traits necessary for commercial production.

There are many possible breeding strategies that can be applied to the development of hybrid poplar woody biomass crops (Riemenschneider et al. 2001). Yet all breeding strategies derive from the need for a commercial variety to possess several attributes simultaneously, such as an adventitious root system, rapid growth, and resistance to pests (US DOE 2011). Elsewhere, inter-specific hybridization may be necessary. For example, in the Upper Midwest, eastern cottonwood cuttings root erratically in the field, and hybridization between that species and another more easily rooted species is necessary to achieve an economical silvicultural system (Zalesny and Zalesny 2009). This need for an aggregate genotype possessing all required commercial attributes gives rise to the several breeding programs found throughout North America and elsewhere in the world. Commercial genotypes in use today have most, if not all, of the important traits affecting production. However, the number of commercial genotypes in use today is relatively low, and diversification, as well as yield improvement, is a goal of breeding programs (Berguson et al. 2010; US DOE 2011; Zalesny et al. 2011). In the United States, genetic improvement research varies considerably across the country (Table 3.2).

3.3.1.4 Biomass Production

Biomass production is one of the criteria of assessing feasibility of *Populus* clones for bioenergy and bioproduct application. The sprouting ability of these species is also necessary to understand yield potential. It is currently understood that more than 20,000 stems per hectare of *Populus* can resprout after first-year thinning from a combination of stump and root sprouts depending on species. *Populus* species have great potential for biomass production. Certain genotypes of the eastern cottonwood (*P. deltoides*) and its hybrids, for example, are capable of producing yields ranging from 9 to 20 Mg ha^{-1} yr^{-1} in the United States (Wright 2006; Berguson et al. 2010; Zalesny et al. 2011; Zamora et al. 2013) and in other parts of the world (Table 3.3) with a wide range of environmental conditions suitable for their production, which differs based on climate, soil properties, and cultural practices.

Table 3.3 Biomass Yield of *Populus* spp. around the Globe

Country	Varieties	Rotation	Biomass (Mg ha⁻¹)	Reference
Ghent, Belgium	*Populus trichocarpa* × *deltoides-Hoogvorst*	Annual	3.5	Vande Walle et al. (2007)
Nanjing, China	*P. deltoides* Bartr. cv. "Lux," *P. euramericana* (Dode) Guinier cv. "San Martino," *P. deltoides* Bartr. cv. "Havard"	Annual	10.3	Fang et al. (1999)
Jiangsu Province, China	*P. deltoides* cv. "35"	Annual	8.8–15.2	Fang et al. (1999)
Bihar, India	*P. deltoides* G3 intercropping system	Annual	12.1	Das and Chaturvedi (2005)
Sweden	*P. balsamifera* L., *P. maximowiczii* × *P. trichocarpa*, *P. trichocarpa* × *P. deltoides*, *P. trichocarpa*	4 years	141.9	Johansson and Karačić (2011)
Quebec, Canada	2 clones of *P. maximowiczii* × *P. nigra* (NM5 and NM6)	4 years	66.5	Labrecque and Teodorescu (2005)
Hungary	*P. trichocarpa* × *deltoides*	Annual	16.3	Marosvölgyi et al. (1999)
Scotland	*P. balsamifera* var. *Michauxii* (Henry) × *P. trichocarpa* var. *Hastata* (Dode) Farwell	Annual	5.5	Proe et al. (1999)
Iowa, USA	7300501	5 years	16.8	Riemenschneider et al. (2001)
Wisconsin, USA	80X00601	5 years	8.5	Riemenschneider et al. (2001)
Minnesota, USA	D121	5 years	6.8	Riemenschneider et al. (2001)
Iowa, USA	Eugenii	7 years	17.0	Goerndt and Mize (2008)
Iowa, USA	Eugenii	7 years	5.4	Tufekcioglu et al. (2003)
Iowa, USA	Cradon	10 years	30.0	Goerndt and Mize (2008)

3.3.2 Shrub Willow (*Salix* spp.)

3.3.2.1 Introduction

Research and development of shrub willows as biomass crops has been occurring since the mid-1970s in Europe because of the multiple environmental and rural development benefits associated with their production and use (Börjesson 1999; Volk et al. 2004; Rowe et al. 2008). At present, there are more than 20,000 ha of willow biomass crops established in Sweden, the United Kingdom, and other countries (Aylott et al. 2008; Volk et al. 2011). In North America, research on willow biomass crops started in the early 1980s, but the system has not been widely deployed—with less than 1000 ha in the region (Volk et al. 2004; Volk and Luzadis 2009).

Shrub willows have several characteristics that make them ideal feedstock for biofuels, bioproducts, and bioenergy: high yields that can be obtained in three- to four-year rotations, ease of propagation from dormant hardwood cuttings, a broad underutilized genetic base, ease of breeding for several characteristics, and ability to resprout after multiple harvests (Figure 3.3). The use of shrub willow for SRWC systems has been extensively evaluated over the past two decades with research centered in northeastern and western Canada, Sweden, and the United States. Information has been provided on sustainability issues (Volk et al. 2004; Rowe et al. 2008), cutting cycle

Figure 3.3   Willow biomass crops resprouting in the spring after being harvested the previous winter at the end of their first rotation. The plants are about six weeks old aboveground on a four-year-old root system. (Courtesy of Tim Volk.)

and spacing effects on biomass yield (Kopp et al. 1997), genetic improvements (Smart et al. 2005), and nutrient management (Aronsson et al. 2000; Adegbidi et al. 2001; Quaye and Volk 2013).

3.3.2.2 Biology and Production Systems

Shrub willow commonly hybridized and evaluated for use in SRWC bioenergy coppice systems includes the North American species *Salix eriocephala* Michx. and *S. discolor* Muhl; the European species *S. viminalis* L., *S. purpurea* L., *S. schwerinii* Wolf, and *S. dasyclados* Wimm.; and the Asian species *S. miyabeana* Seemen, *S. sachalinensis* Schmidt, and *S. koriyanagi* Kimura (Smart et al. 2005; Volk et al. 2011).

The shrub willow cropping system consists of planting genetically improved varieties in fully prepared open land where weeds have been controlled. The varieties of shrub willows that have been bred and selected over the past two decades in New York can be grown successfully on marginal agricultural land across the Northeast, Midwest, parts of the Southeast United States, and in Canada. Furthermore, willows produced through the Swedish breeding program can be grown across parts of Europe (Mola-Yudego 2010). This range could be expanded with the development of new varieties.

Preparing fields prior to planting is an essential step in the biological and economic success of willow biomass crops. Typically field preparation begins in the fall of the year before planting and involves a combination of chemical and mechanical techniques to control weeds. Planting takes place in the spring between the end of April and the beginning of June. In addition, trials incorporating cover crops such as winter rye (*Secale cereale* L.) or white clover (*Trifolium repens* L.) during the establishment year have reduced provided cover that would help reduce erosion potential and that lowered weed pressure without impeding willow establishment and early growth (Adiele and Volk 2011). Studies to evaluate conventional tillage, no tillage, and conservation tillage methods using precision zone tillage equipment are currently underway.

Willows are planted as unrooted, dormant hardwood cuttings at about 15,000 plants per hectare using mechanized planters that are attached to farm tractors and that operate at about

Figure 3.4 Mechanical planting of one-year-old stems of willow biomass crops with an Egedal planter in New York, USA. The planter cuts off 15- to 20-cm-long sections of stem and inserts them 14–18 cm into the soil. (Courtesy of Tim Volk.)

0.8 ha h^{-1} (Figure 3.4). To facilitate the management and harvesting of the crop with agricultural machinery, willows are planted in a double-row system with 1.8 m between double rows, 0.76 m between rows, and 0.51 m between plants within the rows. Following the first year of growth, the willows are cut back close to the soil surface during the dormant season to stimulate coppice regrowth, which increases the number of stems per stool from 1–2 to 8–13 depending on the variety (Tharakan et al. 2005). After an additional three to four years of growth, the stems are mechanically harvested during the dormant season after the willows have dropped their leaves.

Different types of harvesting systems have been developed over the years, including whole stem, cut and bail, and single pass cut and chip systems of different sizes (Berhongaray et al. 2013; Savoie et al. 2013). At present, forage harvesters with a specially designed cutting head cut for woody crops are the most common technique used, although other smaller cut and chip harvesting systems are being developed. These forage harvester cut the willow stems 5–10 cm above the ground, feed the stems into the forage harvester, and produce uniform and consistent sized chips that can be collected and delivered directly to end users with no additional processing (Berhongaray et al. 2013; Eisenbies et al. 2012; Savoie et al. 2013) (Figure 3.5). The chipped material is then delivered to end users for conversion to bioenergy, biofuels, and/or bioproducts. The plants will resprout the following spring when they are typically fertilized with about 100 kg N ha^{-1} (Abrahamson et al. 2002; Adegbidi et al. 2003) of commercial fertilizer or organic sources such as manure or biosolids. Further research is underway to refine these recommendations for new willow varieties across a range of sites. Recent studies (Quaye et al. 2011; Quaye and Volk 2013) have shown that there is no yield response to a range of nutrient amendments in willows across a number of sites, indicating that nutrient additions may not be necessary to maintain production. The willows are allowed to grow for another three- to four-year rotation before they are harvested again. Projections indicate that the crop can be maintained for seven rotations before the rows of willow stools begin to expand to the point that they are no longer accessible with harvesting equipment. At this point, the crop can be replanted by killing the existing stools with herbicides after harvesting, and the killed stools are chopped up with a heavy disk and/or grinding machine followed by planting that year or the following year.

Figure 3.5   **(See color insert.)** A New Holland (NH) forage harvester fitted with a specially designed NH
FB130 coppice header harvesting four-year-old willow biomass crops in central New York.
(Courtesy of D. Angel, SUNY-ESF, Syracuse, NY.)

3.3.2.3 Breeding and Genetic Improvements

The large genetic diversity across the genus *Salix* and the limited domestication efforts to date
provide tremendous potential to improve yield and other characteristics, such as insect and disease
resistance and growth forms of willow biomass crops. Worldwide there are 350–500 species of willow
(Kuzovkina et al. 2008; Smart et al. 2008), with growth forms ranging from prostrate, dwarf species
to trees with heights of greater than 40 m. The species used in woody crop systems are primarily from
the subgenus *Caprisalix* (*Vetrix*), which has more than 125 species worldwide (Kuzovkina et al. 2008).
Although these species have many characteristics in common, their growth habits, life history, and
resistance to pests and diseases vary, which is important in the successful development of woody crops.
Willows' ability for vegetative propagation means that once superior individuals with genetic gains are
identified, they can be maintained and those individuals can be multiplied rapidly for deployment.

Willow breeding in Europe was initiated in the 1980s, but many of the varieties from these
programs that have been tested in the United States have been damaged by the potato leafhopper
(*Empoasca fabae*), making them unsuitable for large-scale deployment. Since the late 1990s, more
than 700 accessions have been collected from around the world and more than 575 controlled
pollinations have been attempted, producing about 200 families (Smart et al. 2008). Selection trials
of new varieties from the initial rounds of the breeding programs in the late 1990s have produced
yields that are up to 15–20% greater in the first rotation than the standard varieties used in early
yield trials. Second rotation results from these same trials indicate that the yield of some of the new
willow varieties is more that 30% greater than the standard varieties. These results indicate that
there is a large potential to make use of the wide genetic diversity of shrub willows to improve yields
with traditional breeding and selection (US DOE 2011).

3.3.2.4 Biomass Production

A rapid growth rate is one of the attributes that make shrub willows an appealing biomass
crop. First rotation, non-irrigated research-scale trials in the U.S. Northeast have produced yields
(oven dry ton, or odt) of 8.4–11.6 Mg ha^{-1} yr^{-1} (Adegbidi et al. 2001, 2003; Volk et al. 2004).
Second rotation yields of the five best-producing varieties in these trials increased by 18–62%

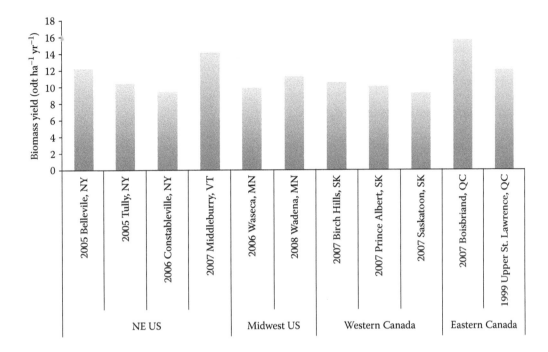

Figure 3.6 First-rotation biomass yield (Mg [oven dry] ha⁻¹ yr⁻¹) of top five clones with biomass crop yield trials established with new willow genotypes between 1999 and 2008. Site labels include the year of planting and location. (Modified from Volk, T. A., et al., *Aspects of Applied Biology*, 112, 67–74, 2011; Nissim, W. G., et al., *Biomass and Bioenergy*, 56, 361–369, 2013; Zamora, D., et al., *Biomass and Bioenergy*, 49, 222–230, 2013.)

compared to first rotations (Volk et al. 2006) and, in subsequent rotations, yields are maintained and largely dependent on weather conditions. The most recent yield trials using improved varieties of willow are showing yield increases of 20–40%. Growth rates of new willow varieties exceed 15 Mg (oven dry) ha⁻¹ yr⁻¹ (Volk et al. 2011) planted in Quebec, Canada (Figure 3.6). However, the mean yield of improved willow varieties grown in Canada and the United States is 11.31 Mg (oven dry) ha⁻¹ yr⁻¹ (Figure 3.6), which is slightly higher than the 10.2 Mg (oven dry) ha⁻¹ yr⁻¹ average yield of willows grown in Europe (Mola-Yudego 2010). Abrahamson et al. (1998) suggested a yield range of 11–18 Mg (oven dry) ha⁻¹ yr⁻¹. However, yields of fertilized and irrigated unimproved varieties of willow grown for three years have exceeded 27 Mg (oven dry) ha⁻¹ yr⁻¹ in the northeastern and midwestern United States (Adegbidi et al. 2001; Labrecque and Teodorescu 2005).

3.3.3 Eucalyptus

3.3.3.1 Introduction

Eucalyptus spp. are successful SRWCs because of their fast growth and environmental tolerance due to attributes such as indeterminate growth, coppicing, lignotubers, drought/fire/insect resistance, and/or tolerance of soil acidity and low fertility, and many have desirable wood properties for bioenergy production. They have been domesticated for various products and have been widely commercialized in the tropics and subtropics. Significant plantings of eucalyptus in the United States began with introductions from Australia as a result of the California Gold Rush in 1849. *Eucalyptus* species were introduced in the southern United States as early as 1878, but no significant commercial plantations were established until the late 1960s (Dougherty and Wright 2012;

Stanturf et al. 2013). Eucalyptus also has a long history in India. Large-scale plantations have been raised in government-owned and private lands (Palanna 2013). In Brazil, eucalyptus was introduced in late 1890s from Uruguay to increase the supply of wood to the railroad industry and is now a primary source of wood for energy (Barnes 2013). Though eucalyptus has naturalized in some areas, it also raises concerns of invasiveness (Booth 2012).

3.3.3.2 Biology and Production System

The native range of eucalyptus is primarily Australia with a few species also native from Indonesia, Philippines, and Papua New Guinea. Eucalypts are grown extensively as exotic plantation species in tropical and subtropical regions throughout Africa, South America, Asia, and Australia and in more temperate regions of Europe, South America, North America, and Australia (Stanturf et al. 2013).

Although there are more than 700 species, less than 15 species are commercially significant worldwide (Gonzalez et al. 2011). Seed introductions of *Eucalyptus* species in the United States were focused primarily in Florida (Zon and Briscoe 1911) due to climatic similarities with Australia. Genetic improvement programs are selected for fast growth, cold tolerance, desirable growth form, and reduced lignin in the southern United States. *Eucalyptus* plantations in the southern United States can be successfully established using freeze tolerant improved seedlings in some specific regions (Gonzalez et al. 2011). Genetic improvement programs aim to improve varieties for various growing conditions (Rockwood et al. 2006; Gonzalez et al. 2011). *Eucalyptus* prefer moderately well-drained soils with some degree of clay content for water retention (Wright and Rosales 2001). *Eucalyptus* could be grown for bioenergy on up to 100,000 ha in Hawaii following guidelines from a research and development program in the 1980s.

Eucalyptus production practices in different parts of the world vary with site conditions, desired products, and scale of commercialization. Genetic selection has led to commercialization of genotypes with unique advantages in different applications. They are commercially propagated by seed and cloning of tissue culture. For conventional pulpwood production, stands are typically established at a planting density of 1482–2471 trees per hectare, and harvested every 6–10 years. They may be replanted at harvest, which can benefit from improved genetic material, or regenerated from coppice growth, which eliminates the cost of replanting. The economically optimum time between harvests may be three to four years, with replanting after two to five harvests, on stands with initial densities of 8401 trees per hectare in Florida (Langholtz et al. 2007).

Silvicultural strategies of *Eucalyptus* continue to evolve with changing markets, genotypes, and applications. Because of high growth rates and tolerance to a range of growing conditions, *Eucalyptus* can be produced in innovative ways, providing nonmarket benefits. For example, research trials demonstrate that *E. grandis* and *E. amplifolia* can be used for restoration of phosphate-mined lands (Rockwood et al. 2006; Langholtz et al. 2007, 2009). *Eucalyptus* spp. has been shown to be effective at phytoremediation of reclaimed wastewater, municipal waste, storm water, and arsenic and trichloroethylene contaminated sites (Langholtz et al. 2005; Rockwood et al. 2008). Eucalyptus plantations that provide these types of environmental services may be viewed more favorably by the public, and compensation for nonmarket environmental services would improve the profitability of these systems.

3.3.3.3 Breeding and Genetic Improvements

A few *Eucalyptus* species and hybrids constitute the majority of these plantations. Most domesticated eucalypts are from the subgenus *Symphyomyrtus*, the largest of the 10 subgenera currently recognized within *Eucalyptus*, containing more than 75% of the species (Rockwood et al. 2008). Four species and their hybrids from this subgenus, *Eucalyptus grandis*, *E. urophylla*, *E. camaldulensis*, and *E. globulus*, account for about 80% of the eucalypt plantations worldwide. *E. grandis* is the most widely used species in plantation forestry worldwide in tropical and subtropical areas not only as a pure species but also as a parental species in hybrid breeding. It has the fastest growth and widest adaptability of all *Eucalyptus* species. The greatest area of plantations of *E. grandis* and its hybrids with other species is in Brazil and several other Central and South American countries. It has been planted extensively in India, South Africa, Zambia, Zimbabwe, Tanzania, Uganda, and Sri Lanka, and is grown in California, Florida, and Hawaii in the United States (US DOE 2011).

Eucalyptus species are ideal for genetic improvement because many typically propagate easily, have short generation intervals and broad genetic variability, and may be genetically engineered. The species/cultivars/hybrids with documented bioenergy potential in the Southeast are described in Zalesny et al. 2011. Due to Florida's challenging climatic and edaphic conditions, much SRWCs emphasis has been on *Eucalyptus* tree improvement for adaptability to infertile soils and damaging freezes. Through a significant advance in the understanding of freezing tolerance in *Arabidopsis*, a freeze-tolerant *Eucalyptus* has been developed (Yamaguchi-Shinozaki and Shinozaki 1993). A renewed effort to identify frost-tolerant species and genetic modifications to increase frost tolerance that will permit expansion of the range of *Eucalyptus* is driven by the potential need for 20 million Mg yr^{-1} of *Eucalyptus* wood for pulp and biofuel production in the southern United States by 2022 (Rockwood et al. 2008). Recent work to select frost-tolerant clones of *E. benthamii* is being done to extend the northern range of the species for biomass production in the southeastern United States (Wright 2006).

3.3.3.4 Biomass Production

Eucalyptus yields are influenced by precipitation, fertility, soil, location, and genetics. In the southern United States, rotation length and yields for biomass for bioenergy can be three to four years with a 12.3–22.4 Mg (oven dry) ha^{-1} yr^{-1} (Gonzalez et al. 2011). *Eucalyptus* spp. yielded 18–35 Mg (oven dry) ha^{-1} yr^{-1} annually after three to five years of growth on clayey soil in central Florida, which is comparable to 21–34 Mg (oven dry) ha^{-1} yr^{-1} estimated for *Eucalyptus* in Australia (Pereira et al. 1989) but higher than the 10–18 Mg (oven dry) ha^{-1} yr^{-1} estimated by Klass (2004), who observed that yields could be improved with SRWC development in the subtropical South. Variability in these estimates is due to soil, climate, and management practices. In traditional pulpwood management systems, this hybrid is predicted to produce 34–43 green Mg ha^{-1} yr^{-1} on a seven-year rotation in a traditional pulpwood management system and with 43–52 green Mg ha^{-1} yr^{-1} in a biomass management system. On the other hand, *Eucalyptus saligna* Smith in five- and six-year rotations and an eight-year *Eucalyptus/Albizia* mix produced 20.2, 18.6, and 26.9 or more Mg (oven dry) ha^{-1} yr^{-1}, respectively (Whitesell et al. 1992). Biomass production of *Eucalyptus* in Brazil (Hall et al. 1993), Spain (San Miguel 1988), and Hawaii (San Miguel 1988) is within this range (Table 3.4).

Table 3.4 Biomass Production of *Eucalyptus* spp. around the Globe at Varying Site Regions

High Yield	Average Yield		
Mg (oven dry) ha^{-1}		Production Region	Reference
30	13–15	NW Spain	San Miguel (1988)
21	5–8	SW Spain	
	12.5	Spain	
35	12–22	Southern USA	Gonzalez et al. (2011)
	9–26	Mid Brazil	Hall et al. (1993)
58	40–50	India	Prasad et al. (2010)
27	13–27	Hawaii	Whitesell et al. (1992)
30	20–24	Australia	Pereira et al. (1989)
	2.9	Ethiopia	Selamyihun et al. (2005)
28	3–21	NE Brazil	Carpentieri et al. (1993)

3.4 CHALLENGES AND LIMITATIONS OF USING SRWC FOR ENERGY

The use of SRWCs for energy is not without challenges and limitations. We have identified the following challenges and limitations.

3.4.1 Inadequate Productivity and Risk of Pests and Diseases

One considerable concern about growing SRWCs for bioenergy is that SRWCs may not produce sufficient biomass as a feasible renewable energy source. In the United States, the DOE projected that a productivity rate of 17–22 Mg (oven dry) ha^{-1} yr^{-1} will be required for the long-term feasibility of renewable production (English et al. 2006). However, the average production of 11.3 Mg (oven dry) ha^{-1} yr^{-1} across site growing conditions in the United States and Canada (Figure 3.6) is well below the required amount of biomass necessary to sustain feasibility of bioenergy systems in the United States (English et al. 2006; Hinchee et al. 2009). Similarly, biomass production of *Populus* (12 Mg [oven dry] ha^{-1} yr^{-1}) (Mercker 2007) falls below the projected rate of SRWCs for sustained bioenergy production. Pests and diseases also hamper optimum production of SRWCs.

Around the globe, several breeding programs exist to improve productivity and to increase SRWCs' resistance to pests and diseases (Table 3.1). Companies and research programs in the United States and Canada, such as ArborGen and the Sungrant Regional Feedstock Partnership Poplar Woody Crops Program, are currently breeding new *Populus* hybrids and varieties with fast growth, high volume increments, and the ability to grow in a wide range of sites that are also resistant to pests and diseases. Studies underway by the Sungrant Poplar Woody Crops Program, for example, have identified superior poplar clones yielding 7.5–11 Mg (oven dry) ha^{-1} yr^{-1} in the Upper Midwest and more than 19 Mg (oven dry) ha^{-1} yr^{-1} in the Pacific Northwest of the United States (Berguson et al. 2010). It should be noted that although breeding programs are underway to improve poplar yield, these programs have only been in operation for a limited time, and successive genetic improvement over a multidecade time period is needed to achieve the genetic gains and to capitalize on the wide genetic diversity present in the species.

Gene insertion is also being employed to improve production. Research on gene insertion of conifer cytosolic glutamine synthetase (GSI) in *Populus* has shown GSI transgenic trees with more than a 100% increase in leaf biomass relative to control trees under low nitrogen conditions (Hinchee et al. 2009). Genomic works are also being carried out in the United States and other parts of the world to identify genome-wide functional gene networks in poplars that are associated with abiotic stress tolerance and bioenergy related traits using combination of computational projects, gene expression analysis, and experimental validation (Berguson et al. 2012). The program investigates developmental of poplar varieties that can thrive under abiotic stress on marginal land that is suitable for food crops.

The willow breeding program in the United States, on the other hand, has been intensified to develop high yielding clones that are also resistant to pests and diseases. In the past, the willow breeding program focused on hybridization. More than 1000 willow clones have been developed through the U.S. breeding program that could be employed in the landscape for biomass production for energy. Genomic work is also underway to investigate how gene expression patterns in willow hybrids are related to yield potential and other traits important for biofuel production (Serapiglia et al. 2013).

Genetic modification of *Eucalyptus* is also expanding to increase yield. In the United States, a genetically engineered hybrid of *E. grandis* × *E. urophylla* S.T. Blake with genes for cold tolerance, lignin biosynthesis, and fertility is currently in the U.S. regulatory approval process (US DOE 2011). In field tests, these trees have survived temperatures as low as 6°C. The variety is well known for its high-quality fiber; it also excels at biomass production and can be planted on marginal lands.

Efforts to identify regions of the *Eucalyptus* genome that regulate biomass growth and wood quality are underway to identify genes of value for bioenergy, particularly those involved in the lignin and carbohydrate/cellulose pathways.

3.4.2 Technological Compatibility

A comprehensive review by Gold and Seuring (2011) details many of the issues involved in the supply chain and logistics of bioenergy production from biomass. Cited technological issues range from harvesting, such as matching biomass supply (natural growing cycles) and continuous bioenergy plant demand, to system design, such as high levels of complexities and inter-dependencies of bioenergy production systems (Gold and Seuring 2011; Yeh 2011). One specific technological challenge in marketing SRWC biomass is providing the biomass in acceptable forms to the end user. For willow SRWCs, biomass is harvested on a three- to four-year cycle using modified agriculture equipment that cuts and chips the biomass in a single operation. Depending on the equipment, biomass chips may be rendered in an inconsistent quality undesirable to the end user. To address this issue, researchers in Canada and the United States have developed a harvesting system for willow biomass crops based on a New Holland (NH) forage harvester fitted with a specially designed willow cutting head to address the issue of harvesting, processing, and storing of willow chips for energy (Figure 3.5). This equipment is now commercially available, and the system continues to be improved. Trials with this system indicated that for three- to four-year-old willow biomass crops with the majority of stems <75 mm in diameter, consistent high-quality chips (>95% of the chips being smaller than 37.5 mm) can be produced with the FR series NH forage harvester and NH FN130 coppice header at a rate of about 60–80 wet Mg h^{-1} (Eisenbies et al. 2012).

Similarly, SRWC hybrid poplars and *Eucalyptus* are not without technological compatibility issues. Harvesting requirements, storage and processing, and quality requirements are among the issues associated with the supply chain of using SRWC hybrid poplars and *Eucalyptus*. Traditional forestry harvesting equipment needs to be retrofitted to allow for the harvesting of these species in a manner that will optimize production. However, several advancements have been made to address logistical needs. Companies associated with producing energy from biomass have made modifications to their operations. Machineries and equipment have been developed to process woods on-site prior to delivery. Materials are chipped on-site to allow for an easy and economical transport of the materials to a facility.

3.4.3 Conversion Efficiency

Two major platforms of converting woody biomass into energy are biochemical (sugar) and thermochemical (pyrolysis or gasification). The deployment of certain conversion processes depends on the wood properties and characteristics. For the biochemical platform of fuel production, SRWCs have been seen by some as a less desirable feedstock because of their high lignin content and recalcitrance to digestion, which is one of the challenges in efficiently converting woody biomass to biofuels. However, other processes such as that proposed by ZeaChem use the lignin in a gasification step to produce hydrogen to increase end product yield. Recalcitrance refers to the resistance of lignocellulosic material in undergoing biofuel conversions; this is due to a variety of factors, such as polysaccharides that are not easily digestible by enzymes and/or microorganisms and compounds that inhibit the fermentation of sugars once they are released (Hinchee et al. 2009). Trees with natural low levels of lignin and high cellulose are ideal for the biochemical process. However, due to the variety of processes currently being considered and the lack of clarity on ultimately commercially successful pathways, wood biomass may be desirable or detrimental, depending on the specific conversion system.

One approach for reducing recalcitrance is to reduce the lignin content of wood or to make it easier to remove. This has been done in a number of transgenic trees, including some *Populus* species, by downregulating genes involved in lignin biosynthesis (Hinchee et al. 2009). Here, however, a challenge exists in avoiding negative pleiotropic effects associated with lignin reduction, such as cavitation and vessel collapse. A number of studies have worked to address these issues (Hinchee et al. 2009). On the other hand, wood biomass such as SRWCs is the preferred feedstock for the pyrolytic conversion production because of high lignin and greater energy density (US DOE 2011). Lignin has less oxygen than carbohydrates (so there is less to remove) and higher energy density, meaning more energy content per Mg of biomass processed. Compared to many proposed herbaceous energy crops, such as switchgrass or miscanthus, the low ash content of woody crops may make SRWCs the preferred source of raw material in thermal conversion systems.

3.4.4 Cost of Production

The commercial production of poplars in the United States has been limited to those locations where the plantations are associated with a specific mill such as the Verso Paper Mill in Sartell, Minnesota (recently closed), or the high-value sawmill near the Green Wood Resources stands in Oregon. In these cases, a higher valued product such as paper or lumber is being produced. Despite the biomass production potential and other environmental and rural development benefits associated with willow biomass crops, their deployment has not yet been widely adopted in the United States due to the high cost of production and the lack of markets (Quaye and Volk 2013). Although projections indicate that *Salix*, *Populus,* and *Eucalyptus* SRWC systems will be important parts of the future biomass supply, only a limited number of hectares have been deployed for that purpose to date in the United States, Canada, and Europe. The high upfront establishment costs, risks associated with the production of this new crop, and the uncertainty of biomass markets over multiple rotations create barriers to large-scale deployment (Buchholz and Volk 2011).

Economics of the system can be improved by increasing yields and reducing production costs because the system has not been widely deployed and can still be improved. Areas for improvement include reducing the cost of planting stock, improving weed control, optimizing harvesting and logistics operations, and reducing input costs such as fertilizer. The use of organic waste materials such as biosolid compost (BC) and digested dairy manure (DM) as nutrient sources for willow biomass production is being explored as an attractive means to decrease fertilization costs, while maintaining nutrient and levels, increasing production, and reducing greenhouse gases associated with the system. Recent studies across a range of sites (Quaye et al. 2011; Quaye and Volk 2013) indicate that that there was no change in willow production when a range of organic and inorganic nutrients were applied. This suggests that it may be possible to produce good yields in willow biomass crops without fertilizer additions in the first and in some cases later rotations at some sites due to the nutrient status of these sites and high internal nutrient cycling in these systems using this cultural practice. It may be possible to remove this input, which is about 10% of total costs over seven rotations, and improve the economics of the systems. Other opportunities such as decreasing planting density and optimizing harvesting operations are being explored.

3.5 SUSTAINABILITY OF SRWC FOR ENERGY

The Brundtland Commission provided one of the most well-known definitions of sustainability, highlighting the equal right of present and future generations to meet their respective needs (WCED 1997). Kiser and Fox (2013) posit the integration of the intensely interrelated economic, ecological, and social aspects of sustainability in a "triple-bottom line," as do Dyllick and Hockerts (2002) who pointed to the three facets of sustainability, conceiving corporate sustainability as the business case (economic), the natural case (environmental), and the societal case (social).

Environmental, economic, and social sustainability of the use of SRWCs is a great concern. Biomass yields with SRWC systems can be improved with proper site selection and silvicultural treatments such as species and provenance choice, fertilization irrigation, tillage, and weed control (Mead 2005; Ranney and Mann 1994). Environmental sustainability issues that arise from the intensive management and inputs used in SRWC systems are likely to fall midway between traditional forestry and agricultural systems. These systems are highly mechanized and require intensive silvicultural inputs from a traditional forestry perspective to achieve desired yields. In SRWC systems, environmental sustainability in its broader context is using the land in a way where future generations can derive the same benefits as the present generations (Andersson et al. 2000). Sustainable management maintains the ecological capacity of the site for tree growth over time (Evans 1997) by avoiding degradation of soil nutrients and water supplies (Andersson et al. 2000; Reijnders 2006). The intensity of silvicultural practices can have a positive effect by increasing yield. Yield increases are required to make SRWC biomass bioenergy cost competitive with fossil fuel energy sources and to also reduce the amount of land required to produce the necessary tonnage. Negative environmental effects of SRWC systems also arise from the intensity of silvicultural practices, resulting in a trade-off between yield increases and environmental impacts. Some factors that negatively impact the environmental sustainability of SRWC systems include tillage and erosion, soil compaction, herbicide, nutrient management, and pests and pathogens (Buchholz et al. 2009; Kiser and Fox 2013).

Concerns about soil erosion potential in SRWC systems occur during the establishment phase because it can take one to three years before sites are fully occupied (Volk et al. 2004; Rowe et al. 2008). Erosion represents possibly the greatest threat to long-term soil productivity due to loss of organic matter and nutrient-rich soil surface (Lal 1998). However, because tillage is only required at establishment, erosion is likely limited to the first few years. Once SRWC crops are established, they are perennial crops with low soil erosion potential. As a result of erosion during the establishment phase of shrub willow SRWC systems, the use of cover crops is being explored to minimize such impacts (Volk et al. 2004; Adiele and Volk 2011). Because tillage is only required at establishment, erosion is likely limited to the establishment year; hence, it should be less in coppice regenerated systems compared to annual tilled systems. Furthermore, during harvest, SRWC systems require the use of machinery with the potential to compact the soil. Compaction may also reduce water infiltration, causing soil erosion. However, the use of machinery in SRWC plantations is usually less than that needed for traditional forestry because fewer passes are needed, making the SRWC systems more environmentally friendly (Ulzen-Appiah 2002; Hamza and Anderson 2005). Established root systems are extensive and diffuse enough to support harvesting equipment when soils are dry (Mitchell et al. 1999).

Herbicide application is a practice used in SRWC systems to control weeds during establishment (Kopp et al. 1997; Volk et al. 2006), and it has the potential to enter groundwater, thus affecting water quality (Tolbert et al. 2002; Spalding et al. 2003). However, herbicides are only used during the first year or two of establishment, so over time these inputs are a fraction of what occurs in agricultural rotations (Volk and Luzadis 2009). SRWCs can be used in the phytoremediation process and can improve water quality compared to traditional agriculture (Mortensen et al. 1998); however, fertilization has the potential to negatively impact the sustainability of SRWC systems when added nutrients are not captured by the crop but instead leached to the groundwater or runoff to surface waters (Kiser and Fox 2013). It appears that fertilization will likely be much less than in traditional agricultural crops, and additional research in growth responses through time in SRWC systems is needed to ensure judicious use of fertilizer from an economic as well as environment point of view.

The perennial nature of SRWC systems and the relatively narrow genetic base in *Salix, Populus,* or *Eucalyptus* SRWCs may promote susceptibility to pests and diseases. Damage of SRWC systems by pests and pathogens has been found to be problematic. Some willow clones are susceptible to potato leafhoppers (Volk et al. 2006), while leaf rust (*Melampsora medusa*) is common disease for *Populus* species affecting performance. Planting multiple cultivars with different genetic backgrounds when large-scale plantings are occurring is an effective way to manage for this concern at this point in time. Further work is needed to understand pest and pathogen dynamics and their potential impact on SRWC systems (Serapiglia et al. 2013).

The perennial nature of SRWCs and their changing structure suggest that across the landscape these systems can enhance biodiversity (Volk et al. 2004; Rowe et al. 2011). A study of bird diversity in willow biomass crops over several years found that these systems provide good foraging and nesting habitat for a diverse array of bird species (Dhondt et al. 2007). For example, an increase in abundance and diversity of ground flora and avian species among commercial SRWC systems was observed in Germany (Schulz et al. 2009).

Although achieving environmental sustainability is critical in SRWC systems, economic and social sustainability of such systems must also be in place. The current yield potential of SRWC systems with typical operational costs and procedures across the inherent site fertility gradient appears to range from 9 to 15 Mg (oven dry) ha^{-1} yr^{-1} (Figure 3.6). Selection of the best genetics and intensive silvicultural practices such as optimal fertilization, irrigation, and weed control have shown the potential to double or triple the yield of these species (Labrecque and Teodorescu 2005). Yield increases can proportionally reduce the area of land required to meet production goals. However, intensive silvicultural practices may raise production costs to the level where SRWC systems may become economically unviable. In addition, these intensive silvicultural practices tend to increase the negative environmental impacts of SRWC systems (Kiser and Fox 2013). Additional research is needed to better understand trade-offs between intensity of management and production to achieve optimal environmental and economic performance across a range of landscapes where SRWCs may be grown.

SRWCs have the potential to provide rural development benefits by diversifying farm crops, creating an alternative source of income for landowners, and by circulating energy dollars through the local economy. A recent study in New York on biofuel production from biomass indicated that, regardless of the scenarios or region that was assessed, more than 80% of the jobs that are created when biofuels are produced from locally produced biomass are associated with the biomass production, harvesting, and transportation portion of the system. As biomass systems develop and are implemented, the feedstock production and delivery portion of the system will create the majority of jobs within this new industry. Willow biomass crop production, harvesting, and transportation in a study by Swenson (2010) created 45 direct and indirect jobs for every 10,000 acres of willow biomass crops established and managed. Induced jobs were not assessed as part of this study because of the difficulty in making these estimates but would increase this figure.

A recent assessment in the United Kingdom indicated that growing and producing willow biomass crops for renewable electricity would create two to four full-time jobs for every MW$_e$ of power (Thornley 2007). Brazil, one of the countries with the highest number of short rotation plantations, has mainly large-scale industrial production, while other countries such as Madagascar have principally small-scale *Eucalyptus* plantations. However, the short rotation wood sector has been criticized for only providing seasonal employment with low security arrangements. Large companies are increasingly trying to reduce costs and make planning more efficient by working on a continuous basis. Although it is difficult to estimate, the plantations are estimated to create one to three jobs per 100 ha (Elbehri et al. 2013).

Overall, by extending the concept of sustainability to bioenergy, sustainable bioenergy production systems "would be environmentally, economically, and socially viable now and for future generations and would move the world away from an unsustainable reliance on fossil fuels" (Friedman et al. 2010).

3.6 PLANTING WILLOWS USING TRADITIONAL FARMING METHODS—A CASE STUDY

Much research has been conducted regarding clonal production of willows using the double-row planting methods of the breeding programs of Sweden, the United Kingdom, and North America; however, biomass production using the traditional planting methods of farmers has not been explored across geographical regions. The double-row planting design of Sweden and State University of New York (SUNY) is compatible with commercial operations; however, it limits the application of such design by farmers because it is not compatible with their existing operations (e.g., spacing of cultivation equipment). This could pose challenges to adoption of the willow as an SRWC energy production system. Efforts must be made to develop willow production systems compatible with existing farming operations.

In an attempt to demonstrate the biomass production potential of willow using farmers' traditional planting arrangements, an experiment was conducted in northern Minnesota to assess willow biomass production using different methods: (1) the SUNY and Sweden double-row planting method and (2) a single-row planting system with an inter-row spacing of 1.5 m. Two intra-row spacings were introduced in the single-row planting method, such as 0.61 m and 0.46 m, representing 10,763 and 14,351 plants per hectare, respectively. The 1.5 m spacing between the rows was used to allow for the passage of weed maintenance equipment. The double-row planting systems of Sweden and SUNY generated a total biomass of 20 Mg (oven dry) ha^{-1} at the end of first rotation (three years of growth), with 31 Mg (oven dry) ha^{-1} and 39 Mg ha^{-1} of biomass using the single planting method with 0.46 m and 0.61 m intra-row spacing, respectively, representing an increase of 27% and 45% more over the double-row planting system (Figure 3.7). The biomass production of willows using the traditional planting method is within the range of a similar planting arrangement

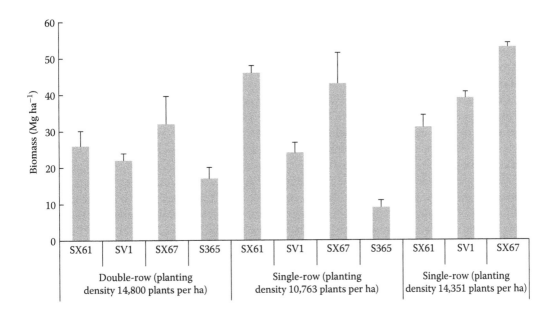

Figure 3.7 Biomass production (Mg oven dry) of willow growing in a clay-loam soil in northern Minnesota, USA. For experiment 1 it employed the SUNY and Sweden double-row system (with spacing 1.5 m between double rows, 0.76 m between rows, and 0.61 m between plants within the rows); experiment 2 used a single-row planting arrangement with 0.61 m plant spacing within the row; and experiment 3 used a single-row planting with 0.46 m spacing within the row. The distance between single rows was 1.5 m.

in Canada (Mirck and Schroeder 2013). Farmers continue to explore options to achieve production cost effectively while end users of biomass continue to demand high-quality feedstock suitable for their production systems. The economics and benefits of the sustainability of utilizing farmers' methods of planting willow is an important consideration, and the compatibility of the SRWC system with the local social landscape is a necessary factor in the long-term success of SRWCs.

Achieving multiple benefits of the bioenergy production system is challenging. Several research projects are underway to understand the potential of SRWCs for energy production in the context of agroforestry while sustaining ecological services such as carbon sequestration, including the integration of willows with agronomic crops in the context of alley cropping systems (e.g., Peichl et al. 2006; Zamora et al. 2014). Such systems are developed to assess water quality improvement, carbon sequestration, and biodiversity as forms of ecosystem services while producing biomass for energy.

3.7 FUTURE PROSPECTS AND CONCLUSIONS

Genetic variation among *Populus* and *Salix* spp. is very high, which presents opportunities to improve yield and nutrient efficiency. Investments must continue to be made to produce new genotypes and to test these clones in replicated trials to minimize risks associated with commercial production. Field testing of new genotypes must be continued, and clone tests at a greater number of sites over the range of sites where *Populus, Salix*, and *Eucalyptus* could grow are necessary. Furthermore, continued development and evaluation of pure species populations are necessary to form a broad foundation for future breeding. Commercial deployment is necessary, and evaluation of improved genotypes in yield blocks must be done to determine biomass and yield gains over current commercial standards. Several other components of the *Salix* system need to be developed to improve the overall economics of willow biomass crop systems. As knowledge increases about the management system required to grow SRWCs and the role they play across the landscape, SRWCs will be more effectively deployed so that other benefits derived from this system, in addition to biomass, can be optimized.

Appropriate silvicultural practices of SRWC must be better understood. The relationship between plant spacing and yield over time is currently unclear particularly as plant populations increase. Understanding farmers' preferences when producing these crops is necessary based on their existing plantation operations. For hybrid poplar, there is evidence that inter-tree competition occurs early in stand development, and the benefits of higher planting density may not be evident or long-lived. This impacts yield and economics. Very little research has been done to evaluate the productivity of poplars under a repeated-harvest, coppiced-based management system, particularly with new clones developed in breeding programs. Due to the high cost of fertilization and the potential for groundwater contamination, determination of nitrogen response to fertilization on a variety of sites must also be explored. Testing methods to assess the nitrogen mineralization potential of sites and correlating this status to growth response must be developed. Also, a more thorough understanding of nutrient budgets within these systems is needed. Economic and land-use impacts of producing SRWCs must be performed across a greater range of landscapes where SRWCs may be grown. If bioenergy becomes economically viable and socially acceptable, some amount of acreage for dedicated biomass production will likely come from the current agricultural land base. In light of this, yields and environmental effects on sites varying in agricultural productivity need to be better understood in order to select sites suitable for energy production while minimizing impacts to the food production system.

In addition to studies on potential yields of different varieties of willow across a range of sites, research must continue on various aspects of the production cycle, including nutrient amendments and cycling, alternative tillage practices, incorporating cover crops, spacing and density studies, harvesting systems development, growth characteristics important for biomass production, use of willow plantations by a variety of macro- and microfauna, changes in soil carbon, economics of the production system, and life cycle assessments of SRWC bioenergy systems. Results from this and other initiatives in North America, Europe, and elsewhere will provide a base from which to begin to expand and deploy willow and other SRWC biomass crops.

Environmental sustainability issues that arise from the intensive silvicultural inputs used in SRWC systems are typically more similar to those in agriculture than in traditional forestry operations. Compared to traditional longer-rotation forestry, these systems are highly mechanized and require intensive silvicultural inputs to achieve desired yields. The trade-offs among management intensity, yield, and environmental effects, positive and negative, are an important research area requiring additional investment.

The promise of SRWCs to produce raw material for the emerging liquid fuel and energy industries is tangible and economically realistic assuming that efficient conversion systems are developed in parallel with the biomass production system. Through the considerable efforts of government, industry, and academia, a foundation of research and commercial application has been laid to form a basic understanding of yield, genetics, needed inputs, and environmental effects. However, development of a new crop for widespread deployment is a long-term effort requiring a consistent investment over a multidecadal time frame. The SRWC production system represents one of a number of potential avenues for producing raw material for renewable energy and for ensuring sustainable production to meet society's demand for energy for generations.

REFERENCES

Abrahamson, L. P., D. J. Robison, and T. A. Volk. 1998. Sustainability and environmental issues associated with willow bioenergy development in New York (USA). *Biomass and Bioenergy* 15: 17–22.

Abrahamson, L. P., T. A. Volk, R. F. Kopp, E. H. White, and J. L. Ballard. 2002. *Willow bioenergy producer's handbook* (rev). State University of New York College of Environmental Science and Forestry (SUNY-ESF), Syracuse, p. 31.

Adegbidi, H. G., R. D. Briggs, T. A. Volk, E. H. White, and L. P. Abrahamson. 2003. Effect of organic amendments and slow-release nitrogen fertilizer on willow biomass production and soil chemical characteristics. *Biomass and Bioenergy* 25: 389–398.

Adegbidi, H. G., T. A. Volk, E. H. White, R. D. Briggs, L. P. Abrahamson, and D. H. Bickelhaupt. 2001. Biomass and nutrient export by willow clones in experimental bioenergy plantations in New York. *Biomass and Bioenergy* 20: 389–398.

Adiele, J. and T. A. Volk. 2011. Developing spring cover crop systems for willow biomass crop establishment. *Aspects of Applied Biology* 112: 113–119.

Al Afas, N., N. Marron, and R. Ceulemans. 2005. Clonal variation in stomatal characteristics related to biomass production of 12 poplar (*Populus*) clones in a short rotation coppice culture. *Environmental and Experimental Botany* 58: 279–286.

Andersson, F. O., G. I. Agren, and E. Fuhrer. 2000. Sustainable tree biomass production. *Forest Ecology and Management* 132: 51–62.

Aronsson, P. G., L. F. Bergström, and S. N. E. Elowson. 2000. Long-term influence of intensively cultured short-rotation Willow Coppice on nitrogen concentrations in groundwater. *Journal of Environmental Management* 58: 135–145.

Aylott, M. J., E. Casella, I. Tubby, N. R. Street, P. Smith, and G. Taylor. 2008. Yield and spatial supply of bioenergy poplar and willow short-rotation coppice in the UK. *New Phytologist* 178: 358–370.

Barnes, J. 2013. *Brazil making use of its* Eucalyptus *resource*. Forestry Investment Blog. http://www.forestry-invest.com/2011/brazil-making-use-of-its-eucalyptus-resources/812 (Accessed October 10, 2013).

Berguson, W. E., J. Eaton, and B. Stanton. 2010. Development of hybrid poplar for commercial production in the United States: The Pacific Northwest and Minnesota experience. In: R. Braun, D. Karlen, and D. Johnson, editors. *Sustainable alternative fuel feedstock opportunities, challenges and roadmaps for six U.S. regions*. Soil and Conservation Society, Ankeny, IA, pp. 282–299.

Berguson, W. E., B. McMahon, B. Stanton, R. Shuren, R. Miller, R. Rousseau, M. Cunningham, and J. Wright. 2012. The sun grant poplar woody crops research program: Accomplishments and implications. http://sungrant.tennessee.edu/NR/rdonlyres/430141C3-38B5-4265-BAC4-E89C9239B62A/3273/BillBerguson_UniversityofMinnesotaDuluth.pdf (Accessed October 5, 2013).

Berhongaray, G., O. El Kasmioui, and R. Ceulemans. 2013. Comparative analysis of harvesting machines on an operational high-density short rotation woody crop (SRWC) culture: One-process versus two-process harvest operation. *Biomass and Bioenergy* 58: 333–342.

Berndes, G., M. Hoogwijk, and R. van den Broek. 2003. The contribution of biomass in the future global energy supply: A review of 17 studies. *Biomass and Bioenergy* 25: 1–28.

Booth, T. H. 2012. Eucalypts and their potential for invasiveness particularly in frost-prone regions. *International Journal of Forest Research* 212: 837165, p. 7.

Börjesson, P. 1999. Environmental effects of energy crop cultivation in Sweden I: Identification and quantification. *Biomass and Bioenergy* 16: 137–154.

Buchholz, T., V. A. Luzadis, and T. A Volk. 2009. Sustainability criteria for bioenergy systems: Results from an expert survey. *Journal of Cleaner Production* 17: S86–S98.

Buchholz, T. and T. A. Volk. 2011. Improving the profitability of willow crops—Identifying opportunities with a crop budget model. *Bioenergy Research* 4: 85–95.

Carpentieri, A. E., E. D. Larson, and J. Woods. 1993. Future biomass-based electricity supply in Northeast Brazil. *Biomass and Bioenergy* 4: 149–174.

Das, D. K. and O. P. Chaturvedi. 2005. Structure and function of *Populus deltoides* agroforestry systems in eastern India: 1. *Dry matter dynamics. Agroforestry Systems* 65: 215–221.

Dhondt, A. A., P. H. Wrege, J. Cerretani, and K. V. Sydenstricker. 2007. Avian species richness and reproduction in short-rotation coppice habitats in central and western New York: Capsule species richness and density increase rapidly with coppice age, and are similar to estimates from early successional habitats. *Bird Study* 54: 12–22.

Dickmann, D. I. 2006. Silviculture and biology of short-rotation woody crops in temperate regions: Then and now. *Biomass and Bioenergy* 30: 696–705.

Dougherty, D. and J. Wright. 2012. Silviculture and economic evaluation of eucalypt plantations in the southern US. *BioResources* 7: 4817–4842.

Dyllick, T. and K. Hockerts. 2002. Beyond the business case for corporate sustainability. *Business Strategy and the Environment* 1: 130–141.

Eisenbies, M. H., L. P. Abrahamson, P. Castellano, C. Foster, M. McArdle, J. Posselius, R. Shurn, et al. 2012. Development and deployment of a short rotation woody crop harvesting system based on a New Holland forage harvester and SRC woody crop header. *Proceedings from Sun Grant National Conference: Science for Biomass Feedstock Production and Utilization*, New Orleans, LA. http://www.sungrant.tennessee.edu/NatConference/ (Accessed on March 15, 2014).

Elbehri, A., A. Segerstedt, and P. Liu. 2013. *Biofuel and sustainability challenge: A global assessment of sustainability issues, trends, and policies for biofuels and related feedstocks.* Rome: FAO.

English, B.C., D.G. de la Torre Ugarte, M.E. Walsh, C. Hellwinket, and J. Menard. 2006. Economic competitiveness of bioenergy production and effects on agriculture of the Southern Region. *Journal of Agricultural and Applied Economics* 38(2): 389–402.

Evans, J. 1997. Bioenergy plantations-experience and prospects. *Biomass and Bioenergy* 13: 189–191.

Fang, S., X. Xu, S. Lu, and L. Tang. 1999. Growth dynamics and biomass production in short-rotation poplar plantations: 6-year results for three clones at four spacings. *Biomass and Bioenergy* 17: 415–425.

Friedman, D., M. Morris, and M. Bomford. 2010. *Sustainability standards for farm energy.* eXtension Sustainable Farm Energy. http://www.extension.org/pages/Sustainability_Standards_for_Farm_Energy (Accessed April 29, 2010).

Goerndt, M. E. and C. W. Mize. 2008. Short-rotation woody biomass as a crop of marginal lands in Iowa. *North Journal of Applied Forestry* 25: 82–86.

Gold, S. and S. Seuring. 2011. Supply chain and logistics issues of bio-energy production. *Journal of Cleaner Production* 19: 32–42.

Gonzalez, R., T. Treasure, J. Wright, D. Saloni, R. Phillips, R. Abt, and H. Jameel. 2011. Exploring the potential of *Eucalyptus* for energy production in the Southern United States: Financial analysis of delivered biomass. Part I. *Biomass and Bioenergy* 35: 755–766.

Hall, D. O., F. Rosillo-Calle, R. Williams, and J. Woods. 1993. Biomass for energy: Supply prospects. In: T. B. Johansson, H. Kelly, A. K. N. Reddy, and R. H. Williams, editors. *Renewable energy—Sources for fuels and electricity*. Island Press, Washington, DC, pp. 589–651.

Hamza, M.A. and W. K. Anderson. 2002. *Soil organic matter in experimental short-rotation intensive culture (SRIC) systems: Effects of cultural factors, season, and age*. State University of New York College of Environmental Science and Forestry, Syracuse, NY.

Hamza, M. A. and W. K. Anderson. 2005. Soil compaction in cropping systems: A review of the nature, causes and possible solutions. *Soil and Tillage Research* 82: 121–145.

Hansen, E. A., M. E. Ostry, W. D. Johnson, D. N. Tolsted, D. A. Netzer, W. E. Berguson, and R. B. Hall. 1994. *Field performance of Populus in short-rotation intensive culture plantations in the north-central U.S.* USDA Forest Service Paper NC-320. USDA Forest Service, St. Paul, MN.

Hinchee, M., W. Rottmann, L. Mullinax, C. Zhang, S. Chang, M. Cunningham, L. Pearson, and N. Nehra. 2009. Short-rotation woody crops for bioenergy and biofuels applications. *In Vitro Cellular and Developmental Biology—Plant* 45: 619–629.

Hoogwijk, M., A. Faaij, B. Eickhout, B. de Vries, and W. Turkenburg. 2005. Potential of biomass energy out to 2100, for four IPCC SRES land-use scenarios. *Biomass and Bioenergy* 29: 225–257.

IEA (International Energy Agency). 2013. Short rotation woody crops systems. http://www.ieabioenergy.com/task/short-rotation-crops-for-bioenergy-systems (Accessed October 5, 2013).

IPC (International Poplar Commission). 2013. Information on the International Poplar Commission. http://www.fao.org/forestry/ipc/en/ (Accessed September 21, 2013).

Jelinski, D. W. and W. M. Cheliak. 1992. Genetic diversity and spatial subdivision of *Populus tremuloides* (Salicaceae) in a heterogeneous landscape. *American Journal of Botany* 79: 728–736.

Johansson, T. and A. Karačić. 2011. Increment and biomass in hybrid poplar and some practical implications. *Biomass and Bioenergy* 35: 1925–1934.

Kauter, D., I. Lewandowski, and W. Claupein. 2003. Quality and quantity of harvestable biomass from *Populus* short rotation coppice for a solid fuel use—A review of the physiological basis and management influences. *Biomass and Bioenergy* 24: 411–427.

Kiser, L. C. and T. R. Fox. 2013. Short-rotation woody crop biomass production for bioenergy. In: B. P. Singh, editor. *Biofuel crop sustainability*. John Wiley, Oxford, UK, pp. 205–237.

Klass, D. L. 2004. Biomass for renewable energy and fuels. *Encyclopedia of Energy* 1: 193–212.

Kopp, R. F., L. P. Abrahamson, E. H. White, K. F. Burns, and C. A. Nowak. 1997. Cutting cycle and spacing effects on biomass production by a willow clone in New York. *Biomass and Bioenergy* 12: 313–319.

Kuzovkina, Y. A., M. Weih, M. A. Romero, J. Charles, S. Hurst, I. McIvor, A. Karp, et al. 2008. *Salix*: Botany and global horticulture. In: J. Janick, editor. *Horticultural reviews*. John Wiley, Hoboken, NJ, pp. 447–489.

Labrecque, M. and T. I. Teodorescu. 2005. Field performance and biomass production of 12 willow and poplar clones in short-rotation coppice in southern Quebec (Canada). *Biomass and Bioenergy* 29: 1–9.

Lal, R. 1998. Soil erosion impact on agronomic productivity and environment quality. *Critical Reviews in Plant Sciences* 17: 319–464.

Langholtz, M., D. Carter, D. Rockwood, and J. R. R. Alavalapati. 2007. The economic feasibility of reclaiming phosphate mined lands with short-rotation woody crops in Florida. *Journal of Forest Economics* 12: 237–249.

Langholtz, M., D. Carter, D. Rockwood, and J. R. R. Alavalapati. 2009. The influence of CO_2 mitigation incentives on profitability of *Eucalyptus* production on clay settling areas in Florida. *Biomass and Bioenergy* 33: 785–792.

Langholtz, M., D. R. Carter, D. L. Rockwood, J. R. R. Alavalapati, and A. Green. 2005. Effect of dendroremediation incentives on the profitability of short-rotation woody cropping of *Eucalyptus grandis*. *Forest Policy and Economics* 7: 806–817.

Marosvölgyi, B., L. Halupa, and I. Wesztergom. 1999. Poplars as biological energy sources in Hungary. *Biomass and Bioenergy* 16: 245–247.

Mead, D. J. 2005. Forests for energy and the role of planted trees. *Critical Reviews in Plant Sciences* 24: 407–421.

Mercker, D. 2007. *Short rotation woody crops for biofuels*. University of Tennessee Agricultural Experiment Station. http://www.utextension.utk.edu/publications/spfiles/SP702-C.pdf (Accessed on March 15, 2014).

Mirck, J. and W. Schroeder. 2013. Composition, stand structure, and biomass estimates of "willow rings" on the Canadian Prairies. *BioEnergy Research* 6: 864–876.

Mitchell, C. P., E. A. Stevens, and M. P. Watters. 1999. Short-rotation forestry—Operations, productivity and costs based on experience gained in the UK. *Forest Ecology and Management* 121: 123–136.

Mola-Yudego, B. 2010. Regional potential yields of short rotation willow plantations on agricultural land in Northern Europe. *Silva Fennica* 44: 63–76.

Mortensen, J., K. H. Nielsen, and U. J. Ørgensen. 1998. Nitrate leaching during establishment of willow (*Salix viminalis*) on two soil types and at two fertilization levels. *Biomass and Bioenergy* 15: 457–466.

Nissim, W. G., F. E. Pitre, T. I. Teodorescu, and M. Labrecque. 2013. Long-term biomass productivity of willow bioenergy plantations maintained in southern Quebec, Canada. *Biomass and Bioenergy* 56: 361–369.

NRC (National Research Council). 2000. *Biobased industrial products: Priorities for research and commercialization.* National Academy Press, Washington, DC.

Palanna, R. M. 2013. *Eucalyptus* in India. http://www.fao.org/docrep/005/ac772e/ac772e06.htm (Accessed October 3, 2013).

Peichl, M., N. Thevathasan, A. Gordon, J. Huss, and R. Abohassan. 2006. Carbon sequestration potentials in temperate tree-based intercropping systems, Southern Ontario, Canada. *Agroforestry System* 66: 243–257.

Perala, D. A. 1977. *Manager's handbook for Aspen in the north-central states.* General technical report NC-36. USDA Forest Service, St. Paul, MN.

Pereira, J. S., S. Linder, M. C. Araújo, H. Pereira, T. Ericsson, N. Borralho, and L. C. Leal. 1989. Optimization of biomass production in *Eucalyptus* globulus plantations—A case study. In: J. S. Pereira and J. J. Landsberg, editors. *Biomass production by fast-growing trees.* NATO ASI. Springer, Dordrecht, The Netherlands, pp. 101–121.

Prasad, J. V. N. S., G. R. Korwar, K. V. Rao, U. K. Mandal, C. A. R. Rao, G. R. Rao, Y. S. Ramakrishna, et al. 2010. Tree row spacing affected agronomic and economic performance of *Eucalyptus*-based agroforestry in Andhra Pradesh, Southern India. *Agroforestry Systems* 78: 253–267.

Proe, M. F., J. Craig, J. Griffiths, A. Wilson, and E. Reid. 1999. Comparison of biomass production in coppice and single stem woodland management systems on an imperfectly drained gley soil in central Scotland. *Biomass and Bioenergy* 17: 141–151.

Quaye, A. K. and T. A. Volk. 2013. Biomass production and soil nutrients in organic and inorganic fertilized willow biomass production systems. *Biomass and Bioenergy* 57: 113–125.

Quaye, A. K., T. A. Volk, S. Hafner, D. J. Leopold, and C. Schirmer. 2011. Impacts of paper sludge and manure on soil and biomass production of willow. *Biomass and Bioenergy* 35: 2796–2806.

Ranney, J. W. and L. K. Mann. 1994. Environmental considerations in energy crop production. *Biomass and Bioenergy* 6: 211–228.

Reijnders, L. 2006. Conditions for the sustainability of biomass based fuel use. *Energy Policy* 34: 863–876.

Riemenschneider, D. E., B. J. Stanton, G. Vallee, and P. Perinet. 2001. Poplar breeding strategies. In: D. I. Dickmann, J. G. Isebrands, J. E. Eckenwalder, and J. Richardson, editors. *Poplar culture in North America.* NRC Research Press, National Research Council of Canada, Ottawa, ON, pp. 43–76.

Rockwood, D. L., D. R. Carter, M. H. Langholtz, and J. A. Stricker. 2006. *Eucalyptus* and *Populus* short rotation woody crops for phosphate mined lands in Florida USA. *Biomass and Bioenergy* 30: 728–734.

Rockwood, D. L., A. W. Rudie, S. A. Ralph, J. Y. Zhu, and J. E. Winandy. 2008. Energy product options for *Eucalyptus* species grown as short rotation woody crops. *International Journal of Molecular Sciences* 9: 1361–1378.

Rowe, R. L., M. E. Hanley, D. Goulson, D. J. Clarke, C. P. Doncaster, and G. Taylor. 2011. Potential benefits of commercial willow Short Rotation Coppice (SRC) for farm-scale plant and invertebrate communities in the agri-environment. *Biomass and Bioenergy* 35: 325–336.

Rowe, R. L., N. R. Street, and G. Taylor. 2008. Identifying potential environmental impacts of large-scale deployment of dedicated bioenergy crops in the UK. *Renewable & Sustainable Energy Reviews* 13: 271–290.

San Miguel, A. 1988. Short rotation biomass plantation. In: F. C. Hummel, W. Pals, and G. Grazzi, editors. *Biomass forestry in Europe: A strategy for the future.* Elsevier Applied Science, London, pp. 540–579.

Sartori, F., R. Lal, M. H. Ebinger, and J. A. Eaton. 2007. Changes in soil carbon and nutrient pools along a chronosequence of poplar plantations in the Columbia Plateau, Oregon, USA. *Agriculture, Ecosystems and Environment* 122: 325–339.

Savoie, P., P. L. Hébert, and F.-S. Robert. 2013. *Harvest of short rotation woody crops with small to medium size forage harvesters*. Paper 131620174. American Society of Agricultural and Biological Engineers, St. Joseph, MI.

Schulz, U., O. Brauner, and H. Gruß. 2009. Animal diversity on short-rotation coppices. *Landbauforschung—vTI Agriculture and Forestry Research* 59: 171–182.

Selamyihun, K., T. Mamo, and L. Stroosnijder. 2005. Biomass production of *Eucalyptus* boundary plantations and their effect on crop productivity on Ethiopian highland Vertisols. *Agroforestry Systems* 63: 281–290.

Serapiglia, M. J., M. C. Humiston, H. Xu, D. A. Hogsett, R. M. de Orduña, A. J. Stipanovic, and L. B. Smart. 2013. Enzymatic saccharification of shrub willow genotypes with differing biomass composition for biofuel production. *Frontiers in Plant Science* 4: 57.

Smart, L. B., K. D. Cameron, T. A. Volk, and L. P. Abrahamson. 2008. *Breeding, selection, and testing of shrub willow as a dedicated energy crop*. NABC Report 19: 85–92. Agricultural Biofuels, National Agricultural Biotechnology Council, Ithaca, NY.

Smart, L. B., T. A. Volk, J. Lin, R. F. Kopp, I. S. Phillips, K. D. Cameron, E. H. White, and L. P. Abrahamson. 2005. Genetic improvement of shrub willow (*Salix* spp.) crops for bioenergy and environmental application in the United States. *Unasylva* 56: 51–55.

Spalding, R. F., M. E. Exner, D. D. Snow, D. A. Cassada, M. E. Burbach, and S. J. Monson. 2003. Herbicides in ground water beneath Nebraska's management systems evaluation area. *Journal of Environmental Quality* 32: 92–99.

Stanton, B., J. Eaton, J. Johnson, D. Rice, B. Schuette, and B. Moser. 2002. Market driven evolution in hybrid poplar management in the Pacific Northwest. *Journal of Forestry* 100: 28–33.

Stanturf, J. A., E. D. Vance, T. R. Fox, and M. Kirst. 2013. *Eucalyptus* beyond its native range: Environmental Issues in exotic bioenergy plantations. *International Journal of Forestry Research* 2013: 1–5.

Stettler, R. F., L. Zsuffa, and R. Wu. 1996. The role of hybridization in the genetic manipulation of Populus. In: R. F. Stettler, H. D. Bradshaw, P. E. Heilman, T. M. Hinckley, editors. *Biology of Populus and its implications for management and conservation*. NRC Research Press, Ottawa, Canada, pp. 87–112.

Swenson, D. 2010. Appendix I: Biofuel and industry economic impacts and analysis. In: *Renewable fuels roadmap and sustainable biomass feedstock supply for New York*. NYSERDA report 10-05, NYSERDA, Albany, NY, p. 59.

Tharakan, P. J., T. A. Volk, C. A. Lindsey, L. P. Abrahamson, and E. H. White. 2005. Evaluating the impact of three incentive programs on cofiring willow biomass with coal in New York State. *Energy Policy* 33: 337–347.

Thornley, P. 2007. *Life cycle assessment of bioenergy systems*. http://www.supergenbioenergy.net/Resources/user/Research%20Output/LCA%20Report%20(P%20Thornley)%20-%20SG%20Research%20Output.pdf (Accessed January 25, 2008).

Tolbert, V. R., D. E. Todd, Jr., L. K. Mann, C. M. Jawdy, D. A. Mays, R. Malik, W. Bandaranayake, A. Houston, D. Tyler, and D. Pettry. 2002. Changes in soil quality and below-ground carbon storage with conversion of traditional agricultural crop lands to bioenergy crop production. *Environmental Pollution* 116: S97–S106.

Tufekcioglu, A., J. W. Raich, T. M. Isenhart, and R. C. Schultz. 2003. Biomass, carbon and nitrogen dynamics of multi-species riparian buffers within an agricultural watershed in Iowa, USA. *Agroforestry Systems* 57: 187–198.

Tuskan, G. A., S. DiFazio, S. Jansson, J. Bohlmann, I. Grigoriev, U. Hellsten, N. Putnam, et al. 2006. The genome of black cottonwood, *Populus trichocarpa* (Torr. & Gray). *Science* 313: 1596–1604.

Ulzen-Appiah, F. 2002. *Soil organic matter in experimental short-rotation intensive culture (SRIC) systems: Effects of cultural factors, season, and age*. State University of New York College of Environmental Science and Forestry, Syracuse, NY.

US DOE (U.S. Department of Energy). 2011. U.S. Billion-ton update: Biomass supply for a bioenergy and bioproducts industry. R.D. Perlack and B. J. Stokes, editors. Oak Ridge National Laboratory, Oak Ridge, TN. p. 227.

US EISA (U.S. Energy Independence and Security Act) of 2007 (Pub. L. 110-140).

Vance, E. D., D. A. Maguire, and R. S. Zalesny, Jr. 2010. Research strategies for increasing productivity of intensively managed forest plantations. *Journal of Forestry* 108: 183–192.

Vande Walle, I., N. Van Camp, L. Van de Casteele, K. Verheyen, and R. Lemeur. 2007. Short-rotation forestry of birch, maple, poplar and willow in Flanders (Belgium) I—Biomass production after 4 years of tree growth. *Biomass and Bioenergy* 31: 267–275.

Verwijst, T. 2001. Willows: An underestimated resource for environment and society. *The Forestry Chronicle* 77: 281–285.

Volk, T. A., L. P. Abrahamson, K. D. Cameron, P. Castellano, T. Corbin, E. Fabio, G. Johnson, et al. 2011. Yields of willow biomass crops across a range of sites in North America. *Aspects of Applied Biology* 112: 67–74.

Volk, T. A. and V. A. Luzadis. 2009. Willow biomass production for bioenergy, biofuels, and bioproducts in New York. In: V. A. Luzadis and B. D. Solomon, editors. *Renewable energy from forest resources in the United States*. Routledge, London, pp. 238–260.

Volk, T. A., T. Verwijst, P. J. Tharakan, L. P. Abrahamson, and E. H. White. 2004. Growing fuel: A sustainability assessment of willow biomass crops. *Frontiers in Ecology and the Environment* 2: 411–418.

WCED (World Commission on Environment and Development). 1997. World commission on environment and development. http://www.un-documents.net/ocf-02.htm#I (Accessed October 3, 2013).

Whitesell, C. D., D. S. DeBell, T. H. Schubert, R. F. Strand, and T. B. Crabb. 1992. *Short rotation management of* Eucalyptus: *Guidelines for plantation in Hawaii*. General Technical Report PSW-137. USDA Forest Service, Albany, CA.

Wright, J. and L. Rosales. 2001. Developing clonal eucalypt forest plantations in Venezuela. *Asian Timber*, p. 8.

Wright, L. 2006. Worldwide commercial development of bioenergy with a focus on energy crop-based projects. *Biomass and Bioenergy* 30: 706–714.

Yamaguchi-Shinozaki, K. and K. Shinozaki. 1993. Characterization of the expression of a desiccation-responsive *rd29* gene of *Arabidopsis thaliana* and analysis of its promoter in transgenic plants. *Molecular and General Genetics* 236: 331–340.

Yeh, B., S. Gold, and S. Seuring. 2011. *Independent assessment of technology characterizations to support the biomass program annual state-of-technology assessments*. NREL (National Renewable Energy Laboratory), Golden, CO. Report, Contract No. DE-AC36-08GO28308.

Zalesny, R. S., Jr., M. W. Cunningham, R. B. Hall, J. Mirck, D. L. Rockwood, J. A. Stanturf, and T. A. Volk. 2011. Woody biomass from short rotation energy crops. In: J. Y. Zhu, X. Zhang, and X. Pan, editors. *Sustainable production of fuels, chemicals, and fibers from forest biomass*. American Chemical Society, ACS Symposium Series, Washington, DC, pp. 27–63.

Zalesny, R. S., Jr. and J. A. Zalesny. 2009. Selecting *Populus* with different adventitious root types for environmental benefits, fiber, and energy. In: K. Niemi and C. Seagel, editors. *Adventitious root formation of forest trees and horticultural plants—From genes to applications*. Research Signpost, Ontario, pp. 359–384.

Zalesny, J. A., R. S. Zalesny, Jr., D. R. Coyle, R. B. Hall, and E. O. Bauer. 2009. Clonal variation in morphology of *Populus* root systems following irrigation with landfill leachate or water during 2 years of establishment. *BioEnergy Research* 2: 134–143.

Zamora, D., G. Wyatt, K. Apostol, and U. Tschirner. 2013. Biomass yield, energy values, and chemical composition of hybrid poplar in short rotation woody crops production and native perennial grasses in Minnesota, USA. *Biomass and Bioenergy* 49: 222–230.

Zamora D. S., K. G. Apostol, G. J. Wyatt. 2014. Biomass production and potential ethanol yields of shrub willow hybrids and native willow accessions after a single 3-year harvest cycle on marginal lands in central Minnesota, USA. *Agroforestry System* 88(4): 593–606.

Zon, R. and J. Briscoe. 1911. *Eucalyptus* in Florida. U.S. Department of Agriculture. *Forest Service Bulletin* 87: 99.

CHAPTER 4

Supply Chain Management of Biomass Feedstock

Sandra D. Ekşioğlu[1] and Cerry M. Klein[2]
[1]Department of Industrial Engineering, Clemson University, Clemson, SC, USA
[2]Department of Industrial and Manufacturing Systems Engineering,
 University of Missouri, Columbia, MO, USA

CONTENTS

4.1 INTRODUCTION

Production of bioenergy, that being biofuels or renewable electricity, is expected to increase in the near future due to legislation enacted by the Energy Independence and Security Act of 2007. The renewable fuel standard (RFS) (EPA 2007) was an outcome of this legislation. Based on the RFS, 36 billion gallons a year (BGY) of biofuels should be produced by 2022. This standard sets a cap on corn-ethanol production at 15 BGY and requires that at least 16 BGY of cellulosic biofuels be produced. In 2013, 1.3 BG of biodiesel (EIA 2013) and 13.3 BG of ethanol were produced (RFA 2013).

A number of policies and incentives at the federal and state level are expected to increase the generation of electricity from renewable resources, such as by using biomass for cofiring. Policies at the federal level—the renewable energy production tax credit (PTC)—provide an income tax credit of 2.2 cents per kilowatt-hour. The Annual Energy Outlook (EIA 2013) projects that electricity production from biomass will increase from 37.26 billion kilowatt-hours in 2011 to 131.89 billion kilowatt-hours in 2040. According to the U.S. Energy Information Administration (EIA), biomass contributes nearly 3.9 quadrillion British thermal units (BTU) and accounts for more than 4% of total U.S. primary energy consumption (EIA 2010). Over the last 30 years, the share of biomass in the total primary energy consumption has averaged about 3.5% (EIA 2010).

Although the public and government officials see bioenergy as a potential energy source, its production and delivery come with major challenges. One of the biggest challenges faced is managing the biomass supply chain. Biomass, in the form of agricultural and forest waste, has low density and poor flowability properties, and thus, it is bulky, heterogeneous, and unstable. Biomass supply is uncertain because its yield—as is the case with other agricultural and forest products and residues—depends on weather conditions, insect populations, plant disease, and so forth. Biomass supply is seasonal, and its supply is constrained by land availability. Biomass deteriorates and loosens dry matter with time. Biomass suppliers are typically small- or medium-sized farms, which are widely dispersed geographically. Due to the large number of suppliers, the assortment of produce, the traditional administrative systems, and less than sophisticated buying options for agricultural products, processes such as collection and storage of biomass and order processing are challenging and expensive. These costs make biofuels less competitive with fossil fuels, and therefore, limit their production quantities.

The challenges listed here have inspired researchers to investigate biomass supply chain–related problems. Within this area of research, most of the existing literature is focused on building models that minimize system-wide costs. These models can be classified based on the managerial decisions they support, those being strategic, tactical, and operational decisions. Thus, some of the models proposed focus on biomass supply chain design, others on biomass supply chain management, and a number of papers focus on coordinated, biomass supply chain design and management problems. These cost minimization problems are modeled either using deterministic or stochastic optimization models. Other researchers propose simulation/optimization models to optimize the biomass supply chain.

The existing literature could also be classified based on the type of product produced or on the type of biomass used. For example, the structure of the supply chain in support of cellulosic ethanol production is different from the structure of the supply chain in support of corn-ethanol, biodiesel, or renewable electricity. The reasons for these differences are product characteristics. Product characteristics impact facility location decisions, transportation mode selection, and inventory replenishment decisions.

The literature on the biomass supply chain has grown recently. The goal of this chapter is to organize and summarize this literature in order to help readers become familiar with the major streams of research within this area. Some related review papers are worth mentioning. Mafakheri and Nasiri (2014) review and classify the existing literature based on the type of model developed and on their relation to strategic supply chain–related decisions. DeMeyer et al. (2014) provide an overview of papers that focus on optimizing the upstream segment of the biomass-for-bioenergy supply chain. They organize the literature based on the optimization model developed and on the decisions supported (such as supply chain design and supply chain management). Sharma et al. (2013) present a review of mathematical programing models developed for biomass supply chains and identify areas that need to be studied further. Although these reviews are of value and cover different aspects of this supply chain, they do not present a complete overview of the problems and challenges related to biomass supply chains. The purpose of this chapter is to give an extensive overview of all aspects related to this supply chain and related decisions.

In addition to the papers cited here, a good resource for related literature is the Bioenergy Knowledge Discovery Framework Web site (KDF 2014) provided by the U.S. Department of Energy. This Web site provides access to a variety of data sets, publications, collaborations, and mapping tools that support bioenergy research, analysis, and decision making. The users can search for information, contribute data, and can use the tools and map interface to synthesize, analyze, and visualize information in a spatially integrated manner.

4.2 CLASSIFICATION OF THE LITERATURE BASED ON THE SCOPE OF DECISIONS IN THE SUPPLY CHAIN

Biomass supply chain–related decisions are classified into strategic, tactical, and operational decisions. Strategic decisions include facility location, inbound and outbound supply chain design, and production capacity decisions. Tactical decisions are those such as transportation mode selection or supplier selection. Operational decisions are related to harvesting schedule, production schedule, inventory control, and transportation schedule. Next, we present current literature in each of these research areas.

4.2.1 Strategic Decisions

Facility location decisions are very important for biofuel plants. These decisions are not easy because they are subject to a number of factors, such as proximity to biomass resources, transportation costs, and highway/railway/waterway accessibility. In the past, plants that produced first-generation biofuels—using corn, soybean, and other edible products—were typically located within a 50-mile radius of their supply in order to minimize inbound transportation costs (Aden et al. 2002). The motivation behind these decisions was the high transportation costs due to biomass being a bulky and low energy density product. The goal of such a supply chain design is to minimize transportation costs. Consequently, the size of these plants was limited by the amount of biomass within this radius. These plants could not benefit from the economies of scale associated with high production volumes (Searcy and Flynn 2008). These supply chain designs do not optimize plant investment and biofuel production costs. To identify facility locations for these supply chain structures, one could use weighted gravity location models. These models are well known in the literature and focus on minimizing the weighted sum of transportation distances and volumes shipped (Chopra and Meindl 2012).

In order to reduce bioenergy production costs, plants should increase their production capacity. Consequently, biomass would be delivered from farms located further away. Eksioglu et al. (2011) discuss the impact that an intermodal facility has on location and transportation model selection of biofuel production plants. An intermodal facility is a facility where two or more transportation modes meet, such as inland ports, seaports, rail ramps.

Increasing production capacity of a bioenergy plant reduces the corresponding production costs. In this case, to obtain the necessary amount of biomass, a plant would receive shipments from farms located further away. Eksioglu et al. (2011) discuss the impact that an intermodal facility has on location and transportation mode selection of biofuel production plants. Modes of transportation such as barge and rail have traditionally been used for transportation of bulk products, such as, corn, soybean, wood chips, and fuel. These transportation modes are cost efficient and would contribute to reducing the total transportation cost of biomass. The model proposed by Eksioglu et al. (2011) indicates that bioenergy plants would reduce the total delivery cost if located near an intermodal facility. This work was inspired by the opening of the first ethanol production plant in Mississippi (Bunge-Ergon, near the port of Vicksburg) in 2008. Later on, in 2013, KiOR started the construction of its first commercial scale renewable fuel production facility along the port of Columbus in Mississippi. Xie et al. (2014) extended this work by developing a fully integrated multimodal transport system for the cellulosic biofuel supply chain network. These models do capture biomass seasonality. Xie et al. (2014) implemented their model using relevant data from California.

A number of papers in the literature present models in support of hub-and-spoke biomass supply chain designs (Roni et al. 2014b; Xie et al. 2014). These models were inspired by practices used in industries such as agricultural, airline, railway, mail carrier, and telecommunication. These industries have used the same concept to design their distribution networks. Such designs are typically used by companies that deal with inbound or outbound distribution networks that are large and geographically dispersed. Hubs serve as transshipment points where shipments are consolidated and disseminated, and transportation modes are changed. Shipment consolidations result in economies of scale in transportation and, as a consequence, decrease transportation costs. (To learn more about the existing literature on hub-and-spoke supply chain design problem see Alumur and Kara 2008; Tunc et al. 2011; Campbell and O'Kelly 2012.)

4.2.2 Tactical Decisions

In recent years, research has been conducted in the field of biofuel supply chains with a prime focus on minimizing the total system costs. These studies focused mainly on developing integrated biofuel supply chain networks in order to deliver biomass at a competitive price to end users. To achieve this goal, Schmidt et al. (2008) propose a mathematical model that considers biomass production, biomass transportation, production of biofuels and byproducts, distribution of biofuels and byproduct, and plant locations. They use this model to identify potential locations for methane and ethanol production plants in Austria that optimize the performance of this supply chain. Studies such as those of Zamboni et al. (2009b) and Eksioglu et al. (2010) develop models that optimize the total of plant location and transportation costs in biofuel supply chain networks. These works have been extended by others (Eksioglu et al. 2009; Huang et al. 2010; Kang et al. 2010; An et al. 2011; Eksioglu et al. 2013; Xie et al. 2014) to better capture system dynamics by considering multiple period optimization frameworks. Xie and Ouyang (2013) developed a mixed-integer programing model for a dynamic, multitype facility colocation problem that minimizes the total costs from facility construction, capacity expansion, and transportation during a fixed planning horizon.

4.2.3 Operational Decisions

The existing literature provides models that estimate the cost of collecting, handling, and hauling biomass to biorefineries and compares different modes of delivering biomass (Mahmudi and Flynn 2006).

The literature on biomass transportation cost analysis is focused on estimating truck, rail, and barge transportation costs. Some of the existing literature analyzes truck transportation costs of densified biomass in the form of pellets, cubes, and bales (Sokhansanj and Turhollow 2004, 2006; Badger and Fransham 2006; Rogers and Brammer 2009). A study by Mahmudi and Flynn (2006) investigates transportation by rail of biomass in the form of forest residues and wood chips. A study by Eksioglu et al. (2010) investigates rail and barge transportation costs of biomass. Rail transportation of biomass is also discussed in Magnanti and Wong (1981), Searcy et al. (2007), Bonilla and Whittaker (2009), Sokhansanj et al. (2009), Ileleji et al. (2010), and Judd et al. (2011).

The aforementioned studies mainly focus on distance-based fuel costs and on loading/unloading costs per ton of biomass. Factors such as railway ownership, shipment volume, and type of rail movements were not taken into consideration. Gonzales et al. (2013) proposed a few regression models to identify the main factors that impact the delivery cost of densified biomass by analyzing transportation costs of similar products, such as grain and wood chips. They derive these equations using rail tariff data publicly available on the Web sites of a few class I railways. The analysis indicates that transportation costs for densified biomass are impacted by transportation distance, volume shipped, transportation mode used, shipment destination, and so on. Results indicate that barges are the least expensive transportation mode for shipments from the Midwest to the Southeast, whenever barges are available. If they are not available, then unit train is the least expensive transportation mode for distances greater than 90 miles. For shipments from the Midwest to the western United States, unit train is the least expensive transportation mode for distances greater than 210 miles. For shorter distances, trucks are the least expensive transportation mode.

Roni et al. (2014a) extend this research by analyzing rail transportation costs for biomass using the carload waybill data, provided by the U.S. Department of Transportation's Surface Transportation Board, for products with similar physical characteristics to densified biomass, such as grain and liquid type commodities, for 2009 and 2011. Other studies analyze pipeline transportation of biomass (Searcy et al. 2007; Ileleji et al. 2010; Judd et al. 2011).

4.3 CLASSIFICATION OF THE LITERATURE BASED ON THE OPTIMIZATION MODELS USED

A number of models are proposed to optimize the performance of the biomass supply chain. Some of the models are deterministic by nature. These models identify bioenergy plant locations, supply chain network designs for bioenergy plants, bioenergy production schedules, biomass transportation schedule, and so forth. Typically, the goal is to minimize costs or to maximize social benefits. Recently, a number of papers have focused on building multiobjective optimization models that optimize costs and the environmental and social impacts of bioenergy.

There are a few stochastic programing models in the literature. The advantage of using these models is the fact that they better represent the random nature of biomass supply. Simulation models are also proposed in order to capture the uncertainties related to supply, demand, and prices in the supply chain.

4.3.1 Deterministic Optimization Models

Some of the deterministic models take an integrated view of plant location, production, and transportation decisions of biomass supply chains (Eksioglu et al. 2009; Zamboni et al. 2009b; Huang et al. 2010; An et al. 2011). Other deterministic models proposed are extensions of the facility location model and are used to identify biorefinery sittings and corresponding production capacities (Eksioglu et al. 2009; Parker et al. 2010; Bai et al. 2011; Kim et al. 2011b; Papapostolou et al. 2011; Roni et al. 2014b). Extensions of these models identify the number, capacity, and location of bioenergy plants in order to make use of the available biomass in a particular region in a cost efficient manner.

The following is a deterministic model that takes an integrated view of the locations, transportation, and inventory management in the corn-ethanol supply chain. The supply chain considered here consists of K harvesting sites, J collection facilities, J potential biorefinery location sites, K blending facilities, and the market.

A biomass harvesting site (such as a cornfield) in this network represents a supply node. Potential locations for building collection facilities, biorefineries, and blending facilities are transshipment nodes. A blending facility is where ethanol and gas are mixed to produce ethanol blends such as

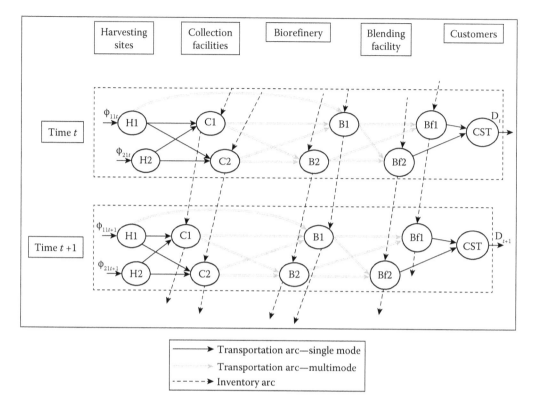

Figure 4.1 Network representation of a supply chain with two harvesting sites, two collection facilities, two biorefineries, two blending facilities, and one market.

E-10 or E-85 that are usable by the consumer. The market represents the demand nodes for ethanol blends. Figure 4.1 gives a graphical representation of a supply chain with two harvesting sites, two collection facilities, two biorefineries, two blending facilities, and the market during time periods t and $t + 1$ of a planning horizon consisting of T periods.

This formulation is an extension of the network design model (Ahuja et al. 1993). It identifies the location of collection facilities and biorefineries that minimize the total cost of locating the facilities; collecting, transporting, processing, and storing biomass; and distributing biofuels to satisfy a regionwide demand for ethanol blends. The model is dynamic in the sense that it considers fluctuations in the supply of biomass and the demand for cellulosic ethanol from one period to another during a time horizon of length T.

In this network, arcs that connect harvesting sites with collection facilities and collection facilities with biorefineries represent biomass transportation arcs. Arcs that connect biorefineries with blending facilities, and blending facilities with the market, represent the transportation of cellulosic ethanol and ethanol blends. Arcs that connect the same facility in two consecutive time periods are the inventory arcs. The problem parameters and decision variables used in the mixed-integer program (MIP) formulation of this problem are presented next.

4.3.1.1 Problem Parameters

The amortized annual cost of constructing and operating a collection facility of size $l \in L$ in location $j \in J$ is Ψ_{jl} ($/yr); Ψ'_{il} ($/yr) is the amortized annual cost of constructing and operating a biorefinery of size $l \in L'$ in location $I \in I$; p_{kb} ($/ton) is the unit price of planting and growing biomass feedstock type $b \in B$ at harvesting site $k \in K$; ρ_b ($/ton) is the cost of harvesting biomass feedstock $b \in B$; h ($/ton) is the unit inventory holding cost; c ($/ton) is the unit transportation cost; ω_b ($/ton) is the unit cost of processing biomass feedstock $b \in B$; L_{kb} is the total number of acres of land at the harvesting site $k \in K$ suitable for planting biomass feedstock type $b \in B$; γ_{kb} is the proportion of available land that can be harvested for production of biofuel in site $k \in K$ ($0 \le \gamma_{kb} \le 1$); δ_{kbt} (tons/acre) is the yield of biomass feedstock $b \in B$ at site $k \in K$ in period $t \in T$; α (%) is the deterioration rate; β_b (gal/ton) is the conversion rate of biomass feedstock $b \in B$; b_t (tons) is the demand for cellulosic biofuel in period $t \in T$; S_{lCF} (tons) is the storage capacity of a collection facility of size $l \in L$; and S_{lBR} (tons) is the storage capacity of a biorefinery of size $l \in L'$.

4.3.1.2 Decision Variables

If a biorefinery of size $l \in L'$ is located at site $i \in I$, x_{il} is a binary variable that equals to 1, and 0 otherwise; x_{jl} is a binary variable that equals to 1 if a collection facility of size $l \in L$ is located at site $j \in J$, and 0 otherwise; ϕ_{kbt} is the quantity of biomass $b \in B$ harvested at site $k \in K$ in period $t \in T$; z_{ibt} is the amount of biomass type $b \in B$ stored at site $i \in I$ in period $t \in T$; z_{it} is the amount of biofuel stored at the biorefinery $i \in I$ in period $t \in T$; y_{ijbt} is the amount of biomass $b \in B$ shipped from facility $i \in I$ to facility $j \in J$ in period $t \in T$; y_{ijt} is the amount of biofuel shipped from the biorefinery $i \in I$ to minimize the cost of satisfying a regionwide demand for biofuel blending facility $j \in J$ in period $t \in T$; and w_{ibt} is the amount of biomass type b processed at biorefinery $i \in I$ in period $t \in T$.

The following is the MIP formulation of this problem.

$$
\begin{aligned}
\min \sum_{t=1}^{T}\Bigg[&\sum_{k=1}^{K}\sum_{b=1}^{B}(\rho_b + p_{kb})\phi_{kbt} + \sum_{b=1}^{B}h_b\left(\sum_{k=1}^{K}z_{kbt} + \sum_{j=1}^{J}z_{jbt} + \sum_{i=1}^{I}z_{ibt}\right) + h^e\sum_{i=1}^{I}z_{it} \\
&+ \sum_{b=1}^{B}\left(\sum_{k=1}^{K}\sum_{j=1}^{J}c_{kj}y_{kjbt} + \sum_{k=1}^{K}\sum_{i=1}^{I}c_{ki}y_{kibt} + \sum_{j=1}^{J}\sum_{i=1}^{I}c_{ji}y_{jibt}\right) + \sum_{i=1}^{I}\sum_{\kappa=1}^{\mathcal{K}}c_{i\kappa}vy_{i\kappa t} \\
&+ \sum_{i=1}^{I}\sum_{b=1}^{B}\omega_b w_{ibt}\Bigg] + \sum_{j=1}^{J}\sum_{l=1}^{L}\Psi_{jl}x_{jl} + \sum_{i=1}^{I}\sum_{l=1}^{L'}\Psi'_{il}x_{il}
\end{aligned}
\tag{P}
$$

$$
\phi_{kbt} \le \delta_{kbt}\gamma_{kb}L_{kb} \qquad k = 1,\ldots,K; \ b = 1,\ldots,B; \ t = 1,\ldots,T
\tag{4.1}
$$

$$
\phi_{kbt} + (1-\alpha)z_{kb,t-1} = \sum_{j=1}^{J}y_{kjbt} + z_{kbt} \qquad k=1,\ldots,K; \ b=1,\ldots,B; \ t=1,\ldots,T
\tag{4.2}
$$

$$
\sum_{k=1}^{K}y_{kjbt} + (1-\alpha)z_{jb,t-1} = \sum_{i=1}^{I}y_{jibt} + z_{jbt} \qquad j=1,\ldots,J; \ b=1,\ldots,B; \ t=1,\ldots,T
\tag{4.3}
$$

$$
\sum_{j=1}^{J}y_{jibt} + (1-\alpha)z_{ib,t-1} = w_{ibt} + z_{ibt} \qquad i=1,\ldots,I; \ b=1,\ldots,B; \ t=1,\ldots,T
\tag{4.4}
$$

$$
\sum_{b=1}^{B}\beta_b w_{ibt} + z_{i,t-1} = \sum_{\kappa=1}^{\mathcal{K}}y_{i\kappa t} + z_{it} \qquad i=1,\ldots,I; \ t=1,\ldots,T
\tag{4.5}
$$

$$\sum_{i=1}^{I}\sum_{\kappa=1}^{K} y_{i\kappa t} = b_t \qquad t = 1,\ldots,T \tag{4.6}$$

$$\sum_{k=1}^{K}\sum_{b=1}^{B} y_{kjbt} + \sum_{b=1}^{B} z_{jb,t-1} \le \sum_{l=1}^{L} S_{lCF} x_{jl} \qquad j = 1,\ldots,J; t = 1,\ldots,T \tag{4.7}$$

$$\sum_{j=1}^{J}\sum_{b=1}^{B} y_{jibt} + \sum_{b=1}^{B} z_{ib,t-1} \le \sum_{l=1}^{L'} S_{lBR} x_{il} \qquad i = 1,\ldots,I; t = 1,\ldots,T \tag{4.8}$$

$$\sum_{l=1}^{L} x_{jl} \le 1 \qquad j = 1,\ldots,J \tag{4.9}$$

$$\sum_{l=1}^{L'} x_{il} \le 1 \qquad i = 1,\ldots,I \tag{4.10}$$

$$\phi_{kbt}, w_{ibt} \ge 0 \quad k = 1,\ldots,K; i = 1,\ldots,I; b = 1,\ldots,B; t = 1,\ldots,T \tag{4.11}$$

$$z_{it}, z_{kbt}, z_{jbt}, z_{ibt} \ge 0 \qquad k = 1,\ldots,K; j = 1,\ldots,J; i = 1,\ldots,I;$$
$$b = 1,\ldots,B; t = 1,\ldots,T \tag{4.12}$$

$$y_{kjbt}, y_{jibt}, y_{i\kappa t} \ge 0 \qquad k = 1,\ldots,K; j = 1,\ldots,J; i = 1,\ldots,I;$$
$$\kappa = 1,\ldots,\mathcal{K}; b = 1,\ldots,B; t = 1,\ldots,T \tag{4.13}$$

$$x_{jl} \in \{0,1\} \qquad j = 1,\ldots,J; l = 1,\ldots,L \tag{4.14}$$

$$x_{il} \in \{0,1\} \qquad i = 1,\ldots,I; l = 1,\ldots,L'. \tag{4.15}$$

The objective here is to minimize the total system costs, which include facility investment costs, transportation and inventory holding costs, and biomass collection and processing costs. Constraints (4.1) show that the total amount of biomass available for delivery depends on the amount of acres of biomass harvested and the production yield. Constraints (4.2), (4.3), (4.4), and (4.5) are the flow conservation constraints at the harvesting sites, collection facilities, biorefineries, and blending facilities, respectively. These constraints ensure that no more biomass or biofuel is shipped from or processed at a location than what is actually available at the time of shipment or processing. Constraints (4.6) enforce that demand for biofuel is satisfied. Constraints (4.7) and (4.8) are capacity constraints. Constraints (4.9) and (4.10) enforce that at most one storage facility and at most one biorefinery of a particular size be open in a given location; (4.11) through (4.13) are the nonnegativity constraints, and (4.14) and (4.15) are the binary constraints.

4.3.2 Stochastic Optimization Models

The main drawback of the deterministic models listed here is that they rely on the assumption that problem parameters are known in advance. However, biomass supply chains are subject to a number of uncertainties, such as biomass supply, biofuel demand, and market price. Therefore, a number of models in the literature account for these uncertainties. Awudu and Zhang (2012) provide an extensive review of papers that incorporate uncertainty in models for biofuel supply chains. The authors point out that there are a limited number of research papers that consider uncertainty in the design and management of biofuel supply chains. There are a number of papers that use extensions of the two-stage, location–transportation stochastic programing model to identify biorefinery sittings (Cundiff et al. 1997; Huang et al. 2010; Kim et al. 2011a; Chen and Fan 2012; Gebreslassie et al. 2012).

The work conducted by Cundiff et al. (1997) is one of the first that discusses the impact of supply uncertainties on the biofuel supply chain. This work is focused on the impact of weather conditions on biomass production yields. The modeling effort is devoted to the design of an efficient biomass delivery system. Chen and Fan (2012) introduce a two-stage stochastic programing model that identifies refinery and terminal sizes and locations, a feedstock resource allocation strategy, an ethanol production and transportation plan under feedstock biomass supply, and biofuel demand uncertainties. Gebreslassie et al. (2012) propose a bicriterion, multiperiod, stochastic mixed-integer linear programing model to address the optimal design of hydrocarbon biorefinery supply chains under supply and demand uncertainties. The aim of this study is to reduce expected annualized total system costs and financial risks simultaneously. Kim et al. (2011a) propose a mixed-integer linear program to determine the processing locations, volumes, supply networks, and the logistics of transporting forest waste to conversion facilities and from conversion to the market. Their model captures the impact of system uncertainties on profits to maximize the overall expected profit. Marufuzzaman et al. (2014b) present a two-stage stochastic programing model used to design and manage biodiesel supply chains. This is an extension of the two-stage stochastic location–transportation model; it captures the impact of biomass supply and technology uncertainty on supply chain–related decisions; and it minimizes costs and emissions in the supply chain.

4.3.2.1 A Two-Stage Stochastic Programing Model

The following is a two-stage stochastic programing model. This is an extension of the well-known two-stage stochastic location–transportation model (Birge and Louveaux 2011). We refer to this as model (SP) because it is the two-stage, stochastic programing version of model (P) presented earlier.

The first-stage decision variables of this model are the ones related to facility location and capacity size: x_{jl} and x_{il}. These decisions are made upfront and prior to observing the variations in supply and demand. Biomass supply and biofuel demand variations impact decisions related to transportation, inventory, and processing in a time period. Consider that there is a set of M different scenarios created to represent the behavior of biomass supply and biofuel demand. Thus, the value of the corresponding second-stage variables depends on the corresponding scenario. The goal is to design a supply chain that minimizes the total of facility location costs and the expected costs associated with biomass harvesting, collection, and transportation as well as biofuel production and delivery.

$$
\begin{aligned}
\min \sum_{j=1}^{J}\sum_{l=1}^{L}\Psi_{jl}x_{jl} + \sum_{i=1}^{I}\sum_{l=1}^{L'}\Psi'_{il}x_{il} + E_m\Bigg[\sum_{t=1}^{T}\Bigg(\sum_{k=1}^{K}\sum_{b=1}^{B}(\rho_b + p_{kb})\phi_{kbt}(m) + h^e\sum_{i=1}^{I}z_{it}(m) \\
+ \sum_{b=1}^{B}h_b\Bigg(\sum_{k=1}^{K}z_{kbt}(m) + \sum_{j=1}^{J}z_{jbt}(m) + \sum_{i=1}^{I}z_{ibt}(m)\Bigg) + \sum_{i=1}^{I}\sum_{b=1}^{B}\omega_b w_{ibt}(m) \\
+ \sum_{b=1}^{B}\Bigg(\sum_{k=1}^{K}\sum_{j=1}^{J}c_{kj}y_{kjbt}(m) + \sum_{k=1}^{K}\sum_{i=1}^{I}c_{ki}y_{kibt}(m) + \sum_{j=1}^{J}\sum_{i=1}^{I}c_{ji}y_{jibt}(m) + \sum_{i=1}^{I}\sum_{\kappa=1}^{K}c_{i\kappa}y_{i\kappa t}(m)\Bigg)\Bigg)\Bigg]
\end{aligned}
\qquad \text{(SP)}
$$

$$\phi_{kbt}(m) \leq \delta_{kbt}(m)\gamma_{kb}(m)L_{kb}(m) \qquad \begin{aligned} & k = 1,\ldots,K; b = 1,\ldots,B; \\ & t = 1,\ldots,T; m = 1,\ldots,M \end{aligned} \qquad (4.16)$$

$$\phi_{kbt}(m) + (1-\alpha)z_{kb,t-1}(m) = \sum_{j=1}^{J} y_{kjbt}(m) + z_{kbt}(m) \qquad \begin{aligned} & k = 1,\ldots,K; b = 1,\ldots,B; \\ & t = 1,\ldots,T; m = 1,\ldots,M \end{aligned} \qquad (4.17)$$

$$\sum_{k=1}^{K} y_{kjbt}(m) + (1-\alpha)z_{jb,t-1}(m) = \sum_{i=1}^{I} y_{jibt}(m) + z_{jbt}(m) \qquad \begin{aligned} & j = 1,\ldots,J; b = 1,\ldots,B; \\ & t = 1,\ldots,T; m = 1,\ldots,M \end{aligned} \qquad (4.18)$$

$$\sum_{j=1}^{J} y_{jibt}(m) + (1-\alpha)z_{ib,t-1}(m) = w_{ibt}(m) + z_{ibt}(m) \qquad \begin{aligned} & i = 1,\ldots,I; b = 1,\ldots,B; \\ & t = 1,\ldots,T; m = 1,\ldots,M \end{aligned} \qquad (4.19)$$

$$\sum_{b=1}^{B} \beta_b w_{ibt}(m) + z_{i,t-1}(m) = \sum_{\kappa=1}^{K} y_{i\kappa t}(m) + z_{it}(m) \qquad \begin{aligned} & i = 1,\ldots,I; t = 1,\ldots,T; \\ & m = 1,\ldots,M \end{aligned} \qquad (4.20)$$

$$\sum_{i=1}^{I}\sum_{\kappa=1}^{K} y_{i\kappa t}(m) = b_t(m) \qquad t = 1,\ldots,T; m = 1,\ldots,M \qquad (4.21)$$

$$\sum_{k=1}^{K}\sum_{b=1}^{B} y_{kjbt}(m) + \sum_{b=1}^{B} z_{jb,t-1}(m) \leq \sum_{l=1}^{L} S_{lCF}x_{jl} \qquad \begin{aligned} & j = 1,\ldots,J; t = 1,\ldots,T; \\ & m = 1,\ldots,M \end{aligned} \qquad (4.22)$$

$$\sum_{j=1}^{J}\sum_{b=1}^{B} y_{jibt}(m) + \sum_{b=1}^{B} z_{ib,t-1}(m) \leq \sum_{l=1}^{L} S_{lBR}x_{il} \qquad \begin{aligned} & i = 1,\ldots,I; t = 1,\ldots,T; \\ & m = 1,\ldots,M \end{aligned} \qquad (4.23)$$

$$\sum_{l=1}^{L} x_{il} \leq 1 \qquad i = 1,\ldots,I \qquad (4.24)$$

$$\sum_{l=1}^{L} x_{jl} \leq 1 \qquad j = 1,\ldots,J \qquad (4.25)$$

$$\phi_{kbt}(m), w_{ibt}(m) \geq 0 \qquad \begin{aligned} & k = 1,\ldots,K; i = 1,\ldots,I; b = 1,\ldots,B; \\ & t = 1,\ldots,T; m = 1,\ldots,M \end{aligned} \qquad (4.26)$$

$$z_{it}(m), z_{kbt}(m), z_{jbt}(m), z_{ibt}(m) \geq 0 \qquad \begin{aligned} & k = 1,\ldots,K;\ j = 1,\ldots,J;\ i = 1,\ldots,I; \\ & b = 1,\ldots,B;\ t = 1,\ldots,T; \\ & m = 1,\ldots,M \end{aligned} \qquad (4.27)$$

$$y_{kjbt}(m), y_{jibt}(m), y_{i\kappa t}(m) \geq 0 \qquad \begin{aligned} & k = 1,\ldots,K;\ j = 1,\ldots,J;\ i = 1,\ldots,I; \\ & \kappa = 1,\ldots,\mathcal{K};\ b = 1,\ldots,B;\ t = 1,\ldots,T; \\ & m = 1,\ldots,M \end{aligned} \qquad (4.28)$$

$$x_{jl}(m) \in \{0,1\} \qquad j = 1,\ldots,J;\ l = 1,\ldots,L' \qquad (4.29)$$

$$x_{il}(m) \in \{0,1\} \qquad i = 1,\ldots,I;\ l = 1,\ldots,L. \qquad (4.30)$$

The objective function of (SP) minimizes the total costs associated with the first-stage decisions plus the expected cost of second-stage decisions. The set of constraints of (SP) is similar to the constraints defined earlier for (P). The main difference from formulation (P) is that in this formulation the constraints, other than (4.24) and (4.25), are defined for each scenario $m = 1,\ldots,M$.

4.3.3 Multiobjective Optimization Models

The main objective of many models developed relating to the biomass supply chain has been minimizing costs. This is a trend observed in the literature related to supply chain optimization in general. Recently, there is a growing interest in incorporating environmental and social objectives in biomass supply chain models. Because the development of the biofuels industry is in its initial stages, there is a great opportunity to influence decisions that will determine the impact that this industry will have on the environment and on society.

There are a few works that present models that optimize the economic, environmental, and social impacts of bioenergy supply chain design and management. For example, Zamboni et al. (2009a) present a mixed-integer linear program (MILP) model that simultaneously minimizes the supply chain operating costs and greenhouse gas (GHG) emissions due to supply chain activities. Perimenis et al. (2011) provide a decision support tool to evaluate biofuel production pathways. This tool integrates technical, economic, environmental, and social aspects along the entire value chain of biofuels starting from biomass production to biofuel end use. Mele et al. (2009) address the problem of optimizing the supply chains for bioethanol and sugar production. Their bicriteria MILP model addresses economic and environmental concerns. The model minimizes the total cost of managing the supply chain network, and it minimizes the environmental impact over the entire product life cycle. El-Halwagi et al. (2013) incorporate safety concerns into the biorefinery location selection and capacity management problem. They establish trade-offs between costs and safety issues using Pareto curves. You and Wang (2011) study the optimal design and planning of biomass-to-liquids (BTL) supply chains under economic and environmental criteria. You et al. (2012) address the optimal design and planning of cellulosic ethanol supply chains under economic, environmental, and social objectives.

In this section, a multiobjective, MILP model in support of biofuel supply chain design and management problems is presented. The model aids biorefineries in achieving their environmental and social goals, while optimizing costs of their supply chain. The environmental impact of this supply chain is evaluated by estimating the CO_2 emissions due to transportation and biorefinery operations.

The model captures the social impact by estimating the number of jobs created due to biomass production, biomass preprocessing, biomass transportation, and biorefinery operations.

The model considers that the raw material used for production of biofuel is densified biomass and the network design in support of biomass delivery has a hub-and-spoke structure. It is assumed that biomass is processed at a preprocessing facility that is located at the farm. Biomass is preprocessed prior to delivery. Preprocessing increases the bulk density of biomass, transforming it into a stable, dense, and flowable commodity, which is easier to load and unload, and cheaper to transport. From the preprocessing facility, biomass is delivered by trucks to a consolidation facility, which is referred to as the depot. If a preprocessing facility is located within 75 miles of a biorefinery, it ships directly to the biorefinery, bypassing the depots. Depots are either rail ramps or ports where small truck shipments are consolidated into large volume shipments. Thus, to deliver biomass to a biorefinery, trucks, barge, and rail can be used. It is expected that a biorefinery will have railway access to handle the large amount of biomass required to operate at high capacity. Thus, depots represent the first hubs and biorefineries represent the second hubs in this supply chain. The final product, cellulosic ethanol, is shipped to a bulk terminal or a redistribution bulk terminal from where it is then delivered to customers. Bulk terminals are typically blending facilities where cellulosic ethanol is stored until it is blended with gasoline. Depending on the volume shipped and the transportation distance, either truck or rail is used for cellulosic ethanol delivery. Typically, rail is used for distances longer than 75 miles. From the bulk terminal, shipments of cellulosic ethanol are delivered by truck and in smaller quantities to gas stations. Such a supply chain structure can provide biorefineries with the necessary amount of biomass to maintain high production volumes without paying high transportation costs. In this model, the use of different modes of transportation is intentionally done to capture the trade-offs that exist between costs and environmental impacts due to transportation and inventory holding. Figure 4.2 gives a network representation of a supply chain consisting of four local preprocessing facilities, two depots, one biorefinery, one terminal for biofuel blending and storage, and two customers.

This problem is modeled using a network design and network flow model (Ahuja et al. 1993). Thus, a supply chain network is defined as $G(N, A)$, where the set of nodes N represents the different members of the supply chain, and the set of arcs A represents the transportation links between these members. The set of nodes N consists of subset J, which represents the set of preprocessing facilities; subset D, which represents the set of depots; subset I, which represents the set of potential

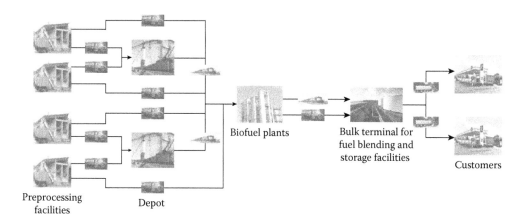

Figure 4.2 Network representation of a supply chain consisting of four local preprocessing facilities, two depots, one biorefinery, one terminal for biofuel blending and storage, and two customers.

biorefinery locations; subset K, which represents the set of bulk terminal locations; and subset C, which represents the set of customers. Set A consists of subset T_1, which represents the set of arcs that connect preprocessing facilities to the depot; T_2, which represents the set of arcs that connect the preprocessing facilities to the biorefinery; subset T_3, which represents the set of arcs that connect biorefinery to the bulk terminal; subset T_4, which represents the set of arcs that connect the bulk terminal to the customer; subset R_1, which represents the set of arcs that connect the depots to the biorefineries; and subset R_2, which represents the set of arcs that connect biorefineries to the bulk terminals. Let $T = T_1 \cup T_2 \cup T_3 \cup T_4$ and $R = R_1 \cup R_2$. The transportation mode used along arcs in T and R are truck and rail, respectively.

4.3.3.1 Cost Objective

The costs along arcs in T are linear, and there are no upper bounds on the amount shipped using these arcs. These arcs represent transportation of biomass and biofuel by truck. Let c_{ij} represent the unit transportation cost per ton shipped along these arcs, and let y_{ij} represent the amount shipped along arc $(i, j) \in T$; then, the total truck transportation costs are equal to: $c_{ij}y_{ij}$.

The total transportation cost along arcs in R is of a multiple-setup structure. Let θ_{ij} be the fixed cost for loading/unloading a unit train along $(i, j) \in R$; v_{ij} be the capacity of a unit train moving along (i, j); and n be the number of unit trains available. Let z_{ij} be an integer variable that represents the number of unit trains moving along arc (i, j). Thus, the total rail transportation costs are equal to: $\theta_{ij}z_{ij} + c_{ij}y_{ij}$. The total transportation costs in this supply chain are

$$T_R C = \sum_{(i,j) \in A} c_{ij} y_{ij} + \sum_{(i,j) \in R} \theta_{ij} z_{ij}. \tag{4.31}$$

Hub location costs represent the investment costs necessary to build the infrastructure in support of loading/unloading unit trains at a depot. Let x_i be a binary variable that takes the value 1 when node $i \in D$ is used as a depot and that takes the value 0 otherwise. Let Ψ_i be the fixed investment cost at node $j \in D$. Total hub location costs are equal to: $\sum_{i \in D} \Psi_i x_i$. Let Ψ_{il} be the fixed investment costs to build a biorefinery of capacity $l(l \in L)$ at node $i \in I$. Let x_{il} be a binary variable that takes the value 1 if node i is selected as the biorefinery location and that takes the value 0 otherwise. Total biorefinery location costs are equal to: $\sum_{i \in I} \sum_{l \in L} \Psi_{il} x_{il}$.

For this formulation, the system is also penalized for not meeting demand. Let π_i represent the demand shortage and let α_i represent the corresponding penalty cost at customer i. Then, expression $\sum_{i \in C} \alpha_i \pi_i$ represents the penalty for not meeting demand. The cost objective function minimizes the total of transportation cost, hub location costs, and a penalty cost for unmet demand and is defined as follows:

$$\text{Minimize: TC} = \sum_{(i,j) \in A} c_{ij} y_{ij} + \sum_{(i,j) \in R} \theta_{ij} z_{ij} + \sum_{i \in D} \Psi_i x_i + \sum_{i \in I} \sum_{l \in L} \Psi_{il} x_{il} + \sum_{i \in C} \alpha_i \pi_i. \tag{4.32}$$

4.3.3.2 Environmental Objective

The model captures CO_2 emissions that result from fuel combustion due to transportation in the supply chain. The model also captures CO_2 emissions due to constructing and operating biofuel plants and to operating the hubs. It is assumed that the emission function is linear with respect to quantities shipped and quantities processed in facilities. Let e_{ij} represent CO_2 emission per ton per mile shipped along arc $(i, j) \in A$. Let ε_{il} represent CO_2 emission per ton processed at the biofuel plant located in $i \in I$. Let μ_i represent CO_2 emission for establishing a hub in $i \in D$. The following environmental objective minimizes total emissions in the supply chain.

$$\text{Minimize: TE} = \sum_{(i,j) \in A} e_{ij} y_{ij} + \sum_{i \in D} \mu_i x_i + \sum_{i \in I} \sum_{l \in L} \epsilon_{il} x_{il}. \tag{4.33}$$

4.3.3.3 Social Objective

The social benefits of this supply chain are measured by the number of accrued local jobs. Jobs are created to support biomass and biofuel transportation, biofuel plant construction and operation, and hub operation. The number of transportation jobs created is linear and depends on the transportation distance and on quantity shipped. The number of jobs created due to biofuel plant construction and operation depends on the production capacity of the plant. The number of jobs created due to hub operation is fixed. Let p_{ij}^T represent the number of transportation jobs created, let p_i^D represent the number of jobs created due to hub operations, and let p_{il}^I represent the number of jobs created due to construction and support operations of biofuel plant i. Then, the social objective function is defined as follows:

$$\text{Maximize: SB} = \sum_{(i,j) \in A} p_{ij}^T y_{ij} + \sum_{(i,j) \in R} p_{ij}^T z_{ij} + \sum_{i \in D} p_i^D x_i + \sum_{i \in I} \sum_{l \in L} p_i^D x_{il}. \tag{4.34}$$

Other notations used in this formulation are: s_j, which represents the amount of biomass available in preprocessing facility j in J; b_j, which represents the demand at customer $j \in C$; u_j, which represents the capacity of depot $j \in D$; and q_{il}, which represents the capacity of a biorefinery of the size $l \in L$ located in $i \in I$.

4.3.3.4 The MILP Model

The following is the multiobjective MILP problem formulation, and this formulation is referred to as (MP).

$$\text{Minimize: } (TC(x,y,z,\pi), TE(x,y))$$

$$\text{Maximize: } (SB(x,y,z))$$

Subject to: (MP)

$$\sum_{i \in (D \cup C)} y_{ji} \leq s_j \qquad \forall j \in J \tag{4.35}$$

$$\sum_{j \in J} y_{ji} - \sum_{j \in I} y_{ij} = 0 \qquad \forall i \in D \tag{4.36}$$

$$\sum_{j \in (J \cup D)} y_{ji} - \sum_{j \in K} y_{ij} = 0 \qquad \forall i \in I \tag{4.37}$$

$$\sum_{i \in I} y_{ij} - \sum_{j \in C} y_{ij} = 0 \qquad \forall i \in K \tag{4.38}$$

$$\sum_{i \in \mathcal{K}} y_{ij} + \pi_j = b_j \qquad \forall j \in C \qquad (4.39)$$

$$y_{ij} - v_{ij}z_{ij} \le 0 \qquad \forall (i,j) \in R \qquad (4.40)$$

$$\sum_{j \in J} y_{ji} - u_i x_i \le 0 \qquad \forall i \in D \qquad (4.41)$$

$$\sum_{j \in (J \cup D)} y_{ji} - q_{il}x_{il} \le 0 \qquad \forall i \in I \qquad (4.42)$$

$$\sum_{l \in L} x_{il} \le 1 \qquad \forall i \in I \qquad (4.43)$$

$$y_{ij} \in R^+ \qquad \forall (i,j) \in A \qquad (4.44)$$

$$\pi_i \in R^+ \qquad \forall i \in C \qquad (4.45)$$

$$x_i \in \{0,1\} \qquad \forall i \in D \qquad (4.46)$$

$$x_{il} \in \{0,1\} \qquad \forall i \in I, l \in L \qquad (4.47)$$

$$z_{ij} \in Z^+ \qquad \forall (i,j) \in R \qquad (4.48)$$

Constraints (4.35) indicate that the amount of biomass shipped from a preprocessing facility is limited by its availability. Constraints (4.36) through (4.38) are the flow balance constraints at depots, biofuel plants, and bulk terminals, respectively. Constraints (4.39) indicate that customer demand could be satisfied through shipments from terminals or from the market. These equations also measure demand shortage. Constraints (4.40) set an upper limit on the amount of biomass shipped using rail cars. Constraints (4.41) set a limit on the storage capacity of a hub. Constraints (4.42) set a limit on the capacity of a biorefinery. Constraints (4.43) set a limit on the number of biofuel plants at a particular location. Constraints (4.44) and (4.45) are the nonnegativity constraints. Constraints (4.46) and (4.47) are binary constraints. Constraints (4.48) are the integer constraints.

4.3.4 Simulation and Simulation/Optimization Models

Biomass supply chain simulation models capture system dynamics, uncertainties, and enable users to perform what-if analysis at a lower cost. The reader is referred to the work of Terzi and Cavalieri (2004) for an extensive review on the use of simulation in the supply chain literature.

A number of simulation models have been developed for biomass supply chains. Typically, discrete-event simulation models are used to capture the interactions among the supply chain entities, changes in the states of the system, and the movement of trucks. Discrete-rate simulation models are typically used to capture the flow of materials and fluctuations in the processing rates of the equipment. Simulation models capture specific qualities of biomass, such as moisture content, bulk density, ash content, and heat value. Incorporating product characteristics into the supply chain model make the model and results realistic. Simulation models capture resource constraints along the supply chain, such as equipment processing capacities and the number and capacity of transportation vehicles. Therefore, simulation models are adequate for modeling and analyzing operational decisions of this supply chain.

Hall et al. (2001) present a simulation model to estimate the logistics cost of delivering forest residues to a power plant in New Zealand. Sims and Venturi (2004) simulate the supply chain of a short rotation coppice crop to a conversion facility in North Island, New Zealand. Sokhansanj and Turhollow (2006) present a simulation model for biomass supply chains referred to as the integrated biomass supply analysis and logistics (IBSAL) model. This model is developed at and supported by Oak Ridge National Laboratory. The model is built in EXTEND and mimics the collection, storage, and transportation of biomass (corn) from the field to a biorefinery. They use this model to present the impact of weather conditions and other operational constraints on the performance of the supply chain. Mahmudi and Flynn (2006) extend this model to simulate the supply chain for forest product and residues. Jacobson and Searcy (2010) present the biomass logistics model (BLM) to estimate logistic costs of biomass and biofuels. This model is developed at and used by the Idaho National Laboratory. Mobini et al. (2011) develop a simulation model to evaluate the delivery cost of forest biomass to a power plant in Quesnel, Canada. This model incorporates three different biomass harvesting systems. Zhang et al. (2012) develop a simulation model to identify the best location for a biomass conversion facility in Michigan. Work by Ileleji (2007) and Ravula et al. (2008) also demonstrate the importance of using simulation tools to evaluate the performance of the biomass supply chain.

Simulation-optimization models in supply chains use simulation models to evaluate/improve the results of an optimization model. Typically, these models would feed each other iteratively until no better solution can be found. In these studies, the simulation model evaluates static solutions obtained by the optimization model in dynamic/stochastic environments. Simulation-optimization models are scarce in the literature of biomass-to-biofuel supply chain design and management. DeMol et al. (1997) develop an optimization-simulation model for biomass-to-biofuel logistics management. Eksioglu et al. (2013) present a simulation-optimization model to optimize the performance of the supply chain for biocrude production from activated sewage sludge in wastewater treatment (WWT) facilities. The optimization model uses a MIP to identify facility locations and assignments. The solution from the optimization model (the structure of the supply chain) is used to build a discrete-event simulation model. The simulation captures the seasonal and random nature of biomass supply. They develop a case study using data from Mississippi.

4.4 CLASSIFICATION OF THE LITERATURE BASED ON RAW MATERIAL USED

The first-generation biofuels were produced using edible products such as corn, soybeans, sunflowers, and so forth. Production of second-generation biofuels relies on using agricultural, forest, animal, and municipal waste. Third-generation biofuels will be generated using algae or some other raw materials. The physical characteristics and energy intensity of these products are different. These differences impact collection, transportation, and processing decisions for biomass. Aligning supply chain strategies with product characteristics is necessary to ensure the practicality of a model.

The literature on supply chain management of corn-based ethanol is scarce. These plants are mainly located in the Midwest in the middle of cornfields. Therefore, the plants receive truck shipments of biomass from farms located nearby. For this reason, research articles on corn ethanol do not focus only on supply chain design and management issues. For example, Zamboni et al. (2009b) propose a multiobjective optimization model to minimize costs and to reduce GHG emissions associated with the future corn-based Italian bioethanol network.

Perlack and Turhollow (2002) discuss the logistics of collecting, handling, and hauling corn stover to an ethanol conversion facility. This work was conducted at the Oak Ridge National Laboratory. The paper provides a very comprehensive cost analysis of two conventional bailing systems—round and rectangular bales, a silage harvesting system, and an unprocessed pickup system. The paper estimates the cost of delivering stover using these systems. Later on, studies conducted at the National Renewable Energy Laboratory (Atchison and Hettenhaus 2004) extend this research to consider a few issues that were observed when removing corn stover from fields to use it as a feedstock. This work provides guidelines for sustainable removal of corn stover for biofuel production. In 2008, the National Renewable Energy Laboratory extended their model to represent the state of technology at the time (Humbird and Aden 2009). From 2004 to 2008, several key biochemical platform accomplishments were achieved and documented that spanned major areas of research related to corn stover as a feedstock (such as biomass collection, pretreatment, conditioning, enzymatic hydrolysis, and fermentation). The goal of this model is to estimate ethanol production cost considering the current state of the technology.

Recently, Cobuloglu and Buyuktahtakin (2014) gave a review of the models developed for supply chain and production of lignocellulosic biofuel and biomass from energy crops, such as switchgrass and miscanthus. They also identified areas of need in terms of research and where gaps exist in better understanding the supply chains issues related to lignocellulosic biomass. Other papers that are focused on the lignocellulosic biomass supply chain are those of Hess et al. (2009), Ileleji et al. (2010), An et al. (2011), and of Cobuloglu and Buyuktahtakin (2014).

Shabani et al. (2013) give a review of the literature related to forest biomass supply chains. They consider deterministic and stochastic mathematical models that use forest biomass for electricity, heat, and biofuel production. They also determined that future work needs to incorporate environmental and social objectives as well as variation in the quality of the forest biomass product.

Biomass, in the form of agricultural and forest waste, has low density and poor flowability properties, and thus, it is bulky, heterogeneous, and unstable. For these reasons, processes such as loading, unloading, and transportation of biomass are challenging and expensive. Densified biomass refers to biomass that has undergone preprocessing to increase the bulk density of the material, such as pelletization and briquetting. Densifying transforms biomass into a stable, dense, and flowable commodity, which is easier to load and unload, and cheaper to transport. A number of papers in the literature discuss characteristics of densified biomass, its performance, and its economic feasibility. The Idaho National Laboratory proposes a commodity-based, advanced biomass supply chain design concept to support the large-scale production of biofuels using densified biomass (Searcy et al. 2007; Hess et al. 2009). This system relies on densifying biomass at local preprocessing facilities before delivering it to a biorefinery and before long distance transportation. This advanced supply system design relies on using long hauls of biomass using high capacity transportation modes, such as rail and barge. Such a system increases biomass availability due to the increase in a biorefinery's supply radius and, as a result, it impacts biorefinery production capacity and reduces feedstock supply risk. Works by Marufuzzaman et al. (2014c) and Roni et al. (2014b) propose inbound supply chain designs to deliver densified biomass-to-biofuel plants or coal plants for cofiring. Both papers rely on the commodity-based, advanced biomass supply chain design concept described here. For additional resources on biomass densification see Clarke and Preto (2011) and REAP (2014).

Eksioglu et al. (2013), Marufuzzaman et al. (2014a), and Marufuzzaman et al. (2014b) present simulation and optimization models for the biodiesel supply chain. Biodiesel is produced via the use of industrial wastes generated by pulp and paper, seafood processing, meat processing, and WWT facilities.

4.5 CLASSIFICATION OF THE LITERATURE BASED ON PRODUCTS PRODUCED

Biomass can be used to produce liquid fuels, such as ethanol, cellulosic ethanol, and biodiesel; electricity; and gases such as biogas and methane. The inbound part of the supply chain for these products has the same challenges discussed thus far because the same material—biomass—is used for their production. However, the outbound supply chain designs are different. For example, liquid fuels are shipped to a blending where they are blended with fuel. Ethanol blends (such as E-10, E-15, and E-85) are distributed to gas stations. The design and management of the supply chain for these products typically integrates the inbound and outbound supply chain decisions. Electricity, on the other hand, is distributed using the existing electricity grid. Therefore, once produced, it is distributed together with nonrenewable electricity generated using coal.

In the previous sections, the use of biomass for production of electricity has not been discussed because the corresponding literature is scarce. Biomass cofiring with coal in coal-fired power plants has drawn attention recently because many researchers and companies view cofiring as a near-term market opportunity for biomass. Cofiring biomass in coal plants is a practice that has been investigated and implemented (Tillman 2000) in order to reduce greenhouse gas emissions; to increase renewable energy production without major capital investments; and to minimize waste (such as wood waste or agricultural waste) and the environmental problems associated with its disposal. Cofiring is indeed a low-risk option for production of renewable energy because the risk associated with major capital investments and uncertain raw material supplies is much smaller compared to other alternative uses of biomass. However, cofiring is a practice that impacts logistics-related costs, capital investments, and efficiency of coal plants. Roni et al. (2014b) propose a framework for designing the supply chain network for biomass cofiring in coal-fired power plants. They present a hub-and-spoke supply chain network design model for long-haul delivery of biomass. Works by Eksioglu et al. (2014) and Eksioglu and Karimi (2014) propose mathematical models to minimize supply chain costs related to delivering biomass to power plants, to minimize corresponding investment costs, and maximize savings due to the 2.3 cents per kilowatt-hour PTC.

4.6 CLASSIFICATION OF THE LITERATURE BASED ON
THE SCOPE OF THE STUDY

Research on using forest biomass as a source of bioenergy has been investigated by a number of researchers around the world. This research has progressed in the United States due to legislation such as the RFS; in Canada, due to the vast availability of forest biomass; in Brazil, due to the vast availability of sugarcane; in Europe, Australia, Ukraine, and so forth because of binding limitations in greenhouse gas emissions due to the ratification of the Kyoto Protocol.

A summary of the existing literature based on the scope of these studies is presented. A distinction is made between the literature that is focused on regional and local supply chains. The difference between these streams of research is on the approach they take in modeling the supply chain. Regional models focus mainly on designing large-scale supply chains. The goal of these supply chains is to identify designs that optimize the regionwide resources and that meet regionwide demands. The literature on local supply chain models focuses on managing small-scale supply chains. This literature identifies good practices and rules of thumb that can be used by small- and some medium-sized facilities to optimize their supply chain.

The following works by Klein et al. (2013) and Yu et al. (2014) are used in order to pinpoint the differences that exist between regional and local models. The model by Yu et al. (2014) is built upon results generated from the model presented in Klein et al. (2013). Although the work by Klein et al. (2013) presents a strategic model that is focused in the state of Missouri, the model by Yu et al. (2014) zooms in within this region. This model provides more details about operational activities of this supply chain. The model by Klein et al. (2013) integrates a mixed-integer linear programing model with a geographic information system (GIS) model. The objective is to help determine optimal facility locations and flow allocations for woody biomass sources. The work by Yu et al. (2014) presents an integrated multiperiod problem of woody biomass operations in a three echelon supply chain system. The problem includes many geographically dispersed harvesting sites, several pretreatment facilities, and a power plant. The objective of the operational planning model is to compute an optimal assignment of harvesting teams at the beginning of each period and an optimal allocation of biomass flows so that the total cost of harvesting, transportation, preprocessing, storage, and conversion over a finite planning horizon is minimized. The constraints include restrictions on supply at each harvesting site, capacity constraints at pretreatment and storage facilities, flow balance between sites and facilities, and demand satisfaction at the power plant. Other considerations affecting operations such as forest regeneration, cutting cycle, and biomass degradation during storage are also modeled. A dynamic programing model is developed to address the problem. A case study is also presented to show the efficacy of the method.

Gronalt and Rauch (2007) develop a model to evaluate regional forest biomass-based supply chain designs and their performance. The model is focused on a federal state of Austria. This state has several forest areas and is home to a number of energy plants. Tursun et al. (2008) present a mathematical programing model to determine the optimal locations and capacities of biorefineries, the amount of bioenergy crops delivered to biorefineries, and the amount of ethanol and coproducts produced and distributed across Illinois. Slade et al. (2009) analyze the role of supply chain design decisions on the viability of commercial cellulosic ethanol projects in Europe. Their work indicates that a supply chain–based analysis can shed light on the major cost contributors in a project. The model proposed by Kim et al. (2011b) identifies an optimal design of the biomass supply chain network for the southeastern region of the United States. This supply chain consists of biomass suppliers, candidate sites for biorefineries, and final markets. Shabani and Sowlati (2013) present a study on using forest biomass as one of the renewable sources of energy for generating electricity. They develop a dynamic optimization model that integrates a number of considerations such as biomass supply and storage, energy production, and management. The developed model is then applied to a real scenario in Canada in which it is shown to produce more profit than the current process in place.

4.7 OTHER RELATED LITERATURE

A few researchers analyze the impact of disruptions to the biofuel supply chain network design and management. Li et al. (2011) proposed one discrete and one continuous model to design reliable bioethanol supply chain networks. They use numerical analysis to evaluate the impact of disruptions on optimal refinery deployment decisions. Wang and Ouyang (2013) proposed a game-theoretical-based, continuous approximation model to locate biorefineries under spatial competition and facility disruption risks. These studies only considered failure risks at biorefineries. Marufuzzaman et al. (2014c) present a mathematical model that designs a reliable multimodal transportation network for a biofuel supply chain system, where intermodal hubs are subject to site-dependent probabilistic disruptions.

4.8 CONCLUSIONS

It is expected that demand for bioenergy will increase in the near future due to legislation such as the RFS and the renewable portfolio standard (RPS). This growth will demand greater amounts of biomass be produced and delivered to bioenergy production plants. Because biomass is a low energy density product, in order to meet the goals set by RFS, the delivery of biomass will require major changes in supply chain infrastructure and supply chain management practices. These changes are expected to reduce biomass delivery costs, and consequently, to reduce the total cost of bioenergy. A number of studies note that processes such as preprocessing and densification are necessary to reduce transportation volumes, and consequently, to reduce biomass transportation costs in the supply chain. Other researchers have studied the role of high-volume and low-cost transportation modes, such as rail and barges, in support of high-volume and long-haul delivery of biomass to large-scale bioenergy production plants. Many researchers propose optimization models that optimize the overall design and management of the biomass supply chain. This chapter summarizes existing models that focus on optimizing the performance of the biomass supply chain. The goal of these models is to build supply chains that are cost efficient, energy efficient, and environmentally friendly. Some of the models developed mimic existing supply chain models for agricultural products that have similar physical characteristics.

The two major U.S. regions that produce biomass are the Midwest and the Southeast. Most of the U.S. population is located in the East and West. Although some of the demand for bioenergy will be satisfied through local facilities of small capacity, it is expected that some of the nationwide demand for fuel will be met through production in large-scale facilities. Such facilities will require long-haul delivery of biomass or bioenergy. To ensure a cost-efficient delivery of bioenergy, the corresponding supply chains would rely on preprocessing biomass prior to delivery, and on using high-volume transportation modes. Therefore, models are needed to optimize the performance of these large-scale supply chains.

Different from the industrial sector, the agricultural sector consists of farms and other agricultural businesses that are of small size. Traditionally, the literature on agricultural products provides models that estimate the cost of collecting, handling, and hauling these products; models that compare different modes of delivering these products; and propose supply chain options for biobased businesses. The literature related to bioenergy production provides models that take an integrated view of the processes involved in collecting, storing, and transporting agricultural products. This literature also emphasizes the importance of building large-scale enterprises to satisfy the nationwide demand for bioenergy. Therefore, it is imperative that we think differently about agriculture, rural economic development, and energy infrastructure.

ACKNOWLEDGMENT

This work was supported by a National Science Foundation grant (CMMI 1052671). This support is gratefully acknowledged.

REFERENCES

Aden, A., M. Ruth, K. Ibsen, J. Jechura, K. Neeves, J. Sheehan, B. Wallace, L. Montague, A. Slayton, and J. Lukas. 2002. *Lignocellulosic biomass to ethanol process design and economics utilizing co-current dilute acid prehydrolysis and enzymatic hydrolysis for corn stover.* Technical Report, NREL/TP-510-32438, National Renewable Energy Laboratory, Golden, CO.

Ahuja, R.K., T.L. Magnanti, and J.B. Orlin. 1993. *Network Flows: Theory, Algorithms, and Applications.* Pearson, Harlow, UK.

Alumur, S. and B.Y. Kara. 2008. Network hub location problems: The state of the art. *European Journal of Operational Research* **190**: 1–21.

An, H., W.E. Wilhelm, and S.W. Searcy. 2011. A mathematical model to design a lignocellulosic biofuel supply chain system with a case study based on a region in central Texas. *Bioresource Technology* **102**: 7860–7870.

Atchison, J.E. and J.R. Hettenhaus. 2004. *Innovative Methods for Corn Stover Collecting, Handling, Storing and Transporting*. Technical Report, NREL/SR-510-33893, National Renewable Energy Laboratory, Golden, CO.

Awudu, I. and J. Zhang. 2012. Uncertainty and sustainability concepts in biofuel supply chain management: A review. *Renewable and Sustainable Energy Reviews* **16**: 1359–1368.

Badger, P.C. and P. Fransham. 2006. Use of mobile fast pyrolysis plants to densify biomass and reduce biomass handling costs—A preliminary assessment. *Biomass & Bioenergy* **30**(4): 321–325.

Bai, Y., T. Hwang, S. Kang, and Y. Ouyang. 2011. Biofuel refinery location and supply chain planning under traffic congestion. *Transportation Research Part B: Methodological* **45**(1): 162–175.

Birge, J.R. and F. Louveaux. 2011. *Introduction to Stochastic Programming*. Springer, New York.

Bonilla, D. and C. Whittaker. 2009. *Freight Transport and Deployment of Bioenergy in the U.K.* Technical Report, Working Paper No. 1043. Transport Studies Unit, University of Oxford, Oxford, UK.

Campbell, J.F. and M.E. O'Kelly. 2012. Twenty-five years of hub location research. *Transportation Science* **46**(2): 153–169.

Chen, C.-W. and Y. Fan. 2012. Bioethanol supply chain system planning under supply and demand uncertainties. *Transportation Research Part E* **48**: 150–164.

Chopra, S. and P. Meindl. 2012. *Supply Chain Management: Strategy, Planning, and Operation*. Prentice Hall, Upper Saddle River, New Jersey.

Clarke, S. and F. Preto. 2011. Biomass densification for energy production. http://www.omafra.gov.on.ca/english/engineer/facts/11-035.pdf

Cobuloglu, H.I. and I.E. Buyuktahtakin. 2014. A review of lignocellulosic biomass and biofuel supply chain models. *Proceedings of the 2014 Industrial and Systems Engineering Research Conference*, Montreal, Canada.

Cundiff, J.S., N. Dias, and H.D. Sherali. 1997. A linear programming approach for designing a herbaceous biomass delivery system. *Bioresource Technology* **59**(1): 47–55.

DeMeyer, A., D. Cattrysse, J. Rasinmäki, and J. Van Orshoven. 2014. Methods to optimize the design and management of biomass-for-bioenergy supply chains: A review. *Renewable and Sustainable Energy Reviews* **31**: 657–670.

DeMol, R.M., M.A. Jogems, P. Van Beek, and J.K. Gigler. 1997. Simulation and optimization of the logistics of biomass fuel collection. *Netherlands Journal of Agricultural Science* **45**: 219–228.

EIA (Energy Information Administration). 2010. *United State Energy Information Administration: Monthly Energy Review*. U.S. Department of Energy, Energy Information Administration, Washington, DC.

EIA (Energy Information Administration). 2013. United States Energy Information Administration: U.S. gasoline demand. http://www.eia.gov/oog/info/twip/twip-gasoline.html

Eksioglu, S.D., A. Acharya, L.E. Leightley, and S. Arora. 2009. Analyzing the design and management of biomass-to-biorefinery supply chain. *Computers & Industrial Engineering* **57**: 1342–1352.

Eksioglu, S.D. and H. Karimi. 2014. An optimization model in support of biomass co-firing decisions in coal fired power plants. In: P. Pawlewski and A. Greenwood, eds., *Process Simulation and Optimization in Sustainable Logistics and Manufacturing*. EcoProduction, Springer, New York, pp. 111–122..

Eksioglu, S.D., H. Karimi, and M. Hu. 2014. A model for analyzing the impact of production tax credit on renewable electricity production. *Proceedings of the Annual ISERC Conference*, Montreal, CA, May 31–June 4.

Eksioglu, S.D., S. Li, S. Zhang, S. Sokhansanj, and D. Petrolia. 2010. Analyzing impact of intermodal facilities on design and management of biofuel supply chain. *Transportation Research Record* **2191**: 144–151.

Eksioglu, S.D., G. Palak, A. Mondala, and A. Greenwood. 2013. Supply chain designs and management for biocrude production via wastewater treatment. *Environmental Progress & Sustainable Energy* **32**(1): 139–147.

Eksioglu, S.D., S. Zhang, S. Li, S. Sokhansanj, and D. Petrolia. 2011. Analyzing the impact of intermodal facilities to the design of the supply chains for biorefineries. *Transportation Research Record, Journal of the Transportation Research Board* **2191**: 144–151.

El-Halwagi, A.M., C. Rosas, J.M. Ponce-Ortega, A. Jimnez-Gutirrez, M.S. Mannan, and M.M. El-Halwagi. 2013. Multiobjective optimization of biorefineries with economic and safety objectives. *Journal of AIChE* **59**: 2427–2434.

EPA. 2007. United states environmental protection agency: Renewable fuels: Regulations & standards. http://www.epa.gov/otaq/fuels/renewablefuels/regulations.htm

Gebreslassie, B.H., Y. Yao, and F. You. 2012. Design under uncertainty of hydrocarbon biorefinery supply chains: Multiobjective stochastic programming models, decomposition algorithm, and a comparison between CVaR and downside risk. *AIChE Journal* **58**(7): 2155–2179.

Gonzales, D., E.M. Searcy, and S.D. Eksioglu. 2013. Cost analysis for high-volume and long-haul transportation of densified biomass feedstock. *Transportation Research Part A* **49**: 48–61.

Gronalt, M. and P. Rauch. 2007. Designing a regional forest fuel supply network. *Biomass & Bioenergy* **31**: 393–402.

Hall, P., J.K. Gigler, and R.E.H. Sims. 2001. Delivery systems of forest arising for energy production in New Zealand. *Biomass & Bioenergy* **21**(6): 391–399.

Hess, J.R., T.C. Wright, L.K. Kenney, and E.M. Searcy. 2009. Uniform-format solid feedstock supply system: A commodity-scale design to produce an infrastructure-compatible bulk solid from lignocellulosic biomass. INL/EXT-09-15423. http://www.inl.gov/technicalpublications/Documents/4408280.pdf

Huang, Y., C.W. Chen, and Y. Fan. 2010. Multistage optimization of the supply chains of biofuels. *Transportation Research Part E* **46**(6): 820–830.

Humbird, D. and A. Aden. 2009. *Biochemical Production of Ethanol from Corn Stover: 2008 State of Technology.* Technical Report, NREL/TP-510-46214, National Renewable Energy Laboratory. Golden, Colorado.

Ileleji, K. 2007. Transportation logistics of biomass for industrial fuel and energy enterprises. *Presented at 7th Annual Conference on Renewable Energy from Organics Recycling*, October 1–3, Indianapolis, IN.

Ileleji, K.E., S. Sokhansanji, and J.S. Cundiff. 2010. Farm-gate to plant-gate delivery of lignocellulosic feedstocks from plant biomass for biofuel production. In: H.P. Blaschek, T. Ezeji, and J. Scheffran. *Biofuels from Agricultural Wastes and Byproducts.* Wiley-Blackwell, Oxford, UK.

Jacobson, J.J. and E.M. Searcy. 2010. Uniform-format feedstock supply system designs for woody biomass. *Presented at the Spring Conference of the AIChE*, March 21–25, Orlando, FL.

Judd, J.S., C. Subhash, and J.S. Cundiff. 2011. Cost analysis of a biomass logistics system. *ASABE Meeting Presentation.* ASABE, St. Joseph, MI.

Kang, S., H. Onal, Y. Ouyang, J. Scheffran, and D. Tursun. 2010. Optimizing the biofuels infrastructure: Transportation networks and biorefinery locations in Illinois. In: *Handbook of Bioenergy Economics and Policy.* Springer, New York, pp. 151–173.

KDF. 2014. Knowledge Discovery Framework. http://bioenergykdf.net/

Kim, J., M.J. Realff, and J.H. Lee. 2011a. Optimal design and global sensitivity analysis of biomass supply chain networks for biofuels under uncertainty. *Computers & Chemical Engineering* **35**: 1738–1751.

Kim, J., M.J. Realff, J.H. Lee, C. Whittaker, and L. Furtner. 2011b. Design of biomass processing network for biofuel production using an MILP model. *Biomass & Bioenergy* **35**(2): 853–871.

Klein, C., W. Jang, and Z. Yu. 2013. A woody biomass based energy logistics system. *Proceedings of the 2013 Industrial and Systems Engineering Research Conference*, May 19–22, 2013, San Juan, Puerto Rico.

Li, X., F. Peng, Y. Bai, and Y. Ouyang. 2011. Effects of disruption risks on biorefinery location design: Discrete and continuous models. *Proceedings of the 90th TRB Annual Meeting*, January 2011, Washington, DC.

Mafakheri, F. and F. Nasiri. 2014. Modeling of biomass-to-energy supply chain operations: Applications, challenges and research directions. *Energy Policy* **67**: 116–126.

Magnanti, T.L. and R.T. Wong. 1981. Accelerating benders decomposition: Algorithmic enhancement and model selection criteria. *Operations Research* **29**: 464–484.

Mahmudi, H. and P. Flynn. 2006. Rail vs. truck transport of biomass. *Applied Biochemistry and Biotechnology* **129**(1): 88–103.

Marufuzzaman, M., S.D. Eksioglu, and R. Hernandez. 2014a. Environmentally friendly supply chain planning and design for biodiesel production via wastewater sludge. *Transportation Science* **48**: 555–574.

Marufuzzaman, M., S.D. Eksioglu, and Y. Huang. 2014b. Two-stage stochastic programming supply chain model for biodiesel production via wastewater treatment. *Computers & Operations Research* **49**: 1–17.

Marufuzzaman, M., S.D. Eksioglu, X. Li, and J. Wang. 2014c. Analyzing the impact of intermodal-related risk to the design and management of biofuel supply chain. *Transportation Research Part E* **69**: 122–145.

Mele, F.D., G. Guillén-Gosálbez, and L. Jiménez. 2009. Optimal planning of supply chains for bioethanol and sugar production with economic and environmental concerns. *Computer Aided Chemical Engineering* **26**: 997–1002.

Mobini, M., T. Sowlati, and S. Sokhansanj. 2011. Forest biomass supply logistics for a power plant using the discrete-event simulation approach. *Applied Energy* **88**(4): 1241–1250.

Papapostolou, C., E. Kondili, and J.K. Kaldellis. 2011. Development and implementation of an optimization model for biofuels supply chain. *Energy* **36**: 6019–6026.

Parker, N., P. Tittmann, Q. Hart, R. Nelson, K. Skog, E. Schmidt, and B. Jenkins. 2010. Development of a biorefinery optimized biofuel supply curve for the western United States. *Biomass & Bioenergy* **34**(11): 1597–1607.

Perimenis, A., H. Walimwipi, S. Zinoviev, F. Mller-Langer, and S. Miertus. 2011. Development of a decision support tool for the assessment of biofuels. *Energy Policy* **39**(3): 1782–1793.

Perlack, R.D. and A.T. Turhollow. 2002. Assessment of options for the collection, handling, and transportation of corn stover. http://bioenergy.ornl.gov/pdfs/ornltm-200244.pdf

Ravula, P., R. Grisso, and J.S. Cudiff. 2008. Cotton logistics as a model for a biomass transportation system. *Biomass & Bioenergy* **32**: 314–325.

REAP. 2014. Resource Efficient Agricultural Production. http://www.reap-canada.com

RFA. 2013. Renewable Fuels Association: Monthly U.S. fuel ethanol production/demand. http://ethanolrfa.org/pages/monthly-fuel-ethanol-production-demand

Rogers, J. and J.G. Brammer. 2009. Analysis of transport costs for energy crops for use in biomass pyrolysis plant networks. *Biomass & Bioenergy* **33**(10): 1367–1375.

Roni, M.S., S.D. Eksioglu, E. Searcy, and J. Jacobson. 2014a. Estimating the variable cost for high-volume and long-haul transportation of densified biomass and biofuel. *Transportation Research Part D: Transport and Environment* **29**: 40–55.

Roni, M.S., S.D. Eksioglu, E. Searcy, and K. Jha. 2014b. A supply chain network design model for biomass co-firing in coal-fired power plants. *Transportation Research Part E: Logistics and Transportation Review* **61**: 115–134.

Schmidt, J., S. Leduc, E. Dotzauer, G. Kindermann, and E. Schmid. 2008. Optimizing the supply chain of biofuels including the use of waste process heat: An Austrian case study. **2191**: 144–151. http://oega.boku.ac.at/fileadmin/userupload/Tagung/2008/ShortPaper2008/Schmi dtetalOEGA2008Tagungsband.pdf

Searcy, E. and P. Flynn. 2008. The impact of biomass availability and processing cost on optimum size and processing technology selection. *Applied Biochemistry and Biotechnology* **154**: 271–286.

Searcy, E., P. Flynn, E. Ghafoori, and A. Kumar. 2007. The relative cost of biomass energy transport. *Applied Biochemistry and Biotechnology* **137–140**(1–12): 639–652.

Shabani, N., S. Akhtari, and T. Sowlati. 2013. Value chain optimization of forest biomass for bio energy production: A review. *Renewable and Sustainable Energy Reviews* **23**: 299–311.

Shabani, N. and T. Sowlati. 2013. A mixed-integer non-linear programming model for tactical value chain optimization of a wood biomass power plant. *Applied Energy* **104**: 353–361.

Sharma, B., R.G. Ingalls, C.L. Jones, and A. Khanchi. 2013. Biomass supply chain design and analysis: Basis, overview, modeling, challenges, and future. *Renewable and Sustainable Energy Reviews* **24**: 608–627.

Sims, R.E.H. and P. Venturi. 2004. All-year-round harvesting of short rotation coppice *Eucalyptus* compared with the delivered costs of biomass from more conventional short season, harvesting systems. *Biomass & Bioenergy* **26**(1): 27–37.

Slade, R., A. Bauen, and N. Shah. 2009. The commercial performance of cellulosic ethanol supply-chains in Europe. *Biotechnology for Biofuels* **2**: 3.

Sokhansanj, S., S. Mani, A. Turhollow, A. Kumar, D. Bransby, L. Lynd, and M. Laser. 2009. Large-scale production, harvest and logistics of switchgrass (*Panicum virgatum* L.)—Current technology and envisioning a mature technology. *Biofuels, Bioproducts and Biorefining* **3**(12): 124–141.

Sokhansanj, S. and A.F. Turhollow. 2004. Biomass densification—Cubing operations and costs for corn stover. *Applied Engineering in Agriculture* **20**: 495–499.

Sokhansanj, S. and A.F. Turhollow. 2006. Economics of producing fuel pellets from biomass. *Applied Engineering in Agriculture* **22**(3): 421–426.

Terzi, S. and S. Cavalieri. 2004. Simulation in the supply chain context: A survey. *Computers in Industry* **53**: 3–16.

Tillman, D.A. 2000. Biomass cofiring: The technology, the experience, the combustion consequences. *Biomass & Bioenergy* **19**: 365–384.

Tunc, H., B. Eksioglu, S.D. Eksioglu, and M. Jin. 2011. Hub-based network design: A review. *International Journal of Networking* **1**(2): 17–24.

Tursun, U., S. Kang, H. Onal, Y. Ouyang, and J. Scheffran. 2008. Optimal biorefinery locations and transportation network for the future biofuels industry in Illinois. In: *Environmental and Rural Development Impacts Conference*, October 15–16, St. Louis, MO.

Wang, X. and Y. Ouyang. 2013. A continuous approximation approach to competitive facility location design under facility disruption risks. *Transportation Research Part B* **50**: 90–103.

Xie, F., Y. Huang, and S.D. Eksioglu. 2014. Integrating multimodal transport into cellulosic biofuel supply chain design under feedstock seasonality with a case study based on California. *Bioresource Technology* **152**: 15–23.

Xie, W. and Y. Ouyang. 2013. Dynamic planning of facility locations with benefits from multitype facility colocation. *Computer-Aided Civil & Infrastructure Engineering* **28**(9): 666–678.

You, F., L. Tao, D.J. Graziano, and S.W. Snyder. 2012. Optimal design of sustainable cellulosic biofuel supply chains: Multiobjective optimization coupled with life cycle assessment and input-output analysis. *AIChE Journal* **58**(4): 1157–1180.

You, F. and B. Wang. 2011. Life cycle optimization of biomass-to-liquid supply chains with distributed-centralized processing networks. *Industrial & Engineering Chemistry Research* **50**(17): 10102–10127.

Yu, Z., C. Klein, and W. Jang. 2014. Multi period operational planning in woody biomass system. *Proceedings of the 2014 Industrial and Systems Engineering Research Conference*, May 31–June 4, 2014, Montreal, Canada.

Zamboni, A., F. Bezzo, and N. Shah. 2009a. Spatially explicit static model for the strategic design of future bioethanol production systems. *Energy & Fuels* **23**(10): 5134–5143.

Zamboni, A., N. Shah, and F. Bezzo. 2009b. Spatially explicit static model for the strategic design of future bioethanol production systems. 1. Cost minimization. *Energy & Fuels* **23**(10): 5121–5133.

Zhang, F., D.M. Johnson, and M.A. Johnson. 2012. Development of a simulation model of biomass supply chain for biofuel production. *Renewable Energy* **44**: 380–391.

PART II

Conversion Processes

CHAPTER 5

Thermochemical Biomass Conversion for Rural Biorefinery

Thallada Bhaskar, Bhavya Balagurumurthy, Rawel Singh, and Priyanka Ohri
Bio-Fuels Division, CSIR–Indian Institute of Petroleum, Dehradun, India

CONTENTS

5.1 INTRODUCTION

Conventional fuels from fossil resources have played the most important role in the rapid technological progress over the past few centuries. It is estimated that more than 85% of the world's energy requirements are obtained from conventional fuels (Weiland 2010). In addition to providing energy, they are also an important feedstock for a majority of commodity products such as plastics and fabrics (Lucian et al. 2007). The recent fluctuating prices and environmental disturbances due to the use of conventional fuels have called for an alternative to conventional fuels among which biomass-derived energy appears to have the maximum potential. Resources such as solar radiation, winds, tides, and biomass are renewable resources, which are (if appropriately managed) in no danger of being over-exploited. The first three resources can be used as renewable sources of energy, while biomass can be used to produce not only energy but also chemicals and materials (Clark and Deswarte 2008).

Biomass feedstocks are energy sources derived from plants, microbial cells, and the wastes and residues associated with their processing (e.g., agricultural residues and forestry wastes).

They are generally formed through photosynthesis, whereby plants (and some microbial cells) garner atmospheric CO_2 and sunlight to produce high energy carbonaceous compounds (i.e., biomass) and oxygen (Klass 1998; Abdeshanian et al. 2010). When the energy constrained within biomass is released, the carbon is oxidized to CO_2, which can be recycled to produce new biomass. Theoretically, no additional greenhouse gas is produced because the emitted CO_2 is part of the current carbon cycle. Therefore, if efficiently utilized, biomass is regarded as an alternative clean and renewable source for energy and other commodities due to its abundance, high energy content, sustainability, and its biodegradability. Although numerous biorefinery schemes and conversion technologies exist for the transformation of biomass into usable energy forms, they are not cost efficient and economically viable in competition with existing petroleum refinery technologies. In particular, the recalcitrant nature of several feedstocks presents a major technological obstacle for their processing and transformation (Srirangan et al. 2012).

Biorefinery ideas and concepts can help to use limited renewable raw materials more efficiently than they are today. With biorefineries, valuable products—such as platform chemicals—can be produced from agricultural feedstock, which can subsequently be further processed into a variety of substances by the chemical industry. Due to the role they play as producers of biomass, rural areas will grow in importance in the decades to come. Parts of the biorefinery process can be relocated to rural areas to bring a high added value to these regions. By refining biomass at the place of production, new economic opportunities may arise for agriculturists, and the industry gets high-grade preproducts. Croplands can provide more renewable raw materials without endangering a sustainable agriculture. To decide if a region can provide adequate amounts of raw material for a biorefinery, new raw material assessment procedures have to be developed. In doing so, involvement of farmers is inevitable to generate a reliable study of biomass refinery potentials (Papendiek et al. 2012).

Once sufficient biomass has been identified at a particular location, processes have to be developed to convert the same to value added products. This chapter focuses on such processes or combination of processes that can valorize the biomass at the rural or decentralized and centralized level.

5.2 FEEDSTOCKS FOR BIOREFINERY

Currently, biomass-derived energy sources supply ~50 exajoules (EJ) of the world's energy, which represents 10% of global annual primary energy consumption and ~75% of the energy derived from alternative renewable energy sources (Haberl et al. 2010). Moreover, it is expected that biomass-derived energy may have to contribute ~1500 EJ by 2050. At this time, only 2% of the biomass-derived energy sources are utilized in the transportation sector; the rest is generally for household uses (Berndes et al. 2003; Alexander et al. 2012). Transportation fuels derived from biomass (i.e., biofuels) can be produced from lignocellulosic crops, and unused agricultural wastes (second generation) (Worldwatch-Institute 2007; Srirangan et al. 2012).

Energy crops reduce the competition for fertile land because they can be grown on land that is not suitable for agricultural crops. Moreover, in comparison with conventional crops that can contribute only a small fraction of the aforementioned standing biomass, biorefineries based on lignocellulosic feedstocks can rely on larger biomass per hectare yields because the whole crop is available as feedstock (Kamm et al. 2006; Katzen and Schell 2006).

Lignocellulosic biomass has three major components: cellulose, hemicellulose, and lignin. Cellulose $(C_6H_{10}O_6)_n$ has a strong molecular structure made by long chains of glucose molecules (C_6 sugar). Starch can be readily hydrolyzed by enzymes or acid attack to the single sugar monomers, while cellulose (30%–50% of total lignocellulosic dry matter) is much more difficult to hydrolyze and set free individual glucose monomers. Hemicellulose $(C_5H_8O_5)_n$ is a relatively amorphous component that is easier to break down with chemicals and/or heat than cellulose, and it contains a mix of C_6 and C_5 sugars. It is the second main component of lignocellulosic biomass (20%–40% of total feedstock dry matter). Lignin ($C_9H_{10}O_2(OCH_3)_n$) is essentially the glue that provides the overall rigidity to the structure of plants and trees and is made of phenolic polymers. Although cellulose and hemicellulose are polysaccharides that can be hydrolyzed to sugars and then fermented to ethanol, lignin cannot be used in fermentation processes, but it may be useful for other purposes (chemical extraction or energy generation). Lignin (15%–25% of total feedstock dry matter) is the largest noncarbohydrate fraction of lignocellulosic biomass. In addition, lignocellulosic biomass contains some amount of inorganic matter and extractives.

5.3 WHAT IS A BIOREFINERY?

A biorefinery is a facility that oversees the sustainable processing of biomass into a spectrum of marketable products (food, feed, materials, chemicals) and energy (fuels, power, heat) (IEA 2009; Papendiek et al. 2012). A biorefinery can be considered as an integral unit that can accept different feedstocks and can convert them to a range of useful products including chemicals, energy, and materials (Clark et al. 2006).

Unlike fossil resources, biomass can be obtained almost all over the world. In addition, biomass typically exhibits a low bulk density and a relatively high water content (up to 90% for grass), which makes its transport much more expensive than the transfer of natural gas or petroleum (Clark and Deswarte 2008). An important stage in a biorefinery system is the provision of a renewable, consistent, and regular supply of feedstock. Initial processing may be required to increase its energy density to reduce transport, handling, and storage costs. Reducing the cost of collection, transportation, and storage of biomass through densification is thus critical to develop a sustainable infrastructure capable of working with significant quantities of raw material (Hess et al. 2003). In addition, the economics of many conversion processes, which are batch operations, would be dramatically improved through an increase in density because the inherent low density of biomass limits the amount of material that can be processed at any one time (Deswarte et al. 2007). This pretreatment also offers the added benefit of providing a much more uniform material (in size, shape, moisture, density, and energy content), which can be easily handled. Preprocessing can be done on the farm during harvesting (Hess et al. 2003; Clark and Deswarte 2008).

Another issue associated with the use of (fresh) biomass is its perishable character or its susceptibility to degradation (Kadam et al. 2000). This issue is of high importance because fossil resources are available all year long but, in many cases, the same kind of biomass might not be available year-round (Thorsell et al. 2004). Hence, strategies have to be developed to have a sustainable supply chain mechanism for biomass feedstock availability and make sure it is not spoilt due to storage.

One way to mitigate the negative effects of local ecosystem services is to convert biomass into a variety of chemicals, biomaterials, and energy, maximizing the value of biomass and minimizing waste. Similar to oil-based refineries, where many energy and chemical products are produced from crude oil, biorefineries will produce many different industrial products from biomass. These will include low-value, high-volume products, such as transportation fuels and commodity chemicals, as well as materials and high-value, low-volume products or specialty chemicals. Energy is the targeted output right now but, as biorefineries become more and more sophisticated with time, other products will be developed. In some types of biorefinery processes, food and feed production may also be incorporated (Clark and Deswarte 2008).

Consumption of nonrenewable energy resources during biorefinery processing should be minimized while the usage of biomass is maximized. Biorefinery industries are expected to develop as dispersed industrial complexes that are able to revitalize rural areas. Unlike oil refinery, which occurs in a very large plant, biorefineries will most probably encompass a whole range of different-sized installations. Several decentralized biorefineries can combine their products in order to completely utilize all biomass components, giving rise to integrated bioindustrial systems or centralized biorefineries. Biomass resources are locally available and their usage will reduce dependence on imported fossil fuels (Cherubini 2010).

5.4 TYPES OF BIOREFINERIES

Several types of biorefineries have been proposed by various researchers all over the world, some of which are discussed below.

Phase I biorefineries use any one feedstock and have single process operation capabilities thereby producing a single major product. Phase II biorefineries also process a single feedstock yet produce several products—which could be energy, chemicals, and/or materials, thus responding to market demand and price. Phase I biorefineries can be converted into phase II biorefineries if the various side streams are upgraded. Production of energy was once the main target of crude oil refining; now sophisticated process chemistry and engineering are being used to develop complex materials and chemicals that "squeeze every ounce of value" from a barrel of oil (Realff and Abbas 2004). Phase III biorefineries are a very developed or advanced type of biorefinery. They produce a variety of energy and chemical products using various types of feedstocks and incorporate several processing technologies to produce multiple products. The diversity of the products provides a high degree of flexibility to suit changing market demands and to offer various options to achieve profitability and to maximize returns. The ability to process multiple feedstocks ensures secure feedstock availability and offers these highly integrated biorefineries the possibility of selecting the most profitable combination of raw materials (de Jong et al. 2006). No commercial phase III biorefineries exist at present, but extensive research work is being carried out. Four types of phase III biorefinery systems have been identified by researchers; they are the lignocellulosic feedstock biorefinery, whole crop biorefinery, green biorefinery, and two-platform concept biorefinery. A lignocellulose feedstock biorefinery uses lignocellulosic biomass consisting of cellulose, hemicellulose, and lignin as feedstock. Through an array of processes, they are converted into a variety of energy and chemical products. In a whole crop biorefinery, the entire plant is converted into energy, chemicals, and/or materials; hence, there is maximum utilization of biomass into various products. A green biorefinery is another form of a phase III biorefinery that has been extensively studied in the European Union (EU). Wet green biomass is converted into useful products through a combination of different technologies (Andersen and Kiel 2000). The various fractions are used for producing multiple products (Clark and Deswarte 2008).

5.5 VARIOUS METHODS OF CONVERSION

The biomass conversion methods are classified into biochemical, physicochemical, or thermochemical processes. Biochemical conversions include a variety of chemical reactions catalytically mediated inside microorganisms as whole-cell biocatalysts and/or enzymes to convert fermentable feedstock substrates (e.g., monosugars) into fuels or other high-value commodities (Balat 2011). Today, bioethanol and biobutanol are being studied because they are considered to be potential replacements for transportation fuels by using biochemical processes. In chemical conversion processes, biomass is converted by means of a chemical reaction into useful products; biodiesel from edible/nonedible oil is an example.

5.6 THERMOCHEMICAL PROCESSES

Thermochemical processes are defined as the processes that are carried out in the presence of heat and at times with a catalyst. In certain cases, high pressures are also maintained during the process using an external source, or the pressure is autogenously generated. In comparison with biochemical processes, thermochemical processes occur faster (in the range of a few seconds, minutes, or hours); whereas the former takes time (in the range of days) to complete. In addition, unlike the biochemical processes, which produce lignin as a by-product and face difficulty in fermenting the C_5 sugars, thermochemical methods of conversion utilize the entire feedstock to produce value added hydrocarbons. The other advantages of thermochemical methods of conversion are that they are not feedstock specific and can mostly process even combinations of feedstocks. Pyrolysis is said to be the basis of all thermochemical methods of conversion, and it is defined as the heating of any material in the absence of oxygen.

Combustion has been carried out by mankind for ages, and it is the most widely practiced form of the thermochemical process all over the world. During combustion, biomass is openly burned with an abundant supply of oxygen, and heat is the main product obtained from the process. Cofiring is mainly followed in thermal power plants where biomass is cofired with coal to produce heat and then in turn electricity. Combustion can be carried out in grate firings, screw systems, and so forth that are fixed bed, and fluidized bed combustion also can be carried out. This process should be used as minimally as possible because it generates huge amount of CO_2 as a by-product. Researchers feel that this CO_2 is used later on by plants for new growth thereby making the process carbon neutral. Despite this fact, it is not advisable to follow this process at the initial stages because the only source of organic carbon from renewable sources of energy would go to waste in terms of CO_2. It would be better if biomass was converted to hydrocarbons, recycled as much as possible, and then—after its end of life—be burned to produce heat. This way the biomass would be efficiently utilized.

The next kind of thermochemical conversion is gasification, which is a partial oxidation process. In certain cases, gasification also is carried out in the presence of steam, but synthesis gas (which is a mixture of CO and H_2) is the product in all cases. The synthesis gas can later be converted into various products such as fuels, waxes, olefins, and so forth by Fischer Tropsch reactions. The challenge in the case of gasification processes is the tar formation during the high temperature process followed by cleaning the gas.

Pyrolysis processes generate bio-oil, biochar, and noncondensable gases as products, and the amount produced varies due to several factors. The various kinds of pyrolysis based on the residence time of vapors are slow, fast, vacuum, and flash pyrolysis. Pyrolysis is generally carried out in an inert atmosphere such as helium or nitrogen; in cases where hydrogen atmosphere is used, the process is termed hydropyrolysis. Depending on the reactor used, the pyrolysis processes are ablative, rotating cone, screw, auger, bubbling fluidized bed, or circulating fluidized bed pyrolysis.

Bio-oil produced by various processes is different owing to the reactions leading to its production due to differences in residence time, atmosphere, and the reactor used. The major components of bio-oils mostly are organic acids, esters, alcohols, ketones, phenols, aldehydes, alkenes, furfurals, sugars, and some inorganic species (Mohan et al. 2006). The bio-oil can be converted into valuable chemicals or fuels depending on the end product requirement. However, there are numerous technical bottlenecks associated with the storage of bio-oils owing to aging problems and utilization as transportation fuels because of their crude and inconsistent nature, thermal instability, and corrosive properties. As a result, several strenuous upgrading steps are required to ensure the applicability of these bio-oils as transportation fuels. Hydrodeoxygenation, catalytic cracking, emulsification, steam reforming, and chemical extraction are relevant techniques developed to improve bio-oil quality (Zhang et al. 2007; Srirangan et al. 2012).

Hydrothermal liquefaction is carried out in the presence of water that is either added externally or by using the inherent moisture of the feedstock at subcritical conditions. As the process uses water, which is an environmentally benign solvent, it is environmentally friendly. At subcritical conditions, the ionic product of water increases and it acts as a reactant as well as a catalyst, cleaving the bonds present in the biomass to yield value added products. Solvents such as methanol or ethanol have also been used in various studies to carry out the hydrothermal liquefaction of biomass at subcritical/supercritical conditions. Alkali-based catalysts such as KOH or K_2CO_3 have been used, and several researchers are designing solid catalysts for this process as well.

Carbonization has been practiced for the longest time to produce charcoal from biomass. It is carried out at temperatures of 300°C–500°C with residence time in the range of many hours to a few days. This process has been gaining interest off late, and several researchers are trying to develop processes that produce biochar efficiently. Biochar research has garnered a lot of interest due to its varied applications. Its utilization is the only way to make biomass conversion processes carbon negative. Carbonization is also being used to produce carbon nanotubes, adsorbents, catalysts/catalyst supports, and so forth from biomass.

5.7 PRODUCTS FROM BIOREFINERIES

The various products that can be obtained from a rural biorefinery using thermochemical methods of conversion have been shown in Figure 5.1.

The products of biorefinery systems can be grouped into two main categories: material products and energy products. Energy products are those products that are used because of their energy content, providing electricity, heat, or transportation service. On the other hand, material products are not used for the purpose of energy generation but for their chemical or physical properties. In some cases, a further distinction for the characterization of products is needed because some products such as biohydrogen or bioethanol might be used either as fuels or as chemical compounds in chemical synthesis. In these cases, it is necessary to identify the addressed markets (for instance, the transportation sector for H_2 and bioethanol). Concerning the chemicals,

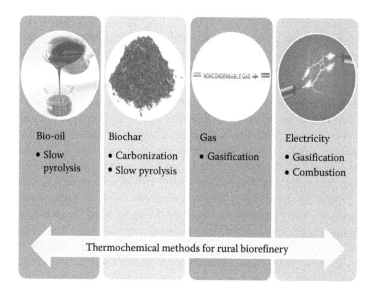

Figure 5.1   Products obtained using thermochemical methods of conversion.

this objective can be met by producing the same chemical species from biomass instead of fossil resources (e.g., phenols) or by producing a molecule having a different structure but an equivalent function. Concerning the fuels, a biorefinery must replace conventional fossil fuels (mainly gasoline, diesel, heavy oil, coal, and natural gas) with biofuels coming from biomass conversion. The most important energy products that can be produced in biorefineries are gaseous biofuels (biogas, syngas, hydrogen, biomethane), solid biofuels (pellets, lignin, charcoal), or liquid biofuels for transportation (bioethanol, biodiesel, FT-fuels, bio-oil). The most important chemical and material products that can be produced in biorefineries are chemicals (fine chemicals, specialty chemicals, bulk chemicals), organic acids (succinic, lactic, itaconic, and other sugar derivatives), polymers and resins (starch-based plastics, phenol resins, furan resins), biomaterials (wood panels, pulp, paper, cellulose), food and animal feed, fertilizers, and petrochemical feedstocks (Cherubini 2010).

5.8 USE OF GREEN CHEMISTRY IN BIOREFINERIES

In order to establish a sustainable production of biofuels and biochemicals, it is of absolute necessity to integrate green chemistry into biorefineries along with the use of low environmental impact technologies. Green chemistry employs a set of principles for the manufacture and application of products that aim at eliminating the use, or generation, of environmentally harmful and hazardous chemicals. It offers a tool kit of techniques and underlying principles that any researcher could, and should, apply when developing the next generation of biorefineries. The overall goal of green chemistry for a biorefinery is the production of genuinely green and sustainable chemical products (Clark et al. 2009). Green chemistry offers a protocol when developing biorefinery processes and may play an important role in facilitating production of commodity chemicals from biomass. Some of the protocols used while producing chemical products in a green process require that during the whole product life cycle energy demands be minimized, safer processes used, and the use of hazardous chemicals be avoided. The final product should be nontoxic, degradable into innocuous chemicals, and should produce minimal waste. Other technologies with great potential as energy efficient extraction methods are microwaves and ultrasounds (Gronnow et al. 2005). Nonconventional energies have not been exploited to their fullest extent and will surely play a key role in improving the sustainability of the process if used extensively. Due to their energy specificity, nonconventional energies have the advantage of being able to selectively cleave the bonds. In addition, the mode of heat transfer enables the elimination of size reduction steps in many cases. Microwave energy can be used to carry out pyrolysis, and some studies on microwave hydrothermal carbonization also exist. Solar gasification can be used to produce synthetic gas using biomass.

In addition, there are numerous natural polymers directly available from biomass with the potential for physical and chemical modifications. These include starches, cellulose, hemicellulose, lignin, proteins, and lipids. Modification of natural polymers is of extreme interest because they can replace fossil-derived polymers such as plastics and textiles. Lignin, which is a highly complex matrix of aromatic units, is a renewable source of aromatic compounds that are widely used in the chemical industry (Ku and Mun 2006; Cherubini 2010).

5.9 RURAL BIOREFINERIES

When production is based more and more on renewable raw materials, the processing is configured to be more efficient and eco-friendly with products that are preferably biodegradable and nontoxic. With the increasing usage of biomass, the chemical industry can produce a variety of chemicals and materials as well as energy with the lowest consumption of fossil resources as much as possible. Industrial plants require a sufficient quantity of high-grade raw materials to manufacture products. The increasing dependence of the industry on renewable raw materials poses new challenges to main producers with regard to altered product requirements because the quality standards for food, energy sources, and industrial feedstock differ considerably. Consequently, the raw materials produced must be adapted to the requirements of the respective buyers. It is necessary to look for new raw material and its potential in addition to exploiting the existing agricultural or forest residues more efficiently.

Refining raw materials at the place of production based on the concept of biorefinery is one opportunity. One such concept has been shown in Figure 5.2 where the biomass produced at the farm is being converted to fuel that is used to run the water pump. The water is being used for subsequent production of biomass. The multiple uses of raw materials, if carried out using biorefineries, will lead to higher product yields and multiple products, thereby increasing the efficiency of plant usage and the generation of fewer waste products. An advancement of this concept is to locate parts of the biorefinery in rural areas and to refine the raw materials on the farm. This advanced concept aims to make optimum use of all exploitable parts of the raw materials cultivated on farms to ensure sustainability and the cost effectiveness of their use. Building small processing plants in rural areas that process the preproducts from the regional farms can reduce the transportation costs of the raw materials and intermediate products even more and can enable the quality of the feedstock for centralized processing to be enhanced. Locating the biorefineries at the heart of rural areas can be achieved if an all-year supply of plant raw material can be supplied within a suitable radius around the farm. Due to its decentralized location, biorefiners would buy products generated at the regional farms. From this, a stable regional sales market would develop for the preliminary products generated by farmers. Industry and agriculture would be linked by this process, and this would open up new entrepreneurial structures for agriculturists. By investing in biomass conversion or biorefineries, agriculturists or farmers could evolve into suppliers of high-grade products to industry, enabling them to tap completely new markets. Refining on the farm allows very fast processing and can deliver high quality preproducts to decentralized biorefinery plants that produce the basic materials. When the biomass refining starts on the farm, the residual materials (eg. press cake)

Water from tubewell is used for farming

Crop is grown and harvested

Fuel generated by the process is used to run water pump

Thermochemical method of conversion

Residues left over on the farm are pyrolyzed/ gasified

Figure 5.2 (See color insert.) Local applications of biomass.

left over after refining can be used as animal feed or as a source of energy and, subsequently, as fertilizer. In other words, nutrients and carbon are directly used on the farm and returned to the soil, which is a vital aspect in the sustainable management of agricultural areas (Kahnt 2008). Usually, all the different users of biomass create individual raw material assessments to see if a region has the potential to supply suitable biomass for their use option. Often the fact that the various biomass potentials overlap, reducing the single use options or even making some of them unavailable, is not considered (Helming et al. 2008; Papendiek et al. 2012).

The major impediment to biomass use is the development of economically viable methods (physical, chemical, thermochemical, and biochemical) to separate, refine, and transform it into energy, chemicals, and materials (European Commission 2005). Indeed, biorefining technologies (some of which are already at a stage of commercialization, while others require further research and technological development) have to compete with processes that have been continuously improved by petrorefineries over the last 100 years (and that have a very high degree of technical and cost optimization). In particular, biorefineries will have to develop clever process engineering to deal with separation—by far the most wasteful and expensive stage of biomass conversion, and currently accounting for 60%–80% of the process cost of most mature chemical processes. The production of chemicals (e.g., succinic acid) and fuels (e.g., bioethanol) through fermentation processes, for example, generates very dilute and complex aqueous solutions, which will have to be dealt with using clean and low-energy techniques. In fact, given that so many of our carefully isolated, functionalized, and purified chemical products end up in formulations, it would also seem wise to seek methods that can convert the multicomponent systems we obtain from biomass into multicomponent formulations with the correct set of properties we require in several applications (Clark 2007).

Slow pyrolysis is carried out in reactors at temperatures of 300°C–500°C with residence times of around one to two hours. The liquid products are obtained as a result of decarboxylation, decarbonylation, dehydration reactions, and so forth. Hence, this crude organic fraction can be used as fuel oil for stationary purposes. Under optimized conditions, with certain forest residues, extraction of bio-oil can also produce several compounds that have high medicinal values. The solid products produced in this process can be used for heating purposes and other applications. When the generation of solid carbon is of utmost importance, carbonization processes may be heavily employed in decentralized units. Gasification is being deployed on a huge scale compared to the other thermochemical methods of conversion currently at the rural level. These processes have high potential in the case of rural biorefineries because they can reduce the dependence of rural population on fossil-based resources that are problematic to obtain at remote locations. This will enable them to utilize the resources available in their vicinity effectively and efficiently. The requirements for these processes are feedstock flexibility and easy operation (even by rural residents).

Among the main drivers for the use of bioenergy and bioproducts are their potential environmental benefits (e.g., carbon dioxide emission reduction, biodegradability). It is thus essential that we assess the environmental impact of all the energy and chemical products we manufacture (across their life cycles) to make sure that they are truly sustainable and that they present real (environmental and societal) advantages compared to their petroleum-derived analogues (Gallezot 2007). A major issue for biomass as a raw material for industrial product manufacture is variability. Questions of standardization and specifications will therefore need to be addressed as new biofuels, biomaterials, and bioproducts are introduced onto the market. Another major challenge associated with the use of biomass is yield. One approach to improve/modify the properties and/or yield of biomass is to use selective breeding and genetic engineering to develop plant strains that produce greater amounts of desirable feedstocks, chemicals, or even compounds that the plant does not naturally produce (Fernando et al. 2006). This essentially transfers part of the biorefining to the plant (Clark and Deswarte 2008).

In contrast with fossil resources, biomass feedstocks are composed of highly oxygenated and/or highly functionalized chemicals. From an energy point of view, this means that the calorific value of biomass is substantially lower than that of fossil fuels. It is therefore preferable to treat biomass before using it as an alternative fuel or as a source of energy. This also means that we must apply significantly different chemistries to such highly functionalized biomass so as to build these up into the valuable chemical products on which our society depends (Clark 2007). In fact, because the production of commodity and specialty functionalized chemicals from fossil resources typically requires highly energy intensive processes, biomass represents a particularly attractive alternative source of these valuable compounds (Sanders et al. 2005). Ragauskas et al. (2006) are of the opinion that the use of carbohydrates as a raw material for chemical production could potentially eliminate the need for several capital-intensive oxidative processes used in the petroleum industry (Clark and Deswarte 2008).

Opinions vary widely on the optimal size of future biorefineries. It is proposed that biorefineries will correspond to a combination of large-scale facilities (which can take full advantage of economies of scale and enjoy greater buying power when acquiring feedstocks) and small-scale plants (which can keep transport costs to an absolute bare minimum and take full advantage of available process integration technologies) (European Commission 2005). Their optimal size would depend on the nature of the feedstock processed, the location of the plant, and the technologies employed corresponding to a balance between the increasing cost of transporting pretreated biomass and the decreasing cost of processing as the size of the biorefinery increases. Proven full-scale technologies and demonstration biorefinery plants are required before commercial scale biorefineries of any size can be built (Clark and Deswarte 2008). As shown in Figure 5.3, small-scale decentralized units can be built in optimal radius and their products can be utilized on site or near the place of generation. Some of the products can also be transported to a centralized unit where they can be collectively processed into bigger units. Similar products from different decentralized refineries can be pooled at the centralized biorefinery. Figure 5.4 shows how the products from the decentralized

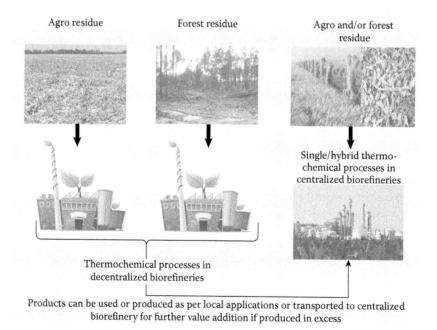

Figure 5.3 Concept of decentralized and centralized biorefineries.

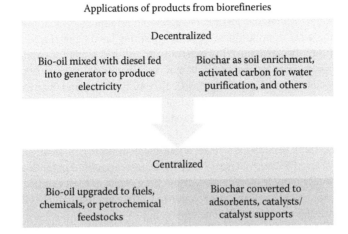

Figure 5.4 Applications of products from decentralized and centralized biorefineries.

biorefineries can be used locally or can be transported and upgraded in a centralized biorefinery to obtain different value added products. The main objective is to design the supply chain to optimize and minimize the transportation costs and capital costs of centralized and decentralized units without compromising on the profits from the various products.

5.10 CHALLENGES AND OPPORTUNITIES

Although biologically derived energy carriers offer an alternative to traditional petroleum derived fuels, these biofuels may be uneconomical and energy insufficient (Solomon 2010). Accordingly, several challenges in sustainable development must be overcome before commercial-scale production of biofuels can be realized. Although the renewable energy encompasses a myriad of different environmentally benign forms (e.g., wind, solar, geothermal, and biohydrogen), the perception is that liquefied biofuels (e.g., bioethanol, biodiesel, and biobutanol) seem to be the most realistic options for large-scale production within the foreseeable future—primarily because these fuels can be readily blended with petroleum-derived fuels and they are more or less compatible with the existing fuel transportation and refueling infrastructures (Borjesson 2009; Amaro et al. 2011; Lutke-Eversloh and Bahl 2011). The necessary push to promote biofuels can only be available if favorable biofuel policies are framed that enable the proliferation of biofuel companies. This will lead to increase in the funding for research and development in this field as well. Several countries have biofuel policies in place such as the U.S. Environmental Protection Agency and the EU biofuel policy (which mandates the blending of biofuels with fossil fuels, enabling a smooth transition into a renewable resources-based economy). Market analysts argue that in order to prevent an unparalleled increase in the number of hungry people, governments should accelerate transition from the first-generation feedstock to the second-generation through policy and regulatory actions (Bindraban et al. 2009; Prabhakar and Elder 2009). Although the applicative potential of several liquid bio-fuels has been extensively explored, all biofuels cost much more than fossil fuels—primarily due to expensive feedstock, which limits their economic feasibility. However, the problem can potentially be alleviated through either producing biofuels in developing nations or by developing production schemes utilizing the second-generation feedstock and their associated waste residues or even the third-generation feedstock, such as microalgae (Srirangan et al. 2012). It is also understood that land and water availability and land-use practices are key limiting factors for the large-scale production of biofuels. Land-use change for the cultivation of energy crops will have significant consequences on the employment and income of regional populations, global food security, and on the biodiversity of ecological communities. Several questions have been raised on the availability of arable land in the future because utilizing existing agricultural lands for energy crop cultivation might result in food shortages. Nevertheless, it may be feasible to convert abandoned, idle, and marginal lands into usable lands for energy crop cultivation (Sagar and Kartha 2007; Bindraban et al. 2009; Phalan 2009). As freshwater is a critical limiting resource (only ~0.6% of the Earth's surface is covered by freshwater), irrigation associated with the extensive farming practices for the production of energy crops may adversely affect the aquatic systems of freshwater. In addition to polluting water bodies, there are also concerns about the availability of freshwater for sustainable production of energy crops. Problems arising from water shortage could potentially be alleviated by selecting energy crops requiring less irrigation, cultivating crops in high rainfall zones, and by utilizing feedstock that can be cultured in saline water (e.g., marine microalgae) (Phalan 2009; Zhou and Thomson 2009; Srirangan et al. 2012). The emergence of biofuel markets is expected to directly affect the livelihood and economy of rural communities, given that almost all feedstocks are cultivated in rural areas (Solomon 2010). Most economists support the notion that global biofuel programs will generally contribute to the sustainable livelihood of agricultural laborers by increasing employment rates in most rural communities as a large portion of feedstock cultivation and refinery processing involves manual labor. Ideally, biofuels should mitigate the environmental impacts associated with the use of petroleum fuels, although several studies indicate that large-scale cultivation of biofuel crops may potentially damage the natural environment. To meet the global demand of biofuels, major strategies should be implemented to significantly increase agricultural productivity without any environmental impacts such as loss of biodiversity, the introduction of invasive energy crops, releasing agro-contaminants (e.g., pesticides, herbicides, and fertilizers) into aquatic systems, and increasing global emissions of NO_x and CO_2 (Sagar and Kartha 2007; Bindraban et al. 2009; Phalan 2009; Zhou and Thomson 2009; Solomon 2010). Some of the concerns associated with monocropping can, however, be mitigated through agrotechnical innovations for improving agricultural practices. For instance, competition with food crops can be avoided through the cultivation of certain second-generation energy crops with low agrochemical demands on marginal lands, such as jatropha (Zhou and Thomson 2009). Large-scale cultivation of microalgae, which can produce third-generation biofuels, can be used to reduce the competition for land. Studies have reported that algae can be grown on both wastewater and saline water. Biofuels derived from certain energy crops may also have emissions similar to petroleum fuels (i.e., NO_x and CO_2), thus questioning their greenhouse gas benefits. The use of crop residues is projected to decrease greenhouse gas emissions by up to ~50% (Cherubini and Ulgiati 2010).

Biorefineries allow more efficient use of biomass to save resources. The opportunities presented by implementation in a rural area are diverse. The cost effectiveness of the biorefinery concept depends to a great extent on the costs involved in transporting the biomass. If a large part of the biorefinery process is implemented in decentralized units, transportation costs are minimized, making biomass processing even more efficient. The products obtained in the decentralized units can be transported over longer distances to the centralized units with minimal transportation costs and associated greenhouse gas emissions. The production of preproducts on the farm using decentralized units results in better quality industrial products and supports the sustainable development of rural areas. Refining biomass can constitute additional income for agriculturists, as is already the case in the use of biomass for energy production. The refining of previously unused or only partially used raw materials can facilitate stable income in a promising market through the close link to biorefineries. After the refining of raw materials, substances are left that can be used to close material cycles on the farm. The end of life materials can also be burned after it is determined that they cannot be used anymore. However, the potential of relocating biomass refineries to rural areas has not yet been sufficiently studied. It is not clear whether the participation of agriculturists will help in the evolution of a new method to describe raw material potentials more reliably. Whether or not on-farm refining can really pay off for the agriculturists still needs to be investigated and also if it is possible to cultivate and refine enough high-grade biomass to meet the future demands of various industries (Papendiek et al. 2012).

For this purpose, long-term strategies that recognize the potential of local renewable resources should be developed. Of paramount importance will be the deployment of biorefineries (of various sizes and shapes) that can convert a variety of biofeedstocks into power, heat, chemicals, and other valuable materials, maximizing the value of the biomass and minimizing waste. These integrated facilities will most likely employ a combination of physical, chemical, biotechnological, and thermochemical technologies, which ought to be efficient and follow green chemistry principles so as to minimize environmental footprints and to ensure the sustainability of all products generated (a cradle-to-grave approach). Local pretreatment of (low bulk density and often wet) biomass will be critical to the development of a sustainable infrastructure capable of working with significant quantities of raw material. Thus, specific attention should be given to the development of these (local) processes. The challenge of the next decade will be to develop demonstration plants, which will require cross-sector collaborations and which will attract the necessary investors required for the construction of full-scale biorefineries (Clark and Deswarte 2008).

The utilization of biomass feedstock appears to be a genuine solution to sustainable production of clean energy carriers in the future. However, various technical issues remain to be addressed—in particular, the recalcitrant nature of the second-generation feedstock and which well-established thermochemical technologies may seem convenient and advantageous to apply. The biorefinery process should be systematically analyzed, modeled, and optimized based on a number of factors—such as feedstock selection, pretreatment method, reaction and separation process, energy integration, water recycling, and coproduct production—to ensure its economical efficacy (Srirangan et al. 2012).

A key driver for the development and implementation of biorefineries is the growth in demand for energy, fuels, and chemicals. Accordingly, the aim of research is in developing new technologies and creating novel processes, products, and capabilities to ensure the growth is sustainable from economic, environmental, and social perspectives. Further research and technology adoption will indicate which new products and processes contribute to more sustainable performances compared to conventional fossil-based systems. The term sustainability itself needs a common definition and criteria for its evaluation. This will be necessary for communication with nongovernmental organizations (NGOs), the general public, regulators, and with policymakers about, for example, CO_2 reductions. When developing chemistry for future biorefineries, it is important that the methods and techniques used minimize impact to the environment and that the final products are truly green and sustainable. The use of sustainable feedstock is not enough to ensure a prosperous future for later generations; protection of the environment using greener methodologies is also required (Cherubini 2010).

The biorefinery industry has developed rapidly in the last few years. In addition to bioethanol and biodiesel, several bio-based products are already or will soon be in commercial production, replacing petroleum-based products in the market. With continuing developments and advances in new energy crops, aquacultures, synthetic biology for cell engineering, and conversion technologies, biorefining will increasingly play a more important role in the supply of energy, fuels, and chemicals for sustainable economic growth with minimal or no negative impact on the environment. However, there are many challenges facing the biorefinery industry. Current infrastructures built on petroleum-based manufacturing and products may not be relevant to bio-based manufacturing and products. For example, a supply of biomass feedstock may be seasonal and limited by a geographical area. Also, the relatively low density of biomass would hinder its storage and transportation, thus severely limiting its ability to support a megascale biorefinery that could benefit from the economy of scale. Not all of the current petroleum-based chemicals can be economically produced from biomass (Yang and Yu 2013). Hence, an optimal strategy has to be developed upon which a biorefinery would be based to keep it cost effective and energy efficient.

ACKNOWLEDGMENTS

The authors thank the Director, CSIR–Indian Institute of Petroleum, Dehradun, for his constant encouragement and support. RS thanks CSIR, New Delhi, for providing senior research fellowship. The authors thank CSIR for funding in the form of XII Five-Year Plan project (CSC0116/ BioEn) and the Ministry of New and Renewable Energy (MNRE), India, for providing financial support.

REFERENCES

Abdeshanian, P., Dashti, M.G., Kalil, M.S., and Yusoff, W.M.W. 2010. Production of biofuel using biomass as a sustainable biological resource. *Biotechnology* 9: 274–82.

Alexander, B.R., Mitchell, R.E., and Gur, T.M. 2012. Experimental and modeling study of biomass conversion in a solid carbon fuel cell. *J Electrochem Soc* 159: B347–54.

Amaro, H.M., Guedes, A.C., and Malcata, F.X. 2011. Advances and perspectives in using microalgae to produce biodiesel. *Appl Energy* 88: 3402–10.

Andersen, M. and Kiel, P. 2000. Integrated utilisation of green biomass in the green biorefinery. *Ind Crops Prod* 11: 129–37.

Balat, M. 2011. Production of bioethanol from lignocellulosic materials via the biochemical pathway: A review. *Energy Convers Manag* 52: 858–75.

Berndes, G., Hoogwijk, M., and van den Broek, R. 2003. The contribution of biomass in the future global energy supply: A review of 17 studies. *Biomass Bioenergy* 25: 1–28.

Bindraban, P.S., Bulte, E.H., and Conijn, S.G. 2009. Can large-scale biofuels production be sustainable by 2020? *Agr Syst* 101: 197–9.

Borjesson, P. 2009. Good or bad bioethanol from a greenhouse gas perspective-what determines this? *Appl Energy* 86: 589–94.

Cherubini, F. 2010. The biorefinery concept: Using biomass instead of oil for producing energy and chemicals. *Energy Convers Manag* 51: 1412–21.

Cherubini, F. and Ulgiati, S. 2010. Crop residues as raw materials for biorefinery systems—A LCA case study. *Appl Energy* 87: 47–57.

Clark, J.H. 2007. Green chemistry for the second generation biorefinery—Sustainable chemical manufacturing based on biomass. *J Chem Technol Biotechnol* 82: 603–9.

Clark, J.H., Budarin, V., Deswarte, F.E.I., Hardy, J.J.E., Kerton, F.M., Hunt, A.J., Luque, R., et al. 2006. Green chemistry and the biorefinery: A partnership for a sustainable future. *Green Chem* 8: 853–60.

Clark, J.H. and Deswarte, F.E.I. 2008. The biorefinery concept—An integrated approach. In *Introduction to Chemicals from Biomass*, eds. J.H. Clark and F.E.I. Deswarte. John Wiley, Chichester, UK.

Clark, J.H., Deswarte, F.E.I., and Farmer, T.J. 2009. The integration of green chemistry into future biorefineries. *Biofuels Bioprod Bioref* 3: 72–90.

de Jong, E., van Ree, R., van Tuil, R., and Elbersen, W. 2006. Biorefineries for the chemical industry—A Dutch point of view. In *Biorefineries—Biobased Industrial Processes and Products. Status Quo and Future Directions*, eds. B. Kamm, M. Kamm, and P. Gruber. Wiley-VCH, Weinheim, Germany.

Deswarte, F.E.I., Clark, J.H., Wilson, A.J., Hardy, J.J.E., Marriott, R., Chahal, S.P., Jackson, C., et al. 2007. Toward an integrated straw-based biorefinery. *Biofuels Bioprod Bioref* 1: 245–54.

European Commission. 2005. Biomass—Green energy for Europe. http://www.managenergy.net/download/ r1270.pdf

Fernando, S., Adhikari, S., Chandrapal, C., and Murali, N. 2006. Biorefineries: Current status, challenges, and future direction. *Energy Fuel* 20: 1727–37.

Gallezot, P. 2007. Process options for converting renewable feedstocks to bioproducts. *Green Chem* 9: 295–302.

Gronnow, M.J., White, R.J., Clark, J.H., and Macquarrie, D.J. 2005. Energy efficiency in chemical reactions: A comparative study of different reaction techniques. *Org Process Res Dev* 9: 516–18.

Haberl, H., Beringer, T., Bhattacharya, S.C., Erb, K.-H., and Hoogwijk, M. 2010. The global technical potential of bio-energy in 2050 considering sustainability constraints. *Curr Opin Environ Sustain* 2: 394–403.

Helming, K., Pérez-Soba, M., and Tabbush, P. 2008. *Sustainability Impact Assessment of Land Use Changes*. Springer-Verlag, Berlin.

Hess, J.R., Thompson, D.N., Hoskinson, R.L., Shaw, P.G., and Grant, D.R. 2003. Physical separation of straw stem components to reduce silica. *Appl Biochem Biotechnol* 105–108: 43–51.

IEA (International Energy Agency). 2009. IEA bioenergy task 42 biorefinery. http://www.iea-bioenergy.task42-biorefineries.com/en/ieabiorefinery/Publications-2.htm#

Kadam, K.L., Forrest, L.H., and Jacobson, W.A. 2000. Rice straw as a lignocellulosic resource: Collection, processing, transportation, and environmental aspects. *Biomass Bioenerg* 18: 369–89.

Kahnt, G. 2008. *Leguminosen im konventionellen und ökologischen Landbau*. DLG Verlags -GmbH, Frankfurt am Main.

Kamm, B., Kamm, M., Gruber, P.R., and Kromus, S. 2006. Biorefinery systems—An overview. In *Biorefineries—Industrial Processes and Products (Status Quo and Future Directions)*, eds. B. Kamm, P.R. Gruber, and M. Kamm, Vol. 1. Wiley-VCH, Weinheim.

Katzen, R. and Schell, D.J. 2006. Lignocellulosic feedstock biorefinery: History and plant development for biomass hydrolysis. In *Biorefineries—Industrial Processes and Products (Status Quo and Future Directions)*, eds. B. Kamm, P.R. Gruber, and M. Kamm, Vol. 1. Wiley-VCH, Weinheim.

Klass, D.L. 1998. *Biomass for Renewable Energy, Fuels, and Chemicals*. Academic Press, San Diego, CA.

Ku, C.S. and Mun, S.P. 2006. Characterization of pyrolysis tar derived from lignocellulosic biomass. *J Ind Eng Chem* 12: 853–6.

Lucian, L.A., Argyropoulos, D.S., Adamopoulos, L., and Gaspar, A.R. 2007. Chemicals, materials, and energy from biomass: A review. In *ACS Symposium Series 954. Materials, Chemicals, and Energy for Forest Biomass*, ed. D.S. Argyropoulos, pp. 2–30. American Chemical Society, Washington, DC.

Lutke-Eversloh, T. and Bahl, H. 2011. Metabolic engineering of *Clostridium acetobutylicum*: Recent advances to improve butanol production. *Curr Opin Biotechnol* 22: 634–47.

Mohan, D., Pittman, C.U., and Steele, P.H. 2006. Pyrolysis of wood/biomass for bio-oil: A critical review. *Energy Fuel* 20: 848–89.

Papendiek, F., Ende, H.P., Steinhardt, U., and Wiggering, H. 2012. Biorefineries: Relocating biomass refineries to the rural area landscape. *Landsc Online* 27: 1–9.

Phalan, B. 2009. The social and environmental impacts of biofuels in Asia: An overview. *Appl Energy* 86(Suppl. 1): S21–9.

Prabhakar, S.V.R.K. and Elder, M. 2009. Biofuels and resource use efficiency in developing Asia: Back to basics. *Appl Energy* 86(Suppl. 1): S30–6.

Ragauskas, A.J., Williams, C.K., Davison, B.H., Britovsek, G., Cairney, J., Eckert, C.A., Frederick, W.J., Jr., et al. 2006. The path forward for biofuels and biomaterials. *Science* 311: 484–89.

Realff, M.J. and Abbas, C. 2004. Industrial symbiosis—Refining the biorefinery. *J Ind Ecol* 7: 5–9.

Sagar, A.D. and Kartha, S. 2007. Bioenergy and sustainable development? *Annu Rev Environ Resour* 32: 131–67.

Sanders, J., Scott, E., and Mooibroek, H. 2005. Biorefinery, the bridge between agriculture and chemistry. *14th European Biomass Conference*. http://www.biorefinery.nl/fileadmin/biorefinery/docs/sanders_br_the_bridge_between_agriculture_and_chemistry.pdf

Solomon, B.D. 2010. Biofuels and sustainability. *Ann N Y Acad Sci* 1185: 119–34.

Srirangan, K., Akawi, L., Moo-Young, M., and Chou, C.P. 2012. Towards sustainable production of clean energy carriers from biomass resources. *Appl Energ* 100: 172–86.

Thorsell, S., Epplin, F.M., Huhnke, R.L., and Taliaferro, C.M. 2004. Economics of a coordinated biorefinery feedstock harvest system: Lignocellulosic biomass harvest cost. *Biomass Bioenerg* 27: 327–37.

Weiland, P. 2010. Biogas production: Current state and perspectives. *Appl Microbiol Biotechnol* 85: 849–60.

Worldwatch-Institute. 2007. *Biofuels for Transport: Global Potential and Implications for Sustainable Energy and Agriculture*. Earthscan, London.

Yang, S.T. and Yu, M. 2013. Integrated biorefinery for sustainable production of fuels, chemicals, and polymers. In *Bioprocessing Technologies in Biorefinery for Sustainable Production of Fuels, Chemicals, and Polymers*, eds. S.T. Yang, H.A. El-Enshasy, and N. Thongchul, 1st ed. John Wiley & Sons, Inc., Hoboken, NJ, USA, 1–26.

Zhang, Q., Chang, J., Wang, T., and Xu, Y. 2007. Review of biomass pyrolysis oil properties and upgrading research. *Energy Convers Manage* 48: 87–92.

Zhou, A. and Thomson, E. 2009. The development of biofuels in Asia. *Appl Energy* 86(Suppl. 1): S11–20.

CHAPTER 6

Conversion of Holocellulose-Derived Polyols to Valuable Chemicals Using High-Temperature Liquid Water and High-Pressure Carbon Dioxide

Masayuki Shirai,[1] **Osamu Sato,**[2] **and Aritomo Yamaguchi**[2,3]
[1]Department of Chemistry and Bioengineering, Faculty of Engineering, IWATE University, Morioka, Japan
[2]Research Center for Compact Chemical System, National Institute of Advanced Industrial Science and Technology (AIST), Sendai, Japan
[3]Japan Science and Technology agency (JST), PRESTO, Sendai, Japan

CONTENTS

6.1 INTRODUCTION

Using numerous industrial processes, which have been developed in the twentieth century, various useful chemical products are manufactured from petroleum. In pursuing the efficient usage of fossil fuel resources, both as a raw material and as an energy source, process technology for their mass production has been successfully developed over the past several years. Their mass consumption was also encouraged for economic development. On the contrary, large-scale disposal of used products and emission of carbon dioxide through incineration have become problematic for the environment and for human beings. To make current activities sustainable, in addition to conventional fossil resources, a technology that uses organic wastes as a supplementary resource for production of raw chemical materials should be developed. Also, new processes would require that the raw materials and their by-products cause less stress to the human body and to the environment, while also having production efficiency equal to or greater than the current petroleum-based system.

Water is the safest chemical for humans and the environment; however, it can be used as a reaction medium under high-temperature and high-pressure conditions with the change in physical properties. The dielectric constant value of water at room temperature is 78, and it decreases with increasing temperature (Figure 6.1), indicating that organic compounds can be soluble in water around 523 K.

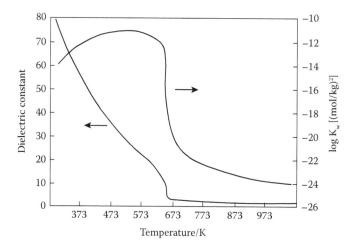

Figure 6.1   Dielectric constant and ionic product of water at 25 MPa.

The ionic product value of water is 10^{-14} at room temperature. It increases with rising temperature to around 10^{-11} at around 523 K (Figure 6.1), and then it decreases with further increase in temperature. Although still neutral, the concentration values of both protons and hydroxide ions in water at around 523 K are about 30 times higher than those at room temperature. These two properties show that acidic and basic reactions of organic compounds are expected to proceed in high-temperature water.

The addition of high-pressure carbon dioxide could provide more acidic reaction media than high-temperature water alone. Carbonic acid is generated when high-pressure carbon dioxide is introduced into high-temperature water (Equation 6.1) and provides protons and carbonic acid anions. The combination of high-temperature liquid water and high-pressure carbon dioxide can be used as a reaction medium with a higher acid catalyst capability than high-temperature water alone due to the protons from carbonic acid. In a typical acid catalyst reaction using an inorganic acid, the usage of base is indispensable for neutralization followed by separation and removal of by-product salts; however, the carbon dioxide molecules in the system can be removed easily by returning the solution to normal pressure after reaction, using high-temperature water and high-pressure carbon dioxide reaction systems. As a result, there is no need to use base for neutralization and no need to treat the by-product salt, providing a distinct advantage of this method.

$$H_2O + CO_2 \rightarrow H_2CO_3 \rightarrow H^+ + HCO_3^- \rightarrow 2H^+ + CO_3^{2-} \tag{6.1}$$

This chapter reports the development of environmentally benign techniques to convert biomass-related compounds, especially hollocellulose-derived polyols, to valuable cyclic ethers using high-temperature liquid water and high-pressure carbon dioxide [1] (Figure 6.2).

6.1.1 Dehydration of Polyalcohol in High-Temperature Liquid Water and in High-Pressure Carbon Dioxide

Herbaceous biomass is an organic polymer made up of carbon, oxygen, and hydrogen, and its partial hydrolysis can produce various polyalcohols. Cyclic ethers obtained by dehydrating polyalcohols are useful as raw functional polymers, intermediates for food additives and drugs. Dehydration of alcohols typically progresses with inorganic acid; however, we report here that the dehydration of polyalcohols can also be accomplished using high-temperature water and that the rate of reaction increases under high-pressure carbon dioxide.

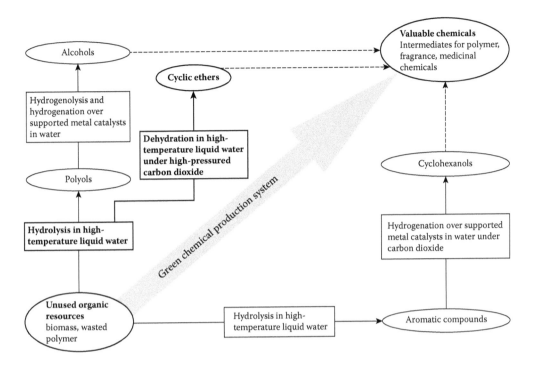

Figure 6.2   Synthesis of valuable chemicals from unused resources using water and carbon dioxide media.

Dehydration was carried out in a batch reactor made of a SUS 316 tube and the reaction temperature was controlled by salt or sand baths [2].

Figure 6.3 shows the results of the treatment of an aqueous solution of 1,2,5-pentanetriol (1.0 mol/dm^3) [3]. Tetrahydrofurfuryl alcohol and 3-hydroxytetrahydropyran were generated when an aqueous solution of 1,2,5-pentanetriol was treated in the reactor at 573 K without adding any acid. The material balance was more than 95%, indicating that intramolecular dehydration was the main reaction, and polymerization reactions rarely proceeded. The initial rate of the dehydration reaction increased with an introduction of carbon dioxide at high pressure; however, the final yield values did not depend on the carbon dioxide pressure. Also, the aqueous solution of tetrahydrofurfuryl alcohol was treated at 573 K, generating 1,2,5-pentanetriol and 3-hydroxytetrahydropyran [4]. These results indicate that carbon dioxide acted as an acid catalyst for the dehydration reactions and shortened the time to reach chemical equilibrium.

The sorbitol dehydration in water at 523 K without adding any acid catalyst is shown in Figure 6.4; 1,4-anhydrosorbitol and 1,5-anhydrosorbitol were produced by the monomolecular dehydration of sorbitol [5]; isosorbide was produced by the stepwise dehydration of 1,4-anhydrosorbitol. The dehydration rate of sorbitol to 1,4-anhydrosorbitol was faster than that of 1,4-anhydrosorbitol to isosorbide; therefore, 1,4-anhydrosorbitol could be obtained as an intermediate product.

We also studied the reaction profiles of several types of diols and triols in high-temperature liquid water. Acetol was obtained from the aqueous solution of glycerol [6], and cyclic ethers having a five-member ring were mainly obtained from polyols having more than three carbon atoms. Figure 6.5 summarizes the dehydration rates for various polyalcohols in water at 573 K and under 17.7 MPa of carbon dioxide pressure [7]. Dehydration of polyalcohols proceeded in high-temperature liquid water and their initial reaction rates were enhanced by the addition of carbon dioxide. Here, it should be noted that the enhancement in reaction rates induced by carbon dioxide was dependent on the structure of polyalcohol in terms of degree (primary, secondary, and tertiary) of alcohols.

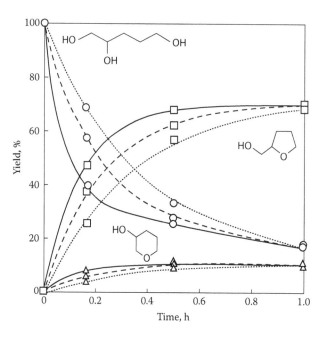

Figure 6.3   Dehydration of 1.0 mol/dm³ 1,2,5-pentanetriol (○) to tetrahydrofurfuryl alcohol (□) and 3-hydroxytetrahydropyran (△) in water at 573 K under 0 (dotted line), 17.7 (dashed line), and 26.6 MPa (solid line) of carbon dioxide pressure.

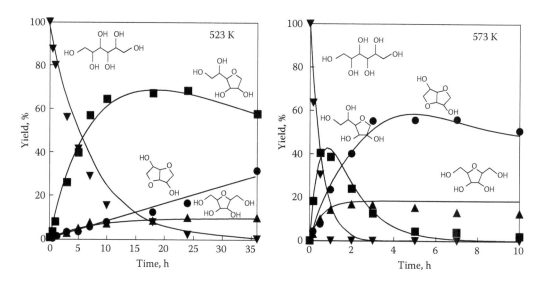

Figure 6.4   Dehydration of 1.0 mol/dm³ sorbitol (▼) to 1,4-anhydrosorbitol (■), 2,5-anhydrosorbitol (▲), and isosorbide (●) in water at 523 and 573 K.

Figure 6.6 shows the initial dehydration rates under argon and high-pressure carbon dioxide for an aqueous solution of 1,4-butanediol, with two primary hydroxyl groups, a secondary 2R,5R-hexanediol, and tertiary 2,5-dimethyl-2,5-hexanediol [8]. The initial rate of reaction for dehydration in high-temperature water greatly increased for all the substrates with the addition of carbon dioxide and in the following order: 1,4-butanediol < 2R,5R-hexanediol < 2,5-dimethyl-2,5-hexanediol.

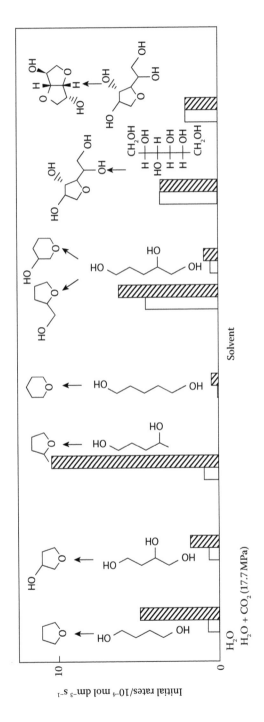

Figure 6.5 Initial intramolecular dehydration rate of 1.0 mol/dm³ polyalcohol at 573 K in water under 0 (□) and 17.7 MPa (▨) of carbon dioxide.

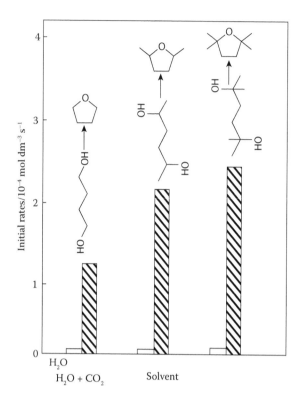

Figure 6.6 Initial intramolecular dehydration rates of polyalcohol for dehydration of 1,4-butanediol and 2R,5R-hexanediol and 2,5-dimethyl-2,5-hexanediol. For dehydration of 1,4-butanediol and 2R,5R-hexanediol, reaction conditions were: initial concentration, 1.0 mol/dm³, temperature, 523 K, carbon dioxide pressure, 0 MPa (□), 16.2 MPa (▨). For dehydration of 2,5-dimethyl-2,5-hexanediol, initial concentration, 0.5 mol/dm³, temperature 473 K, carbon dioxide pressure 0 MPa (□), 14.6 MPa (▨).

The intramolecular dehydration rate of 2R,5R-hexanediol, with two secondary hydroxyl groups, was greater than that of primary 1,4-butanediol. This is because the secondary alcohol has a greater basicity than the primary alcohol; protons from carbonic acid would attach more easily to the secondary alcohol in high-temperature liquid water in the presence of a carbonic acid. Dehydration reactions also could take place rapidly for 2,5-dimethyl-2,5-hexanediol, of which contain tertiary hydroxyl groups because a carbocation is generated by the proton from carbonic acid formed from carbon dioxide and water.

In intramolecular dehydration of a polyalcohol with two secondary hydroxyl groups in high-temperature water, a stereoselectivity could be shown for the production of the corresponding cyclic ether. The reaction profiles of 2R,5R-hexanediol and 2S,5S-hexanediol in liquid water at 523 K as well as with added carbon dioxide were studied [9]; 2,5-dimethyltetrahydrofuran was obtained from 2R,5R-hexanediol in 523 K water, and cis-selectivity was 90% (Figure 6.7, left). By the addition of 16.2 MPa of carbon dioxide, the dehydration rate increased and the cis-selectivity was maintained. We also checked the dehydration of 2S,5S-hexanediol and confirmed that the behavior was nearly equal to 2R,5R-hexanediol, and 2,5-dimethyltetrahydrofuran is obtained at 90% with cis-selectivity (Figure 6.7, right). The cis-trans isomerization reaction of 2,5-dimethyltetrahydrofuran did not take place at 523 K in water with 16.2 MPa carbon dioxide, indicating that the high cis-selectivity for 2,5-hexanediol dehydration reaction would be explained by the S_N2 reaction mechanism in which a proton from carbonic acid adds to one of the hydroxyl groups of hexanediol (without forming

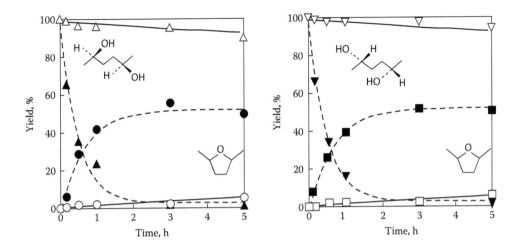

Figure 6.7 Dehydration of 1.0 mol/dm³ of 2R,5R-hexanediol (△, ▲) and 2S,5S-hexanediol (▽, ▼) to 2,5-dimethyltetrahydration (●, ○, ■, □) in water (△, ▽, ○, □) and in water with 16.2 MPa carbon dioxide (▲, ▼, ●, ■) at 523 K.

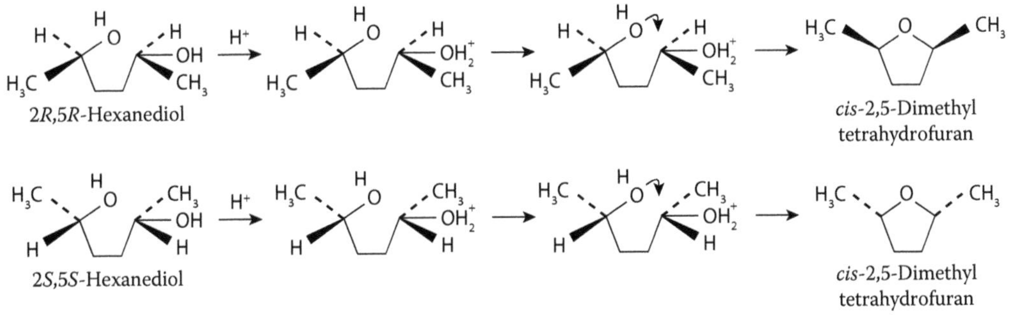

Figure 6.8 Reaction mechanism of 2,5-hexanediol in liquid water under carbon dioxide at 523 K.

a carbocation), and the other hydroxyl group attacks the carbon atom with the added proton from behind, causing the water molecule to detach as a leaving group (Figure 6.8). The dehydration of 2R,5R-hexanediol proceeded at room temperature when sulfuric acid was used as an inorganic catalyst; however, the *cis*-selectivity was only 50% because S_N1 dehydration proceeded in the presence of sulfuric acid. Dehydration of hexanediol using a combination of high-pressure carbon dioxide and high-temperature water was slower than when using an inorganic acid, but it is advantageous for stereoselectivity.

6.2 CONCLUSION

We showed that high-temperature liquid water around 523 K is an acidic reaction medium for the dehydration of polyols, which could be obtained by partial decomposition of hollocellulose, and the initial dehydration rates of polyols to the corresponding cyclic ethers were enhanced by the addition of high-pressure carbon dioxide in a batch reaction system. The development of flow-type reactors could provide continuous production of commercially useful cyclic ethers.

REFERENCES

1. M. Shirai. Recovery of chemicals from unused resources using water and carbon dioxide (Japanese). *Kagaku to Kyoiku*, 60(9) (2012): 386–387.

2. O. Sato, K. Arai, and M. Shirai. Decomposition behavior of poly(ethylene-2,6-naphthalene dicarboxylate) in high temperature water. *Fluid Phase Equilibria*, 228–229 (2005): 523–525.

3. A. Yamaguchi, N. Hiyoshi, O. Sato, and M. Shirai. Dehydration of triol compounds in high-temperature liquid water under high-pressure carbon dioxide. *Topics in Catalysis*, 53(7–10) (2010): 487–491.

4. A. Yamaguchi, N. Hiyoshi, O. Sato, K. K. Bando, Y. Masuda, and M. Shirai. Thermodynamic equilibria between polyalcohols and cyclic ethers in high-temperature liquid water. *Journal of Chemical & Engineering Data*, 54(9) (2009): 2666–2668.

5. A. Yamaguchi, N. Hiyoshi, O. Sato, and M. Shirai. Sorbitol dehydration in high temperature liquid water. *Green Chemistry*, 13(4) (2011): 873–881.

6. A. Yamaguchi, N. Hiyoshi, O. Sato, C. V. Rode, and M. Shirai. Enhancement of glycerol conversion to acetol in high-temperature liquid water by high-pressure carbon dioxide. *Chemistry Letters*, 37(9) (2008): 926–927.

7. A. Yamaguchi, N. Hiyoshi, O. Sato, K. K. Bando, and M. Shirai. Enhancement of cyclic ether formation from polyalcohol compounds in high temperature liquid water by high pressure carbon dioxide. *Green Chemistry*, 11(1) (2009): 48–52.

8. A. Yamaguchi, N. Hiyoshi, O. Sato, and M. Shirai. Cyclization of alkanediols in high-temperature liquid water with high-pressure carbon dioxide. *Catalysis Today*, 185(1) (2012): 302–305.

9. A. Yamaguchi, N. Hiyoshi, O. Sato, and M. Shirai. Stereoselective intramolecular dehydration of 2,5-hexanediol in high-temperature liquid water with high-pressure carbon dioxide *ACS Catalysis*, 1(1) (2011): 67–69.

CHAPTER 7

Utilization of Ionic Liquids for the Processing of Biopolymers

Tushar J. Trivedi[1] and Arvind Kumar[1,2]
[1]AcSIR, CSIR–Central Salt and Marine Chemicals Research Institute, Bhavnagar, India
[2]Salt and Marine Chemicals Division, CSIR–Central Salt and Marine
 Chemicals Research Institute, Bhavnagar, India

CONTENTS

7.1 INTRODUCTION

The vast majority of industrial chemical reactions are performed either in the gas phase, with a solid catalyst, or in the solution phase, where the solvent fulfills several functions during a chemical reaction. Solvents dissolve the reactant, which facilitates collisions that must occur in order to transform the reactants to products. The solvent also provides a means of temperature control, either to increase the energy of the colliding particles so that they will react more quickly or to absorb heat that is generated during an exothermic reaction [1]. The selection of the appropriate solvent is guided by theory and experience. Generally, a good solvent should meet the following criteria: (a) it should dissolve the reactants; (b) it should be inert to reaction conditions; (c) it should have an appropriate boiling point so that the solvent does not vaporize but should still be removed by distillation or vaporization; and (d) it should be easy to remove the product from the solvent.

In general, solvents can be classified according to their type of intermolecular interaction. There are three main group of solvents arranged according to the persisting intermolecular interaction, namely: (1) molecular liquids (solid melts with molecules having intermolecular interactions such as a dipole–dipole interaction); (2) ionic liquids (solid melts that contain only ionic interactions); and (3) atomic liquids (low-melting metals such as liquid mercury or liquid sodium, where there are only weak "metallic" bonds between the atoms) [2]. The majority of chemical reactions are performed exclusively in molecular solvents. The chemical industry is under considerable and ever-increasing pressure to reduce the detrimental impact of volatile organic compounds (VOCs), which are hazardous and cause environment pollution because of their volatile and flammable nature and, in some cases, due to the formation of explosive peroxides. Chlorinated hydrocarbons in conjunction with crucial environmental concerns, such as atmospheric emissions and contamination of water, cause dangerous chronic exposure. All these concerns restricted the use of VOCs and raised the importance of "green" solvents as alternatives to these environmentally hazardous solvents that can drastically reduce waste solvent production and hence reduce their environmental impact. The most prevalent of among these new green systems are supercritical fluids (such as supercritical water and carbon dioxide) [3,4], solvent-less processes [5,6], fluorous techniques, such as the use of perflurocarbons and fluorous biphasic catalysis [7,8], and ionic liquids (ILs) [9]. ILs (salts composed entirely of ions which are liquid below 100°C) [10] are a relatively new class of solvents that are becoming increasingly popular as viable replacements for molecular solvents because of their favorable inherent properties. The reputation of ILs being "green" solvents is largely because of their physicochemical properties and their nonvolatility in relation to classic VOCs. More detailed information about the physical properties of ILs can be found in *Ionic Liquids in Synthesis* by Wasserscheid and Welton [10]. The important properties of ILs, such as high thermal stability and high solvating ability, are useful for biopolymer processing.

Biopolymers are polymers produced by living organisms. Similar to synthetic polymers, biopolymers contain monomeric units that are covalently bonded to form larger structures. There are three main classes of biopolymers, which are classified according to the monomeric units used and the structure of the biopolymer formed: polynucleotides, polypeptides, and polysaccharides. Biopolymers, mainly the polysaccharides, can be extracted from the inedible biomass that is reproduced in abundance, year after year, by means of natural solar energy. Natural polysaccharides, such as cellulose, starch, chitosan, chitin, agarose, and so on, have recently received much attention for use as new green and sustainable materials because of their ecofriendly properties and economic viability. These biopolymers can be converted into hydrocarbon fuel for energy [11]. For such conversions, it is necessary for the biopolymer to dissolve/depolymerize adequately under mild reaction conditions. However, the natural biopolymers have often exhibited difficulties in solubilization and processing due to numerous inter- or intrahydrogen bonds in their polymeric chains, which restrict their employment for a wide variety of material applications. The processing of biopolymers using common organic solvents creates difficulties in terms of limited solubility, especially at high temperatures. Besides solubility limitations, there are several other problems—such as solvent handling, volatility, generation of poisonous gas or waste, and solvent recovery—that are generally encountered while processing the biopolymers in common organic solvents. Therefore, when considering green chemistry concerns, sustainability, and ecoefficiency, it is imperative to use greener solvents with high solvating ability for developing novel and efficient processing of biopolymers with minimum energy consumption.

7.2 BIOPOLYMER PROCESSING USING ILs—A BRIEF LITERATURE REVIEW

The solvation ability of ILs is the most important prerequisite in the present context. Some of the ILs have special solvation abilities to dissolve wood, lignocelluloses biomass, and other biopolymers, under relatively mild conditions. This particular ability of some ILs, accompanied by a series of concurrent advantages, has enabled the development of improved processing strategies for the manufacturing of a plethora of biopolymer-based advanced materials. The properties of biomaterials can be modified in a chemical, physical, or enzymatic way after dissolution in an IL media. The following section gives a brief literature review on the processing of biopolymers and biopolymer-based materials using ILs.

7.2.1 Dissolution Studies

In 1934, Graenacher [12] first reported dissolution of cellulose in a mixture of 1-ethylpyridinium chloride and a nitrogen-containing base when heated at 120°C. Much later, in 2002, Swatloski et al. [13] reported that alkyl imidazolium-based ILs, such as [C_4mim][Cl], can dissolve cellulose up to 25 wt% and can be easily regenerated by precipitation upon addition of antisolvents without severe degradation of the cellulose. A high chloride concentration in the IL was found to be responsible for breaking the extensive hydrogen-bonding network promoting dissolution. After this report, Rogers and coworkers [14] investigated the cellulose solubilizing mechanism in ILs using ^{13}CNMR spectroscopy. Another group also showed mechanisms of cellulose dissolution in ILs; Zhang et al. [15] did NMR spectroscopic studies to understand the solvation and dissolution of cellobiose in the IL [Emim][OAc] and reported that hydrogen bonding among hydroxyl groups of cellobiose with both anion and cation of the IL drives the dissolution (Figure 7.1) [15]. Remsing et al. [16] have reported the mechanism of cellulose dissolution in chloride-based ILs using NMR spectroscopy. Their studies revealed the hydrogen bonding among a hydroxyl group of biopolymers and small anions of ILs in a 1:1 stoichiometry. Apart from the experimental dissolution studies, a molecular dynamics simulation study performed by Liu et al. [17] on cellulose oligomers and [C_2mim][OAc] have shown that in addition to the anion forming strong hydrogen bonds with hydroxyl groups of the cellulose, the cations are in close contact with the cellulose through hydrophobic interactions. Very recently, it has been shown that nature of cations also plays a significant role in the dissolution of cellulose [18].

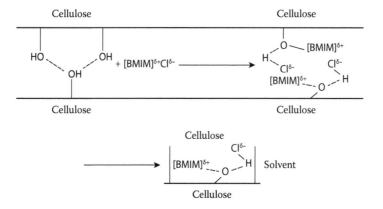

Figure 7.1 Dissolution mechanism of cellulose in [BMIM][Cl]. (Adapted from Zhang, J. et al., *Phys. Chem. Chem. Phys.*, 12, 1941–7, 2010.)

Effects of the nature of cations or anions of ILs and additives in ILs have been studied for improvement in the dissolution of cellulosic materials. Xu et al. [19] have investigated the effect of the addition of lithium salt on cellulose dissolution in ILs and found that lithium salt increases the solubility of cellulose owing to the interactions of Li^+ with the hydroxyl oxygen atoms of cellulose. Zhang et al. [20] have demonstrated that a short alkyl chain in the cation (i.e., [Amim]) enhances the solubility of cellulose when compared to [C_4mim] cation. Ohno and his coworkers have reported a series of alkyl imidazolium salts containing different phosphate anions that have the potential to solubilize cellulose under mild conditions. This group also published a review article describing the design of ILs for dissolution, depolymerization, and energy conversion of cellulose and their derivatives [21,22]. Vitz et al. [23] have investigated cellulose solubility in a variety of imidazolium ILs and have shown the extended solubility of cellulose in phosphate-containing ILs. Abe et al. [24] have reported the fast dissolution of 20 wt% cellulose in tetrabutylphosphonium hydroxide [TBPH] with the mixture of water (40 wt%) in milder conditions at 25°C and in short times (5 min) where the solution remained stable throughout the process. Ohira et al. [25] have designed biobased halogen-free ILs containing amino acid moiety as anions that could efficiently dissolve cellulose. Recently, Wang et al. [26] have reported a critical review on ILs for the processing of cellulose; in this review they did a detailed literature survey on cellulose dissolution with the different role of cations and anions of ILs. From the literature, it has been concluded that dissolution is greatly affected by the source of cellulose and the degree of polymerization (DP) of cellulose, and that the dissolution conditions (heating method, irradiation, heating temperature, time, etc.) also affect the dissolution process. Generally, with the same cation, the solubility of cellulose in ILs with different anions decreases in the order: $[(CH_3CH_2)_2PO_4]^- \approx [OAc]^- > [SHCH_2COO]^- > [HCOO]^- > [Cl]^- > [Br]^- \approx [SCN]^-$ [26], whereas with the same anion, the solubility of cellulose in ILs with different cations follow the order: $[C_4mim] > [C_1OC_2mim] > [C_4dmim] > [C_2OHmim] \approx [phC_1mim] > [C_2mmor] > [C = C_2mmor] > [C = C_2mpip] > [C_4mpip] > [C_4mpyr]$ [18].

Besides cellulose, chitin and chitosan have also been investigated for dissolution in an IL media by several research groups. Yamazaki et al. [27] have reported dissolving 10 wt% of chitin in [Amim][Br]. Xie et al. [28] have reported the dissolution of chitin and its *N*-deacetylated analogue (chitosan) up to 10 wt% in [C_4mim][Cl]. Wang et al. [29] have studied the dissolution behavior of chitin in different ILs containing different anions. He has concluded that dissolution of chitin in ILs depends mainly on its degree of acetylation, molecular weight, and anions of ILs used. Kadokawa et al. [30] have reported that chitin nanofibrils can be easily formed by gelation of commercial chitin powders with [Amim][Br] followed by the regeneration with methanol. Chen et al. [31] have studied the influence of anionic structure on the dissolution of chitosan in [C_4mim]$^+$-based ILs. Muzzarelli [32] have reviewed the biomedical exploitation of chitosan/chitin dissolution, regeneration, and electrospinning of chitosan/chitin threads in imidazolium-based ILs. Recently, Sharma et al. [33] have observed dissolution of α-chitin biobased deep eutectic mixtures. Maximum dissolution of the biopolymer (9% w/w) was observed in the deep eutectic solvent (DES) consisting of choline chloride–thiourea. The solubility behavior of other biopolymers (pectin, guar gum, and keratin) and carbohydrates (amylose, dextrin, inulin, pectin, and xylan) also have been studied in different ILs by many research groups [34–39].

7.2.2 Ionogels

An ionogel is defined as "a solid interconnected network spreading throughout a liquid phase" and these have been exploited for a variety of applications. One can produce ionogels by dissolution of organic/inorganic solutes or biopolymers in ILs and cooling the solution; this novel, soft, and flexible hybrid "ionogel" material may be useful for many electrochemical devices. In 2008, Vidinha et al. [40] prepared gelatin-based ion-gels and built an electrochromic window based on Prussian blue (PB) and poly(3,4-ethylenedioxythiophene) (PEDOT) as electrochromic layers showing the utility of biopolymer-based ion-gels for electrochemical devices. Ning et al. [41] have reported the solid biopolymer electrolytes based on [Amim][Cl] plasticized cornstarch. Mine et al. [42] prepared a biopolymer-based gel electrolyte in ILs, a guar-gum-based functional material prepared in IL. Prasad et al. [43] have reported polysaccharide-ILs-based functional materials using carrageenans (k-, i-, and l-) and their composite gels with cellulose using ILs; k-carrageenan formed a hard gel while the i- and l-carrageenan formed softer gels with [C$_4$mim]Cl. This group also reported weak gels of chitin in the IL; [C$_4$mim][Br] have also been reported by the same group. Chitin (with 7%, w/w) in IL has been reported to form a viscous gel-like material [44]. Guar-gum-based temperature-induced shapeable films in [C$_4$mim]Cl ILs, formed by soaking and compression techniques, have also been reported. This film was found to exhibit good mechanical and conducting properties that are comparable to semiconductors [45]. Pimenta et al. [46] has reported electrospinning of ion jelly materials containing gelatin in an IL that combines the chemical and the conducting versatility of an IL with the morphological versatility of the biopolymer. These ion jellies exhibit a high conductivity and high thermostability and have been used successfully to design electrochromic windows. Agarose-based ionogels in ILs have scarcely been studied—except Shamsuri and Diak [47], who have demonstrated the plasticizing effect of choline chloride-/urea-based deep eutectic mixture on agarose film.

7.2.3 Functionalization

Because of their high solvating ability, certain ILs have been found suitable to dissolve different types of biomaterials without derivatization and degradation. The ease of dissolution of biomaterials in ILs under mild conditions has led to their efficient functionalization in numerous ways; for example, esterification, etherification, acetylation, and carbanilation of cellulose have been carried out by the Heinze research group [48–52]. Abbott et al. [53,54] have reported the cationic functionalization of cellulose in choline chloride-based ILs and the O-acetylation of cellulose and monosaccharide using zinc-based ILs. Wu et al. [55] have been the initiators of a work on biopolymer functionalization in ILs. They have investigated the homogeneous acetylation of cellulose in [Amim][Cl] medium without any catalyst and have achieved a high degree of substitution (DS) in lesser time. Tomé et al. [56] have reported the surface hydrophobization through chemical modification of bacterial and vegetable cellulose fiber with several anhydrides in an IL, while Gericke et al. [57] have suggested ILs as promising but challenging solvents for cellulose derivatization. Apart from cellulosic materials, the functionalization of other biomaterials, such as starch, sugarcane bagasse, sucrose, cyclodextrin, zein protein, chitosan, or chitin, has also been carried out in an IL medium with or without catalyst in a short time with a controllable DS [36,58–62].

7.2.4 Extraction

Because of high solvating ability, ILs are promising solvents for an extraction process. As far as extraction of biopolymers from biomass utilizing ILs is concerned, various biomass fractions such as lignocellulose, lignin, and so forth have been obtained by direct dissolution of wood biomass in ILs [63–66]. Fort et al. [66] have reported that carbohydrates can be gradually released to the IL phase from the wood particles or other lignocellulosic materials. ILs have also been proposed as an efficient extracting solvents for the extraction of valuable biomolecules such as DNA, RNA from aqueous phase systems, and proteins from yeast or biological fluids [67,68]. Besides the extraction of cellulosic materials, there are other biopolymers, such as keratin [69], pectin [70], and suberin [71], that have been successfully isolated from the bioresources using ILs. Qin et al. [72] have extracted chitin from crustacean shells using ILs, 1-ethyl-3-methylimidazolium acetate of a high purity, and high molecular weight chitin powder and to fibers and films that can be spun directly from the extract solution. Very recently, Barber et al. [73] reported a one-pot process for the extraction of high molecular weight chitin fibers directly from dried shrimp shells using ILs, 1-ethyl-3-methylimidazolium acetate. Cláudio et al. [74] have recently investigated the extraction of caffeine from bioresources (guaraná seeds) using several ILs composed of imidazolium or pyrrolidinium cations and chloride, acetate, and tosylate anions. The extraction ability of ILs has also been employed for removing other valuable chemicals from different bioresources [75–80]. Marine microalgae is another category of biomass that contains many useful and valuable chemicals such as lipid, oil, polysaccharides, fuels, chemical feedstock, and so forth in abundance; the ILs have been used as an extraction medium for the recovery of such chemicals from various algae sources [81–84].

7.2.5 Composite/Nanocomposite Materials and Their Ionogels

Composites are materials that are made up of two or more constituents with significantly different physical or chemical properties and with the components retaining their macroscopic properties distinctly within the overall structure. Blending biopolymers to improve their chemical and physical properties in ILs has received substantial attention in the past few years. Xie et al. [85] have published the first report on blending of wool keratin/cellulose in IL, $[C_4mim][Cl]$ by a simple dissolution and regeneration method. Park et al. [86] have prepared ternary biocompatible and blood compatible composites of heparin–cellulose–charcoal in an IL medium. Prepared composites might be useful for direct hemoperfusion to remove free-diluted and protein-bound toxins of small size or as potential oral agents in the cases when strict preservation of large molecules (proteins) is necessary for drug detoxification. Sun et al. [87] have reported the magnetite-embedded cellulose fibers prepared from the solution of cellulose in ILs, 1-ethyl-3-methylimidazolium chloride, $[C_2mim][Cl]$ where suspended magnetite Fe_3O_4 (particles) were incorporated in the cellulose matrix by a dry-jet wet spinning process in an aqueous bath. Yu et al. [88] have investigated the formation of blend films from cellulose and Konjac glucomannan (KGM) in IL [Amim][Cl] by coagulation with water. Kuzmina et al. [89] have investigated the cellulose (CE)/silk fibroin (SF) blends preparation in $[C_4mim][Cl]$ using thermo gravimetric analysis (TGA), infra-red (IR)-spectroscopy. Wu et al. [90] have reported green composite film preparations that are composed of three different biopolymers—cellulose, starch, and lignin—in different proportions in an IL medium. Sun et al. [91] have prepared chitosan/cellulose composites as biodegradable biosorbents for heavy metal removal. Huang et al. [92] has introduced a new composite preparation approach in ILs, where they have fabricated a unique biosensor. Liu et al. [93] have also investigated the enzyme immobilization on magnetic cellulose–chitosan hybrid gel. Very recently, Tran et al. [94] have reported ILs-mediated chitosan–cellulose composite materials (as green solvents) for the removal of microcystin. The prepared composites have superior mechanical strength, excellent adsorbent capacity, and recyclability as microcystin can be easily desorbed. Agarose-based composites have also been reported recently [95].

Besides composite preparation, it is also possible to confine ILs as gel matrices by cooling composite material–IL solutions at ambient temperatures to form "composite ionogels." This composite–ILs sol can be cast in different forms such as films, fibers, or gels. There are many research groups that are working on biopolymer-based composite materials. Kadokawa et al. [96] have reported composite ionogels of cellulose (10% w/w) and starch (5%w/w) in $[C_4mim][Cl]$. Gels of three types of carrageenans (κ, ι, and λ) as well as their composite gels with cellulose were prepared using $[C_4mim][Cl]$; κ-Carrageenan formed a hard gel while the other two carrageenans formed softer gels with $[C_4mim]$ [Cl]. Shang et al. [97] have reported the composite film based on SF/CE using $[C_4mim][Cl]$. Takegawa et al. [98] have investigated chitin/cellulose composite ionogels. Luo et al. [99] prepared cellulose/curcumin composite films by casting the solution mixture obtained after dissolution of 0–5 wt% in [Amim][Cl]. These composite films possess good mechanical properties and thermal stability, both of which are comparable to the pure cellulose films.

Nanocomposites based on a biopolymer matrix and metal nanoparticles using ILs are scarcely noted in current literature. Li et al. [100] were the first to report that the colloidal gold nanoparticles/cellulose hybrid precipitates by reducing $HAuCl_4$ to gold with $NaBH_4$ as a reducing agent in cellulose solution in IL, $[C_4mim][Cl]$. Brondani et al. [101] have reported the preparation of chitin/gold nanoparticles in a biopolymeric matrix of chitin in an IL as a biosensor for rosmarinic acid. Gelesky et al. [102] have reported nanocomposite materials of metal formed of nanoparticle–IL–cellulose that are catalytically active membrane materials for hydrogenation reactions. Ma et al. [103,104] have prepared biomass-based hybrid nanocomposites of cellulose/$CaCO_3$ using IL as a green solvent. This group has also reported the preparation of microwave-assisted cellulose/CuO nanocomposite in IL by an environmentally friendly method. Bagheri and Rabieh [105] have reported the nanocomposite based on cellulose/zinc oxide (ZnO) in IL, $[C_4mim][Cl]$ by microwave-assisted dissolution of cellulose in IL followed by addition of a premixed ground of $Zn(CH_3COO)_2 \cdot H_2O$ and NaOH. Singh et al. [106] have investigated IL-based processing of electrically conducting chitin nanocomposite scaffolds for stem cell growth. A nanocomposite was prepared by chitin/Carbon Nano Tube (CNT) dissolution in imidazolium-based ILs. This method has enabled the fabrication of chitin-based advanced multi-functional biocompatible scaffolds where electrical conduction is critical for tissue functioning. The nanocomposites of polysaccharide guar gum and Multi Walled Carbon Nano Tubes (MWCNTs) in ILs having self-healing properties have been reported very recently by the Prasad group [107]. Also, very recently, Liu et al. [93] have investigated a new strategy to prepare a hybrid gel, and they have reported magnetic Fe_3O_4–cellulose–chitosan hybrid gel microsphere prepared via sol–gel transition using IL as a solvent. Though rare, the formation and characterization of biopolymer nanocomposite ionogels has also been reported by cooling hot nanocomposite–IL solution [93,108–111].

7.3 AGAROSE PROCESSING IN ILs

The present section will focus on the processing details for a gelling polysaccharide "agarose" using ILs. Agarose is composed of linear chains of agarobiose, a disaccharide made up of D-galactose and 3,6-anhydro-L-galactopyranose units, linked by α-(1→3) and β-(1→4) glycosidic bonds (Figure 7.2). Agarose is extracted from a class of marine red algae (*agarophyta*) and is known for its thermoreversible gelling property. Over the years, agarose gel has been exploited for its biomedical applications, such as in DNA electrophoresis, drug delivery, cell therapy, and in molecular and tissue engineering because of its high mechanical strength, biocompatibility, and bio inert nature [112,113].

7.3.1 Extraction from Algae

Trivedi [114] reported ILs as a medium for efficient extraction of agarose from microalgae (Rhodophyta)—mainly *Gracilaria dura* obtained from western coastal regions of India—via dissolution under varying conditions of heating or microwave irradiation. Three different ILs have been

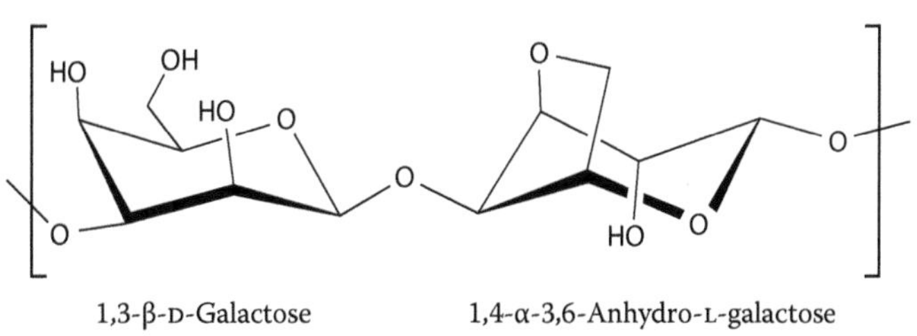

1,3-β-D-Galactose   1,4-α-3,6-Anhydro-L-galactose

Figure 7.2 Repeating unit of agarose.

used as extracting medium: 1-ethyl-3-methylimidazolium acetate, [Emim][OAc]; choline acetate, [Ch][OAc]; and 1-ethyl-3-methyl imidazolium diethyl phosphate, [Emim][Dep]. As compared to conventional methods, a very high extraction yield of good quality agarose (39 wt%) could be achieved depending upon the nature of the IL used and on the applied experimental conditions. Purity of extracted agarose was confirmed from various spectral and analytical techniques. ILs were recovered after the extraction process and were reused for the further extraction process. Recovered ILs were characterized for their purity by NMR technique; the percentage of recycling and extraction ability of recycled ILs in different cycles have been measured.

Extraction of agarose from red algae using ILs was carried out according to the representative steps shown in Figure 7.3 [114]. The extraction process was performed in 10 ml beakers with controlled conditions of temperature or microwave (MW) or both. Typically, 0.5 g of dried and powdered algae was treated with 10.0 g of preheated ILs in beakers. The mixture (containing algae + ILs) was stirred at 80/100°C for 2 h or MW treatment for 2 min; at 3 s, a pulse was given prior to the normal heating. In order to reduce the viscosity of the IL–algae solution, 10% boiled water was added. The hot water treated mixture was stirred until free flow homogenization, and it was filtered using thin cotton under a vacuum. Filtrate was treated with methanol to precipitate agarose dissolved in ILs. Several methanol washings were given to ensure the complete removal of ILs from extracted agarose. Agarose extracted from different experiments was dried under the vacuum and percent of yield was calculated on the basis of used dried algae weight. Imidazolium-based ILs having high basicity anions such as [OAc]- or [Dep]- have been found to have good dissolution ability for biomass [84,115].

Table 7.1 [114] shows the percentage of yield of extracted agarose under applied different conditions of heating or combination of microwave heating followed by normal heating. Various parameters such as increases in temperature or microwave heating before normal cooking have been found to assist in rapid and higher dissolution of algae and efficient agarose extraction.

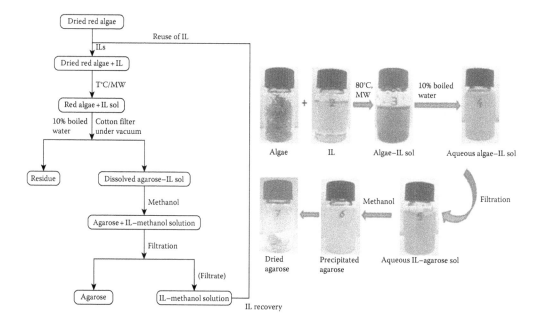

Figure 7.3 Schematic process and representation of different steps carried out for efficient extraction of agarose using ionic liquids. (From Trivedi, T. J., *Studies on utilization of room temperature ionic liquids for processing of biopolymer agarose*, Academy of Scientific and Innovative Research, India PhD thesis, 2013.)

Table 7.1   Used Ionic Liquids, Applied Condition, and Percentage of Yield for Agarose Extraction Using Ionic Liquids

ILs	Treatment Condition	% Yield
[Emim][OAc]	80°C; 2 h	28.5
	MW; 80°C; 2 h	39
	100°C; 2 h	35
[Ch][OAc]	80°C; 2 h	2.5
	MW; 80°C; 2 h	12.5
	100°C; 2 h	7.2
[Emim][Dep]	80°C; 2 h	7.6
	MW; 80°C; 2 h	18.5
	100°C; 2 h	17

Source: Trivedi, T. J., *Studies on utilization of room temperature ionic liquids for processing of biopolymer agarose*, Academy of Scientific and Innovative Research, India PhD thesis, 2013.

Table 7.2   Extraction Ability and Recovery of 1-Ethyl-3-Methyl Imidazolium Acetate in Different Cycles

Recycled IL	Yield (%)	IL Recovery (%)
First	28.5	87.56
Second	28	79.31
Third	24.97	65.30
Fourth	21.66	46.23

Source: Trivedi, T. J., *Studies on utilization of room temperature ionic liquids for processing of biopolymer agarose*, Academy of Scientific and Innovative Research, India PhD thesis, 2013.

Nearly 2.5 and 7.6 wt% of agarose could be extracted using [Ch][OAc] and [Emim][Dep], respectively, by simple heating at 80°C for 2 h whereas the yield was dramatically higher (28.5 wt%) for [Emim][OAc] under similar conditions. It was observed that by increasing reaction temperature from 80°C to 100°C, the percentage yield for agarose extraction increased from 2.5 to 5% in [Ch][OAc], 7.6% to 17% in [Emim][Dep], and 28.5% to 35% in [Emim][OAc]. Application of microwave irradiation for 2 min at 5 s pulse prior to the conventional heating at 80°C increased the percentage yield of agarose in all the ILs. Under coupled conditions of microwave and normal heating, the agarose yield enhanced from 2.5% to 12.5% in [Ch][OAc], 7.6% to 18.5% in [Emim][Dep], and 28.5% to 39% in [Emim][OAc]. The extraction efficiency of used ILs followed the order: [Emim][OAc] > [Emim][Dep] > [Ch][OAc], suggesting that the extraction ability depends on viscosity and the nature of cation/anion of ILs. It is well known that strong hydrogen-bonding basicity is effectively weakening the hydrogen-bonding network of the polymer chain, which concludes that the higher the β value, the higher is the dissolution and extraction efficiency [83,116]. The higher extraction efficiency of imidazolium-based ILs containing OAc and Dep anions is due to their high solvating ability, hydrogen bond basicity (β), and low viscosity, while the low extraction in [Ch][OAc] may be due to the nature of the cation of the IL.

Recycling of used solvents from the reaction mixture after materials extraction is important for development of an energy saving process. After the extraction of agarose, as shown in Figure 7.3, the filtrates containing a mixture of ILs, methanol, and water were evaporated under reduced pressure using a rotary evaporator to obtain pure ILs. Extraction reactions were performed at 80°C (2 h for up to four cycles), and results of extraction are illustrated in Table 7.2 [114]. Results show that the extracting of agarose was reduced from 28.5% to 21% after the fourth cycle, indicating a reduction of about

27% efficiency of extraction. Percentage of recovery of ILs decreased from 87.5% to 46.23% in the fourth cycle, which is mainly because of experimental and handling error after each experiment as well as because of the rigid process.

7.3.2 Dissolution and Regeneration

Kumar and coworkers [117] have reported the agarose dissolution in imidazolium- or pyridinium-based ILs at 70°C and have determined the effect of anions and alkyl chain length effect in the ILs toward dissolution. The solubility of agarose in ILs depends on the nature of cation and anion. For the same alkyl chain length, ILs containing chloride anion show better solubility for agarose as compared to the methylsulfate anion. High efficiency of ILs containing chloride anion for agarose solubilization is due to strong interaction of chloride ions with the hydroxyl groups because of the very high hydrogen-bonding basicity of these ionic liquids. The interactions among chloride ions and the hydroxyl groups leads to disruption of the hydrogen-bonding network of the polymer and results in higher solubility. A remarkable decrease in solubility was observed with the increase in alkyl chain length of the cation. Dissolved agarose was regenerated using methanol, and ILs were recovered and recycled for different experiments. Regenerated agarose largely maintained the features of native agarose in terms of molecular weight, polydispersity, thermal stability, and crystallinity but varied slightly in conformation preferences (Table 7.1). In order to examine the possible degradation of the material, the molecular weight was determined from gel permeation chromatography. Results show that agarose regenerated from $[C_4mim][C_1OSO_3]$, $[C_8mim][Cl]$, and $[C_4mim][Cl]$ had comparable molecular weights to that of native agarose, whereas the material regenerated from $[C_4mpy][Cl]$ had a comparatively lower molecular weight. Regenerated agarose was characterized via spectral techniques such as Fourier Transform Infra-Red (FTIR), circular dichroism (CD) spectroscopy, TGA, and X-ray Diffraction (XRD), which are presented in Figure 7.4. FTIR spectra show that regenerated agarose largely maintains the native structure. Characteristic IR bands at 773, 894, and 932 cm^{-1} (because of 3,6-anydro-β-galactose skeletal bending in agarose) remain unaltered in the material regenerated from water, $[C_4mim][C_1OSO_3]$ and $[C_8mim][Cl]$, but appear slightly diminished when regenerated from $[C_4mim][Cl]$ and $[C_4mpy][Cl]$. The IR bands at 1158 and 1071 cm^{-1} corresponding to –C–O–C– and glycosidic linkage are also altered in the regenerated material. No peak corresponding to the ILs moiety indicates that these are a fairly noncoordinating media for dissolution of agarose. Circular dichroism (CD) spectra of regenerated agarose from various ILs indicate their overall secondary structure, and in the case of agarose, a characteristic CD spectrum with a positive band centered at ~185 nm is observed. The material regenerated from solutions of $[C_4mim][C_1OSO_3]$ shows slight loss in intensity, and agarose regenerated from $[C_8mim][Cl]$ and $[C_4mim][Cl]$ appears to be red shifted with a loss in intensity indicating less conformational change in the regenerated material. In the case of agarose regenerated from $[C_4mpy][Cl]$, a very large decrease in the CD band intensity is observed, and the band is highly red shifted, which is indicative of change in conformation and disruption of ordered structure. From TGA profiles, it is seen that the decomposition temperature (T_{dec}) is characterized by a narrow temperature range from 275°C to 300°C. No significant weight loss was observed in TGA curves of the agarose regenerated from $[C_4mim][Cl]$, $[C_4mpy][Cl]$, and $[C_4mim][C_1OSO_3]$ before the decomposition temperature. No sharp or narrow peak in XRD and low crystallinity indicate that the regenerated material, such as native agarose, is amorphous in nature.

Dissolution of agarose in mixed IL systems containing low viscosity alkyl or hydroxyalkyl ammonium formate ILs and imidazolium- or pyridinium-based ILs (scheme shown in Figure 7.5.) has also been carried out to improve the dissolution [118]. Results show that higher dissolution can be achieved in mixed ILs systems. Dissolved agarose was regenerated using methanol as a precipitating solvent and further characterized for degradation/conformational changes using various analytical technique such as FTIR, CD spectroscopy, TGA, and so forth. Spectral characterization of regenerated materials shows no degradation in regenerated agarose. Dissolution results of agarose

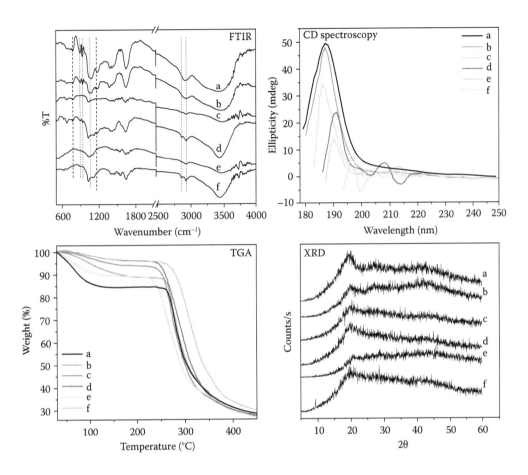

Figure 7.4 **(See color insert.)** Spectral analysis of agarose regenerated from various solutions: (a) native agarose, (b) water, (c) [C$_4$mim][C$_1$OSO$_3$], (d) [C$_8$mim][Cl], (e) [C$_4$mim][Cl], and (f) [C$_4$mpy][Cl].

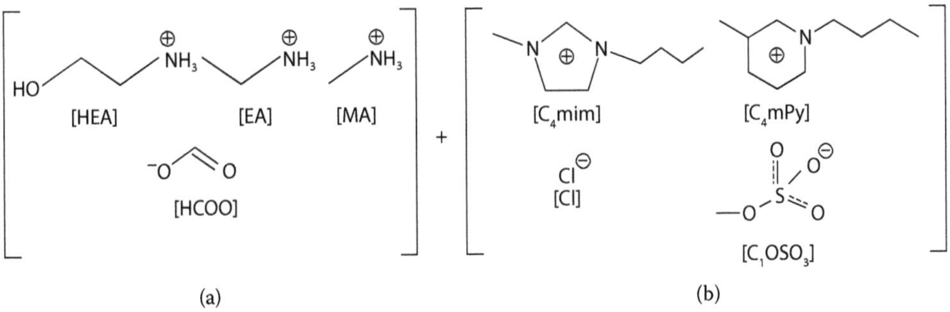

Figure 7.5 Constituents of ionic liquids (ILs) used: (a) ammonium-based ILs and (b) imidazolium-based ILs.

in the ILs: [HEA][HCOO], [MA][HCOO], [EA] [HCOO], and their mixtures with [C$_4$mim][Cl] or [C$_4$mpy][Cl] or [C$_4$mim][C$_1$OSO$_3$] at 70°C are listed in Table 7.3. Dissolution results indicate that the agarose has very good solubility in [MA][HCOO] and [EA] [HCOO], whereas the solubility in [HEA][HCOO] is less than that in water. Exceptionally higher solubility was observed in the systems ([MA][HCOO] or [EA][HCOO] + [C$_4$mpy][Cl] or [C$_4$mim][C$_1$OSO$_3$]). The solubility

Table 7.3   Solubility of Agarose in Ionic Liquids/Mixed Ionic Liquids at 70°C and Molecular Weight (M_n and M_w), Polydispersity (P), and Degree of Polymerization (D_p) of Regenerated Agarose

Entry	Solvent System	s (wt%)	M_n	M_w	P	D_p
1	[C$_4$mim][Cl]	16	77,850	152,000	1.95	497
2	[C$_4$mim][C$_1$OSO$_3$]	5	81,100	162,400	2.00	530
3	[C$_8$mim][Cl]	4.5	70,250	142,300	2.03	465
4	[C$_4$mPy][Cl]	13	39,500	98,200	2.49	320
5	[HEA][HCOO]	5	81,531	160,326	2.31	527
6	[MA][HCOO]	14	61,879	130,259	3.16	432
7	[EA][HCOO]	20	44,933	127,908	2.74	420
8	[HEA][HCOO] + [C$_4$mim][Cl]	8	73,907	160,470	2.82	527
9	[MA][HCOO] + [C$_4$mim][Cl]	18	72,681	148,760	2.71	489
10	[EA][HCOO] + [C$_4$mim][Cl]	7	69,398	159,217	2.91	523
11	[HEA][HCOO] + [C$_4$mPy][Cl]	7	66,020	138,124	2.89	454
12	[MA][HCOO] + [C$_4$mPy][Cl]	21	47,147	140,822	2.98	463
13	[EA][HCOO] + [C$_4$mPy][Cl]	25	46,192	115,630	2.75	380
14	[HEA][HCOO] + [C$_4$mim][C$_1$OSO$_3$]	14	82,387	161,787	1.97	532
15	[MA][HCOO] + [C$_4$mim][C$_1$OSO$_3$]	23	43,490	98,073	2.63	322
16	[EA][HCOO] + [C$_4$mim][C$_1$OSO$_3$]	26	68,451	156,960	2.29	450

Table 7.4 E_T (30), E_T^N, Kamlet–Taft Parameters of Ionic Liquids or Mixed Ionic Liquids (1:1 w/w) at 70°C

No	Solvent System	E_T (30) kcal mol^{-1}	E_T^N	α	β	π^*
1	[HEA][HCOO]	59.6	0.882	1.11	0.59	0.97
2	[MA][HCOO]	59.3	0.873	1.02	0.83	1.05
3	[EA][HCOO]	56.4	0.787	1.10	0.87	0.77
4	[HEA][HCOO] + [C$_4$mim][Cl]	55.7	0.765	0.77	0.67	1.10
5	[MA][HCOO] + [C$_4$mim][Cl]	57.0	0.802	0.87	0.86	1.07
6	[EA][HCOO] + [C$_4$mim][Cl]	58.9	0.861	1.04	0.85	1.01
7	[HEA][HCOO] + [C$_4$mPy][Cl]	56.0	0.772	0.75	0.62	1.14
8	[MA][HCOO] + [C$_4$mPy][Cl]	54.2	0.718	1.02	0.88	0.71
9	[EA][HCOO] + [C$_4$mPy][Cl]	55.5	0.758	0.91	0.94	0.92
10	[HEA][HCOO] + [C$_4$mim][C$_1$OSO$_3$]	57.8	0.827	0.97	0.84	1.01
11	[MA][HCOO] + [C$_4$mim][C$_1$OSO$_3$]	57.0	0.802	1.02	0.89	0.90
12	[EA][HCOO] + [C$_4$mim][C$_1$OSO$_3$]	57.2	0.809	1.13	1.00	0.79

of agarose in ILs depends on the nature of the cation and anion. IL systems containing [C$_1$OSO$_3$]$^-$) anions dissolved higher amounts of agarose, whereas the ammonium ILs having a hydroxyl group in cation are not very effective in solubilizing agarose. Similar to cellulose, the high efficiency of ILs containing [HCOO]$^-$, [Cl]$^-$, or [C$_1$OSO$_3$]$^-$ toward agarose solubilization is due to strong interactions of these ions with the hydroxyl groups of agarose, which eventually leads to disruption of the hydrogen-bonding network of this polymer.

Specific physicochemical properties, such as polarity of a solvent, strongly influence the dissolution of materials. The polarity parameters [α, β, π^*, $E_T(30)$, and E_T^N] of these IL systems were measured to explain the dissolution ability for agarose (Table 7.4) [114]. As indicated by large $E_T(30)$ values, the ILs used are of reasonably good polarity, and the mixing of ILs does not alter the $E_T(30)$ values very significantly. These parameters provide a quantitative measure of solvent polarizability (π^* parameter), hydrogen bond donor capacity (α parameter), and hydrogen bond acceptor capacity

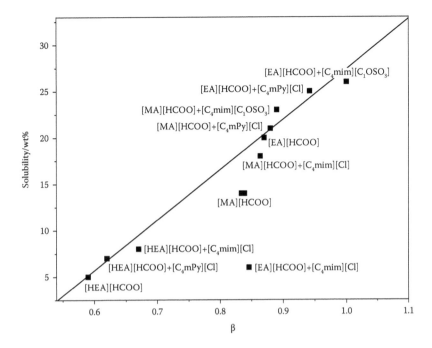

Figure 7.6   Linear correlation between solubility of agarose and β parameter of ILs/mixed ILs at 70°C.

(β parameter) [119]. As can be seen from Table 7.2, α parameter slightly decreases when the ILs [C_4mim][Cl] or [C_4mpy][Cl] are added to the alkyl or hydroxyalkyl ammonium ILs, although it hardly changes upon addition of [C_4mim][C_1OSO$_3$]. Recent studies have revealed that ILs with higher β values are more capable of dissolving biomaterials, such as cellulose [21,120]. Higher dissolution of agarose in alkyl ammonium ILs can be attributed to their very high β values. Also, the β values raised considerably when the ammonium ILs were mixed with imidazolium (particularly, [C_4mim] [C_1OSO$_3$]) or pyridinium ILs and possibly were the reason for higher dissolution of agarose. β values of ILs/mixed ILs systems versus agarose solubility is plotted in Figure 7.6, and it is seen that a near linear correlation exists in most cases. Deviation from this correlation for the system [EA][HCOO] + [C_4mim][Cl] is perhaps due to the formation of ionogel at working temperature and, hence, no more agarose could be dissolved.

7.3.3 Ionogels in ILs/Mixed ILs

For ion-gel preparation, agarose was dissolved in ILs/mixed ILs at 70°C [117,118]. Viscous clear solutions were allowed to cool for gelling. Gelling and melting temperatures of various ion-gels were monitored through visual inspection of liquid and gel states. The physicochemical properties of ionogels from different ILs/mixed ILs are recorded in Table 7.5). Least gelling temperature was observed for the hydrogel prepared using agarose regenerated from [C_4mim][C_1OSO$_3$] indicating some degradation of material. The formation of ionogels hydrogen-bonding between ILs and hydroxyl groups of agarose molecules must play an important role in gelling. Large ions will interact with agarose strands and prevent the formation of double helical structures and will finally lead to a comparatively weaker gel. Exceptionally high strength ionogels were obtained from the agarose solutions in the N-(2-hydroxy ethyl) ammonium formate or in its mixture with 1-butyl-3-methylimidazolium chloride. Viscoelastic conducting measurements and the gelling melting property of ionogels prepared from IL/mixed ILs are shown in Table 7.3. Results show that mixing

Table 7.5 Ionic Liquids/Mixed Ionic Liquids, Gelling Temperature (T_{gel}), Melting Temperature (T_m), Gel Strength, and Ionic Conductivity of Ionogels ($\kappa_{ionogel}$) and Ionic Liquid Systems (κ)

No	ILs/Mixed ILs	T_{gel} (°C)	T_m (°C)	Gel Strength (g cm^{-2}) at 30°C	κ (mS cm^{-1}) at 25°C	$\kappa_{ionogel}$ (mS cm^{-1}) at 25°C
1	[HEA][HCOO]	36	56	>1850	3.274	3.458
2	[MA][HCOO]	21	23	Liquid	35.98	31.91
3	[EA][HCOO]	24	26	Liquid	19.20	16.83
4	[C$_4$mim][Cl]	40	53	—	0.250	0.290
5	[C$_8$mim][Cl]	30	25	Weak gel	0.034	0.049
6	[C$_4$mim][C$_1$OSO$_3$]	17	35	Weak gel	0.802	0.854
7	[HEA][HCOO] + [C$_4$mim][Cl]	51	59	590	3.326	3.326
8	[MA][HCOO] + [C$_4$mim][Cl]	35	34	330	12.73	11.97
9	[EA][HCOO] + [C$_4$mim][Cl]	45	45	540	7.902	7.158
10	[HEA][HCOO] + [C$_4$mPy][Cl]	37	42	490	2.611	2.541
11	[MA][HCOO] + [C$_4$mPy][Cl]	22	37	Weak gel	10.84	9.353
12	[HEA][HCOO] + [C$_4$mim][Cl]	51	59	590	3.326	3.326
13	[EA][HCOO] + [C$_4$mPy][Cl]	34	46	140	6.078	5.682
14	[HEA][HCOO] + [C$_4$mim][C$_1$OSO$_3$]	28	44	<100	3.424	4.215
15	[MA][HCOO] + [C$_4$mim][C$_1$OSO$_3$]	18	19	Liquid	16.45	13.79
16	[EA][HCOO] + [C$_4$mim][C$_1$OSO$_3$]	22	25	<100	11.93	11.13

of [HEA][HCOO]-based ILs in imidazolium ILs increases the ionogel strength, although low viscous ammonium-based ILs such as [MA][HCOO] or [EA][HCOO] increase conductivity as well as solubility.

For preparation of ionogel, only a small quantity of agarose is sufficient; hence, the conducting properties of native ILs can be retained in gels. Conductivity depends mainly on the nature of constituent ions of ILs. In the case of imidazolium-based ILs, the conductivity of [C$_4$mim][C$_1$OSO$_3$] is very high as compared to [C$_8$mim][Cl] and [C$_4$mim][Cl]. Increasing temperature introduces thermal motion and may lead to a cooperative mechanism of ion conduction between cation and anion at higher temperatures. Low viscosity of [C$_4$mim][C$_1$OSO$_3$] is also responsible for its higher conductivity. As compared to imidazolium-based ILs, the ammonium-based ILs had very high conductivity. Addition of low viscosity ammonium-based ILs to imidazolium ILs increased the conductivity of ionogels significantly (Table 7.5).

Agarose–IL sol, ionogel and its different shapes, stability of ionogels, or hydrogel open atmosphere are represented in Figure 7.7. When connected to a power supply, ionogels lit up the LED, exhibiting a mixed electronic and ionic conduction. Ionogel—particularly when based on hydroxyalkyl ammonium formate ILs or their mixtures—remained unaltered even under large strains. To demonstrate the suitability of ionogels in electronic devices, we used a reasonably good strength and ionic conductivity ionogel prepared in the mixed ILs system, [HEA][HCOO] + [C$_4$mim][Cl] and poly(3,4-ethylenedioxythiophene) (PEDOT) to build an electrochromic window (Figure 7.8). The electrochromic window built with ionogel was found to have a capacity for charge transport and electrical conduction, and it changed the color of the PEDOT on glass-ITO plate from a bleached state to slight blue to brownish with the variations in voltage.

7.3.4 In Situ Functionalization and Ionogels of Modified Materials

In situ functionalization (acetylation and carbanilation) of agarose (AG) in an IL (1-butyl 3-methylimidazolium acetate, [C$_4$mim][OAc]) medium show a high DS: 3.35 for acetylation and 1.35 for carbanilation, respectively, which is higher than that reported in literature for the common

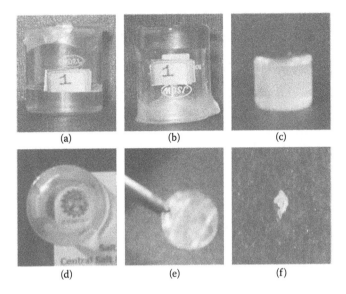

Figure 7.7 Representative (a) agarose–IL sol, (b) agarose ionogel, (c) ionogel cuboid, (d) transparent ionogel film, (e) stable ionogel thin film, and (f) shrunken hydrogel.

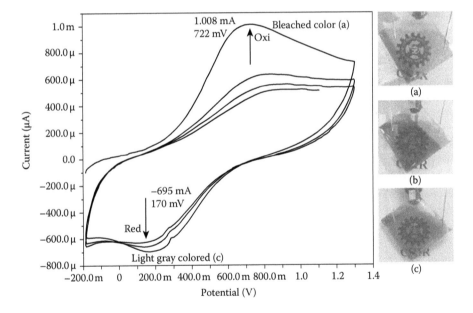

Figure 7.8 Cyclic voltammograms obtained for PEDOT/ITO-glass/ionogels devices for 10 cycle at 100 mV S^{-1} scan rate between –0.2 V and 1.3 V and electrochromic smart window of PEDOT/glass-ITO/ionogels: (a) bleached state, (b, c) unbleached state.

organic solvents [114]. Functionalization of agarose was confirmed in the various analytical techniques. Acetylated agarose was found sparingly soluble in water, although carbanilated agarose could dissolve in hot water and form hydrogels. Reaction schemes for agarose functionalization in ILs are shown in Figure 7.9 [114].

For acetylated agarose (AAG), the DS values were measured from ^{1}H NMR and varied from 0.46 to 3.35 under different experimental conditions. In the presence of pyridine, the DS increased

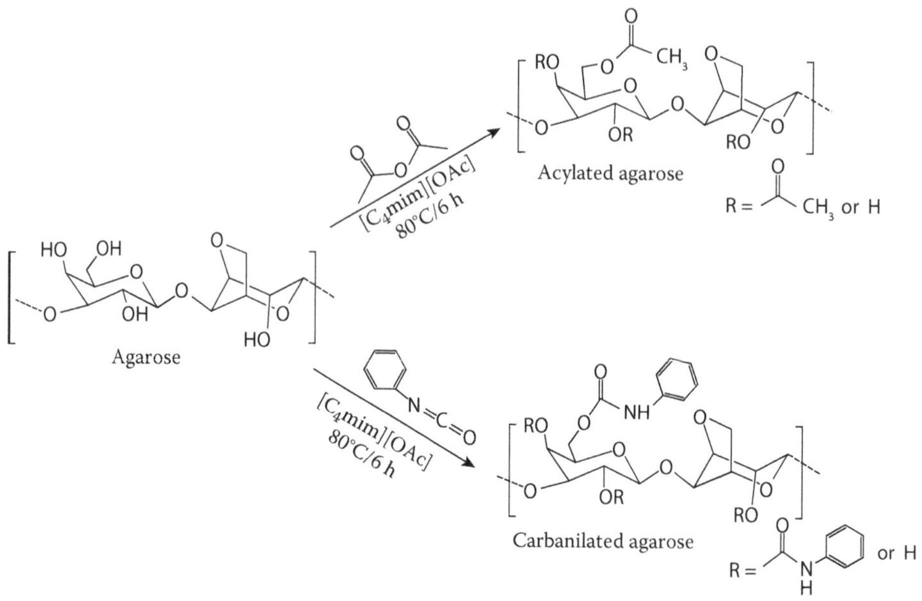

Figure 7.9 Reaction scheme for agarose functionalization in ionic liquid. (From Trivedi, T. J., *Studies on utilization of room temperature ionic liquids for processing of biopolymer agarose*, Academy of Scientific and Innovative Research, India PhD thesis, 2013.)

from 0.46 to 2.02 when acetic anhydrite was used as reagent, whereas not much of an increase in DS was observed when acetyl chloride was used as a reagent, indicating that the former reagent is more effective. At a molar ratio of 16 mol of acetic anhydride per ARU for 6 h, showed maximum acetylation (3.35), whereas with acetyl chloride, a maximum DS of 2.95 could be achieved. In the case of carbanilation, a higher amount of phenyl isocyanate and more reaction time were required. For carbanilated agarose (CAG), the DS was calculated using elemental analysis (CHNS) of the sample, and it varied from 0.07 to 1.35 according to reaction time and applied conditions. The variation of reaction conditions and the addition of pyridine did not affect the conversion significantly. This may be due to high sensitivity of phenyl isocyanate toward inherent water (moisture) present in AG moiety. The reaction between phenyl isocyanate and water produces aniline and carbon dioxide, which leads to less conversion. Morphology (SEM analysis) of the native and modified AG has been characterized in Figure 7.10 [114]. AAG having a high DS was found porous with a flake type structure, although the CAG had fibrous structure with very high porosity.

Dissolution of modified agarose (AAG and CAG) in different ILs (1-butyl-3-methylimidazolium chloride [C₄mim][Cl]; and *N*-(2-hydroxyethyl) ammonium formate [HEA][HCOO]/mixed ILs: [C₄mim][Cl] + [HEA][HCOO]) have been tested (the solubility is reported in Table 7.6), and ionogels have been prepared. It can be seen that the agarose solubility is higher compared to the modified agarose in neat ILs, whereas, in the mixed ILs ([HEA][HCOO] + [C₄mim][Cl]), the solubility is higher for modified agarose. High solubility of biopolymers in ILs depends on the nature of cation and anion, whereas the ILs containing small anions such as [Cl]⁻ or [HCOO]⁻ are more effective for solubilization. The dissolution of agarose in ILs/mixed ILs can be attributed to their very high hydrogen bond acceptor (HBA) capacity (β) values. Higher solubility of modified agarose in mixed ILs systems is due to the synergic effect of anions as well as to the presence of hydrogen-bonding sites in modified AG, which lead to stronger interactions with the anions of ILs. Such interactions eventually lead to disruption of the hydrogen-bonding network among biopolymers and enhance the solubility.

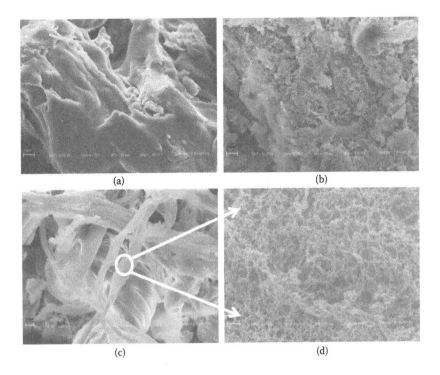

Figure 7.10 Scanning electron microscopy (SEM) images of (a) agarose, (b) acetylated agarose, (c) carbanilated agarose, and (d) carbanilated agarose fiber with high magnification. (From Trivedi, T. J., *Studies on utilization of room temperature ionic liquids for processing of biopolymer agarose*, Academy of Scientific and Innovative Research, India PhD thesis, 2013.)

Ionogels properties (gelling and melting temperatures) of native and modified agarose in ILs/mixed ILs were monitored through visual inspection of liquid and gel states and are noted in Table 7.6. Surprisingly, the CAG ionogels formed in the [HEA][HCOO] + [C$_4$mim][Cl] system showed a remarkable self-healing property. Visual inspection showed that after it was cut in two separate pieces with blade, the CAG ionogel healed spontaneously and returned back to original conditions after a certain time without any treatment or external stimuli. Photographs taken before and after the time of damage clearly show the self-healing process (Figure 7.11) [114]. As can be seen from photographs, CAG ionogels self-healed in 60 min, whereas the AG ionogels remained unchanged with damage. The introduction of functionality in the AG skeleton led to the insertion of new HBA or donor sites that helped in the self-healing process of CAG ionogels via multiple interactions. Such interactions, which may include π–π stacking (between IL, imidazolium, and the phenyl ring of the CAG), electrostatic or van der Waals, and the charge transfer induced between CAG moiety and mixed ILs reinforce the reversible bond formation.

7.3.5 Agarose–Chitosan Composite Materials/Ionogels

Simultaneous dissolution of agarose (AG) and chitosan (CH) in varying proportions in [C$_4$mim][Cl] has been performed [121]. Composite materials were constructed from AG–CH–IL solutions using antisolvent methanol, and IL was recovered from the solutions. Structure, stability, and physiochemical property of AG–CH composite materials were characterized via several analytical techniques, such as FTIR, CD, DSC, TGA, GPC, and SEM. The result shows that composite materials have good thermal and conformational stability, compatibility, and strong hydrogen-bonding interactions between AG–CH complexes.

Table 7.6 Solubility (s, wt%) of Agarose (AG), Acetylated Agarose (AAG), and Carbanilated Agarose (CAG) in Ionic Liquids/Mixed Ionic Liquids at 70°C, Gelling Temperature (T_{gel}), Melting Temperature (T_m), Gel Strength (g.cm^{-2}), and Conductivity of Ionogels ($\kappa_{ionogel}$) (3 wt%)

Type	[C$_4$mim][Cl]					[HEA][HCOO]					[HEA][HCOO] + [C$_4$mim][Cl]				
	s	T_{gel}	T_m	$\kappa_{ionogel}$ (mA)	Gel Strength	s	T_{gel}	T_m	$\kappa_{ionogel}$ (mA)	Gel Strength	s	T_{gel}	T_m	$\kappa_{ionogel}$ (mA)	Gel Strength
AG	16	40	53	0.0653	140	5	28	40	1.737	680	8	49	61	0.869	510
AAG	6.5	33	37	0.0994	<100	2.5	No gel formation		NM	9.5	27	35		1.205	500
CAG	11	38	42	0.0754	130	4.5	21	33	1.321	380	14	45	56	0.848	550

Source: Trivedi, T. J., Studies on utilization of room temperature ionic liquids for processing of biopolymer agarose, Academy of Scientific and Innovative Research, India PhD thesis, 2013.

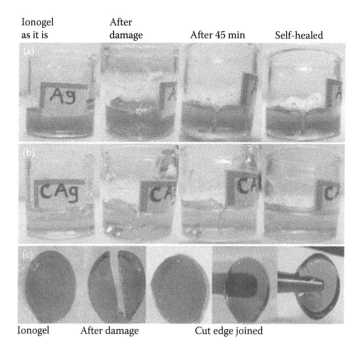

Figure 7.11 Photographs of ionogels during self-healing process: (a) agarose (no self-healing), (b) carbanilated agarose (self-healing), and (c) self-healing process of carbanilated agarose in open atmosphere. (From Trivedi, T. J., *Studies on utilization of room temperature ionic liquids for processing of biopolymer agarose*, Academy of Scientific and Innovative Research, India PhD thesis, 2013.)

FTIR and CD spectra of native, regenerated biopolymers, and composites having different proportions of AG and CH are presented in Figure 7.12 (left); FTIR spectrum of native CH shows a broad band at 3200–3500 cm^{-1} is attributed to the –NH$_2$ and –OH stretching vibrations, peak at 1560 cm^{-1} for the NH bending (amide II) (NH$_2$), 1647 cm^{-1} is due to the C = O stretching (amide I) O = C–NHR, 2927, 2884, 1411, 1321, and 1260 cm^{-1} correspond to CH$_2$ bending due to pyranose ring, 1078 cm^{-1} for saccharide structures. FTIR spectrum of AG shows an absorption band at about 3400 cm^{-1} that is associated with O–H stretching, bands at 773, 894, and 932 cm^{-1} because of 3,6-anydro-β-galactose skeletal bending, and a band at 1072 cm^{-1} that is attributed to the deformation mode of the C–O groups [117,118]. As can be seen in Figure 7.12, the spectra of composites show a combination of functional groups originated from both CH and AG. An increase in the AG content in the composite decreases the intensity of the band arising from the NH bending (amide II) at 1560 cm^{-1} and increases the band absorbance at 1380 cm^{-1}. The peak 1647 cm^{-1}, attributed to the C = O stretching (amide I) O = C–NHR of CH is shifted toward a lower frequency as the concentration of AG in the composite is increased. A gradual shift of the characteristic absorption bands of CH and AG indicate AG–CH complex formation either by formation of hydrogen bonds between the –OH/–NH$_2$ groups in CH molecules with the –OH groups in AG or by the reciprocal entanglement between the macromolecular chains. CD spectra of native and composite materials with different proportions are shown in Figure 7.12 (right). In the case of AG, a characteristic CD spectrum with a positive band centered at ~190 nm is observed, whereas in CH, a negative band centered near ~210 nm is observed. Analysis of CD spectra in Figure 7.12, right reveals that AG and CH regenerated from IL solutions largely maintain their native confirmation, whereas in AG–CH composites, the spectra is red shifted with a decrease in intensity from positive to negative (i.e. toward the native CH with the increase in CH amount in composite indicating the disruption of ordered structure of AG due to complexation/intermolecular interactions). CD spectra of AG–CH (50:50) gave equal intensity in positive and negative bands, showing a homogeneous blending of the biopolymers in IL.

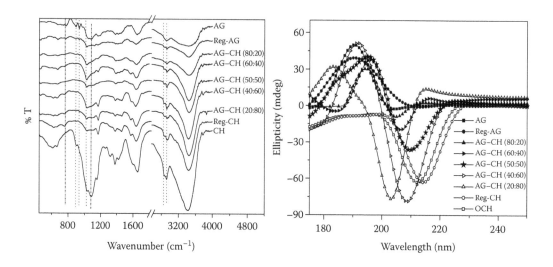

Figure 7.12 FTIR (left) and CD spectrum (right) of native, regenerated biopolymers, and agarose–chitosan (AG–CH) composites having different proportions.

Composite ionogels (3 wt%) were obtained upon cooling AG, CH, or AG–CH (50:50) dissolved viscous solutions of IL. Prepared ionogels have been characterized by gelling, melting, gel strength, conducting, and viscoelastic measurement. Gelling and melting temperatures of various ionogels were monitored through visual inspection of gel and liquid state and are noted in Table 7.7. All the ionogels, except that of pure AG-[C_4mim][Cl], indicate thermal hysteresis. The formation of ionogels is expected to occur in a slightly different fashion than the formation of water-based gels. Here, hydrogen bonding between IL ions and hydroxyl groups of AG molecules must play an important role in gelling. The presence of a large organic cation will interact with AG strands and prevent the formation of double helical structures and finally will lead to a comparatively weaker gel. On the contrary, CH do not produce temperature-induced hydrogels because of a relatively rigid molecular conformation given by the beta glycosidic links of chitosan, which are further stabilized by hydrogen bonds among consecutive units. However, once dissolved in ILs, the CH (being a charged polymer and having an ionic character) interacts with IL ions electrostatically and through hydrogen bond formation. The IL is immobilized in the gel matrix through strong hydrogen bonding between protons of imidazolium and oxygen of AG/CH as well as because of the hydrogen bonding between –OH/–NH$_2$ protons of AG/CH and Cl ions.

Rheological measurements performed for ionogels are shown in Figure 7.13. Frequency dependences of dynamic storage (G') and loss moduli (G'') (Figure 7.13a) show G' is larger than G'' and is nearly frequency independent, showing a solid-like behavior of the ionogels. Dynamic strain sweep experiments on various ionogels over a wide range of strains (0.1%–10%) at a frequency of 100 rad s^{-1} are shown in Figure 7.13b. It has been observed that, for most of the ionogels, a linear viscoelastic regime is maintained indicating no deformation of the gel microstructure under the experimental conditions. In the case of AG ionogels G' and G'' gap slightly decreases with the increase in strain, showing a decrease in solid-like behavior with continuous change in gel microstructure wherein the molecular connections in the gel network are disrupted under the large strains; although in CH and AG–CH composite ionogels, G' remains quite higher than G'' within the region of linear viscoelasticity, indicating persistence of solid-like behavior. Temperature dependences of G' and G'' measured at a frequency of 1 s^{-1} is shown in Figure 7.13c,d. A transition for both G' and G'' as a function of temperature is indicative of melting temperature (T_m). Before the crossover of G' and G'', IL–biopolymer solutions are optically transparent ionogels. T_m obtained from temperature dependent rheology measurements are close to that observed from visual inspection reported in Table 7.7.

Table 7.7 Gelling Temperature (T_{gel}), Melting Temperature (T_m), Gel Strength, and Conductivity of Composite/Nanocomposite Ionogels ($\kappa_{ionogel}$)

Composition	T_{gel} (°C)	T_m (°C)	Gel Strength (g cm^{-2}) at 30°C	$\kappa_{ionogel}$ (mS cm^{-1}) 25°C
AG	40	53	140	0.065
CH	67	96	600	0.099
AG–CH (50:50)	56	84	670	0.075
Ag NPs–AG–CH	38	80	590	0.675

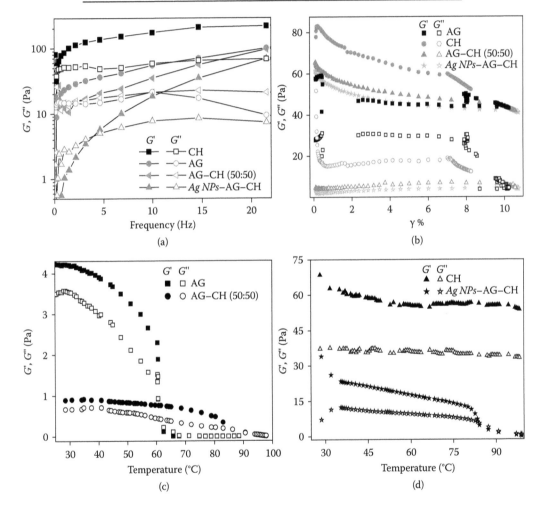

Figure 7.13 (a) Frequency dependence, (b) strain dependence, and (c, d) temperature dependence of dynamic storage (solid symbols) and loss moduli (open symbols) (G' and G'') of different ionogels.

7.3.6 Nanocomposite Materials/Ionogels in ILs

Composite materials has been uniformly decorated with silver oxide (Ag_2O) nanoparticles (*Ag NPs*) to form nanocomposites in a single step by in situ synthesis of *Ag NPs* in AG–CH–IL sols, wherein the biopolymer moiety acted as a reducing and stabilizing agent [121]. For preparation of nanocomposites, 100 mM silver nitrate was added in AG–CH (50:50)–IL solution. The mixture was heated at 100°C under continuous stirring for 6 h. The appearance of color change from slight yellow to dark yellow to light brownish confirmed formation of silver oxide nanoparticles in situ (Figure 7.14). Nanocomposite materials isolated by addition of methanol to the resultant solution enabled precipitation of nanocomposite materials. Decoration of *Ag NPs* in nanocomposite composite materials was confirmed from UV-V is spectroscopy, SEM, TEM, EDAX, and XRD. SEM images (Figure 7.15a,b) of the nanocomposite show that *Ag NPs* are nearly spherical and uniformly decorated on the AG–CH surface. TEM of nanocomposite (Figure 7.15c) showed 10–20 nm size *Ag NPs*; the lattice fringes in high resolution and electron diffraction pattern (SAED) indicated their crystalline and polycrystalline nature. The crystalline nature of nanocomposite is also confirmed from comparison of XRD patterns of regenerated AG–CH and *Ag NPs*–AG–CH composites (Figure 7.15d). AG–CH displayed only a single broad band

Figure 7.14   Photographs of IL-biopolymer sols from left to right: agarose (OAg), chitosan (OCh), agarose–chitosan composite (Ag–Ch), and nanocomposite (AgCh–*Ag NPs*).

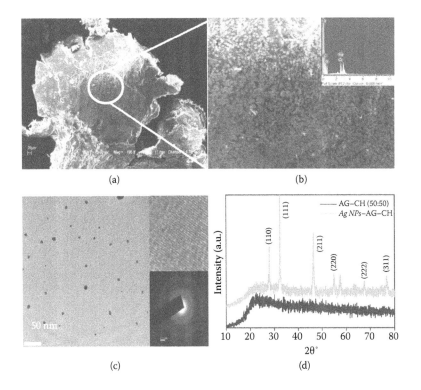

Figure 7.15   (a, b) SEM, (c) TEM image *Ag NPs*–AG–CH composite, and (d) comparison of XRD pattern of AG–CH and *Ag NPs*–AG–CH composite.

showing the amorphous nature of the composite, whereas the *Ag NPs*–AG–CH composite showed two intense peaks at 27.94 and 32.27, which corresponds to (110) and (111) of Ag_2O. Apart from this, diffraction peaks at 46.34, 54.92, 67.48, and 76.81 can be indexed to (211), (220), (222), and (311) planes of face-centered cubic silver, respectively.

Hot solutions of 100 mM $AgNO_3$ in AG–CH (50:50)-[C_4mim][Cl] (at 100°C, under continuous stirring for 6 h) upon cooling to room temperature resulted in the formation of homogeneous nanocomposite ionogels with well dispersed *Ag NPs* (Figure 7.15). Cooling of *Ag NPs*–AG–CH–IL sols to room temperature resulted in high conductivity and high mechanical strength nanocomposite ionogels. High gel strength of nanocomposite ionogel is possibly due to the bridging role of *Ag NPs* among polymer chains and to forming a network structure along with the usual electrostatic and hydrogen bond interactions between polymer groups and IL ions similar to that observed by Shen and Xu [122]. Rheological investigations of nanocomposite ionogels are shown in Figure 7.13. Unlike other ionogels, in nanocomposite ionogels, G' increased continuously and G'' increased slightly initially and then becomes constant with the increase in frequency. Temperature dependences of G' and G'' (Figure 7.13c,d) showed a gel melting temperature (T_m) ~83°C that is very close to that observed from visual inspection (Table 7.7). Strain experiments (Figure 7.13b) indicated no deformation of the gel microstructure. Because *Ag NPs* show good antimicrobial activity, the *Ag NPs*–AG–CH composite materials have the potential to be used in biotechnology and biomedical applications, whereas the nanocomposite ionogels will be suitable as precursors for applications, such as quasisolid dye sensitized solar cells, actuators, sensors, or electrochromic display.

7.4 CONCLUSION

ILs show good potential as a novel green medium for the processing of biopolymers. In literature, several ILs have been used for efficient extraction of various biopolymers from different bioresources. It has been shown that ILs can dissolve various biopolymers under comparatively milder conditions without degradation of materials. Functionalization of biopolymers, which otherwise is difficult, uneconomical, and environment unfriendly in conventional organic solvents, can be done with ease in ionic liquid medium. A case study on processing of a gelling biopolymer "agarose" utilizing ILs as medium is presented. It has been shown that the agarose can be easily extracted from algae in an economical manner. Agarose can be dissolved to a good extent in imidazolium-based ILs, and the dissolution ability can be tremendously increased in mixed ionic liquid systems because of synergic effects and lowering of viscosity. Regeneration of agarose without degradation can be done through the addition of an antisolvent. In addition, one can obtain composite materials just by simultaneous dissolution of different biopolymers. Here we have shown composite/nanocomposite preparation of agarose and chitosan as a representative system. Ionogels are another category of novel materials that can be easily obtained by cooling the biopolymers-ionic liquid solutions. Such ionogels can be promising material for quasisolid dye sensitized solar cells, actuators, sensor-based materials, or for electronic devices.

ACKNOWLEDGMENTS

The authors thank the director of CSIR-CSMCRI for providing the facility to carry out their research work. The authors also thank the Department of Science and Technology (DST), Government of India, for financial support through various sponsored projects (SR/S1/PC-55/2008 and SR/S1/PC-04/2010).

REFERENCES

1. Reichardt, C. 2003. *Solvent and solvent effect in organic chemistry.* Wiley-VCH, Weinheim.
2. Gutmann, V. 1968. *Coordination chemistry in non-aqueous solution.* Springer, New York.
3. Mchugh, M.A. and Krukonis, V.J. 1994. *Supercritical fluid extraction.* Butterworth-Heinemann, Newton, MA.
4. Jessop, P.G. and Leitner, W. 1999. *Chemical synthesis using supercritical fluids.* Wiley-VCH, Weinheim.
5. Rothenberg, G., Downie, A.P., Raston, C.L., and Scott, J.L. 2001. Understanding solid/solid organic reactions. *J. Am. Chem. Soc.*, 123: 8701–8.
6. Jessop, P., Wynne, D.C., Dehaai, S., and Nakawatase, D. 2000. Carbon dioxide gas accelerates solventless synthesis. *Chem. Commun.*, 8: 693–4.
7. Curran, D. and Lee, Z. 2001. Fluorous techniques for the synthesis and separation of organic molecules. *Green Chem.*, 3: G3–G7.
8. Fish, R.H. 1999. Fluorous biphasic catalysis: A new paradigm for the separation of homogeneous catalysts from their reaction substrates and products. *J. Chem. Eur.*, 5: 1677–80.
9. Welton, T. 1999. Room-temperature ionic liquids. Solvents for synthesis and catalysis. *Chem. Rev.*, 99: 2071–84.
10. Wasserscheid, P. and Welton, T. 2008. *Ionic liquids in synthesis.* 2nd ed. Wiley- VCH Verlag GmbH, Weinheim, Germany.
11. Ohno, H. and Fukaya, Y. 2009. Task specific ionic liquids for cellulose technology. *Chem. Lett.*, 38: 2–7.
12. Graenacher, C. 1934. Cellulose solution. U.S. Patent 1943176.
13. Swatloski, R.P., Spear, S.K. John, D., Holbrey, J.D., and Rogers, R.D. 2002. *J. Am. Chem. Soc.*, 124: 4974–5.
14. Moulthrop, J.S., Swatloski, R.P., Moyna, G., and Rogers, R.D. 2005. High-resolution 13C NMR studies of cellulose and cellulose oligomers in ionic liquid solutions. *Chem. Commun.*, 12: 1557–9.
15. Zhang, J., Zhang, H., Wu, J., Zhang, J., He, J., and Xiang, J. 2010. NMR spectroscopic studies of cellobiose solvation in EmimAc aimed to understand the dissolution mechanism of cellulose in ionic liquids. *Phys. Chem. Chem. Phys.*, 12: 1941–7.
16. Remsing, R.C., Swatloski, R.P., Rogers, R.D., and Moyna, G. 2006. Mechanism of cellulose dissolution in the ionic liquid 1-n-butyl-3-methylimidazolium chloride: A 13C and 35/37Cl NMR relaxation study on model systems. *Chem. Commun.*, 1271–3.
17. Liu, H., Sale, K.L., Holmes, B.M., Simmons, B.A., and Singh, S. 2010. Understanding the interactions of cellulose with ionic liquids: A molecular dynamics study. *J. Phys. Chem. B.*, 114: 4293–301.
18. Lu, B., Xu, A., and Wang, J. 2013. Cation does matter: How cationic structure affects the dissolution of cellulose in ionic liquids. *Green Chem.*, 16: 1326–35.
19. Xu, A., Wang, J., and Wang, H. 2010. Effects of anionic structure and lithium salt addition on the dissolution of cellulose in 1-butyl-3-methylimidazolium-based ionic liquid solvents systems. *Green Chem.*, 12: 268–75.
20. Zhang, H., Wu, J., Zhang, J., and He, J. 2005. 1-Allyl-3-methylimidazolium chloride room temperature ionic liquid: A new and powerful non-derivatizing solvent for cellulose. *Macromolecules*, 38: 8272–7.
21. Fukaya, Y., Hayashi, K., Wadab, M., and Ohno, H. 2008. Cellulose dissolution with polar ionic liquids under mild conditions: Required factors for anions. *Green Chem.*, 10: 44–7.
22. Fukaya, Y., Sugimoto, A., and Ohno, H. 2006. Superior solubility of polysaccharides in low viscosity, polar, and halogen-free 1,3-dialkylimidazolium formates. *Biomacromolecules*, 7: 3295–7.
23. Vitz, J., Erdmenger, T., Haensch, C., and Schubert, U.S. 2009. Extended dissolution studies of cellulose in imidazolium based ionic liquids. *Green Chem.*, 11: 417–24.
24. Abe, M., Fukaya, Y., and Ohno, H. 2012. Fast and facile dissolution of cellulose with tetrabutylphosphonium hydroxide containing 40% water. *Chem. Commun.*, 48: 1808–10.
25. Ohira, K., Abe, Y., Kawatsura, M., Suzuki, K., Mizuno, M., Amano, Y., and Itoh, T. 2012. Design of cellulose dissolving ionic liquids inspired by nature. *ChemSusChem.*, 5: 388–91.
26. Wang, H., Gurau, G., and Rogers, R.D. 2012. Ionic liquid processing of cellulose. *Chem. Soc. Rev.*, 41: 1519–37.
27. Yamazaki, S., Takegawa, A., Kaneko, Y., Kadokawa, J.-i., Yamagat, M., and Ishikawa, M. 2009. An acidic cellulose–chit in hybrid gel as novel electrolyte for an electric double layer capacitor. *Electrochem. Commun.*, 11: 68–70.

28. Xie, H., Zhang, S., and Li, S. 2006. Chitin and chitosan dissolved in ionic liquids as reversible sorbents of CO_2. *Green. Chem.*, 8: 630–3.
29. Wang, W.-T., Zhu, J., Wang, X.-L., Huang, Y., and Wang, Y. 2010. Dissolution behavior of chitin in ILs. *J. Macromol. Sci. Phys.*, 49: 528–41.
30. Kadokawa, J.-i., Takegawa, A., Mine, S., and Prasad, K. 2011. Preparation of chitin nanowhiskers using an ionic liquid and their composite materials with poly (vinyl alcohol). *Carbohydr. Polym.*, 84: 1408–12.
31. Chen, Q., Xu, A., Li, Z., Wang, J., and Zhang, S. 2011. Influence of anionic structure on the dissolution of chitosan in 1-butyl-3-methylimidazolium-based ionic liquids. *Green Chem.*, 13: 3446–52.
32. Muzzarelli, R.A.A. 2011. Biomedical exploitation of chitin and chitosan via mechano-chemical disassembly, electrospinning, dissolution in imidazolium ionic liquids, and supercritical drying. *Mar. Drugs.*, 9: 1510–33.
33. Sharma, M., Mukesh, C., Mondal, D., and Prasad, K. 2013. Dissolution of α-chitin in deep eutectic solvents. *RSC Adv.*, 3: 18149–55.
34. Kimizuka, N. and Nakashima, T. 2001. Spontaneous self-assembly of glycolipid bilayer membranes in sugar-philic ionic liquids and formation of ionogels. *Langmuir*, 17: 6759–61.
35. Liu, Q., Janssen, M.H.A., Rantwijk, F.V., and Sheldon, R.A. 2005. Room-temperature ionic liquids that dissolve carbohydrates in high concentrations. *Green. Chem.*, 7: 39–42.
36. Zakrzewska, M.E., Bogel-Łukasik, E., and Bogel-Łukasik, R. 2010. Solubility of carbohydrates in ionic liquids. *Energ. Fuels*, 24: 737–45.
37. Seoud, O.A.E., Koschella, A., Fidale, L.C., Dorn, S., and Heinze, T. 2007. Application of ionic liquids in carbohydrate chemistry: A window of opportunities. *Biomacromolecules*, 8: 2629–47.
38. Mine, S., Prasad, K., Izawa, H., Sonoda, K., and Kadokawa, J.-i. 2010. Preparation of guar gum-based functional materials using ionic liquid. *J. Mater. Chem.*, 20: 9220–5.
39. Idris, A., Vijayaraghavan, R., Rana, U.A., Patti, A.F., and MacFarlane, D.R. 2013. Dissolution of feather keratin in ionic liquids. *Green Chem.*, 15: 525–34.
40. Vidinha, P., Lourenço, N.M.T., Pinheiro, C., Brás, A.R., Carvalho, T., Santos-Silva, T., Mukhopadhyay, A., et al. 2008. Ion jelly: A tailor-made conducting material for smart electrochemical devices. *Chem. Commun.*, 44: 5842–5.
41. Ning, W., Xingxiang, Z., Haihui, L., and Benqiao, H. 2009. 1-Allyl-3-methylimidazolium chloride plasticized-corn starch as solid biopolymer electrolytes. *Carbohydr. Polym.*, 76: 482–4.
42. Mine, S., Prasad, K., Izawa, H., Sonodaa, K., and Kadokawa, J.-i. 2010. Preparation of guar gum-based functional materials using ionic liquid. *J. Mater. Chem.*, 20: 9220–5.
43. Prasad, K., Kaneko, Y., and Kadokawa, J.-i. 2009. Novel gelling systems of κ-, ι- and λ-carrageenans and their composite gels with cellulose using ionic liquid. *Macromol. Biosci.*, 9: 376–82.
44. Prasad, K., Murakami, M., Kaneko, Y., and Kadokawa, J.-i. 2009. Weak gel of chitin with ionic liquid, 1-allyl-3-methylimidazolium bromide. *Int. J. Biol. Macromol.*, 45: 221–5.
45. Prasad, K., Izawa, H., Kaneko, Y., and Kadokawa, J.-i. 2009. Preparation of temperature—shapeable film material from guar gum-based gel with an ionic liquid. *J. Mater. Chem.*, 19: 4088–90.
46. Pimenta, A.F.R., Baptist, A.C., Carvalho, T., Brogueirae, P., Lourençoc, N.M.T., Afonsod, C.A.M., Barreirosb, S., Vidinhab, P., and Borgesa, J.P. 2012. Electrospinning of Ion jelly fibers. *Mater. Lett.*, 83: 161–4.
47. Shamsuri, A.A. and Diak, R. 2012. Plasticizing effect of choline chloride/urea eutectic based ionic liquid on physicochemical properties of agarose films. *Bioresources*, 7: 4760–75.
48. Schlufter, K., Schmauder, H., Dorn, S., and Heinze, T. 2006. Efficient homogeneous chemical modification of bacterial cellulose in the ionic liquid 1-N-Butyl-3-methylimidazolium chloride. *Macromol. Rapid Commun.*, 27: 1670–6.
49. Heinze, T., Schwikal, K., and Barthel, S. 2005. Ionic liquids as reaction medium in cellulose functionalization. *Macromol. Biosci.*, 5: 520–5.
50. Barthel, S. and Heinze, T. 2006. Acylation and carbanilation of cellulose in ionic liquids. *Green Chem.*, 8: 301–6.
51. Heinze, T., Dorn, S., Schöbitz, M., Liebert, T., Köhler, S., and Meister, F. 2008. Interactions of ionic liquids with polysaccharides—2: Cellulose. *Macromol. Symp.*, 262: 8–22.
52. Liebert, T. and Heinze, T. 2008. Interaction of ionic liquids with polysaccharide 5. Solvents and reaction media for the modification of cellulose. *Bioresources*, 3: 576–601.

53. Abbott, A.P., Bell, T.J., Handa, S., and Stoddart, B. 2005. O-Acetylation of cellulose and monosaccharides using zinc based ionic liquid. *Green Chem.*, 7: 705–7.

54. Abbott, A.P., Bell, T.J., Handa, S., and Stoddart, B. 2006. Cationic functionalisation of cellulose using a choline based ionic liquid analogue. *Green Chem.*, 8: 784–6.

55. Wu, J., Zhang, J., He, J., Ren, Q., and Guo, M. 2004. Homogeneous acetylation of cellulose in a new ionic liquid. *Biomacromolecules*, 5: 266–8.

56. Tomé, L.C., Freire, M.G., Rebelo, L.P.N., Silvestre, A.J.D., Neto, C.P., Marruchoab, I.M., and Freire, C.S.R. 2011. Surface hydrophobization of bacterial and vegetable cellulose fibers using ionic liquids as solvent media and catalysts. *Green Chem.*, 13: 2464–70.

57. Gericke, M., Fardim, P., and Heinze, T. 2012. Ionic liquids—Promising but challenging solvents for homogeneous derivatization of cellulose. *Molecules*, 17: 7458–502.

58. Biswas, A., Shogren, R.L., Stevenson, D.G., Willett, J.L., and Bhowmik, P.K. 2006. Ionic liquids as solvents for biopolymers: Acylation of starch and zein protein. *Carbohydr. Polym.*, 66: 546–50.

59. Liu, C.F., Sun, R.C., Zhang, A.P., Ren, J.L., Wang, X.A., Qin, M.H., Chao, Z.N., and Luo, W. 2007. Homogeneous modification of sugarcane bagasse cellulose with succinic anhydride using a ionic liquid as reaction medium. *Carbohydr. Res.*, 342: 919–26.

60. Liu, Z., Shen, L., Liu, Z., and Lu, J. 2009. Acetylation of β-cyclodextrin in ionic liquid green solvent. *J. Mater. Sci.*, 44: 1813–20.

61. Kadokawa, J.-i. 2013. Ionic liquid as useful media for dissolution, derivatization, and nanomaterial processing of chitin. *Green Sustain. Chem.*, 3: 19–25.

62. Liu, L.S., Zhou, B., Wang, F., Xu, F., and Sun, R. 2013. Homogeneous acetylation of chitosan in ionic liquids. *J. Appl. Polym. Sci.*, 129: 28–35.

63. Hamada, Y.Y., Yoshida, K., and Asai, R.-i. 2013. A possible means of realizing a sacrifice-free three component separation of lignocellulose from wood biomass using an amino acid ionic liquid. *Green Chem.*, 15: 1863–8.

64. Xin, Q., Pfeiffer, K., Prausnitz, J.M., Clark, D.S., and Blanch, H.W. 2012. Extraction of lignins from aqueous–ionic liquid mixtures by organic solvents. *Biotechnol. Bioeng.*, 109: 346–52.

65. Prado, R., Erdocia, X., and Labidi, J. 2013. Lignin extraction and purification with ionic liquids. *J. Chem. Technol. Biotechnol.*, 88: 1248–57.

66. Fort, D.A., Remsing, R.C., Swatloski, R.P., Moyna, P., Moyna, G., and Rogers, R.D. 2007. Can ionic liquids dissolve wood? Processing and analysis of lignocellulosic materials with 1-n-butyl-3-methylimidazolium chloride. *Green Chem.*, 9: 63–9.

67. Du, Z., Yu, Y.-L., and Wang, J.-H. 2007. Extraction of proteins from biological fluids by use of an ionic liquid/aqueous two-phase system. *Chem. Eur. J.*, 13: 2130–7.

68. Wang, J.H., Cheng, D.H., Chen, X.W., Du, Z., and Fang, Z.L. 2007. Direct extraction of double-stranded DNA into ionic liquid 1-butyl-3-methylimidazolium hexafluorophosphate and its quantification. *Anal. Chem.*, 79: 620–5.

69. Wang, Y.-X. and Cao, X.-J. 2012. Extracting keratin from chicken feathers by using a hydrophobic ionic liquid. *Process Biochem.*, 47: 896–9.

70. Guolin, H., Jeffrey, S., Kai, Z., and Xiaolan, H. 2012. Application of ionic liquids in the microwave-assisted extraction of pectin from lemon peels. *J. Anal. Methods Chem.*, 2012: 1–8.

71. Ferreira, R., Garcia, H., Sousa, A.F., Petkovic, M., Lamosa, P., Freire, C.S.R., Silvestre, A.J.D., Rebeloa, L.P.N., and Pereira, C.S. 2012. Suberin isolation from cork using ionic liquids: Characterisation of ensuing products. *N. J. Chem.*, 36: 2014–24.

72. Qin, Y., Lu, X., Sun, N., and Rogers, R.D. 2010. Dissolution or extraction of crustacean shells using ionic liquids to obtain high molecular weight purified chitin and direct production of chitin films and fibers. *Green Chem.*, 12: 968–71.

73. Barber, P.S., Griggs, C.S., Bonner, J.R., and Rogers, R.D. 2013. Electrospinning of chitin nanofibers directly from an ionic liquid extract of shrimp shells. *Green Chem.*, 15: 601–7.

74. Cláudio, A.F.M., Ferreira, A.M., Freire, M.G., and Coutinho, J.A.P. 2013. Enhanced extraction of caffeine from *Guaraná* seeds using aqueous solutions of ionic liquids. *Green Chem.*, 15: 2002–10.

75. Manic, M.S., Visak, V.N., and Ponte, M.N.D. 2011. Extraction of free fatty acids from soybean oil using ionic liquids or poly (ethyleneglycol)s. *AIChE J.*, 57: 1344–55.

76. Usuki, T., Yasuda, N., Fujita, M.Y., and Rikukawa, M. 2011. Extraction and isolation of shikimic acid from *Ginkgo biloba* leaves utilizing an ionic liquid that dissolves cellulose. *Chem. Commun.*, 47: 10560–2.

77. Ressmann, A.K., Gaertner, P., and Bica, K. 2011. From plant to drug: Ionic liquids for the reactive dissolution of biomass. *Green Chem.*, 13: 1442–7.

78. Yansheng, C., Zhida, Z., Qingshan, L., Peifang, Y., and Biermann, U.W. 2011. Microwave-assisted extraction of lactones from *Ligusticum chuanxiong* Hort. using protic ionic liquids. *Green Chem.*, 13: 666–70.

79. Chowdhury, S.A., Vijayaraghavan, R., and MacFarlane, D.R. 2010. Distillable ionic liquid extraction of tannins from plant materials. *Green Chem.*, 12: 1023–8.

80. Bica, K., Gaertner, P., and Rogers, R.D. 2011. Ionic liquids and fragrances-direct isolation of orange essential oil. *Green Chem.*, 13: 1997–9.

81. Teixeira, R.E. 2012. Energy-efficient extraction of fuel and chemical feedstocks from algae. *Green Chem.*, 14: 419–27.

82. Kim, Y.-H., Choi, Y.-K., and Park, J. 2012. Ionic liquid-mediated extraction of lipids from algal biomass. *Bio. Technol.*, 109: 312–15.

83. Abe, M., Fukaya, Y., and Ohno, H. 2010. Extraction of polysaccharides from bran with phosphonate or phosphinate-derived ionic liquids under short mixing time and low temperature. *Green Chem.*, 12: 1274–80.

84. Fujita, K., Kobayanshi, D., Nakamura, N., and Ohno, H. 2013. Direct dissolution of wet and saliferous marine microalgae by polar ionic liquids without heating. *Enzyme Microb. Technol.*, 52: 199–202.

85. Xie, H.B., Li, S.H., and Zhang, S.B. 2005. Ionic liquids as novel solvents for the dissolution and blending of wool keratin fibers. *Green Chem.*, 7: 606–8.

86. Park, T.-J., Lee, S.-H., and Simmons, T.J. 2008. Heparin–cellulose–charcoal composites for drug detoxification prepared using room temperature ionic liquids. *Chem. Commun.*, 40: 5022–4.

87. Sun, N., Swatloski, R.P., Maxim, M.L., Rahman, M., Harland, A.G., Haque, A., Spear, S.K., Daly, D.T., and Rogers, R.D. 2008. Magnetite-embedded cellulose fibers prepared from ionic liquid. *J. Mater. Chem.*, 18: 283–90.

88. Yu, Z., Jiang, Y., Zou, W., Duan, J., and Xiong, X. 2009. Preparation and characterization of cellulose and Konjac glucomannan blend film from ionic liquid. *J. Polym. Sci. B. Polym. Phys.*, 47: 1686–94.

89. Kuzmina, O.G., Sashina, E.S., Novoselov, N.P., and Zoboroski, M. 2009. Blends of cellulose and silk fibroin in 1-butyl-3-methylimidazolium chloride-based solutions. *FIBRES TEXTILES East. Eur.*, 17: 36–9.

90. Wu, R.-L., Wang, X.-L., Li, F., Li, H.-Z., and Wang, Y.-Z. 2009. Green composite films prepared from cellulose, starch and lignin in room-temperature ionic liquid. *Bioresour. Technol.*, 100: 2569–74.

91. Sun, X., Peng, B., Ji, Y., Chen, J., and Li, D. 2009. Chitosan(chitin)/cellulose composite biosorbents prepared using ionic liquid for heavy metal ions adsorption. *AIChE J.*, 55: 2062–9.

92. Huang, K.-J., Miao, Y.-X., Wang, L., Gan, T., Yu, M., and Wang, L.-L. 2012. Direct electrochemistry of hemoglobin based on chitosan–ionic liquid–ferrocene/graphene composite film. *Process Biochem.*, 47: 1171–7.

93. Liu, Z., Wang, H., Li, B., Liu, C., Jiang, Y., Yu, G., and Mu, X. 2012. Biocompatible magnetic cellulose–chitosan hybrid gel microspheres reconstituted from ionic liquids for enzyme immobilization. *J. Mater. Chem.*, 22: 15085–91.

94. Tran, C.D., Duri, S., Delneri, A., and Franko, M. 2013. Chitosan-cellulose composite materials: Preparation, characterization and application for removal of microcystin. *J. Hazard. Mater.*, 252–253: 355–66.

95. Shamsuri, A.A. and Daik, R. 2013. Utilization of an ionic liquid/urea mixture as a physical coupling agent for agarose/talc composite films. *Materials*, 6: 682–98.

96. Kadokawa, J.-i., Murakami, M., Takegawa, A., and Kaneko, Y. 2009. Preparation of cellulose–starch composite gel and fibrous material from a mixture of the polysaccharides in ionic liquid. *Carbohydra. Polym.*, 75: 180–3.

97. Shang, S., Zhu, L., and Fan, J. 2011. Physical properties of silk fibroin/cellulose blend films regenerated from the hydrophilic ionic liquid. *Carbohydr Polym.*, 86: 462–8.

98. Takegawa, A., Murakami, M., Kaneko, Y., and Kadokawa, J.-i. 2010. Preparation of chitin/cellulose composite gels and films with ionic liquids. *Carbohydr. Polym.*, 79: 85–90.

99. Luo, N., Varaprasad, K., Reddy, G.V.S., Rajulu, A.V., and Zhang, J. 2012. Preparation and characterization of cellulose/curcumin composite films *RSC Adv.*, 2: 8483–8.

100. Li, Z., and Taubert, A. 2009. Cellulose/gold nanocrystal hybrids via an ionic liquid/aqueous precipitation route. *Molecules*, 14: 4682–8.

101. Brondani, D., Zapp, E., Vieira, I.C., Dupontb, J., and Scheeren, C.W. 2011. Gold nanoparticles in an ionic liquid phase supported in a biopolymeric matrix applied in the development of a rosmarinic acid biosensor. *Analyst*, 136: 2495–505.

102. Gelesky, M.A., Scheeren, C.W., Foppa, L., Pavan, F.A., Dias, S.L.P., and Dupont, J. 2009. Metal nanoparticle/ionic liquid/cellulose: new catalytically active membrane materials for hydrogenation reactions. *Biomacromolecules*, 10: 1888–93.

103. Ma, M.-G., Dong, Y.-Y., Fu, L.-H., Li, S.-M., and Shung, R.-C. 2113. Cellulose/CaCO3 nanocomposites: Microwave ionic liquid synthesis, characterization, and biological activity. *Carbohydr. Polym.*, 92: 1669–76.

104. Ma, M.-G., Quing, S.-J., Li, S.-M., Zhu, J.-F., Fu, L.-H., and Shung, R.-C. 2013. Microwave synthesis of cellulose/CuO nanocomposites in ionic liquid and its thermal transformation to CuO. *Carbohydr. Polym.*, 91: 162–8.

105. Bagheri, M. and Rabieh, S. 2013. Preparation and characterization of cellulose-ZnO nanocomposite based on ionic liquid ([C4mim]Cl). *Cellulose*, 20: 699–705.

106. Singh, N., Koziol, K.K.K., Chen, J., Patil, A.J., Gilman, J.W., Trulove, P.C., Kafienah, W., and Rahatekar, S.S. 2013. Ionic liquids-based processing of electrically conducting chitin nanocomposite scaffolds for stem cell growth. *Green Chem.*, 15: 1192–202.

107. Sharma, M., Mondal, D., Mukesh, C., and Prasad, K. 2013. *Carbohydr. Polym.*, 98: 1025–30.

108. Mahmoudian, S., Wahit, M.U., Ismail, A.F., and Yussuf, A.A. 2012. Preparation of regenerated cellulose/montmorillonite nanocomposite films via ionic liquids. *Carbohydr. Polym.*, 88: 1251–7.

109. Xiao, W., Wu, T., Peng, J., Bai, Y., Li, J., Lai, G., Wu, J., and Dai, L. 2013. Preparation, structure, and properties of chitosan/cellulose/multiwalled carbon nanotube composite membranes and fibers. *J. Appl. Polym. Sci.*, 128: 1193–9.

110. Sun, W., Li, X., Liu, S., and Jiao, K. 2009. Electrochemistry of hemoglobin in the chitosan and TiO2 nanoparticles composite film modified carbon ionic liquid electrode and its electrocatalysis. *Bull. Korean Chem. Soc.*, 30: 582–8.

111. Gayet, F., Viau, L., Leroux, F., Monge, S., Robin, J.-J., and Vioux, A. 2010. Polymer nanocomposite ionogels, high-performance electrolyte membranes. *J. Mater. Chem.*, 20: 9456–62.

112. Selby, H.H. and Whistler, R. L. 1993. *Agar. Industrial gums*. Academic Press, New York, pp. 87–103.

113. Novak, R., Zeng, Y., Shuga, J., Venugopalan, G., Fletcher, D.A., Smith, M.T., and Mathies, R.A. 2011. Single-cell multiplex gene detection and sequencing with microfluidically generated agarose emulsions. *Angew. Chem. Int. Ed. Engl.*, 123: 410–15.

114. Trivedi, T. J. 2013. *Studies on utilization of room temperature ionic liquids for processing of biopolymer agarose*. PhD Thesis, Academy of Scientific and Innovative Research, India.

115. Sun, N., Rahman, M., Qin, Y., Maxim, M.L., Rodrigues, H., and Rogers, R.D. 2009. Complete dissolution and partial delignification of wood in the ionic liquid 1-ethyl-3-methylimidazolium acetate. *Green Chem.*, 11: 646–55.

116. Brandt, A., Hallett, J.P., Leak, D.J., Murphy, R.J., and Welton, T. 2010. The effect of the ionic liquid anion in the pretreatment of pine wood chips. *Green Chem.*, 12: 672–9.

117. Singh, T., Trivedi, T.J., and Kumar, A. 2010. Dissolution, regeneration and ion-gel formation of agarose in room temperature ionic liquids. *Green Chem.*, 12: 1029–35.

118. Trivedi, T.J., Srivastava, D.N., Rogers, R.D., and Kumar, A. 2012. Agarose processing in protic and mixed protic–aprotic ionic liquids: Dissolution, regeneration and high conductivity, high strength ionogels. *Green Chem.*, 14: 2831–9.

119. Kamlet, M.J., Abboud, J.L.M., Abraham, M.H., and Taft, R.W. 1983. Linear solvation energy relationships. 23. A comprehensive collection of the solvatochromic parameters, pi. alpha, and beta, and some methods for simplifying the generalized solvatochromic equation. *J. Org. Chem.*, 48: 2877–87.

120. Rinaldi, R. 2011. Instantaneous dissolution of cellulose in organic electrolyte solutions. *Chem. Commun.*, 47: 511–13.

121. Trivedi, T.J., Rao, K.S., and Kumar, A. 2014. Facile preparation of agarose-chitosan hybrid materials and nanocomposite ionogels using an ionic liquid via dissolution, regeneration and sol-gel transition. *Green Chem.*, 16: 320–30.

122. Shen, J.-S. and Xu, B. 2011. In situ encapsulating silver nanocrystals into hydrogels. A "green" signaling platform for thiol-containing amino acids or small peptides. *Chem. Commun.*, 47: 2577–9.

CHAPTER 8

Economically Viable Biochemical Processes for Advanced Rural Biorefinery and Downstream Recovery Operations

Patrick Dube* and Pratap Pullammanappallil
Department of Agricultural and Biological Engineering, University of Florida, Gainesville, FL, USA

CONTENTS

8.1 INTRODUCTION

Rural biorefineries offer an alternative to traditional ethanol production by providing the opportunity to produce fuel on site to reduce costs associated with biomass transportation thus making the fuel economically viable. Widespread installation of rural biorefineries could lead to increased uptake of biofuels and could help meet biofuel standards in the upcoming years. Unfortunately, even with reduced transportation costs at a rural biorefinery, biofuel is still not competitive with traditional gasoline fuel due to high processing costs and inefficient processes. Various technologies and processes that are available for conversion of biomass to biofuels and bioproducts can be grouped under two platforms, namely syngas and sugar. Under the syngas platform, the biomass is first thermochemically converted to syngas (which is primarily a mixture

---

* Present affiliation: USDA-ARS, Florence, South Carolina.

of carbon monoxide and hydrogen) at high temperatures. The syngas is a precursor for the production of fuels and chemicals using catalytic processes. In the sugar platform, the polysaccharides in biomass are first saccharified to sugars by employing a combination of chemical, thermal, and enzymatic pretreatments. The sugars are then biochemically converted to fuels and chemicals using microbial biocatalysts. Biochemical options in industrial use require expensive pretreatment and hydrolysis costs that drive up final prices and that are not feasible in a rural setting. This chapter aims to identify biochemical options that can be employed in a rural locale and still produce an economically viable biofuel product.

First, feedstocks will be investigated in order to identify options that can viably use a biochemical platform to produce ethanol. Both lignocellulosic and cellulosic feedstocks will be considered with ease of growth, convertibility, and simplicity of system design being major factors that will determine how viable they are. Once feedstocks are chosen, their conversion to biofuels will be determined. Because only ethanol has emerged as the biofuel that can be commercially produced employing biochemical pathways, this chapter will focus on ethanol production. Topics will include conversion technologies, fermentation, and economic analysis of processes that can efficiently convert feedstocks to ethanol fuel. In order to help drive down costs and make this process more profitable, the conversion of the spent residues (that is residues remaining after ethanol extraction) using anaerobic digestion will be examined. Biogas produced from this step can generate energy that can be used to run the facility. A combined integrated system will be proposed in which as much of the feedstock as possible will be converted to a valuable product.

8.2 FEEDSTOCKS

When choosing feedstocks for a rural biorefinery, there are many characteristics to consider in order to choose a feedstock that can be biochemically converted to a valuable product in a cost-effective manner. Lignocellulosic feedstocks, although the most abundant raw material on the planet, are made up of cellulose, hemicellulose, and lignin. These carbohydrate polymers combined are resistant to degradation due to cross-linking among the polysaccharide cellulose and hemicellulose and lignin. These cross-linking ester and ether links must be eliminated before the sugar in the feedstock can be readily accessible. In a typical biochemical ethanol pathway, lignocellulose is broken down using a pretreatment step followed by hydrolysis. These steps, although effective at releasing sugars, are expensive processes that add costly equipment, supply and process costs, inhibiting widespread use of this technology.

In a rural setting, choosing a feedstock that can be easily processed is of utmost importance. The optimal feedstock for a rural biorefinery would be one that does not require extensive processing before sugar extraction could occur. Sugarcane, sugar beet, and sweet sorghum are all feedstocks from which sugar can be extracted by pressing, milling, or similar simple means. Once pressed, the juice contains six carbon hexose sugars that can be readily converted to ethanol via fermentation. The residue remaining after extracting the juice can be further processed via anaerobic digestion or can be burned to help offset energy costs.

Sugarcane is not commonly used for ethanol in the United States due to limited tropical areas in which to grow it, but it is common in other countries with extensive tropical regions, such as Brazil. The sugarcane-to-ethanol model has been proven in Brazil where, in 2012–2013, 6.1 billion gallons of fuel was produced from 588 million tons of sugar (UNICA 2013). With a chemical composition of 20% lignin, 24% hemicellulose, and 43% cellulose in sugarcane bagasse, ethanol from sugarcane bagasse does not favor a simple pathway of conversion due to the high lignin content (Kim and Day 2011). Containing 10%–15% sucrose, sugar beets can provide a large amount of sugars for ethanol production. From sucrose, the theoretical yield of

Table 8.1 Sugar-Based Crop Yields and Potential Ethanol Yields

	Yield (tons/ac)[a,b]	Total Production (million tons)[a]	Sugar Yield (kg/ha)	Ethanol Yield (L/ha)
Sugarcane	34.6	29.7		6900
Sugar beet	28.5	32.8	9525[d]	5060[d]
Sweet sorghum	28.75		1345[c]	704[c]

Sources: [a]USDA, 2014; [b]Amosson, A., et al., *Economic Analysis of Sweet Sorghum for Biofuels Production in the Texas High Plains*, Texas A&M Agrilife Extension, Amarillo, TX, 2011; [c]Rutto, L. K., et al., *Journal of Sustainable Bioenergy Systems*, 3, 113–8, 2013; [d]Panella, L., *Sugar Technology*, 12, 288–293, 2010.

Note: Sugarcane—40–50 kg juice per 100 kg cane; 17%–22% sugar in juice.

ethanol is 163 gallons of ethanol per ton of sucrose with the realistic expected recovery taking into consideration plant operations being 141 gallons per ton of sucrose. One ton of sugarcane typically yields 19.5 gallons of ethanol.

Sugar beets are produced in the United States in the Northwest, Great Plains, and Great Lakes regions. There were more than 32.8 million tons of sugar beets produced in 2013–2014 from 1.15 million acres (USDA 2014). Sugar beets consist of 74% total sugar with 21.1% of that being glucose. After being pressed, the tailings and pulp that remain are typically sold at low cost for affordable cattle feed.

Sweet sorghum is attractive for ethanol production mainly because the juice in the stalk is readily available, full of sugar, and highly fermentable (Montross et al. 2009). Using nitrogen requirements as a reference to determine the energy potential of sweet sorghum, it can be found that 60–90 lb/ac is required to grow sweet sorghum. Considering that one pound of N requires 18,000 Btu for production, sorghum requires approximately 1.35 million Btu/ac in nitrogen alone. Assuming that sweet sorghum can yield 400 gal/ac and that ethanol has an energy content of 76,000 Btu/gal, there is potential for 30.4 million Btu/gal of fuel to be produced. Table 8.1 lists crop and ethanol yields for sugar based feedstocks.

8.3 ETHANOL

In order to produce economically competitive ethanol in a rural setting, lowering costs must be taken heavily into consideration. The first step to lowering costs is to choose a viable feedstock, as outlined earlier. Sugar-based feedstocks such as sugar beets, sugarcane, and sweet sorghum are optimal because they eliminate a costly step in ethanol production—hydrolysis. When dealing with a cellulosic feedstock such as corn or wheat, following a pretreatment step, enzymatic hydrolysis must be utilized to convert the starches to sugars that can then be fermented into ethanol. With a sugar-based feedstock, enzymatic hydrolysis is not necessary because, after a simple milling or pressing, sugars are ready for fermentation.

The first step in turning a sugar-based feedstock into ethanol comes after harvesting. In order to maximize the amount of sugar available, it must be utilized as soon as possible. A rural biorefinery is an excellent choice for sugar biomass due to the proximity of the ethanol production facility to the crop. For sugarcane, the next step is for the sugar to be washed clean of debris then sent to the roll mills for extraction. This step separates the sugarcane juice containing sugars from the bagasse. Sugar beets are processed differently than sugarcane. Following harvesting, they are cleaned, sliced into fine slices, and then the sugar is extracted using hot water in a diffuser. This step separates the spent sugar beet from a sugar-based liquid. Sweet sorghum is very similar to sugarcane in which cleaned stems are processed using mills to create sweet sorghum syrup and spent sweet sorghum biomass. More details on ethanol production can be found in USDA (1982).

Following sugar extraction, the resulting juice is taken through a screening process and then limed, heated, and decanted. Next, evaporation occurs in order to achieve a specific sugar concentration. Depending on the system design, molasses can be added to form a sugarcane mash. Once the juice is concentrated, it needs to be fermented by yeast. *Saccharomyces cerevisiae* can ferment this concentrated sugary syrup over a period of 8–12 hours, resulting in a liquid with 7%–10% ethanol. In order to concentrate this liquid, it must be further processed beginning by using distillation. Distillation alone creates a hydrated ethanol product, which then must be further dehydrated using a dehydrating column or by adsorption using molecular sieves.

8.3.1 Economics

Much of the challenge in producing ethanol is a result of the high energy input needed for the different steps during conversion of a biomass into fuel. The energy demand comes from three different areas: thermal (steam for heaters, evaporators, and distillation), mechanical (preprocessing and pumping), and electric (engines and lighting). The thermal energy demand to produce hydrated bioethanol ranges from 370–410 kg of steam/tonne of cane to 500–580 kg of steam/tonne of cane for anhydrous bioethanol. Mechanical energy demands are much lower at 16 kWh/tonne of cane for hydrated and anhydrous bioethanol. Both hydrated and anhydrous bioethanol require 12 kWh/tonne of cane for electric requirements.

8.3.2 By-Products

During the processing of sugar biomass to ethanol, there are a variety of side products that are currently disposed of or sold for minimum value. In an effort to produce more revenue and make a rural biorefinery more economically feasible, processing these products can result in a valuable commodity.

Bagasse is the discarded biomass after sugar juice is extracted from sugarcane; it is similar to sugar beet pulp and spent sweet sorghum. Although there is not much readily available sugar left in this spent biomass, it is full of cellulose. This cellulose can be transformed into paper or cardboard, used for animal feed, burned for energy, or further treated with enzymes for additional ethanol production. Similarly, it can be anaerobically digested and converted into biogas to help offset facility costs.

Molasses is a by-product of the processing of sugarcane for crystalline sugar. It can be further converted to ethanol or added as a supplement to some animal feeds. Some facilities are currently using it to grow bacteria and fungi that are used in the chemical and pharmaceutical industries, including yeast production. During the fermentation process, yeast produces ethanol as well as growing more cells for further use—as much as 15–30 g of dry yeast per liter of ethanol produced. This yeast can be separated from the liquid components or extracted from the bottom of the fermenting vessels and used as a source of protein for animal feed.

Another by-product of ethanol production is carbon dioxide. Most facilities must first treat this off-gas before venting it to the atmosphere to reduce global warming. It can be washed and used to recover bioethanol before being released but has higher value elsewhere. By purifying, liquefying, deodorizing, and storing it under pressure, it can be sold for use in the beverage industry, sodium bicarbonate manufacturing, or in other industries in need of carbon dioxide. Other by-products, such as citric acid and amino acids, can be made for the chemical industry but require additional processing steps and therefore are unfit for a rural biorefinery.

8.3.3 Challenges

Current challenges facing widespread bioethanol production stem from making the process economically competitive with current fuel sources. There is a large focus on increasing the positive energy balances displayed in bioethanol production while maintaining environmental standards (Amorim et al. 2011). Current technologies have improved preprocessing of feedstocks in order to maximize the quantity and quality of sugar collected. This has improved milling, pressing, and separation technologies seen on a rural biorefinery. Future challenges that must be implemented to improve the energy balance on a rural biorefinery will look at yeast for conversion of sugars to ethanol. Researching new strains and combining this work with genetic engineering of current strains will improve yields and help create a more effective process.

8.4 ANAEROBIC DIGESTION

Anaerobic digestion is the process of breaking down organic material to methane and carbon dioxide by a mixed culture of microorganisms in the absence of oxygen. This process takes place in four distinct steps: hydrolysis, acidogenesis, acetogenesis, and methanogenesis. To begin, the proteins, carbohydrates, and lipids present in the organic matter are hydrolyzed into amino acids, sugars, and fatty acids (Gujer and Zehnder 1983). The next two steps happen in concert with each other. Some of the long-chain fatty acids and alcohols that have been produced are oxidized into acetate and hydrogen while the rest are converted to volatile fatty acids before transforming into acetate and hydrogen. The final methanogenic step converts the resulting acetate and hydrogen into methane and carbon dioxide (Gujer and Zehnder 1983).

In order to optimize biogas production, there are certain parameters that must be monitored and adjusted to achieve maximum results. These parameters include temperature (mesophilic vs. thermophilic), loading rates, pH, pretreatment, retention time, reactor design, and system design (Mata-Alvarez et al. 2000).

Temperature is a very important factor in anaerobic digestion. Bacteria that drive methane production fall into two different temperature ranges: mesophilic bacteria work best around 35°C while thermophilic bacteria thrive around 55°C. Mesophilic anaerobic digestion is more widely used due to the lower temperature, lower operations costs, and stable effluent quality but takes longer to achieve a maximum methane yield (Song et al. 2004). Thermophilic anaerobic digestion achieves a maximum yield quicker, can destroy certain pathogens, and has an increased organic solid reduction, but it has a poor effluent quality, requires more energy input to the system, and needs special equipment to deal with the higher temperatures (Kim et al. 2002).

One of the most important considerations when designing an anaerobic digestion system is system design. A simple design includes the entire process in a one-stage system, but during acidogenesis, the volatile fatty acids and other acids that are formed can cause a drop in pH that can inhibit the entire system unless closely monitored (Bouallagui et al. 2005). In order to combat this problem, many systems are designed in two separate reactors or stages. In the first reactor, organic matter is loaded along with the necessary bacteria where hydrolysis and acidification occur. Once this process is complete, it can be transferred to an anaerobic filter where methanogenesis occurs, producing carbon dioxide and methane. In a two-stage system, each reactor has its own optimal pH, growth, and nutrient uptake rates and over time develop an efficient bacterial community that is optimized for its job. The low operating and start-up costs for one-stage systems can be useful for feedstocks that are not prone to overacidification (De Baere 2000).

Equally important as system design, system operation has a large role in optimal anaerobic digestion as well. Many systems are run by continuously loading and removing organic matter from the system. In larger scale reactors, by the time the organic matter has made it through the system, it has been fully digested and is no longer beneficial. This process limits acidification by reducing hydraulic retention time but can lead to the lack of a strong microbial community as well as possible methanogenic bacteria inhibition due to overfeeding (Bouallagui et al. 2005).

High loading rates in the system can inhibit the system with accumulation of volatile fatty acids and/or a microbial population shift, leading to low methane yields (Salminen and Rintala 2002; Mohan et al. 2007).

8.4.1 Biogas Cleanup

For use on site, biogas that is created must first be cleaned up in order to be used. Typically, biogas straight from an anaerobic digester contains 55%–70% methane (CH_4), 30%–45% carbon dioxide (CO_2), as well as hydrogen sulfide (H_2S), hydrogen (H_2), and water vapor (Frazier et al. 2014). Biogas is an odorless and colorless gas akin to natural gas. Due to the impurities present in the biogas, the energy content is 20–26 MJ/m^3 compared to the 39 MJ/m^3 of commercial natural gas.

Cleaning up biogas allows for more efficient burning and increased power and heat output. In a simple system similar to what would be on a rural biorefinery, CO_2 does not need to be removed because it can move through an engine without harm. Alternatively, removing CO_2 increases the heating value and maintains a constant gas quality. High pH, low temperature, and high pressures in water can result in the scrubbing of CO_2. Organic solvents (polyethyleneglycol and alkanolamines) can also efficiently remove CO_2 and H_2S but are very costly and not sensible on a small scale. Similarly, using activated carbon or molecular sieves can be very efficient but not practical due to the high temperatures and pressures needed. Water vapor is another impurity in biogas, but it can be removed relatively easily. Thermal heat piping can help condense out the water as can compressing the gas before cooling, for simple removal. More expensive methods such as adsorption are an option but can be cost prohibitive. Hydrogen sulfide (H_2S) must be removed from biogas for all but the simplest burning device. When combined with water vapor, the two react and form sulfuric acid, which is corrosive and can destroy engine parts. Activated carbon removes H_2S by oxidizing it to sulfur in the presence of oxygen. Scrubbing the gas with sodium hydroxide, water, or iron salt is also an option. The simplest solution might be to introduce oxygen to the gas which oxidizes H_2S to sulfur and water that can then be removed. There are more expensive methods that include membrane filtration and biofiltration, but on a simple rural biorefinery, these are not economically feasible.

There are a variety of ways to utilize biogas for energy production. Without cleaning up the biogas, it can be burned in a simple gas burner by adjusting the air-to-gas ratio. Hydrogen sulfide levels must be below 1000 ppm and the pressure of the gas at the burner orifice must be between 8 and 25 mbar in order for this to be successful. Although a simple and cheap option, this unit must be replaced periodically due to H_2S corrosion of the equipment. An internal combustion engine may be used, but the biogas must first be cleaned up. H_2S must be removed to below 1000 ppm and water vapor must also be entirely removed in order to use in an internal combustion gas engine that can produce both head and power (25–100 kW). Depending on the size of the rural biorefinery, a fuel cell operation can be used. The gas must be meticulously cleaned to remove sulfides and siloxanes but power and head can be produced at more than 60% efficiency.

8.4.2 Challenges

The challenges faced when using anaerobic digestion lie in improving the total methane yield from specific feedstocks in order to increase the energy that can be produced. Manipulating loading rates, microbial community populations, process design, system design, and monitoring the entire system closely can increase methane yields. Different feedstocks have different requirements and thus must be individually tested and researched. In addition to the challenges faced in improving yields, environmental concerns involving dumping of spent feedstocks, biogas cleanup, and conversion to energy must be taken into account.

8.5 INTEGRATING ANAEROBIC DIGESTION OF WASTES INTO A BIOCHEMICAL RURAL BIOREFINERY

The main goal of a biorefinery is to streamline the process of biofuel and bioenergy production with minimal environmental impacts. Generating fuels by choosing feedstocks that can be readily processed and by combining this with anaerobic digestion that will turn waste into further energy, the system can be set up to run as efficiently as possible. By utilizing wastes and residues produced in the process of making the primary product, the biorefinery may be able to offset fossil fuels. A case study outlining the impact of integrating anaerobic digestion into a sugar beet ethanol process facility is described here.

Sugar beets are an annual crop that can be grown over a wide range of climatic conditions. Currently, sugar beets are grown in eleven states in the United States, from the hot climate of California's Imperial Valley to the colder climates of Michigan. Sugar beets are rich in sugars that can be readily fermented to ethanol. A process block diagram depicting the conversion process is shown in Figure 8.1. Sugar beets are brought into the facility for processing from outdoor stockpiles and indoor storage. These are first washed, which produces a "tailings" waste stream. This stream mainly consists of pieces of beets (10%–30%), weeds, sugar beet tops, debris, and soils held by sugar beets when harvested. Typically about 4%–5% of the weight of sugar beets processed is discarded as tailings. The washing process also produces a wastewater stream. The washed beets are then sliced into long, thin strips called cossettes. Sugar is extracted from these cossettes in a two-step process. In the first step, the sugar is extracted by a diffusion process (the equipment is called a diffuser), which brings the cossettes in contact with hot water (temperature of around 80°C). This step produces a sugar-rich solution. In the second step, the cossettes leaving the diffuser are pressed to expel any remaining sugar. In a sugar factory, the pressed pulp may be dried and sold as cattle feed. The sugar solution from the press can be combined with the sugar-rich solution from the diffuser to produce a juice stream. The juice is then concentrated and fermented to ethanol. The fermentation broth is then distilled to recover ethanol. The distillation process produces a stillage waste stream. The wastewater and tailings from washing process, spent pulp, and the stillage can be anaerobically digested to produce methane. Figure 8.1 also shows typical daily mass flow rates of various streams.

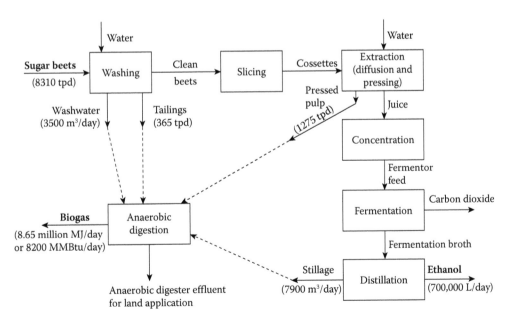

Figure 8.1 A process block diagram for production of ethanol from 8310 metric tonnes per day of sugar beets.

Data shown in the figure, up to the spent pulp stream flow rate, is actual data from the American Crystal Sugar Company sugar factory in Minnesota (Polematidis et al. 2008). It is expected that the processes and process conditions for extracting the sugar will be the same irrespective of whether the sugar beets are used for sugar or ethanol production. Subsequent calculations are performed for a biorefinery processing 8310 metric tonnes per day (tpd) of sugar beets. The washing process would generate 3500 m³/day of wastewater and 365 tpd of tailings. About 1275 tpd of pressed pulp would be produced. The facility would produce 700,000 L/day of ethanol (i.e., 185,000 gallons/day). In North America, a sugar beet campaign lasts 270 days in a year. So such a biorefinery operating on sugar beets will produce 50 million gallons of ethanol annually, which is similar to the output of large corn ethanol plants. The volume of stillage waste produced from distillation is about 11.3 times the volume of ethanol in a sugar beet ethanol process (Wilkie et al. 2000).

The washwater stream, tailings, pressed pulp, and stillage are rich in organic constituents and can be fed to an anaerobic digester to produce biogas. Table 8.2 shows the typical composition of these waste and residue streams (compiled from Wilkie et al. 2000; Polematidis et al. 2008). In Table 8.2, the organic content of liquid (with low particulate matter content) streams is reported in terms of chemical oxygen demand (COD), whereas the content of solid streams is reported in terms of its volatile matter (measured as weight loss of a dried sample at 550°C). The dry matter content is measured as weight remaining when a sample is heated at 105°C. The specific methane potential of these waste streams was measured using biochemical methane potential assays. In these assays, a known mass of sample is incubated after inoculating with an anaerobic digester inoculum and provided with a more than adequate pH buffer and all macro- and micronutrients. The biogas produced from the assay is periodically measured and its methane content analyzed. The assay is continued until biogas production ceases. The total methane produced is then calculated and reported per g volatile matter or g COD of sample incubated. Based on this specific methane yield value, the methane that could potentially be produced from an anaerobic digester treating all waste and residuals from a sugar beet ethanol plant is estimated. It should be noted that anaerobic digestion of these streams produces biogas, which is a mixture of methane and carbon dioxide. Table 8.2 lists only the production of the methane portion of biogas because it is the methane in the biogas that contributes to energy (or heat of combustion) content. About 221,000 m³ of methane will be produced daily from an anaerobic digester processing all waste and residues generated in a sugar beet ethanol plant. Typically, biogas contains about 60% methane; therefore, the digester will produce about 370,000 m³ of biogas daily. Heat of combustion of methane is 37 MJ/m³ at standard temperature and pressure (STP), this translates to around 8.2×10^6 MJ/day (or 7700 MMBtu/day). Ethanol fermentation and the recovery process is energy intensive. Energy in the form of process heat is required for distillation, concentration of juice, and for supplying hot water for diffusers. Electrical energy is required for pumps, pressing, and generally for conveying materials from one unit to the next. Krochta (1980) provides a breakdown of energy requirements in various sections of a sugar beet ethanol biorefinery. Distillation for

Table 8.2   Methane Potential of Sugar Beet Residuals and Wastewater

Residuals/ Wastewater	Dry Matter Content	Volatile Matter (% dry matter)	Chemical Oxygen Demand (gCOD/L)	Specific Methane Potential at STP[a] Conditions	Biogas (Methane) Production from Anaerobic Digestion (m³ at STP[a]/day)
Washwater	2.1 g/L	58	28.5	286 L/kg COD	28,200
Tailings	15% (by wet weight)	90		285 L/kg VM	14,000
Pressed pulp	22% (by wet weight)	96		330 L/kg VM	88,700
Stillage			65	175 L/kg COD	90,000
				Total	220,900

[a] STP standard temperature (0°C) and pressure (1 atm).

recovery of ethanol is the most energy intensive step, which consumes about 50% of the total energy input into a plant (Krochta 1980). Energy requirements for by-product recovery, which in the case of sugar beet would be the energy utilized to dry the pressed pulp for sale as animal feed, is neglected in this example because all the pressed pulp is assumed to be used in anaerobic digestion. Typical energy requirements for distillation ranges from 5.6 to 8.4 MJ/L of anhydrous ethanol (Krochta 1980; Brown and Brown 2014). Therefore, distillation energy requirements for the ethanol plant in this case study would be between 4×10^6 and 6×10^6 MJ/day. It can be seen that biogas fuel from anaerobic digestion would be more than sufficient to meet the energy requirements for distillation. This leaves between 4.2×10^6 and 2.2×10^6 MJ/day of energy for other purposes. The electrical energy requirement in a sugar beet ethanol plant is around 2.8 MJ/L (Krochta 1980). Another approach to utilize the biogas would be in a combined heat and power (CHP) system. CHP systems have an overall efficiency of 80%. If the electrical power generation efficiency is 30% and if the waste heat is utilized at 70% efficiency (which will equate to an overall efficiency of 80%), then 680 MWh$_e$ of electricity can be generated daily with an excess 4.0×10^6 MJ of thermal energy available for distillation. This thermal energy would meet at least 68% of the energy required for distillation of the ethanol. In addition electricity can be sold to the grid. To put the electricity generated in perspective, typical average household consumption of electricity in the United States ranged between 15,046 kWh/year in Louisiana (highest) and 6367 kWh/year in Maine (lowest) for the year 2012. The average annual electricity consumption for a U.S. residential utility customer was 10,837 kWh, an average of 903 kWh per month (USEIA 2014). Based on this consumption, the electricity generated in the sugar beet biorefinery would be sufficient to supply about 22,000 households. Figure 8.2 pictorially represents the options for biogas usage and includes the process flow rates, process heat generated, and the electricity generation. It should be noted that for both scenarios the biogas has to be cleaned. In the case of utilization directly in a boiler (Figure 8.2a), the extent of cleanup need not be as intensive as for utilization for power generation (Figure 8.2b). For boiler usage, the biogas is dried and hydrogen sulfide is removed. For power generation, in addition to the aforementioned, it may be necessary to remove other impurities such as ammonia, Volatile Organic Carbons (VOCs), and siloxanes. The cleanup step for power generation may be more expensive.

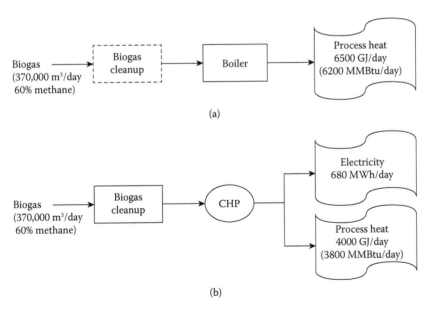

Figure 8.2 Biogas utilization options. (a) Directly fed to boilers after some cleanup for steam generation and (b) used as a fuel for combined heat and power generation.

An integrated rural biorefinery will be designed such that feedstocks produced in the fields will be directly sent to preprocessing; then, once the feedstocks have been processed, the sugar will be extracted and the waste will go immediately into anaerobic digesters. Biogas produced from anaerobic digestion would be utilized on site as fuel for process heat or as fuel for CHP generation. The effluent from the anaerobic digesters will be trucked away for land application locally on the farms. By closing the entire loop of process and doing most of it on site, costs will be reduced and the biofuels will be a step closer to competitiveness in the marketplace.

8.6 CONCLUSION

The biochemical platform offers advantages over other ethanol production methods for a rural biorefinery. With the right feedstock selection, there is no need for pretreatment or enzymatic hydrolysis that is currently to blame for high costs in the ethanol industry. By using simple, proven technologies along with the correct feedstock selection, ethanol can be produced in a rural setting in a cost-effective way.

Sugar-based feedstocks are ideal for biochemical conversion in a rural setting due to the ease of extraction of sugars. A simple milling, grinding, or hot water soak process can release the sugars from sugarcane, sugar beets, or sweet sorghum in a concentrated syrup form. Once the six carbon sugars have been extracted, a simple fermentation to ethanol can be performed. The remaining biomass left over from sugar extraction still has value and is an excellent source for anaerobic digestion. Biogas produced from anaerobic digestion can be used on site or converted to electricity to power the biorefinery and to sell, driving down costs and making the final ethanol product more economically attractive.

REFERENCES

Amorim, H. V., Lopes, M. L., Oliveria, J. V., Buckeridge, M. S., and Goldman, G. H. (2011). Scientific challenges of bioethanol production in Brazil. *Applied Microbiology and Biotechnology*, 91(5), 1267–1275.

Amosson, A., Girase, J., Bean, B., Rooney, W., and Becker, J. (2011). *Economic Analysis of Sweet Sorghum for Biofuels Production in the Texas High Plains*. Amarillo, TX: Texas A&M Agrilife Extension.

Bouallagui, H., Touhami, Y., Ben Cheikh, R., and Hamdi, M. (2005). Bioreactor performance in anaerobic digestion of fruit and vegetable wastes. *Process Biochemistry*, 40, 989–995.

Brown, R. C. and Brown, T. R. (2014). *Biorenewable Resources—Engineering New Products from Agriculture*. 2nd ed. Chichester, UK: Wiley, p. 193.

De Baere, L. (2000). Anaerobic digestion of solid waste: State-of-the-art. *Water Science and Technology*, 41(3), 283–290.

Frazier, R. S., Hamilton, D., and Ndegwa, P. M. (2014). *Anaerobic Digestion: Biogas Utilization and Cleanup*. Stillwater, OK: Oklahoma Cooperative Extension Service Fact Sheet. BAE-1752. pp. 1–4.

Gujer, W. and Zehnder, A. J. B. (1983). Conversion processes in anaerobic digestion. *Water Science & Technology*, 15(8–9), 127–167.

Kim, M., Ahn, Y.-H., and Speece, R. E. (2002). Comparative process stability and efficiency of anaerobic digestion; mesophilic vs. thermophilic. *Water Research*, 36, 4369–4385.

Kim, M. and Day, D. F. (2011). Composition of sugarcane, energy cane, and sweet sorghum suitable for ethanol production at Louisiana sugar mills. *Journal of Industrial Microbiology and Biotechnology*, 38(7), 803–7.

Krochta, J. M. (1980). Energy analysis for ethanol. *California Agriculture*, 34(6), 9–11.

Mata-Alvarez, J., Mace, S., and Llabres, P. (2000). Anaerobic digestion of organic solid wastes. An overview of research achievements and perspectives. *Bioresource Technology*, 71(1), 3–16.

Mohan, S. V., Babu, V. L., and Sarma, P. N. (2007). Anaerobic biohydrogen production from dairy wastewater treatment in sequencing batch reactor (AnSBR): effect of organic loading rate. *Enzyme and Microbial Technology* 41(3), 506–515.

Montross, M. D., Pfeiffer, T. W., Crofcheck, C. L., Shearer, S. A., and Dillon, C. R. (2009). Feasibility of ethanol production from sweet sorghum in Kentucky. http://www.uky.edu/Ag/CCD/sorghumethanol.pdf (Accessed September 1, 2014).

Panella, L. (2010). Sugar beet as an energy crop. *Sugar Technology*, 12(3–4), 288–293.

Polematidis, I., Koppar, A., Pullammanappallil, P., and Seaborn, S. (2008). Biogasification of sugarbeet processing by-products. *Sugar Industry/Zuckerindustrie*, 133(5), 323–329.

Rutto, L. K., Xu, Y., Brandt, M., Ren, S., and Kering, M. K. (2013). Juice, ethanol and grain yield potential of five sweet sorghum (*Sorghum bicolor* [L.] Moench) cultivars. *Journal of Sustainable Bioenergy Systems*, 3, 113–118.

Salminen, E., and Rintala, J. (2002). Anaerobic digestion of organic solid poultry slaughterhouse waste-a review. *Bioresource Technology*. 83(1), 13–26.

Song, Y., Kwon, S., and Woo, J. (2004). Mesophilic and thermophilic temperature co-phase anaerobic digestion compared with single-stage mesophilic-and thermophilic digestion of sewage sludge. *Water Research*, 38, 1653–1662.

UNICA. (2013). Final Report of 2012/2013 Harvest Season South-Central Region. Department of Economics and Statistics, Brazilian Sugarcane Industry Association, UNICA, São Paulo, Brazil.

USDA (United States Department of Agriculture). (1982). *A Guide to Small-Scale Ethanol Production*. 2nd ed. Golden, CO: United States Department of Agriculture.

USDA (United States Department of Agriculture). (2014). Sugar and Sweeteners Yearbook Tables. http://www.ers.usda.gov/datafiles/Sugar_and_Sweeteners_Yearbook_Tables/US_Sugar_Supply_and_Use/TABLE14.XLS (Accessed September 1, 2014).

USEIA (United States Energy Information Administration). (2014). How much electricity does an American home use? http://www.eia.gov/tools/faqs/faq.cfm?id=97&t=3 (Accessed September 1, 2014).

Wilkie, A. C., Riedesel, K. J., and Owens, J. M. (2000). Stillage characterization and anaerobic treatment of stillage from conventional and cellulosic feedstocks. *Biomass and Bioenergy*, 19, 63–102.

CHAPTER 9

Effect of Cometals in Copper Catalysts for Hydrogenolysis of Glycerol to 1,2-Propanediol

Rasika B. Mane and Chandrashekhar V. Rode
Chemical Engineering and Process Development Division, CSIR–National Chemical Laboratory, Pune, India

CONTENTS

9.1 INTRODUCTION

For the last 70 years or so, crude oil (a nonregenerative feedstock) was the source of fuel and a host of synthetic materials in modern society. Increasing demand for fuels and chemicals in the developing world for power, housing, clothing, and agriculture use has led to increases in crude oil prices that are more than five times those at the beginning of the twentieth century. In addition, CO_2 emission by intensive use of fossil resources is seriously threatening the climate of the Earth [1]. The outcome of the recent research for alternate sustainable feedstocks for energy and chemicals is the concept of biorefineries, which provide various pathways for biomass conversion [2]. Although the biorefinery concept is analogous to the conventional petrorefinery that produces multiple fuels and products in an integrated complex, the chemical transformations involved in both are fundamentally different from each other. The catalysts developed in petrorefinery were mainly for the selective functionalization of hydrocarbons; for biorefinery, there is a need to design new catalysts for the removal of some functionalities because biomass-derived molecules are highly functionalized. This scope widens further for taking advantage of the functional groups already present in different types of biomass [3].

Selective deoxygenation is one of the common unit processes in biomass conversion [4], which is exemplified here by demonstrating our own work on hydrodeoxygenation (hydrogenolysis) of glycerol to 1,2-PDO. Glycerol is the smallest, highly functionalized polyol obtained from biomass either directly or as a by-product of (i) industrial conversion of lignocelluloses into ethanol [5,6] and (ii) biodiesel production [7,8]. With wide industrial applications of 1,2-PDO, its production

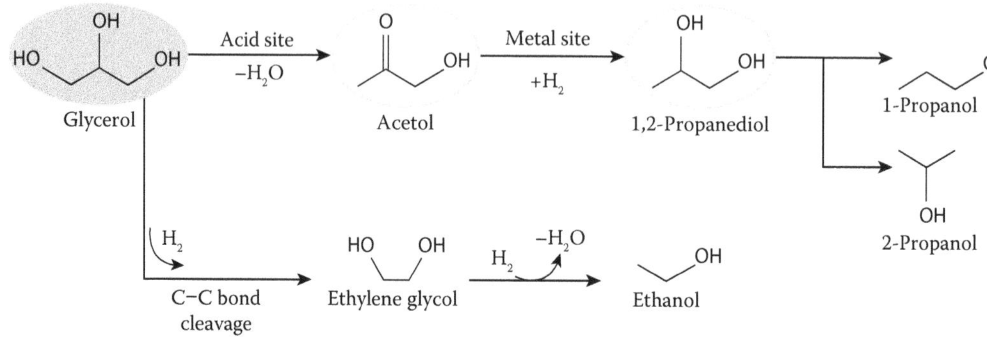

Scheme 9.1   Glycerol hydrogenolysis pathways.

from glycerol becomes a viable sustainable process as is evident from the first commercial plant (Archer Daniels Midland Co., Chicago, Illinois, United States) having a capacity of 0.1 million tpa [9–11]. Glycerol hydrogenolysis involves parallel and series reactions producing several products, as shown in Scheme 9.1. However, 1,2-PDO formation via dehydration is more acceptable due to simple preparation methods of catalyst systems having inherent acidic properties. Between the noble and nonnoble metal catalysts reported for glycerol hydrogenolysis, the choice of the latter—in spite of lower activity—is obvious due to their (i) much lower prices, (ii) higher resistance to poisoning by trace impurities, and (iii) their suppression of C–C cleavage [9]. The pioneering work of Suppes on Cu–Cr catalysts for glycerol hydrogenolysis was subsequently studied extensively by several other researchers for further improvement in catalyst stability, activity, and selectivity [12–16]. In all these studies, Cr containing catalysts have been shown to be the robust and most efficient systems for glycerol hydrogenolysis. However, active research efforts are ongoing to develop new catalysts without chromium [17–19]. In continuation of our efforts to understand the structure activity correlation of catalysts in glycerol hydrogenolysis, here we discuss the effect of combination of alkaline earth (Mg, Ba) and transition (Zn, Cr, Al) metals with Cu on activity and product selectivity. The activity of these catalysts was studied in an aqueous as well as in an organic medium. One of the best catalysts, Cu–Al, was tested for the in situ glycerol hydrogenolysis without use of external hydrogen.

9.2 EXPERIMENTAL

Glycerol (98%), ethylene glycol (EG), and 30% aqueous ammonia were purchased from Merck Specialities, Mumbai, India; acetol and 1,2-peopanediol were procured from Sigma-Aldrich, Bangalore, India. Nitrate salts of aluminum, barium, zinc, and ammonium dichromate of Guaranteed reagent (GR) grade, extra pure copper nitrate, and magnesium nitrate were purchased from Loba Chemie, Mumbai, India. Hydrogen and nitrogen of high purity were obtained from Inox-I.

Copper chromite catalysts with different cometals were prepared by the coprecipitation method according to which already prepared ammonium chromate solution was added to the aqueous nitrate solutions of copper and the respective cometals Al, Zn, Mg, and Ba under constant stirring. A reddish brown precipitate thus formed and was filtered, washed with deionized water, and then dried in a vacuum oven (100°C, 8 h) followed by calcination (400°C, 6 h) and activation under H_2 flow at 200°C for 12 h. Nonchromium Cu-based catalysts with different cometals (Zn, Mg, Ba, Al) were also prepared by the coprecipitation method reported elsewhere [19].

Prepared catalysts were characterized by X-ray powder diffraction, and patterns were recorded on a Rigaku, D-Max III VC model, using nickel filtered CuKα radiation. The samples were scanned in the 2θ range of 10°–80°. X-ray photoelectron spectroscopy (XPS) data were collected on a VG

Scientific ESCA-3000 spectrometer using a non-monochromatized Mg Kα radiation (1253.6 eV). The binding energy values were charge-corrected to the C1s signal (285.0 eV).

Catalyst activity testing was carried out in an autoclave of 300 mL capacity supplied by Parr Instruments Co., USA. The hydrogenolysis experiments were carried out using 100 mL of 20 wt% glycerol solution along with 1 g of catalyst. The reaction was monitored by observing the pressure drop as a function of time as well as GC analysis of the liquid samples withdrawn at regular time intervals. Continuous runs were carried out in a high-pressure, fixed-bed reactor set-up supplied by M/s Geomechanique, France. GC analysis was done using a Varian 3600 model equipped with flame ionization detector (FID) and a capillary column (HP-FFAP; 30 m, 0.53 mm, 1 μm). Gas analysis was carried out using Chemito 8610 GC fitted with packed Porapac-Q column (2 m) connected to the thermal conductivity detector (TCD).

9.3 RESULTS AND DISCUSSION

Although Cu–Cr catalysts were proved to give optimum performance for glycerol hydrogenolysis, the addition of a third metal may lead to further improvement in the catalyst performance. Because glycerol hydrogenolysis proceeds via acid catalyzed first step dehydration, the selection of the third metal depends mainly on its Lewis acidity. Therefore, in this work, a systematic study was undertaken with different metals having such properties along with Cu–Cr. In order to understand the role of acid sites of Al and Zn, metals with basic nature (namely Ba and Mg) were also purposely incorporated in Cu–Cr catalysts. The same metals incorporated in Cu–Cr catalysts were also used in preparing nonchromium Cu catalysts. The performance of such catalysts was found to be better than Cu–Cr catalysts, which is discussed in detail here.

X-ray diffraction (XRD) patterns of the reduced Cu–Cr catalysts without and with cometals (Al, Zn, Ba, and Mg) are shown in Figure 9.1. In all these samples, the content of Cu as well as that of different promoters was kept constant at 52% and 30%, respectively. Copper chromite catalyst without any promoter showed predominant peaks at 2θ values of 43.2° (111) and 50.1° (200) attributed to metallic Cu phase [20] along with two small peaks at 2θ values of 35.4° (111) and 38.4° (111), which showed the presence of CuO [21]. Al- and Mg-promoted Cu–Cr showed a major peak at 2θ of 36.4° (101), which was due to Cu_2O state, two minor broad peaks at 2θ values of 35.4° (111) and 38.4° (111) for CuO, and very low intensity peak of metallic Cu at $2\theta = 43.3°$ (111). Although Ba-promoted copper chromite catalyst showing new peaks at 2θ values of 22.4° (111), 25.4° (210), 28.2° (211), 30.9° (112), 41.6° (113), 41.9° (203), and 43° (410) confirms the formation of $BaCrO_4$ phase with a very low intensity single peak of Cu^0 at $2\theta = 43°$ [22]. However, it did not show peaks corresponding to CuO or Cu_2O phases. Cu–Cr catalyst with Zn showed the predominant presence of unreduced CuO phase with only a small amount of metallic copper, which leads to the lower activity, as will be discussed later in this chapter.

The same cometals (Al, Zn, Ba, and Mg) studied earlier were also incorporated in nonchromium catalysts in which the ratio of Cu to cometal was kept same at 1:1. Figure 9.2, showing the XRD patterns of reduced Cu–Al and Cu–Mg catalysts, revealed the predominant diffraction peaks at $2\theta = 43.5°$ (111), 50.7° (200), and 74.2° (220) attributable to metallic Cu (JCPDS file no. 85-1326), which was unlike the Cu–Cr catalysts in which only a single small peak at $2\theta = 43.3°$ was observed. These catalysts also showed the presence of Cu_2O [$2\theta = 36.6°$ (101); 29.5° (110); 61.4° (220)] (JCPDS file nos. 78-2076) and CuO [$2\theta = 35.7°$ ($\bar{1}$11) and 38.8° (111)] (JCPDS file no. 80-1268) phases, respectively. CuO phase was more predominant in Cu–Mg samples as evidenced by higher intensity peaks as compared to those observed in the Cu–Al sample in which metallic copper phase appeared to be predominant, indicating that magnesium suppressed the reduction of CuO during the catalyst reduction protocol [23]. The complete suppression of formation of metallic copper phase was observed in the case of reduced Cu–Ba and Cu–Zn catalysts. The distinct peaks at $2\theta = 23.9°$ (111),

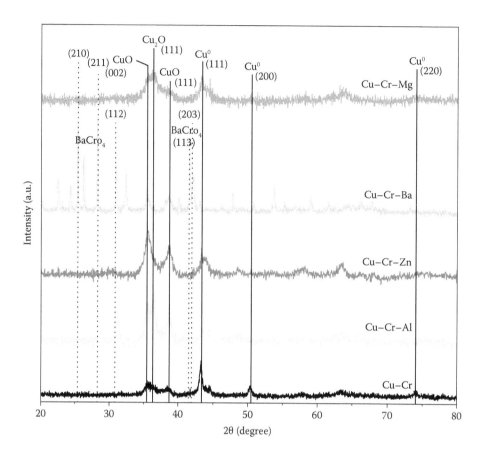

Figure 9.1   XRD patterns of chromium containing Cu catalysts.

27.7° (002), 33.7° (200), 42.8° (041), 44.9° (132), and 46.8° (113) (JCPDS file no. 71-2394) shown in the XRD pattern of Cu–Ba (Figure 9.2) were attributed to the formation of barium carbonate ($BaCO_3$) phase. Barium carbonate is a highly stable phase [24] adversely affecting the glycerol hydrogenolysis activity, as will be discussed later. The copper phases identified in reduced Cu–Ba catalyst were Cu_2O and CuO from the corresponding low intensity peaks at $2\theta = 29.5°$ (110) and $2\theta = 35.7°$ ($\bar{1}11$) and 38.8° (111), respectively. Likewise for reduced Cu–Zn catalyst, only Cu_2O (major) and CuO phases were observed while the Cu^0 phase was completely absent. Further, ZnO phase was also distinct as evidenced from the peaks at $2\theta = 31.7°$ (100), 34.4° (002), 47.4° (111), 56.7° (110), 62.8° (103), and 67.9° (112) [JCPDS file no. 80-0075]. The absence of metallic Cu and the abundance of ZnO led to lower hydrogenolysis activity, as discussed here.

The activity of Cu catalysts with and without chromium was found to be dependent on the crystallite size of active metallic Cu ($2\theta = 43.2°$) phase in the respective catalysts, as shown in Figures 9.3 and 9.4. In both cases, the increasing crystallite size led to a decrease in glycerol conversion, whereas the 1,2-PDO selectivity was not affected considerably in the case of Cr containing catalysts. Chromium containing, as well as nonchromium catalysts with Mg and Al as cometals, showed the smaller crystallite sizes of 13.5, 95 and 5.5, 14.4 nm with higher glycerol conversion of 65%, 31% and 87%, 47%, respectively, compared to those catalysts with other cometals. This is because the presence of Cu_2O species in Mg and Al containing catalysts helps to avoid the aggregation of Cu metal particles [19]. The presence of Ba in Cu–Cr catalysts caused a decrease in crystallite size up to 69 nm due to inhibition of aggregation of Cu particles that also has been reported earlier [25].

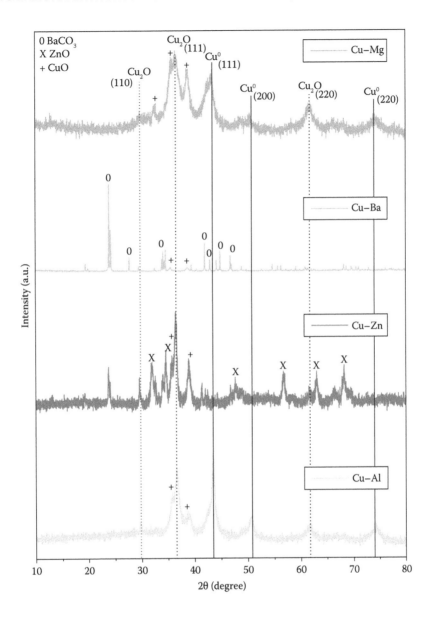

Figure 9.2 XRD patterns of nonchromium Cu catalysts.

This has contributed immensely to the catalyst stability as seen from the time on stream activity of 800 h for continuous hydrogenolysis of glycerol [13]. In contrast to this, nonchromium Cu–Ba and Cu–Zn catalysts showed the absence of metallic copper phase, which led to a decrease in glycerol conversion and 1,2-PDO selectivity compared to other cometals (Al and Mg). The fact that both Cu–Ba and Cu–Zn catalysts showed some measurable hydrogenolysis activity indicate that not only metallic Cu phase and lower crystallite size but also other physicochemical properties such as acidity, surface area, and so forth, also contribute to the catalyst activity.

Because the glycerol hydrogenolysis in aqueous phase is more beneficial than that in organic solvents, both types of catalyst were evaluated in aqueous medium; only nonchromium catalysts showed excellent activity. Therefore, these were further characterized in detail by X-ray photoelectron spectroscopy (XPS) in order to know the oxidation states of Cu (Figure 9.5), which also corroborated with

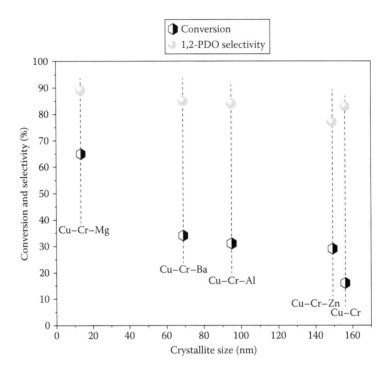

Figure 9.3 Conversion and selectivity versus crystallite size plot of Cu–Cr catalysts for glycerol hydrogenolysis in 2-propanol.

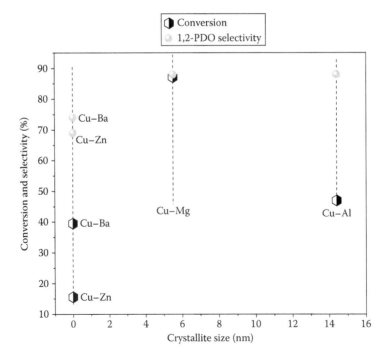

Figure 9.4 Conversion and selectivity versus crystallite size plot of nonchromium Cu catalysts for glycerol hydrogenolysis in 2-propanol.

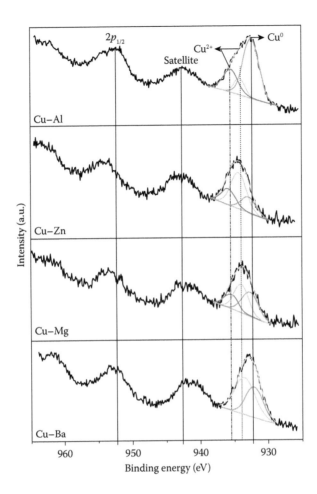

Figure 9.5   XPS spectra of different reduced nonchromium Cu catalysts.

the XRD characterization. Cu $2p$ XPS spectra of reduced copper catalysts are shown in Figure 9.5 in which a broad peak of Cu $2p_{3/2}$ observed in the range of 932 to 935 eV was due to the presence of various Cu species. All the samples showed a peak at about 932 eV that could be attributed to either Cu^{1+} or Cu^0 species [26]. Other peaks at higher binding energies at about 933.5 and at about 935 eV were assigned to Cu^{2+} of CuO, which was also supported by the satellite peak centered at 943 eV. Because the XRD patterns of Cu–Zn and Cu–Ba did not show any metallic Cu phase, the peak at 932.3 eV could be assigned to only Cu^{1+} species—although in the case of the Cu–Al catalyst, the peak at 935 eV assigned to Cu^{2+} was due to the formation of CuO as well as of $CuAl_2O_4$ phases [27].

Various Cu catalysts with and without Cr prepared in this work were evaluated for glycerol hydrogenolysis in aqueous and organic (2-propanol) medium. Figure 9.6 shows the activity results of Cr containing Cu catalysts with various other cometals for hydrogenolysis of glycerol to 1,2-PDO in a 2-propanol solvent. Cu–Cr catalyst without any other cometal showed the least activity with glycerol conversion of 16% due to least acidity (0.3414 mmol NH_3 g^{-1}) and higher crystallite size as discussed earlier; although addition of various cometals Al, Zn, Ba, along with Cr enhanced the catalytic activity in a range of 31%–34%. However, Zn-containing catalysts showed the highest selectivity to dehydration product, acetol (21%), by eliminating the formation of other by-products. Copper chromite catalysts having Mg as a cometal gave the highest activity with 65% glycerol conversion and 89% selectivity to 1,2-PDO and 8% to the undesired products

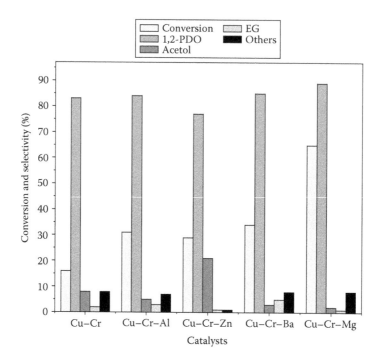

Figure 9.6   Activity of Cu–Cr catalysts for hydrogenolysis of glycerol in 2-propanol. Reaction conditions: temperature, 220°C; glycerol, 20 wt% in 2-propanol; H_2 pressure, 51.7 bar; catalyst, 1 g; reaction time, 5 h.

such as 2-propanol, methanol, EG, and so forth. The highest activity of Mg-containing catalysts was mainly due to the presence of Cu^0, Cu_2O, and CuO phases and the smallest crystallite size of 13.5 nm. With water as reaction medium, Cu–Cr catalysts with and without promoters showed very poor hydrogenolysis activity (6%–10% conversion) and lower 1, 2-PDO selectivity in the range of 54%–65% while remaining selectivity to acetol without any other by-product formation. The least activity of Cu–Cr catalysts with any of the cometals—for hydrogenolysis of glycerol in water—could be probably due to the oxo/hydroxy inactive species formed with water and/or type of acid sites present.

The nonchromium Cu catalysts also have been tested for glycerol hydrogenolysis in organic (2-propanol) as well as aqueous medium. Figure 9.7 shows the catalytic activity of Cu with different cometals, namely Al, Zn, Ba, and Mg for selective hydrogenolysis of glycerol in 2-propanol solvent at 220°C and 51.7 bar H_2 pressure. It was observed that the Cu catalyst having Mg as a cometal showed the highest glycerol conversion of 87% with 88% selectivity to 1,2-PDO, and the remaining was acetol and EG. The catalytic activity of Cu, with different cometals, for glycerol hydrogenolysis varied in the following order: Cu–Mg > Cu–Al > Cu–Ba > Cu–Zn showing the similar trend to that of the Cr containing catalysts. Also, Cu catalysts with Ba and Zn as cometals showed less selectivity to 1,2-PDO (74% and 69%, respectively) with higher selectivity to dehydration product acetol (18% and 28%, respectively) indicating the incomplete hydrogenation as compared to that with the Cu–Mg catalyst. The lower activity and 1,2-PDO selectivity of the Cu–Ba and Cu–Zn catalysts was mainly due to the lower surface area, absence of metallic Cu, and the least acidity [23].

Similarly, activity testing of these catalysts was also carried out in an aqueous medium under reaction conditions the same as those for 2-propanol solvent; the results are shown in Figure 9.8. In aqueous medium, the catalyst screening studies showed a different trend than that observed

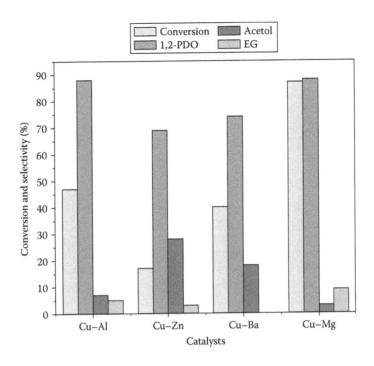

Figure 9.7 Activity of nonchromium Cu catalysts for hydrogenolysis of glycerol in 2-propanol. Reaction conditions: temperature, 220°C; glycerol, 20 wt% in 2-propanol; H_2 pressure, 51.7 bar; catalyst, 1 g; reaction time, 5 h.

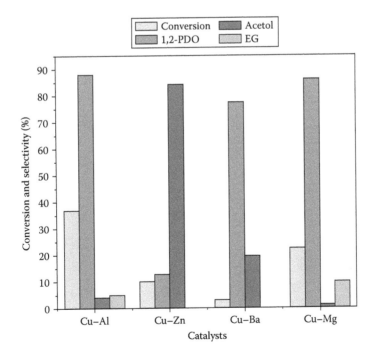

Figure 9.8 Activity of nonchromium Cu catalysts for hydrogenolysis of glycerol in water. Reaction conditions: temperature, 220°C; glycerol, 20 wt% aq.; H_2 pressure, 51.7 bar; catalyst, 1 g; reaction time, 5 h.

for 2-propanol solvent. As can be seen from Figure 9.8, Cu catalyst with Al as a cometal showed the highest glycerol conversion of 38% with maximum selectivity of 91% to 1,2-PDO; Cu with Mg gave comparatively much less conversion (23%) and increased selectivity to C–C cleavage product EG (10%). Cu with Ba showed the least glycerol conversion of 3% with 80% selectivity to 1,2-PDO and 10% to acetol while the Zn-promoted Cu catalyst gave the highest selectivity of 87% to acetol. Nevertheless, activity was two- to three-fold less than that observed in an organic medium, which was due to the lower solubility of H_2 in aqueous medium than that in the former. This confirms that the catalysts having a combination of active metallic Cu, Brønsted acid sites and higher surface area are active for the aqueous phase hydrogenolysis of glycerol. Highly active Cu–Al catalysts were further tested in a continuous fixed-bed reactor for aqueous phase hydrogenolysis of glycerol (Figure 9.9) and showed increased glycerol conversion of 67% and 1,2-PDO selectivity in the range of 92%–93%. This catalyst was also found to be highly stable, giving time on stream activity of 400 h [28].

Because our Cu–Al catalyst showed excellent activity for aqueous glycerol hydrogenolysis, it was further tested for in situ glycerol hydrogenolysis utilizing H_2 generated by glycerol aqueous phase reforming (APR) under inert conditions. The performance of the Cu–Al catalyst along with other catalysts for autogeneous hydrogenolysis under inert atmosphere of nitrogen is shown in Figure 9.10. The results are expressed in terms of carbon selectivity of liquid products, which was calculated as follows.

$$\text{C selectivity } (\%) = \frac{\text{Moles of individual product formed} \times \text{number of C atoms}}{\text{Moles of glycerol feed} \times \text{conversion} \times 3} \times 100$$

The Cu–Al (3 g) catalyst showed 51% conversion with higher product selectivities of 36% and 24% to acetol and 1,2-PDO, respectively, at 220°C. Total liquid product selectivity was 60%, indicating formation of gaseous products (Table 9.1). Under conditions devoid of hydrogen, only dehydration product acetol was expected; however, 1,2-PDO was also formed substantially. On the contrary, much lower glycerol conversions of 10 and <2% were obtained for Cu/Al_2O_3 and Al_2O_3 alone, respectively, even at 230°C. In both the cases, selective acetol formation was observed. Thus, the formation of

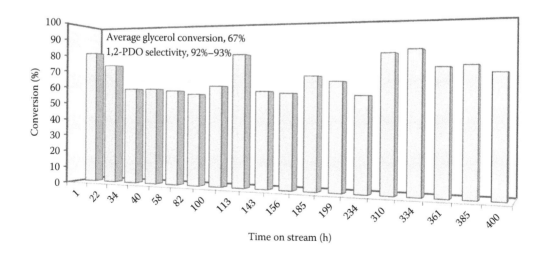

Figure 9.9 TOS of nonchromium Cu–Al catalysts for hydrogenolysis of glycerol in water. Reaction conditions: glycerol, 20 wt% aq.; Cu–Al, 20 g; temperature, 220°C; feed flow rate, 30 mL h⁻¹; H_2 pressure, 27.5 bar; H_2 flow rate, 10 NL/h; time, 400 h.

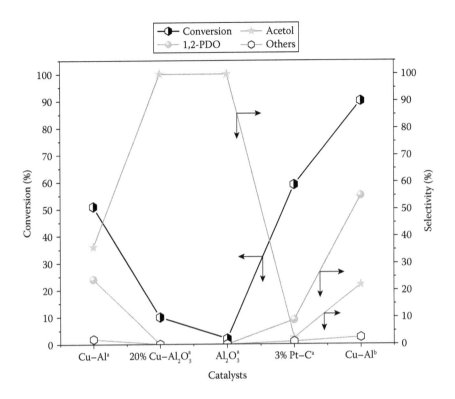

Figure 9.10   Catalyst screening for autogeneous glycerol hydrogenolysis under inert atmosphere. Reaction conditions: [a]batch operation, glycerol, 20 wt% aq.; reaction time, 3 h; [b]continuous operation at GHSV, 513 h⁻¹; LHSV, 1.53 h⁻¹, temperature, 220°C, catalyst weight, 20 g.

Table 9.1   Gas Phase Analysis

Catalysts	Temperature (°C)	Gas Phase Composition (%)		
		H_2	CO_2	CH_4
Cu–Al (1:1)[a]	220	80	20	0.0
20% Cu/Al$_2$O$_3$[a]	230	0.0	0.0	0.0
Al$_2$O$_3$[a]	230	0.0	0.0	0.0
3%Pt/C[a]	220	86	8	6
Cu–Al (1:1)[b]	220	76	24	0.0

[a] batch operation, reaction time, 3 h.
[b] continuous operation at GHSV = 513 h⁻¹, LHSV = 1.53 h⁻¹, temperature, 220C, catalyst weight, 20 g.

1,2-PDO under inert dehydration conditions over Cu–Al indicated hydrogen generation was followed by in situ glycerol hydrogenolysis (Scheme 9.2). In situ hydrogenolysis, also called autogeneous hydrogenolysis, is possible due to H_2 generation either by glycerol reforming or by dehydrogenation of glycerol to glyceraldehyde followed by its further dehydration to pyruvaldehyde [27]. Because glyceraldehyde or pyruvaldehyde was not detected during this reaction and only acetol was the only intermediate detected, glycerol APR was solely contributing to the hydrogen generation for producing 1, 2-PDO from acetol under an inert atmosphere. We propose that in glycerol APR, Cu and Al sites of the catalyst play distinct roles to produce hydrogen and CO_2 [29]. CO_2 and the extent of hydrogen formation were enhanced through water-gas shift (WGS) reaction (Scheme 9.2) on Cu sites

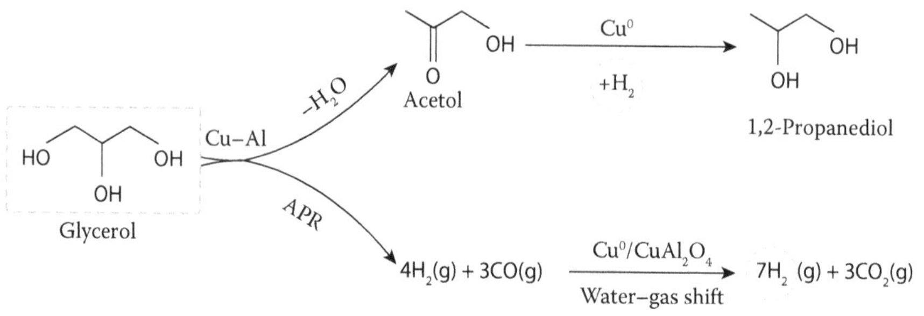

Scheme 9.2 Autogeneous hydrogenolysis of glycerol under inert atmosphere.

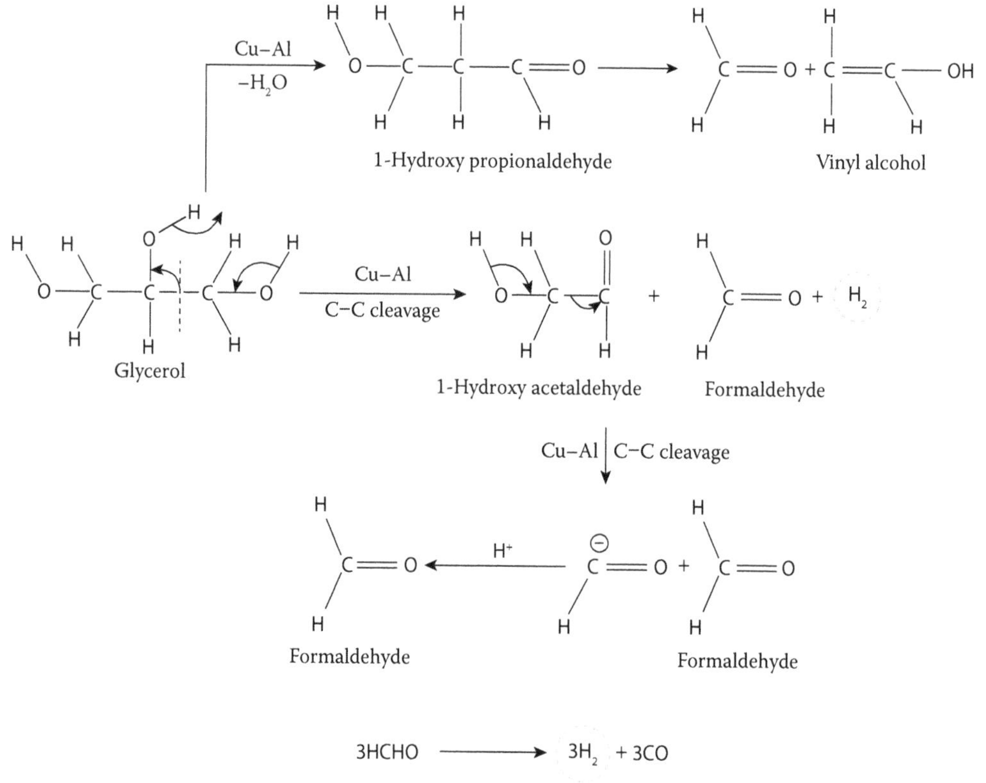

Scheme 9.3 Glycerol aqueous phase reforming for hydrogen generation.

under low temperature conditions (220°C) of the present work. For WGS reaction, H_2 and CO were initially formed via formaldehyde intermediate (Scheme 9.3) [30] as a result of Cu catalyzed dehydrogenation [31] and C–C cleavage of glycerol. Glycerol is capable of undergoing APR because it has C–O ratio of 1:1, which was also found to be facile over Cu catalysts under lower temperature conditions, as observed in our work [32]. Gas phase analysis showed only H_2 and CO_2 without any CO and alkanes (Figure 9.11), confirming glycerol APR over Cu–Al catalyst for in situ hydrogenolysis of glycerol. $CuAl_2O_4$ phase present in our catalyst actively catalyzed the water gas shift reaction even at a low temperature (220°C), thus suppressing the formation of CO and CH_4 [33–35]. This unique

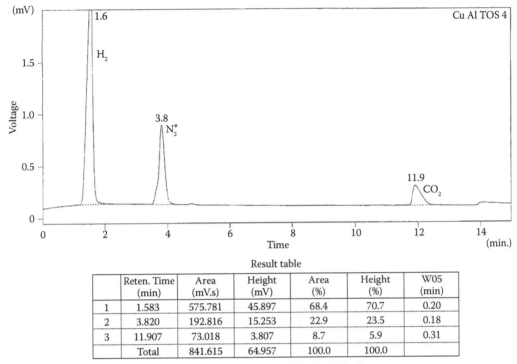

Result table

	Reten. Time (min)	Area (mV.s)	Height (mV)	Area (%)	Height (%)	W05 (min)
1	1.583	575.781	45.897	68.4	70.7	0.20
2	3.820	192.816	15.253	22.9	23.5	0.18
3	11.907	73.018	3.807	8.7	5.9	0.31
	Total	841.615	64.957	100.0	100.0	

* N_2 was used as an inert in the reaction hence the compositions of other gases were calculated by excluding N_2.

Figure 9.11 Gas analysis of Cu–Al catalyst.

ability of Cu–Al catalyst in contrast to Ru and Pt catalysts, contributes to the higher concentration of hydrogen required for in situ hydrogenolysis of glycerol. A control experiment with 3% Pt/C catalyst gave liquid product selectivity as low as ~11% along with methane in the gaseous products (Table 9.1). In a continuous operation over Cu–Al catalyst, the carbon selectivity of total liquid products substantially increased by almost three-fold to 77% (Figure 9.10), due to a several times higher catalyst to substrate ratio (0.67) than that in a batch operation (0.05). The simultaneous increase in 1,2-PDO selectivity was also observed in a continuous mode because H_2 evolution by APR and hydrogenolysis of remaining glycerol proceeded at higher rates under these conditions. The major by-products formed up to the extent of 2.5% include EG, 1-propanol and trace amount of propionic and acetic acids, acetaldehyde, and propionaldehyde.

9.4 CONCLUSIONS

Among various cometals (Al, Zn, Mg, and Ba) in Cu catalysts studied for glycerol hydrogenolysis in 2-propanol, Mg in Cu catalysts with and without Cr were found to enhance the glycerol conversion to 84% and 65%, respectively, in spite of the lower extent of Cu^0 species. Cr-containing catalysts showed very poor activity in a water medium as compared to that of the nonchromium catalysts. This is because Cr may form oxo/hydroxy inactive species with water and/or may be due to the type of acid sites present. On the contrary, Cu–Al catalyst in water medium gave higher glycerol conversion (38%) and 1,2-PDO selectivity (91%) than that of Cu–Mg catalyst. Cu–Al catalyst also showed an excellent activity for autogeneous glycerol hydrogenolysis to 1,2-PDO under inert conditions. Glycerol hydrogenolysis with and without external hydrogen involves various reactions such as dehydration, APR, WGS, hydrogenation, and so forth, which could be possible due to a multifunctional role of Cu–Al catalyst. This was deduced from the coexistence of Cu^0, Cu^{1+}, and Cu^{2+} species in the reduced Cu–Al Brønsted acid sites and the comparatively lower crystallite size. Thus, mixed metal oxides favored glycerol hydrogenolysis rather than Cu^0 alone.

REFERENCES

1. A. Martin, U. Armbruster, and H. Atia, Recent developments in dehydration of glycerol toward acrolein over heteropolyacids. *Eur. J. Lipid Sci. Technol.*, 2012, **114**, 10–23.
2. U.S. Department of Energy, *Top Value Added Chemicals from Biomass.* Oak Ridge, TN: U.S. Department of Energy; 2004.
3. J. ten Dam and U. Hanefeld, Renewable Chemicals: Dehydroxylation of Glycerol and Polyols. *ChemSusChem.*, 2011, **4**, 1017–1034.
4. M. Schlaf, M. E. Thibault, D. DiMondo, D. Taher, E. Karimi, and D. Ashok, Bioenergy II: Group 8 Metal Complexes as Homogeneous Ionic Hydrogenation and Hydrogenolysis Catalysts for the Deoxygenation of Biomass to Petrochemicals – Opportunities, Challenges, Strategies and the Story so Far. *Int. J. Chem. React. Eng.*, 2009, **7**, A34.
5. C. S. Gong, J. X. Du, N. J. Cao, and G. T. Tsao, Coproduction of Ethanol and Glycerol. *Appl. Biochem. Biotechnol.*, 2000, **84**, 543–560.
6. P. L. Rogers, Y. J. Jeon, and C. J. Svenson, Application of Biotechnology to Industrial Sustainability. *Process Saf. Environ. Prot.*, 2005, **83**, 499–503.
7. J. D. Hooker, Apparatus and method for the production of fatty acid alkyl ester. US Pat. 027137, 2005.
8. F. J. Luxem, Method of making alkyl esters. US Pat. 254387, 2004.
9. Y. Nakagawa and K. Tomishige, Heterogeneous catalysis of the glycerol hydrogenolysis. *Catal. Sci. Technol.*, 2011, **1**, 179–190.
10. C. W. Chiu, M. J. Goff, and G. J. Suppes, Distribution of methanol and catalysts between biodiesel and glycerin phases. *AIChE J.*, 2005, **51**, 1274–1278.
11. L. Bournay, D. Casanave, B. Delfort, G. Hillion, and J. A. Chodorge, New heterogeneous process for biodiesel production: A way to improve the quality and the value of the crude glycerin produced by biodiesel plants. *Catal. Today*, 2005, **106**, 190–192.
12. M. A. Dasari, P.-P. Kiatsimkul, W. R. Sutterlin, and G. J. Suppes, Low-pressure hydrogenolysis of glycerol to propylene glycol, *Appl. Catal. A*, 2005, **281**, 225–231.
13. R. B. Mane, A. A. Ghalwadkar, A. M. Hengne, Y. R. Suryawanshi, and C. V. Rode, Role of promoters in copper chromite catalysts for hydrogenolysis of glycerol. *Catal. Today*, 2011, **164**, 447–450.
14. C. V. Rode, A. A. Ghalwadkar, R. B. Mane, A. M. Hengne, S. T. Jadkar, and N. S. Biradar, Selective Hydrogenolysis of Glycerol to 1,2-Propanediol: Comparison of Batch and Continuous Process Operations. *Org. Process Res. Dev.*, 2010, **14**, 1385–1392.
15. Z. Xiao, J. Xiu, X. Wang, B. Zhang, C. T. Williams, D. Su, and C. Liang, Controlled preparation and characterization of supported CuCr2O4 catalysts for hydrogenolysis of highly concentrated glycerol. *Catal. Sci. Technol.*, 2013, **3**, 1108–1115.
16. N. D. Kim, S. Oh, J. B. Joo, K. S. Jung, and J. Yi, The Promotion Effect of Cr on Copper Catalyst in Hydrogenolysis of Glycerol to Propylene Glycol. *Top. Catal.*, 2010, **53**, 517–522.
17. Z. Huang, F. Cui, H. Kang, J. Chen, and C. Xia, Characterization and catalytic properties of the CuO/SiO2 catalysts prepared by precipitation-gel method in the hydrogenolysis of glycerol to 1,2-propanediol: Effect of residual sodium. *Appl. Catal. A*, 2009, **366**, 288–298.
18. L. Guo, J. Zhou, J. Mao, X. Guo, and S. Zhang, Supported Cu catalysts for the selective hydrogenolysis of glycerol to propanediols. *Appl. Catal. A*, 2009, **367**, 93–98.
19. R. B. Mane, A. M. Hengne, A. A. Ghalwadkar, S. Vijayanand, P. H. Mohite, H. S. Potdar, and C. V. Rode, Cu:Al Nano Catalyst for Selective Hydrogenolysis of Glycerol to 1,2-Propanediol. *Catal. Lett.*, 2010, **135**, 141–147.
20. J. Pike, S.-W. Chan, F. Zang, X. Wang, and J. Hanson, Formation of stable Cu_2O from reduction of CuO nanoparticles. *Appl. Catal. A*, 2006, **303**, 273–277.

21. S. Wang and H. C. Liu, Selective hydrogenolysis of glycerol to propylene glycol on Cu–ZnO catalysts. *Catal. Lett.*, 2007, **117**, 62–67.
22. Y. Yan, Q.-S. Wu, L. Li, and Y.-P. Ding, Simultaneous Synthesis of Dendritic Superstructural and Fractal Crystals of $BaCrO_4$ by Vegetal Bi-templates. *Cryst. Growth Des.*, 2006, **6**, 769–773.
23. R. B. Mane, A. Yamaguchi, A. Malawadkar, M. Shirai, and C. V. Rode, Active sites in modified copper catalysts for selective liquid phase dehydration of aqueous glycerol to acetol. *RSC Adv.*, 2013, **37**, 16499–16508.
24. B. Sreedhar, Ch. SatyaVani, D. Keerthi Devi, M. V. B. Rao, and C. Rambabu, Shape Controlled Synthesis of Barium Carbonate Microclusters and Nanocrystallites using Natural Polysachharide – Gum Acacia. *Am. J. Mater. Sci.*, 2012, **2**, 5–13.
25. V. R. Choudhary and S. G. Pataskar, Thermal decomposition of ammonium copper chromate: effect of the addition of barium. *Thermochimica. Acta.*, 1985, **95**, 87–98.
26. J. Agrell, H. Birgersson, M. Boutonnet, I. Melián-Cabrera, R. M. Navarro, and J. L. G. Fierro, Production of hydrogen from methanol over Cu/ZnO catalysts promoted by ZrO_2 and Al_2O_3. *J. Catal.*, 2003, **219**, 389–403.
27. R. B. Mane and C. V. Rode, Simultaneous glycerol dehydration and in situ hydrogenolysis over Cu–Al oxide under an inert atmosphere. *Green. Chem.*, 2012, **14**, 2780–2789.
28. R. B. Mane and C. V. Rode, Continuous Dehydration and Hydrogenolysis of Glycerol over Non-Chromium Copper Catalyst: Laboratory-Scale Process Studies. *Org. Process Res. Dev.*, 2012, **16**, 1043–1052.
29. P. D. Vaidya and A. E. Rodrigues, Glycerol Reforming for Hydrogen Production: A Review. *Chem. Eng. Technol.*, 2009, **32**, 1463–1469.
30. J. Deleplanque, J.-L. Dubois, J.-F. Devaux, and W. Ueda, Production of acrolein and acrylic acid through dehydration and oxydehydration of glycerol with mixed oxide catalysts. *Catal. Today*, 2010, **157**, 351–358.
31. J. Schaferhans, S. Gomez-Quero, D. V. Andreeva, and G. Rothenberg, Novel and Effective Copper–Aluminum Propane Dehydrogenation Catalysts. *Chem. Eur. J.*, 2011, **17**, 12254–12256.
32. R. R. Davda, J. W. Shabaker, G. W. Huber, R. D. Cortright, and J. A. Dumesic, A review of catalytic issues and process conditions for renewable hydrogen and alkanes by aqueous-phase reforming of oxygenated hydrocarbons over supported metal catalysts. *Appl. Catal. B*, 2005, **56**, 171–186.
33. B. L. Kniep, F. Girgsdies, and T. Ressler, Effect of precipitate aging on the microstructural characteristics of Cu/ZnO catalysts for methanol steam reforming. *J. Catal.*, 2005, **236**, 34–44.
34. R. R. Soares, D. A. Simonetti, and J. A. Dumesic, Glycerol as a Source for Fuels and Chemicals by Low-Temperature Catalytic Processing. *Angew. Chem. Int. Ed.*, 2006, **45**, 3982–3985.
35. M. Slinn, K. Kendall, C. Mallon, and J. Andrews, Steam reforming of biodiesel by-product to make renewable hydrogen. *Bioresour. Technol.*, 2008, **99**, 5851–5858.

CHAPTER 10

Energy Harvest
A Possible Solution to the Open Field Stubble Burning in Punjab, India

Sudhakar Sagi,[1] **Amit R. Patel,**[2] **Robert F. Berry,**[1] **and Harpreet Singh**[2]

[1]European Bioenergy Research Institute, School of Engineering and Applied Science,
 Aston University, Birmingham, UK
[2]School of Mechanical, Materials and Energy Engineering, Indian Institute of Technology Ropar,
 Rupnagar, India

CONTENTS

10.1 INTRODUCTION

Processing of agricultural crops for harvesting produces a substantial amount of residues. Crop residues are natural resources that can have tremendous value to farmers. These residues are used for different purposes such as animal feed, composting, thatching for rural homes, and fuel for domestic and industrial use. However, a large portion of these residues is burned in the field primarily to clear the field from straw and stubble after the harvest of the preceding crop. The problem is severe in irrigated agriculture, particularly in the mechanized rice-wheat system. The main reasons for burning crop residues in the field include unavailability of labor, the high cost of removing the residues, and the use of combines in the rice-wheat cropping system. Primary crop types with residues that typically are burned include rice and wheat; farmers in northwest India's Punjab region dispose of a large amount of rice straw by burning in situ.

Burning crop residues leads to adverse effects such as (1) release of soot particles and smoke causing human health problems; (2) emission of greenhouse gases (GHGs) such as carbon dioxide, methane, and nitrous oxide causing global warming; (3) loss of plant nutrients such as nitrogen (N), phosphorus (P), carbon (C), and potassium (K); (4) adverse impact on soil properties; and (5) wastage of valuable C and energy rich residues. There are several options that can be practiced—such as composting, generation of energy, production of biofuels, and recycling in soil—to manage the residues in a productive manner.

According to the Ministry of New and Renewable Energy [1], the Government of India estimated that about 500 Mt of crop residue is generated every year. The cereal crops (rice, wheat, maize, millets) contribute 70%; the rice crop alone contributes 34% of crop residues, and wheat ranks second with 22% of residues. The share of residues generated from various crops is shown in Figure 10.1 (calculated from Ref. [1]).

Increased mechanization, particularly the use of combine harvesters, declining numbers of livestock, long periods required for composting, and no economically viable alternate use of residues are some of the reasons for residues being burned in the field. Because combined harvesters leave a major portion of the residue—including some uncut straw—in the field (because it cuts only at a certain height), about 1.7–1.8 tons per acre of straw is left unused in the field [1].

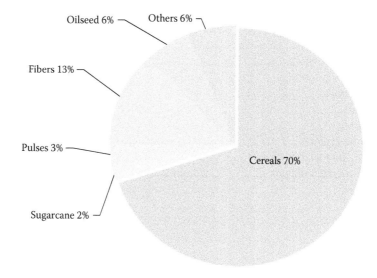

Figure 10.1   Residue share of various crops in India. (From Ministry of New and Renewable Energy Resources, 2009, http://www.mnre.gov.in/relatedlinks/biomassresources.)

Figure 10.2   Open field burning of agricultural crop residues.

Collection and disposal of this residue remains a practical problem, and all available options of disposal lack economical feasibility.

Other reasons for intentional burning include clearing of fields, fertility enhancement, and pest and pasture management. It is believed that burning provides a fast way to clear the agricultural field of residual biomass and facilitates further land preparation and planting. It also provides a fast way of controlling weeds, insects, and diseases, by eliminating them directly or by altering their natural habitat. The time gap between rice harvesting and wheat sowing in northwest India is 15–20 days. In this short duration, farmers prefer burning the rice stalk in the field instead of harvesting it for fodder. A farmer's perception is also that burning boosts soil fertility, although burning actually has a differential impact on soil fertility. It increases the short-term availability of some nutrients (e.g., P and K) and reduces soil acidity, but it leads to a loss of other nutrients (e.g., N and S) and of organic matter. Figure 10.2 shows the typical burning of agricultural fields in Punjab, India.

10.2 RESIDUES AS A FEEDSTOCK FOR ENERGY GENERATION

In recent times, there has been a great interest in biomass as a renewable energy source throughout the world. The major reasons for this are as follows. First of all, technological developments relating to this conversion, to crop production, and so forth, promise the application of biomass at lower cost and with higher conversion efficiency than was possible previously. For example, when low-cost biomass residues are used for fuels, the cost of electricity is often competitive with fossil-fuel-based power generation [2]. The potential threat posed by climate change, due to high emission levels of GHGs—the most important being CO_2—has become a major stimulus for renewable energy sources in general. When produced by sustainable means, biomass emits roughly the same amount of carbon during conversion as is taken up during plant growth. The use of biomass therefore does not contribute to a buildup of CO_2 in the atmosphere and can be treated as CO_2 neutral [3]. It is possible to generate energy from biomass agricultural residue without affecting existing crop production. A recent report released by UK Energy Research Centre (UKERC) suggests that up to one-fifth of global energy could be provided by biomass from plants without damaging food production [4].

Large amounts of agriculture residue are produced in an agrarian country such as India. These can be converted to generate energy. Currently in India, these residues either remain unused or

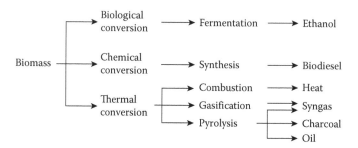

Figure 10.3   Possible energy conversion routes of biomass.

are utilized inefficiently. Biomass residues and their by-products can be converted into energy by means of thermal, biological, and chemical processes. The possible conversion routes of biomass to energy are shown in Figure 10.3. In this chapter, biological and chemical processes, which occupy a lot of space in the literature regarding biomass treatment, are not considered. There are three main thermal processes available for converting biomass into a more useful energy form, namely combustion, gasification, and pyrolysis.

10.3 COMBUSTION

Combustion of biomass and related materials is widely practiced on a commercial basis in order to provide heat and power. The product is heat, which must be used immediately for generating power because storage is not a viable option. Overall efficiency of conversion to power tends to be rather low: typically, 15% for small plants and up to 30% for larger and newer plants. In these plants, some technical problems involving emissions and ash handling remain. The technology is commercially available, and it presents a minimum risk for investors. There are also many producers on the Italian market who have suggested economic solutions in particular for heating houses and connected uses. Denmark and Germany are leaders in this commercial field. In India as well there are biomass power plants coming up recently using this technology.

10.4 GASIFICATION

Gasification can be defined as the thermochemical conversion of a usually solid or liquid fuel into a combustible gaseous product. Integrated gasification combined cycle (IGCC) processes are considered to be among the most attractive technologies for future power production. Research projects in this field, both at bench and at large plant scale and focused primarily on fossil fuel feedstock (coal), have been pursued over the past two decades. The gasifier operates at high temperatures and high pressure to produce a fuel gas that is then burned in a turbine to make electric power.

Biomass gasification is a complex thermochemical process that consists of a number of elementary chemical reactions, beginning with the partial oxidation of a lignocellulosic fuel with a gasifying agent (usually air, oxygen, or steam). Volatile matter released as the biomass fuel is heated partially oxidizes to yield the combustion products H_2O and CO_2 plus heat to continue the endothermic gasification process. Water vaporizes and biomass pyrolysis continues as the fuel is heated.

Thermal decomposition and partial oxidation of the pyrolysis vapors occur at higher temperatures and yield a product gas composed of CO, CO_2, H_2O, H_2, CH_4, and other gaseous hydrocarbons.

10.5 PYROLYSIS

Pyrolysis is the thermal decomposition occurring in the absence of oxygen. Also, it is always the first step in the combustion and gasification process, where it is followed by total or partial oxidation of the primary products. The pyrolysis process can be described with the transformation of complex organic compounds and materials into more simple products. The pyrolysis process consists of a very complex set of reactions involving the formation of radicals. The increase of temperature causes the leak of some hydrocarbons, forming the instable radicals, which break into lower, more stable molecules and into other radicals. The process becomes a chain reaction. The increase of temperature favors the formation of radicals and their instability. Therefore, the reactions are accelerated, and they lead to the formation of products with low molecular weight and gases. With gasification, where a partial oxidation takes place, the pyrolysis is a process of hydrocarbons cracking, which produces compounds with lower volume and high quality of products. Generally, the pyrolysis is divided mainly into two types—slow (conventional) pyrolysis and fast pyrolysis.

10.6 SLOW PYROLYSIS

Conventional pyrolysis or slow pyrolysis is defined as the pyrolysis that occurs under a slow heating rate (typically <1 K/s). In slow pyrolysis, the reactions taking place are always in equilibrium. The heating period is sufficiently slow to allow equilibration during the pyrolysis process. In this case, the ultimate yield and product distribution are limited by the heating rate. The first stage of biomass decomposition is called pre-pyrolysis. During this stage, some internal rearrangement such as water elimination, bond breakage, appearance of radicals, and the formation of carbonyl, carboxyl, and hydroperoxide groups takes place. The second stage of the solid decomposition corresponds to the main pyrolysis process. It proceeds with a high rate and leads to the formation of pyrolysis products. During the third stage, the char decomposes at a very slow rate and forms a carbon-rich residual solid [5].

10.7 FAST PYROLYSIS

Fast pyrolysis is a high temperature process in which the feedstock is rapidly heated in the absence of air, vaporizes, and condenses into a dark brown unstable liquid that has a heating value of about half that of conventional fuel oil. Although it is related to traditional (slow) pyrolysis processes used for making charcoal, fast pyrolysis is a more advanced process that can be carefully controlled to give high yields of liquid. The main product, bio-oil, is obtained in yields of up to 75 wt% on a dry feed basis, together with by-product char and gas. The essential features of a fast pyrolysis process for producing liquids are

- Very high heating and heat transfer rates at the reaction interface, which usually requires a finely ground biomass feed, carefully controlled pyrolysis reaction temperature of around 500°C, and vapor phase temperature of 400°C– 450°C.
- Short vapor residence times of typically less than 2 s and rapid cooling of the pyrolysis vapors to give the bio-oil product. The heating SDM rate is high (50–1000 K/s).

10.8 INTERMEDIATE PYROLYSIS

This kind of pyrolysis operates between the reaction conditions of slow and fast pyrolysis, offering much different product qualities. Intermediate pyrolysis offers reaction conditions that prevent the formation of high molecular tars thereby offering dry and brittle chars that are suitable for different applications. This type of pyrolysis works with a new reactor system, the "Pyroformer™" [6], which is specially designed to separate the ash-rich residues from the fuel. The Pyroformer was developed by the European Bioenergy Research Institute at Aston University, Birmingham, U.K. This process can treat different kinds of biogenic feedstock such as agricultural residues, algae, residues from biogas plants, energy grass, wood residues, and other residues from forestry (leaves, twigs, etc.). The products of the process are pyrolysis liquids (bio-oil), pyrolysis gas (noncondensed vapors), and biochar.

10.9 PROJECT ENERGY HARVEST

Aston University and the Indian Institute of Technology (IIT), Ropar, have started this project together as part of a joint research activity. This project aims solely at stopping the open burning of straw in fields. The work presented here focuses on the need for stopping such burning, its impact on the environment as practiced today (open burning), and on the economics of the pyrolysis route as a viable alternative, as well as its technical realization. Various aspects concerning the application of intermediate pyrolysis technology for conversion of waste to energy as applied in the leading agro driven state, Punjab, India, are discussed here. The work was carried out in two phases. The first phase dealt with the test runs in laboratories at IIT Ropar and the product application. The second phase dealt with community engagement and demonstration of the technology in villages and with the utilization of products by the local community for their daily needs.

10.10 PHASE 1 LAB SCALE TESTING

Phase 1 of the project comprised testing Pyroformer technology in the laboratory by running the feedstock straw and converting the straw to stable products that could be used for energy generation. The Pyroformer basically is a reactor with a coaxial screw system to which the pellets made from different biomass are fed. Pellets are conveyed through the inner screw. The residence time of the pellets can be carefully controlled by the screw's speed. Changes in the rate of feed can be obtained by varying the rotational speed of the screw. Pyrolysis vapor is discharged from the top. The reaction temperature is carefully monitored by electrical heaters with temperature controls. The feed is then passed from the feed inlet to the outlet. The char can then be collected separately in a container. However, part of the char from the process is transported back into the reactor with the outer screw. This helps in better heat transfer to the fresh feed coming into the system, and the char also acts as a catalyst for reforming reactions. The process flow sheet is shown in Figure 10.4.

The first step in the whole process is the preparation of the feed; for better handling and storage purposes the feedstock straw has been pelletized. The pelletization process is a standard one that is very commonly used in the region for the preparation of cattle feed. The machine is unique in that it can change the size of pellets required with slight adjustments. There are two rollers placed on a rotating die that presses the straw with the help of the rollers in to a perforated die. The pellets produced are around 8 mm in diameter and 15 mm in length and can be seen in Figure 10.5.

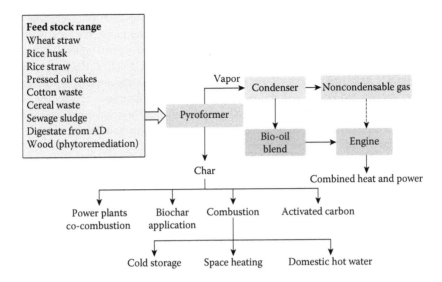

Figure 10.4 The intermediate pyrolysis process for energy conversion.

Figure 10.5 Straw pellets produced for thermal conversion.

10.11 PYROLYSIS PROCESS

Once the straw pellets are produced, they are fed into the reactor, which is set for the desired process temperature. The feed, when conveyed through the reactor from one end to the other end through the inner screw, undergoes the pyrolysis process for the desired residence time. Changes in the rate of feed can be obtained by varying the rotational speed of the screw. The vapor produced during the pyrolysis process is then condensed in a condensing tank using a quenching process. The condenser designed was a special quenching unit where the vapors produced from pyrolysis were directly quenched using either biodiesel or diesel as quench media. The condensable fraction of the vapor, which is roughly 50% of the total vapor produced, is condensed using the quenching process; the remaining noncondensable gas fraction is flared. The char is collected in the char pot.

During the pyrolysis process, monitoring of various parameters takes place through system control by checking on the indicated pressure value inside the reactor, screw rotation, flare operation, condenser operation, and feeding rate. Initially, a known amount of biodiesel/diesel is fed to the quench tank, and this is recirculated in the tank and used to quench the incoming pyrolysis vapor. Once the pyrolysis vapors are condensed, the biodiesel/diesel inside the tank is mixed with the pyrolysis liquid and a blend is created. The level indicator on the quench tank indicates the exact amount of the pyrolysis liquid produced. Once the desired blend is reached, the process is stopped to collect the liquids and then a fresh charge of quench medium is charged again into the quench tank. The collected blended liquid is stored to be used later on the engine.

Several experiments were carried out to optimize the residence time and the correct temperature for the pyrolysis of straw. The products of the process are the vapor and the char. The vapors are quenched using biodiesel/diesel as a direct contact medium without any hot gas filtration. The feedstock straw had a heating value of 16 MJ/kg with a moisture content of 5%. The straw was pelletized prior to pyrolysis. This took place around 120°C. The pellets produced were very hard and highly densified. This pretreatment step is similar to torrefaction.

The pyrolysis runs were carried out on the Pyroformer at various temperatures from 325°C to 400°C. The temperatures compared were 325°C, 350°C, 375°C, and 400°C. By looking at the yields and the quality of the products, 375°C was fixed as the operating temperature, and the residence time was in the range of 2–3 min for solids. At this temperature, the product yields were 34%, 33%, and 33% for char, liquid, and gas, respectively. The yield of the gas fraction was obtained by difference. Hence, this cannot be taken as a true yield of the gas fraction. This remained practically constant; it could be inferred that, within the range of the temperature we carried out the experiments, there was no real significant change in the yield. However, the gas when flared was burning very well. The overall process is shown in Figure 10.6.

The char has a heating value of 20 MJ/kg, which can further be used for co-combustion and/ or as biochar for fertilizer application. The pyrolysis oil was condensed in a quenching unit using biodiesel. The blend produced was 30% pyrolysis liquid and 70% biodiesel. The blend was V/V%. The blended oil has a heating value of 34 MJ/kg and an acidity of 6.3; no separate water fraction or aqueous phase appeared. The pure oil sample had a heating value of 24 MJ/kg. The blended

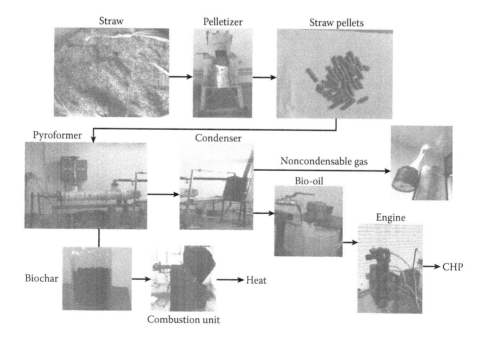

Figure 10.6   (See color insert.) Conversion of straw to bio-oil and biochar.

(pyro-oil + biodiesel) product can be stored for a long period with its direct usage toward combined heat and power. The oils were stored over a six-month period and still no effect on the quality of oils was observed in terms of engine application. The char produced is still in pellet form and has kept its structure as feed material apart from shrinking in size. The char had to be ground fine to be distributed on the field for biochar trials.

10.12 BIOCHAR TRIALS

The char yield as well as the chemical and physical characteristics of biochar depend on the nature of the feedstock used (woody versus herbaceous) and on the operating conditions of the pyrolysis unit (temperature, residence time, heating rate, and feedstock preparation). The wide range of process parameters leads to the formation of biochar products that vary considerably in their elemental and ash composition, density, porosity, pore size distribution, surface area, surface chemical properties, water and ion adsorption, and release. The biochar used in this study is produced out of wheat straw grown in the Punjab region. Prior to the pyrolysis process, the straw had been pelletized into pellets with a size in the range of 8 mm diameter and 12–15 mm length. The pyrolysis temperature was fixed at 375°C and the resultant char yields were around 34%. The pictures of the feed and the char before and after pyrolysis are shown in Figure 10.7.

The biochar produced from wheat straw pellets was also in pellet form as can be seen in Figure 10.7. This is one of the salient features of the intermediate pyrolysis process—that the pellets are still in shape after pyrolysis. The pellets had to be milled to fine powder in order to mix them with the soil on the field. Milling was done by crushing the pellets and then sieving the crushed material. This process had to be repeated several times until we obtained finely ground biochar. The finely grounded biochar can be seen in Figure 10.8. Once the char has been made into fine powder, it was ready to be spread on field.

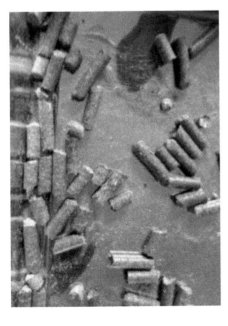 

Figure 10.7   Straw pellets and biochar.

Figure 10.8 Finely grounded biochar.

10.13 FIELD PREPARATION

The field chosen for cultivation is located at the IIT Punjab campus. The field used for this study has not been used previously for any type of farming or any irrigation activity. This is essential for the reason that the application of fertilizer or any other chemical use may influence the study. Further, because the fertility of soil also depends on the previous crop pattern, if the field is previously used in any farming activity, this may affect the yield of the biochar trial. The size of the field chosen for this study has an area of 320 square feet and is divided into four eight-by-ten plots. Two plots (B and D) will be planted with capsicum and the two other plots (A and C) with onion. Field preparation is the first and foremost stage and is very much necessary for the purpose of uniform farming activity. Field preparation will also ensure the proper distribution of water and biochar and will help in efficient harvesting. The number of plants per plot can be found in Table 10.1. The field preparation steps are plowing by tractor, drying the field for one day, manual scooping of the field, drying the field for one day, and dressing and leveling the field. After the field preparation, the finely ground biochar is distributed uniformly in the designated plots. The biochar is weighed prior to being spread on the field. Char weighing 2.785 kg is spread in 80 sq ft (per plot, 10 ft × 8 ft), which is equivalent to 3.75 ton/ha.

After spreading the biochar uniformly over the field, it is essential to mix the biochar with the soil. After a few watering cycles, the large porous surface of biochar will be filled with water and soil, which adsorbs permanently; this will increase its density and will settle the biochar deeply into soil. Crop selection is a very crucial aspect for with respect to the field and determining the plant yield. The soil type for the field considered in the present case is soil with a high bonding strength. Therefore, once the water is dried up, the small masses will have a higher strength, resulting in a tighter grip over the root of the plant. Two crops considered are a root type (onion) and a nonroot type (capsicum). Planting of both is carried out using standard practices. Once the plants are in, it essential to water at regular intervals, but also it is essential to scoop the land with trowel a few inches deeper one day before watering. This will rotate and allow aeration for the soil. This will also break the soil that has formed lumps into small pieces, resulting in a reduction of the pressure around the root of the plants (very essential in the case of onions).

The biochar trials are carried out for a period of six months, which includes the field preparation, biochar distribution on the field, cropping, watering, as well as harvesting. The onion crop is harvested only once, and the capsicum is harvested at regular intervals. It is worth noting that the application of biochar increased the moisture retaining capacity of the field; this is evident from the fact that time required for water to drain in the fully watered field is different for both the fields.

Table 10.1 Summary of the Plantation Scheme

			Amended Field	Controlled Field
Crop Considered	**Plants in Each Field**	**Size (in Feet)**	**Biochar Added**	**No Biochar**
	(row × column)	Length (↓)		
Capsicum	5 × 9	8	B	D
Onion	8 × 14	8	A	C
	Size in foot	Width (→)	10	10

(a)

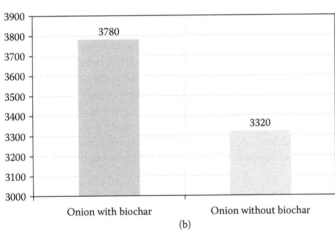

(b)

Figure 10.9 (a) Crop comparison and (b) yield of onions in grams with and without biochar.

This could be one of the reasons for the increase of yield for the field where biochar was added. Another important fact that is observed is that weed growth or growth of unwanted plants (such as grass, etc.) is also substantially less in the biochar added field, compared to that of the field where biochar was not added. From observations, it may be concluded that the cost of fertilizer and chemicals to treat weed growth is a direct saving that is obtained by adopting the biochar route of farming.

Following the harvest of the onions, after a brief period of six months, results showed that the yield of the onions from plot A (the field with biochar) is higher than the yield of the onions in plot C (without biochar). A look at the size of onions reveals that the size is much smaller. The onions grown in plots A and C (with and without biochar, respectively) are shown in Figure 10.9.

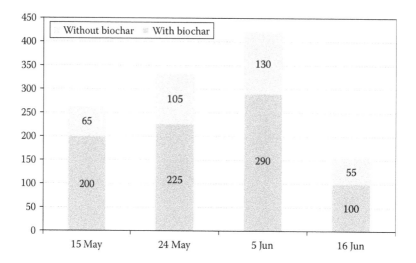

Figure 10.10   Yields of capsicum on various dates in grams.

All the growth conditions for the two plants were absolutely similar except for the biochar aspect. Also the plant with biochar had been given less water compared to the plant without biochar.

The total yields of the onions from both plots after complete growth cycle have also showed considerable differences. Yield from the biochar field is higher compared to that of the field without biochar. A look at the onion size reveals that it is much smaller. This is because the species of onion chosen is a particular variety. The results of the onion yields are shown as a bar graph in Figure 10.9. The overall increase in the yield of onions is around 14%.

The yield of capsicum from the plots with and without biochar on the various days is plotted as a bar graph shown in Figure 10.10. There is substantial difference in the yields. The large yield of capsicum with biochar shows the biochar effectiveness for the crop. A productivity increase of almost 130% is observed in case of capsicum. However, there is only 12% improvement observed in the case of the onions. The reason for this variation can be attributed to the type of the soil because this soil may not be that suitable for the cultivation of root crops. It is also worth mentioning that the present yield is without the addition of any fertilizer and any other chemicals that are generally used in farming while protecting the products from insects and unwanted crops.

10.14 PHASE 2 OF THE PROJECT

As part of the project planning, the technology was installed as one mobile unit that can be moved from village to village for demonstration purposes. The technology was successfully demonstrated in the village of Khuaspura, in the Ropar District of Punjab. The project was conceptualized to be multifaceted in that apart from meeting rural energy needs and reducing CO_2 emissions, it would galvanize self-reliance, local employment, and gender and health-related issues in addition to land reclamation. The focus was to meet the energy needs of the rural population and to stop open burning. The demonstration aimed at studying aspects such as

- Assessment of the potential of bioenergy for rural development and socioeconomic improvement
- Partial replacement of traditional fossil fuels through the application of bio-oil generated from the straw for electricity and the bio char for cooking
- Studying the feasibility of application of biochar as a fertilizer
- Demonstrate technical feasibility and financial viability of bioenergy technologies

- Build capacity and develop mechanisms to implement, manage, and monitor such projects
- Develop strategies to overcome technical, financial, institutional, and market barriers for bioenergy packages
- Disseminate bioenergy technologies across Punjab
- Evaluating barriers to the use of bioenergy technology such as social acceptance and rural engagement (active participation, etc.)
- Sustainability issues

The plan is to achieve these objectives through

- Testing of technical and also, in particular, the economic and financial feasibility of selected bioenergy technology on the basis of new business and financing models and developing further the financial, institutional, and market strategies for their medium- to large-scale replication
- Building the capacity of the supply side to market, finance, and deliver rural bioenergy services
- Institutionalizing the support provided by the project to facilitate sustainable growth of the market after the end of the project

In the second phase of the project, a dual fuel engine is used for power generation with a capacity of 20 kW. The mobile plant has been modified and automated to such an extent that it is completed with all associated auxiliary systems and works as a turnkey solution. All the necessary auxiliary facilities of the power plant include diesel quenched condenser, pelletization plant to convert the straw into pellets for better handling and storage, feed hopper, electrical heaters, mechanized automatic valves, and a flaring system for noncondensable gas flaring (if we choose not to give the gas to the engine). The entire plant is completely automated as a turnkey solution with instrumentation and control systems.

10.15 MOBILE UNIT DEMONSTRATION

As part of the project planning, the technology has been made mobile so that it could be demonstrated in the villages. The intention is also to train the villagers using the technology so that they can operate the plant. The mobile pyrolysis plant is shown in Figure 10.11. The mobile unit was

Figure 10.11 Mobile pyrolysis unit in the village of Khuaspura.

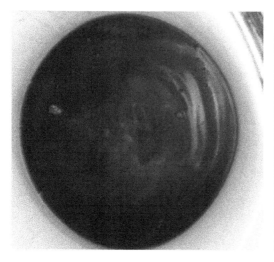

Figure 10.12   Blended oil and biochar.

stationed in Khuaspura, which has a population of around 3000. The agricultural land area under rice and wheat straw cultivation was around 200 acres. The village is very well connected to the electric grid, but the shortage of power during the peak summer season is enormous.

The engine was tested for blends up to 30% (oil from wheat straw) and with a running time close to 250 hours. The engine has been inspected on a regular basis, and it showed no signs of deterioration. It was also tested on a dual fuel mode of operation by injecting the noncondensable gas fraction as well. The results showed that we could displace the liquid fuel up to 20%. This could also possibly be raised with an increase in feedstock throughput. However, we did experience some difficulties with rice straw blends up to 30%. This was due to viscosity and pH increase. We could easily run the engine with 20% blends.

As mentioned earlier, we processed the straw as a part of our demonstration. The feedstock was converted into stable products such as blended oil (pyrolysis oil blended with conventional diesel), biochar, and noncondensable gas fraction. The products are shown in Figure 10.12. The produced blends were run on the dual fuel engine to produce combined heat and power. During the technology demonstrations, the dual fuel engine was run on blended fuel. Tests were also carried out with the noncondensable gas running the engine in the dual fuel mode. The products produced during the demonstration were partly used for running the engine, and the rest were distributed among villagers for them to run the liquid blended fuel on the Lister engines in their fields. Char, one of the products of the process, was also widely distributed among the villagers who used it as a replacement for the firewood traditionally burned for cooking.

10.16 FEEDBACK FROM THE VILLAGERS

The villagers were all very interested in the whole demonstration and participated willingly. The school to which power was provided as part of the demonstration was very thankful for the project. The villagers also expressed interest in actively participating if the unit would be stationed there much longer. During the harvest season, we did observe that at least four or five farmers did stop burning the field when rice was harvested this season. Because we were providing electricity to the community center, the villagers were even using it for their weekly meetings as well.

10.17 KEY MILESTONES

- The demonstration was quite successful in the village, and the villagers actively participated during the whole demonstration.
- Working with Punjab Agricultural University (PAU) Ludhiana on the trials of biochar for fertilizer application.
- The villagers' feedback was very encouraging and could bring about change for at least a few farmers.
- Rotary Club International has presented a vocational award in recognition of our work toward a better society.

ACKNOWLEDGMENTS

The authors thank Prof. M. K. Surappa, director of IIT Ropar, and Dr. Himasnhu Tyagi of the School of Mechanical, Materials and Energy Engineering, for their support of the project. Last but not the least we also thank Oglesby Charitable Trust U.K. and Aston University for their generous financial support.

REFERENCES

1. Ministry of New and Renewable Energy Resources. Biomass power and cogeneration programme. 2009. http://www.mnre.gov.in/relatedlinks/biomassresources
2. McKendry P. Energy production from biomass (part 1): Overview of biomass. *Bioresour Technol* 2002; 83: 37–46.
3. GHG accounting misses the mark on biofuels, biomass—Power and thermal magazine. Posted October 22, 2009, patent filed on 2008.
4. Slade R, Saunders R, Gross R, Bauen A. *Energy from biomass: The size of the global resource*. Imperial College Centre for Energy Policy and Technology, London, 2011.
5. Shafizadeh F. Introduction to pyrolysis of biomass. *J Anal Appl Pyrol* 1982; 3; 283–305.
6. Hornung A, Apfelbacher A. *Combined pyrolysis reformer*, GB 0808739.7—"Thermal treatment of biomass." 2008.

CHAPTER **11**

Algal Biorefinery

S. Venkata Mohan, M. V. Rohit, Rashmi Chandra, and Kannaiah R. Goud
Bioengineering and Environmental Centre, CSIR–Indian Institute of Chemical Technology,
Hyderabad, India

CONTENTS

11.1 INTRODUCTION

The direct conversion of solar energy into green fuels using photosynthesis-enabled microorganisms is one of the alternatives to produce renewable energy. Employing these organisms, especially microalgae, for biofuel generation can alternate the ongoing conflict between the use of land for food or fuel production (Machado and Atsumi 2012). Considering the diversity and potential of microalgae, they could be very important in a future mix of third-generation biofuel technologies. Algal biofuels have become a focal point in academic and industrial research in recent years (Jones and Mayfield 2012). These photosynthetic organisms are known to produce high oil and biomass yields, can be cultivated with nonfreshwater sources, can be grown on nonarable land, do not compete with common food resources, and very efficiently use water and nutrients for growth, making them an exciting addition to the sustainable fuel portfolio (Hannon et al. 2010).

Microalgae adopt various routes of metabolism for their growth and survival, namely autotrophic, heterotrophic, and mixotrophic modes. They are efficient in shifting their metabolism with response to changes in the environmental stress conditions (Devi et al. 2012). Algae are comprised of a diverse group of organisms including prokaryotes and eukaryotes. They occur in the form of single cells, colonies, and multicellular plants (Yu et al. 2011). These microbial cells (particularly green algae) are responsible for more than half of the world's primary production of oxygen. They can harness the energy from sunlight, allowing them to double their biomass rapidly (Seibert et al. 2008). They can be employed to produce various types of energy for transportation including biodiesel, biohydrogen, jet fuel, electric power, and ethanol.

They are known for the presence of high-value compounds such as polysaccharides, proteins, and pigments. Coproducing these products together with biofuels makes the process economically feasible and less dependent on fossil fuel imports (Mata et al. 2010). The real advantage of photosynthetic microorganisms lies in their metabolic flexibility, which offers the possibility of modification in their biochemical pathways (Tredici 2010). Photosynthetic microorganisms, including cyanobacteria, are currently being engineered for platforms to convert solar energy to biochemicals (Ducat et al. 2011; Ruffing 2011). This approach can be applied to produce valuable chemicals that the cyanobacteria host strains do not produce naturally by constructing new biosynthetic pathways.

In a biorefinery concept, the extraction of more than one type of biofuel from algal biomass or an additional coproduct increases the value of the biomass and offers additional offsets to the environmental impacts (Mussgnug et al. 2010; Ehimen et al. 2011; Yang et al. 2011). This strategy can be used to produce a wide range of biofuels, such as alcoholic fermentation of deoiled algae biomass, after extraction of lipids, synthesis of biogas, and combustion of this gas to produce bioelectricity. Integration of a biorefinery concept with wastewater treatment will provide efficient utilization of algae biomass, and this reduces the overall residual waste component of biomass and favors sustainable economics (Harun et al. 2010). In this regard, wastewater treatment integration with biofuels production is gaining attention in the current scenario of sustainable fuel synthesis (Venkata Mohan et al. 2011).

Scientific research, commercialization initiatives, and media coverage have generated a wave of renewed interest in algae cultivation over the last few years. In most cases, the main driver of this interest is its high potential as a renewable energy source, mainly algae-based biofuels and a range of value-added products (Shariati and Hadi 2011). This chapter gives a brief overview of how algae can be cultivated and how biofuels can be produced through conventional and metabolic engineering approaches. The chapter also tries to assess the viability of bioenergy generation and wastewater treatment as an important strategy for integration with the biorefinery concept.

11.2 MODE OF NUTRITION IN MICROALGAE

Microalgae can be classified mainly into autotrophs, mixotrophic and heterotrophs based on light and carbon source utilization. Autotrophic organisms can convert light and chemical energy (CO_2 and H_2O) into carbohydrates, which form the base for all other carbon containing biomolecules (Yoo et al. 2010). Mostly, the energy is stored as a reduced form (carbohydrates), that is the source of energy for the entire needs of the cell. Autotrophic organisms are relatively self-sufficient and self-sustainable because they obtain their energy from sunlight (Nelson et al. 1994; Eberhard et al. 2008). Heterotrophic organisms utilize organic carbon (mainly glucose) as energy for their metabolic functions because they cannot utilize atmospheric CO_2. Nutritional modes significantly influence the carbon assimilation and lipid productivity of the microalgae (Xu et al. 2006). Depending on the available sources, three types of nutritional modes—namely autotrophic, heterotrophic, and mixotrophic—are reported to produce algal biofuel.

11.2.1 Autotrophic

The most common mode of cultivation of microalgae is autotrophic mode. In photoautotrophic nutrition, algae use sunlight as the energy source and inorganic carbon (CO_2) as the carbon source to form biochemical energy through photosynthesis (Figure 11.1). This is one of the most favorable environmental conditions for the growth of microalgae (Chen et al. 2011). The simpler form of photosynthates, such as carbohydrates, serves as the sole energy source for carrying out the metabolic activities of the algal cells (Chang et al. 2011). These carbohydrates under nutrient-limiting and stress conditions will favor lipid biosynthesis (Figure 11.2a), (Gouveia and Oliveira 2009). Lipid productivity greatly depends on the photosynthetic activity in terms of atmospheric CO_2 fixation, microalgae species, operating conditions and species diversity (Ohlroggeav and Browseb 1995; Murata and Siegenthaler 2004; Mata et al. 2010; Chen et al. 2011). An advantage of the autotrophic nutritional mode is the algal oil production at the expense of atmospheric CO_2 and sunlight.

11.2.2 Heterotrophic

In the heterotrophic nutritional mode, microalgae utilize glucose as carbon sources for their growth (Kaplan et al. 1986). In this nutritional mode, the carbohydrates enter the cell, are subsequently converted to lipids, and participate in other metabolic pathways (Venkata Mohan et al. 2013).

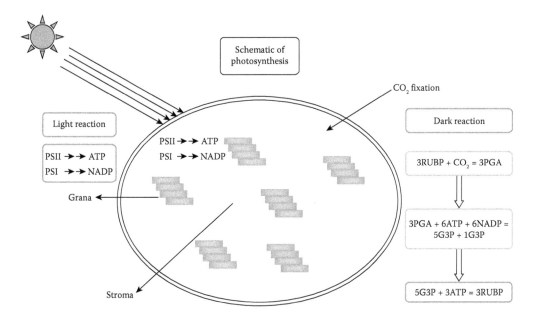

Figure 11.1 Photosynthetic conversion of CO_2 to various metabolic precursors.

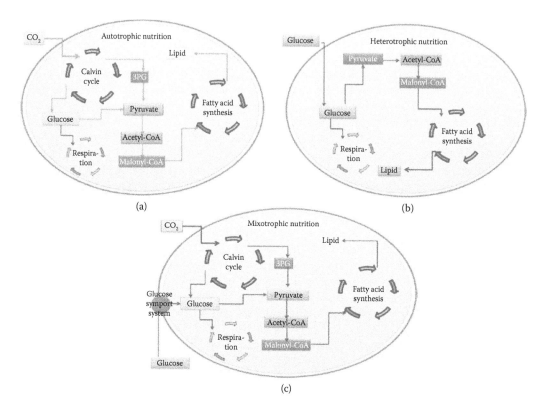

Figure 11.2   (a) Autotrophic, (b) heterotrophic, and (c) mixotrophic.

Heterotrophic nutrition takes place in the presence and absence of light. In photoheterotrophic nutrition, light acts as an energy source, but the source of carbon remains organic only (Figure 11.2b). Heterotrophic growth in a dark condition is supported by a carbon source replacing the light energy (Perez-Garcia et al. 2011). The oxidative assimilation takes place in algae through two pathways, the Embden–Meyerhof pathway and the pentose phosphate pathway (Neilson and Lewin 1974). Both pathways are carried out in the cytosol and are functional in microalgae. However, the PP pathway might have a higher flux rate than the other, depending on the carbon source and the presence of light. Light is not required for the transport of glucose inside the cell during dark heterotrophic operation. The glucose transport system in the algal cell becomes inefficient in the presence of light because of the higher availability of photosynthates inside the cell due to photosynthesis and downregulation of hexose transport protein.

The carbon is obtained as glucose from outside the cell and is converted to the acetyl-CoA via pyruvate. Acetyl-CoA carboxylase (ACC) is the important enzyme that catalyzes the conversion of acetyl-CoA to metabolic precursors such as malonyl coA and subsequently enters the lipid biosynthetic pathway. The photoheterotrophic nutritional mode avoids the limitations of light dependency, which is the major obstruction to gaining high cell density in large-scale photobioreactors (Huang and Wang 2004). *Chlorella protothecoides* showed higher lipid content (40%) during heterotrophic growth (Xu et al. 2006). The major advantage of the heterotrophic nutritional mode is the facilitation of wastewater treatment along with lipid productivity, which gives an edge to its application in the present state of increasing pollution loads (Venkata Mohan et al. 2014). Moreover, cost effectiveness, relative simplicity of operation, and easy maintenance are the main attractions of the heterotrophic growth approach.

11.2.3 Mixotrophic

Microalgae can be cultivated as a combination of autotrophic and heterotrophic mechanisms which is called the mixotrophic condition (Figure 11.2c). It facilitates atmospheric CO_2 sequestration as well as consumption of organic molecules and micronutrients from the growing environment. The CO_2 released by microalgae via respiration will again be trapped and reused in the mixotrophic nutritional mode (Chandra et al. 2014). The mixotrophs have the ability to utilize organic carbon; therefore, light energy is not a limiting factor for biomass growth (Chang et al. 2011). Mixotrophic cultures show reduced photo inhibition and improved growth rates compared with autotrophic and heterotrophic cultures (Chojnacka and Noworyta 2004). Algae have the flexibility to switch their nutritional mode based on substrate availability and light conditions (Chandra et al. 2014). The advantages of mixotrophic nutrition are its independence in terms of photosynthesis and growth substrates (Kong et al. 2012). Water ecosystems generally consist of nutrients and organic carbon as integral parts (Venkata Mohan et al. 2009) where microalgae, along with other living components, function together symbiotically. Some microalgal species are not truly mixotrophs but have the ability to switch between phototrophic and heterotrophic metabolisms, depending on environmental conditions (Kaplan et al. 1986).

Microalgae accumulating lipids generally grow in natural water bodies; therefore, ecological water bodies enriched with diverse microalgae species can be considered as potential reservoirs for harnessing biodiesel. Mixotrophic cultivation has shown to be a good strategy to obtain a large biomass and high growth rates (Ogawa and Aiba 1981; Lee and Lee 2002), with the additional benefit of producing photosynthetic metabolites (Chen 1996; Perez-Garcia et al. 2011). Among the various types of microalgae, mixed microalgae are highly diverse and ecologically dominant species that thrive in eutrophicated wastewater bodies in different ecological habitats (Venkata Mohan et al. 2011). Mixed microalgae are highly efficient in biological CO_2 fixation apart from nutrient removal and simultaneous wastewater treatment of these streams.

11.3 BIOFUELS FROM MICROALGAE

11.3.1 Biodiesel

Biodiesel can be defined as monoalkyl esters of plant oils or animal fats produced via transesterification reactions of lipids. These neutral lipids serve as energy stores in plants and animals having a common structure of triple esters, usually three long-chain fatty acids (LCFA) coupled to glycerol. The transesterification reaction turns the triple esters to single ones by displacing glycerol with small alcohols (e.g., methanol). Algae fix CO_2 during the day via photophosphorylation (thylakoid) and produce carbohydrates during the Calvin cycle (stroma), which gets converted into various products, including triacylglycerides (TAGs), depending on the species of algae or specific conditions pertaining to cytoplasm and plastid (Liu and Benning 2012). Microalgae are capable of surviving and functioning under photoautotrophic, heterotrophic, and mixotrophic conditions.

In photoautotrophic systems, biosequestration of CO_2 from industrial flue gases is one option for biodiesel production as well as CO_2 mitigation. Studies have shown that CO_2 sparging period and time interval have a significant influence on lipid accumulation of microalgae when cultivated in domestic wastewaters (Devi and Venkata Mohan 2012). The organic carbon source is the major factor governing lipid production, when mixotrophic systems are taken into account. Fatty acid–rich effluents generated from the acidogenic biohydrogen production process contain a high amount of organic carbon in the form of volatile fatty acids (VFAs), which can be used for lipid synthesis by mixed microalgae. A VFA platform offers more advantages in terms of inexpensiveness and the ease of converting them to liquid fuels more rapidly (Venkata Mohan and Devi 2012). Apart from microalgae cultivation, the major nutrients, namely, nitrogen (N), phosphorus (P), carbon (C), and potassium (K) also play an important role in biomass growth and lipid accumulation. The nutrient-deprived condition during starvation phase (SP) operation documented increments in lipid productivity due to acceleration of TAG formation. Previous studies indicate that higher biomass growth was observed in N + P conditions while higher lipid productivity was observed in C condition (Devi et al. 2013). In all these conditions, fatty acid methyl ester (FAME) composition is different in terms of saturation content, and this depends on the substrate and mode of cultivation.

Heterotrophic nutrition is again light-dependent and light-independent. However, carbon assimilation is more favorable in the case of light-independent processes (dark heterotrophic) over the light-dependent ones (photoheterotrophic). Carbon uptake depends upon the inducible active hexose symport system from outside the cell, and this process invests energy in the form of ATP (Tanner 1969). In dark heterotrophic algae, light inhibits the expression of the hexose/H^+ symport system (Kamiya and Kowallik 1987; Perez-Garcia et al. 2011) which decreases glucose transport inside the cell. In mixotrophic nutrition, the biochemical processes of autotrophs and heterotrophs occur simultaneously, and the preference of substrate uptake depends on the substrate availability in addition to other environmental conditions (Devi et al. 2012). Carbohydrates provide a source of acetyl-CoA by activated acetyl-CoA synthase in the stroma due to cytosolic conversion of glucose to pyruvate during glycolysis (Somerville et al. 2000; Schwender and Ohlrogge 2002). This acetyl-CoA is transported from the cytosol to the plastid, where it is converted to the fatty acid and subsequently to TAG, which again moves to the cytosol and forms the lipid bodies. Diversity studies have shown dominance of lipid accumulating microalgae such as *Scenedesmus* sp., *Diatoms*, and *Chlorella* spp.

11.3.2 Biohydrogen

Biohydrogen is a natural and transitory by-product of various microbial-driven biochemical reactions. Various biological routes are available for biohydrogen production pertaining to anaerobic fermentation, photobiological, enzymatic, and electrogenic mechanisms. On the basis of light dependency, the biological H_2 production processes can be further classified into light-dependent or photofermentation (photosynthetic process) and light-independent dark fermentation. Light-dependent processes are through biophotolysis of water using green algae and cyanobacteria via direct and indirect biophotolysis or via photofermentation mediated by photosynthetic bacteria (Chandra and Venkata Mohan 2011). Photofermentative hydrogen production using microalgae can be integrated with dark fermentation to achieve higher yields of hydrogen when compared to dark fermentation. Studies have shown that integration of photofermentation with the acidogenic biohydrogen process resulted in maximum H_2 production along simultaneous substrate degradation. Also, the photofermentative process showed higher H_2 yield when compared to dark fermentation. Two-stage process integration can be used for bioenergy production and treatment of organic wastewater generated in preliminary stage. (Mohanakrishna and Venkata Mohan 2013). Cyanobacteria and microalgae undergo direct and indirect biophotolysis to produce H_2 by utilizing inorganic CO_2 in the presence of sunlight and water, whereas photosynthetic bacteria (PSB) manifest H_2 production through photofermentation by consuming a wide variety of substrates ranging from inorganic to organic acids in the presence of light (Venkata Mohan et al. 2013).

11.3.3 Bioelectricity

Microbial fuel cell (MFC) application for bioelectricity generation from various substrates, including waste/wastewater, has gained prominence in the recent bioenergy scenario due to its sustainable nature. In this context, photosynthetic fuel cells (PhFC) are similar to the MFC in operation, but instead of a chemotrophic mechanism, green algae or PSB will act as a biocatalyst. PhFC utilizes sunlight as the electron donor, and the electron will pass through a cascade of proteins, generating a proton motive force similar to the MFC, which provides the feasibility to harness bioelectricity (Chandra et al. 2012). Studies on electrogenic activity of oxygenic PhFC fuel cells using microalgae under the mixotrophic mode was evaluated for harnessing bioelectricity (Venkata Subhash et al. 2013). Dissolved oxygen produced during the photolysis of water (oxygenic photosynthesis) was found to be the major limiting factor lowering performance. The electrogenic activity of the anoxygenic PhFC was relatively higher than microalgae-based oxygenic PhFC due to the release of oxygen that consumes the released electrons responsible for electricity and lowers current generation. Anoxygenic conditions are more favorable for electrogenesis, which can be achieved by using PSB. Nevertheless, microalgae-based oxygenic PhFC is gaining importance in the current scenario and has a vast scope for research in coming years. Applications of PhFC can be extended to the utilization of acidogenic effluents rich in VFAs generated during H_2 production to convert them to either lipids, carbohydrates, or other value-added products along with power generation (Strik et al. 2008).

11.4 VALUE ADDITION

Various species of algae are used in the human health market (e.g., *Chlorella*, *Dunaliella*, and *Spirulina*) as animal and aquaculture feed, antioxidants, with 3 fatty acids, anti-inflammatory products and specialty biochemicals. There is growing interest in new natural products from microalgae (Andlaeuer and Furst 2002), these compounds could change from the low status of a by-product to the high revenue stream of raw materials for nutraceuticals which can provide health benefits (Arnaiz et al. 2011; Andlaeuer and Furst 2002).

11.4.1 Polysaccharides

Algae (macroalgae) contain large contents of polysaccharides, which are polymers of simple sugars (monosaccharides) linked by glycosidic bonds and which contribute to cell-wall structure as well as storage polysaccharides (Holdt and Kraan 2011). These polysaccharides contain numerous commercial applications as stabilizers, thickeners, and emulsifiers in food and feed (Tseng 2001). Alginates are polymers extracted from the cell walls of various brown algae, particularly the species *Laminaria*, *Saccharina*, *Macrocystis*, and *Ascophyllum*. They are composed of D-mannuronic acid and L-glucuronic acid monomers, available in both acid and salt forms; the latter constitutes 40–47% of the dry weight of this brown algal biomass (Arasaki and Arasaki 1983). Alginates are commonly applied as intermediate feedstock in the food and pharmaceutical industries as stabilizers for the preparation of emulsions and suspensions in ice cream, jam, cream, custard, lotions, and toothpaste but also as coatings for pills. Carrageenans consist of linear polysaccharide chains with sulfate half-esters attached to the sugar unit. They are also used as suspension agents and stabilizers in drugs, lotions, and medicinal creams. An illustrative medical application is treatment of bowel problems—such as diarrhea, constipation, or dysentery. They are also used to make internal poultices to control stomach ulcers (Morrissey et al. 2001). Agar is a mixture of polysaccharides typically extracted from the cell walls of red algae. It is composed of agarose and agropectin and exhibits structural and functional properties. *Gelidium* sp. and *Gracilaria* sp. are the major commercial sources of agar (Carlsson 2007).

11.4.2 Protein Compounds

Algal proteins may play structural as well as nutritional roles. The proteins, peptides, and amino acids vary with the algal species as well as the habitat and the season (Arasaki and Arasaki 1983).

Proteins may indeed represent 35%–45% of dry matter in macroalgae (Holdt and Kraan 2011) and even 60%–70% in microalgae (Babadzhanov et al. 2004; Samarakoon and Jeon 2012). Most algal species contain all essential amino acids and are, in particular, a rich source of aspartic and glutamic acids (Fleurence 1999). Bioactive proteins and peptides have been found in micro- and macroalgae that possess a nutraceutical potential (DeFelice 1995), as is the case of their role in reducing the risk of cardiovascular diseases (Erdmann et al. 2008). Bioactive peptides usually contain 3–20 amino acid residues, and their activities stem from their amino acid composition and sequence (Leppälä 2000). Usually such short chains of amino acids are inactive within the sequence of the parent protein, but they become active upon release during gastrointestinal digestion or during food processing, including fermentation.

11.4.3 Chlorophylls

Chlorophylls are lipid-soluble pigments present in microalgae, higher plants, and cyanobacteria as light-harvesting systems, and they carry out photosynthesis (Rasmussen and Morrissey 2007). Chlorophyll is converted into pheophytin, pyropheophytin, and pheophorbide in processed vegetable foods following ingestion by humans. These bioactive compounds show antioxidant and antimutagenic effects and thus play a significant role in cancer prevention (Chernomorsky et al. 1999).

11.4.4 Carotenoids

Carotenoids are the most widespread pigments in nature, and they appear in all algae, higher plants, and many PSB. In general, green algae contain β-carotene, lutein, violaxanthin, astaxanthin, neoxanthin, and zeaxanthin, whereas, red species contain mainly α- and β-carotene, lutein, and zeaxanthin; β-carotene, violaxanthin, and fucoxanthin are present chiefly in brown species (Haugan and Liaaen-Jensen 1994). The major large-scale applications are food and health. Carotenoids' antioxidant properties have been shown to play a role in preventing pathologies linked to oxidative stress (Yan et al. 1999). Few carotenoids are part of vitamins, which have diverse biochemical functions, including hormones, antioxidants, mediators of cell signaling, and regulators of cell and tissue growth and differentiation (Holdt and Kraan 2011).

11.4.5 Phycobiliproteins

These proteins are major photosynthetic accessory pigments in algae and cyanobacteria, which include phycoerythrin, phycocyanin, allophycocyanin, and phycoerythrocyanin (Feng et al. 2006). Various combinations of the two major phycobilins, phycoerythrobilin (red) and phycocyanobilin (blue), can absorb at distinct spectral regions (Lobban and Harrison 1994). These proteins have been used as natural colorants for food and cosmetic applications, for example, chewing gum, ice sherbets, jellies, and dairy products in addition to lipsticks and eyeliners (Sekar and Chandramohan 2008). Several phycobiliproteins have been shown to exhibit antioxidant, anti-inflammatory, neuroprotective, hypocholesterolemic, hepatoprotective, antiviral, antitumoral, serum lipid reducing, and lipase-inhibiting activities (Sekar and Chandramohan 2008). Therefore, such health products as tablets, capsules, or powders that include phycocyanin have successfully reached the market recently (Guil-Guerrero et al. 2004).

11.5 METABOLIC ENGINEERING

Intensive global research efforts are currently targeting manipulation of the native metabolic pathways of lipids, alcohols, hydrocarbons, polysaccharides, and other energy storage compounds in photosynthetic organisms through genetic engineering (Radakovits et al. 2010). Both cyanobacteria and eukaryotic microalgae possess several unique metabolic attributes with respect to biofuel production. There are several advantages to using these organisms, such as their readily available genetic tools, sequenced genomes, and higher growth rates (Machado and Atsumi 2012). Various gene transformation techniques are available for incorporation of a desired gene into the genome of microalgae. The present focus is on potential avenues of genetic engineering that may be undertaken in order to improve microalgae as a biofuel platform for the production of hydrocarbon derived alcohols, isoprenes, diesel fuel alternatives, and alkanes. Induction of desired fluxes in metabolism can redirect cellular function toward the synthesis of preferred products and can even expand the processing capabilities of microalgae (Rosenberg et al. 2008).

Biological conversion of CO_2 into hydrocarbon based fuels could greatly simplify the overall production process and reduce the cost of biofuel production. The exploration of autotrophic microorganisms capable of performing this CO_2 to fuel conversion started in the late 1970s with the U.S. Department of Energy's Aquatic Species Program (ASP) (Sheehan et al. 1998). The ASP isolated and screened more than 3000 species of microalgae from a diverse range of environmental habitats. The main focus of the program was to naturally produce significant amounts of TAGs from eukaryotic algae. Recombinant DNA technology was developed at that time but, due to the infancy of this technology, it was not applied to microalgae for fuel applications until final stages of the ASP (Dunahay et al. 1996). With the development of recombinant DNA technology, prokaryotic microalgae (cyanobacteria) were recognized as potential hosts for fuel production, and the successful engineering of cyanobacteria for ethanol production confirmed their potential (Deng and Coleman 1999). A novel approach allows cells to secrete chemicals without overexpression of transporters or additional genetic modifications, which makes the system simple and inexpensive (Oliver et al. 2013). With the overwhelming prices of crude oil, there is renewed interest in biofuel recovery from microalgae developed for fuel production and in the application of metabolic engineering to enhance fuel yields.

Metabolic precursors such as glyceraldehyde-3-phosphate (G3P) and pyruvate are common in photosynthetic and fermentative metabolisms, making it possible to link both in one organism (Anemaet et al. 2010). The combination of these two metabolisms is referred to as the "Photanol" (Hellingwerf and Teixeira de Mattos 2009) concept (Figure 11.3). This is a new approach wherein photosynthetic production by algae and fermentative pathways occurring in bacteria can be combined to minimize

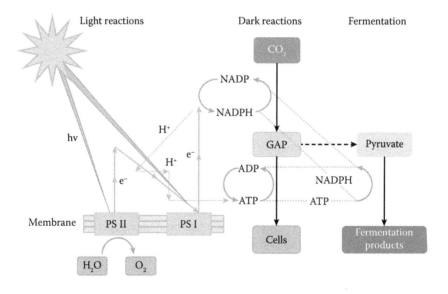

Figure 11.3 Integration of phototrophic and heterotrophic pathways leading to photanol concept.

losses incurred during individual metabolic pathways. Using metabolic engineering, a new enzyme or even an entire pathway can be cloned into cyanobacteria. This approach for biofuel production, where the capacity of cyanobacteria to use sunlight efficiently as an energy source for the reduction of the greenhouse gas CO_2 into organic compounds, (Hellingwerf and Teixeira de Mattos 2009; Anemaet et al. 2010) is gaining importance in current biofuels research. *Synechococcus elongatus* has been studied for isopropanol production (Atsumi et al. 2009). Using a similar method, fatty acid production has also been achieved in the cyanobacterium *Synechocystis* sp. *PCC6803*, using only light and carbon dioxide as feedstocks (Liu et al. 2011).

11.6 MICROALGAE-BASED WASTEWATER TREATMENT

Successful implementation of combining microalgal biomass with biofuel production and wastewater treatment would allow the minimization of freshwater usage, which is precious resource, especially for dry or populous countries. Extensive works have been conducted to explore the feasibility of using microalgae for wastewater treatment, specifically for the removal of nitrogen and phosphorus from effluents (Mallick 2002; Hernandez et al. 2006), which would otherwise result in eutrophication if dumped into lakes and rivers (Janssen et al. 2001). Consumption of nitrogen and phosphorus by microalgae in a controlled manner benefits the environment rather than causing its deterioration. Mixotrophic cultivation is beneficial due to the advantage of using wastewater as a substrate and its flexibility toward photosynthesis and carbohydrate utilization (Devi et al. 2013). The syntrophic association of microalgal cultures in eutrophic microenvironments can be considered as a potential biocatalyst for bioenergy production and can diminish wastewater pollution. Levels of several contaminant heavy metals have also been shown to be reduced by the cultivation of microalgae (Munoz and Guieysse 2006). A major concern associated with using wastewater for microalgae cultivation is contamination (De la Noue and De Pauw 1988). This can be managed by using appropriate pretreatment technologies to remove sediment and to deactivate the wastewater (Tamer et al. 2006). This problem of contamination and pathogen influence can be eliminated by using antibiotics during microalgal cultivation. The most important aspect of using wastewater is that it can be integrated with the biorefinery process for cutting down the investment on substrates used in the conventional biorefining process.

11.7 ALGAL BIOREFINERY

The term biorefinery was coined to describe the production of a wide range of chemicals and biofuels from biomass. This is achieved by the integration of bioprocessing and environmentally low impact chemical technologies in a cost-effective and sustainable manner (Chisti 2007). Examples include photofermentative biohydrogen, bioelectricity, and fermentation of deoiled biomass to ethanol, bio-oils, and biogas production by pyrolysis of deoiled algae cake (Mohan et al. 2006) (Figure 11.4). VFA accumulation during acidogenic biohydrogen production is a major problem. Photosynthetic biohydrogen production can provide a solution to this problem by utilizing the VFA as substrate via a bioaugmentation strategy. Recent reports have shown that augmentation of the acidogenic bacteria with photosynthetic cultures has improved the hydrogen production significantly (Chandra and Venkata Mohan 2014). Selective enrichment of microalgae for photofermentative biohydrogen production is also an important strategy because the substrate composition significantly influences the final product. Acetate-rich VFA contents in dark fermentation effluents can be used as a substrate to microalgae, which in turn generate photofermentative hydrogen (Chandra and Venkata Mohan 2011). This strategy can be directly integrated with the biorefinery concept for recycling the nutrients in generated wastes.

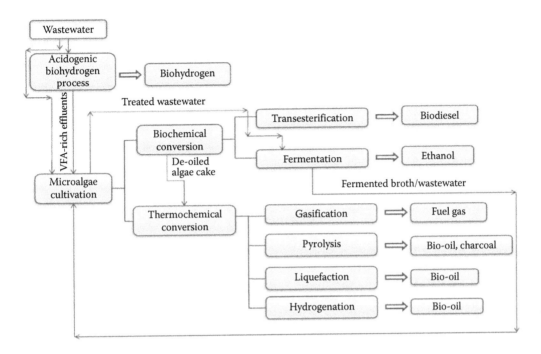

Figure 11.4 Microalgae based biorefinery approach.

Microalgae have the capacity of producing a vast array of high-value bioactive compounds that can be used as pharmaceutical compounds, nutritional supplements, and natural pigments along with biofuels. The economic feasibility of microalgal biofuel production should be significantly enhanced by a high-value coproduct strategy that would sequentially involve the cultivation of microalgae in a microalgal farming facility (CO_2 mitigation), extracting bioactive products from harvested algal biomass, thermal, and hydrothermal processing (pyrolysis, liquefaction, or gasification) (Sarkar et al. 2014); extracting high-value chemicals from the resulting liquid, vapor, or solid phases; and reforming biofuels for different applications (Li et al. 2008). For instance, this concept can be used in a two-stage biohydrogen and biogas production. In the first stage, biohydrogen and biogas are produced by anaerobic digestion of deoiled algae biomass. In the second stage, electricity generated (Venkata Subhash et al. 2014) from biogas can be used to offset the energy requirements for anaerobic digestion of microalgae. Agriculturally derived biogas can be used to provide a CO_2 stream for algae growth and biogas can be used to power the cultivation and lipid extraction process for algae biodiesel (Douskova et al. 2010; Collet et al. 2011; Harun et al. 2011). The employment of a high-value coproduct strategy through the integrated biorefinery approach can significantly enhance the overall cost effectiveness of microalgal biofuel production.

11.8 SUMMARY AND CONCLUSIONS

Microalgae possess several unique metabolic attributes with respect to biofuel production and value-added products. Nutritional modes significantly influence the carbon assimilation and lipid productivity of the microalgae, depending on the available sources. Heterotrophic cultivation is known to produce high biomass and lipids that can be extracted to produce oil. Biohydrogen production by photofermentation is a novel area for hydrogen production through biological routes that can be integrated with wastewater treatment. Various species of algae are used in the human health market for producing antioxidants, anti-inflammatory products and specialty biochemicals such as nutraceuticals. The building of an algae-based bioeconomy has the capacity to not only move the world through present difficulties but also result in industries with very small environmental footprints.

ACKNOWLEDGMENT

The authors wish to thank the director CSIR-IICT, Hyderabad, for encouragement. Funding from CSIR in the form of XII five-year plan project on BIOEN (CSC-0116) is greatly acknowledged.

REFERENCES

Andlaeuer, W. and Furst, P. 2002. Nutraceuticals: A piece of history, present status and outlook. *Food Res. Int.* 35: 171–176.

Anemaet, I.G., Bekker, M., and Hellingwerf, K.J. 2010. Algal photosynthesis as the primary driver for a sustainable development in energy, feed and food production. *Mar. Biotechnol.* 12: 619–629.

Arasaki, S. and Arasaki, T. 1983. *Vegetables from the Sea: Low calorie, high nutrition to help you look and feel better.* Tokyo: Japan Publications, p. 196.

Arnaiz, E., Bernal, J., Martin, M.T., Viguera, C.G., Bernal, J.L., and Toribio, L. 2011. Supercritical fluid extraction of lipids from broccoli leaves. *Eur. J. Lipid Sci. Technol.* 113: 479–486.

Atsumi, S., Higashide, W., and Liao, J.C. 2009. Direct photosynthetic recycling of carbon dioxide to isobutyraldehyde. *Nat. Biotechnol.* 27: 1177–1180.

Babadzhanov, A.S., Abdusamatova, N., Yusupova, F.M., Faizullaeva, N., Mezhlumyan, L.G., and Malikova, M.K. 2004. Chemical composition of *Spirulina platensis* cultivated in Uzbekistan. *Chem. Nat. Comp.* 40: 276–279.

Carlsson, A.S. 2007. *Micro- and macro-algae: Utility for industrial applications: Outputs from the EPOBIO project.* Newbury, UK: CPL Press.

Chandra, R. and Venkata Mohan, S. 2011. Microalgal community and their growth conditions influence biohydrogen production during integration of dark-fermentation and photo-fermentation processes. *Int. J. Hydrogen Energ.* 36: 12211–12219.

Chandra, R., Subhash, G.V., and Venkata Mohan, S. 2012. Mixotrophic operation of photo-bioelectrocatalytic fuel cell under anoxygenic microenvironment enhances the light dependent bioelectrogenic activity. *Bioresour. Technol.* 109: 46–56.

Chandra, R. and Venkata Mohan, S. 2014. Enhanced bio-hydrogenesis by co-culturing photosynthetic bacteria with acidogenic process: Augmented dark-photo fermentative hybrid system to regulate volatile fatty acid inhibition. *Int. J. Hydrogen Energ.* 39: 7604–7615.

Chandra, R., Rohit, M.V., Swamy, Y.V., and Venkata Mohan, S. 2014. Regulatory function of organic carbon supplementation on biodiesel production during growth and nutrient stress phases of mixotrophic micro-algae cultivation. *Bioresour. Technol.* 165: 279–287.

Chandra, R., Arora, S., Rohit, M.V., and Venkata Mohan, S. 2015. Lipid metabolism in response to individual short chain fatty acids during mixotrophic mode of microalgae cultivation: Influence on biodiesel saturation and protein people. *Bioresour. Technol.* doi. 10.1016/j.biortech.2015.01.088.

Chang, R.L., Ghamsari, L., Manichaikul, A., Hom, E.F.Y., Balaji, S., Fu, W., Shen, Y., et al. 2011. Metabolic network reconstruction of *Chlamydomonas* offers insight into light-driven algal metabolism. *Mol. Syst. Biol.* 7: 518.

Chen, C.Y., Yeh, K.L., Aisyah, R., Lee, D.J., and Chang, J.S. 2011. Cultivation, photobioreactor design and harvesting of microalgae for biodiesel production: A critical review. *Bioresour. Technol.* 102: 71–81.

Chen, F. 1996. High cell density culture of microalgae in heterotrophic growth. *Trends Biotechnol.* 14: 421–426.

Chernomorsky, S., Segelman, A., and Poretz, R.D. 1999. Effect of dietary chlorophyll derivatives on mutagenesis and tumor cell growth. *Teratog. Carcinog. Mutagen.* 19: 313–322.

Chisti, Y. 2007. Biodiesel from microalgae. *Biotechnol. Adv.* 25: 294–306.

Chojnacka, K. and Noworyta, A. 2004. Evaluation of *Spirulina* sp. growth in photoautotrophic, heterotrophic and mixotrophic cultures. *Enzyme Microb. Technol.* 34: 461–465.

Collet, P., Helias, A., Lardon, L., Ras, M., Goy, R.A., and Steyer, J.P. 2011. Life cycle assessment of microalgae culture coupled to biogas production. *Bioresour. Technol.* 102: 207–214.

DeFelice, S.L. 1995. The nutraceutical revolution: Its impact on food industry R&D. *Trends Food Sci. Technol.* 6: 59–61.

De la Noue, J. and De Pauw, N. 1988. The potential of microalgal biotechnology: A review of production and uses of microalgae. *Biotechnol. Adv.* 6: 725–770.

Deng, M.D. and Coleman, J.R. 1999. Ethanol synthesis by genetic engineering in cyanobacteria. *Appl. Environ. Microbiol.* 65: 523–528.

Devi, M.P., Subhash, G.V., and Venkata Mohan, S. 2012. Heterotrophic cultivation of mixed microalgae for lipid accumulation and wastewater treatment during sequential growth and starvation phases. Effect of nutrient supplementation. *J. Renew Energ.* 43: 276–283.

Devi, M.P., Swamy, Y.V., and Venkata Mohan, S. 2013. Nutritional mode influences lipid accumulation in microalgae with the function of carbon sequestration and nutrient supplementation. *Bioresour. Technol.* 142: 278–286.

Devi, M.P. and Venkata Mohan, S. 2012. CO_2 supplementation to domestic wastewater enhances microalgae lipid accumulation under mixotrophic microenvironment: Effect of sparging period and interval. *Bioresour. Technol.* 112: 116–123.

Douskova, I., Kastanek, F., Maleterova, Y., Kastanek, P., Doucha, J., and Zachleder, V. 2010. Utilization of distillery stillage for energy generation and concurrent production of valuable microalgal biomass in the sequence: Biogas-cogeneration microalgae products. *Energ. Convers. Manage.* 51: 606–611.

Ducat, D.C., Way, J.C., and Silver, P.A. 2011. Engineering cyanobacteria to generate high-value products. *Trends Biotechnol.* 29: 95–103.

Dunahay, T., Jarvis, E., Dais, S., and Roessler, P. 1996. Manipulation of microalgal lipid production using genetic engineering. *Appl. Biochem. Biotechnol.* 57: 223–231.

Eberhard, S., Finazzi, G., and Wollman, F.A. 2008. The dynamics of photosynthesis. *Annu. Rev. Genet.* 42: 463–515.

Ehimen, E.A., Sun, Z.F., Carrington, C.G., Birch, E.J., and Eaton-Rye, J.J. 2011. Anaerobic digestion of microalgae residues resulting from the biodiesel production process. *Appl. Energ.* 88: 3454–3463.

Erdmann, K., Cheung, B.W.Y., and Schroder, H. 2008. The possible roles of food-derived bioactive peptides in reducing the risk of cardiovascular disease. *J. Nutr. Biochem.* 19: 643–654.

Feng, N.J., Guang-Ce, W., and Tseng, C.K. 2006. Method for large-scale isolation and purification of R-phycoerythrin from red alga *Polysiphonia urceolata* Grev. *Protein Expr. Purif.* 49: 23–31.

Fleurence, J. 1999. Seaweed proteins: Biochemical, nutritional aspects and potential uses. *Trends Food Sci. Technol.* 10: 25–28.

Gouveia, L. and Oliveira, A.C. 2009. Microalgae as a raw material for biofuels production. *J. Ind. Microbiol. Biotechnol.* 36: 269–274.

Guil-Guerrero, J.L., Navarro-Juárez, R., Lopez-Martinez, J.C., Campra- Madrid, P., and Rebolloso-Fuentes, M.M. 2004. Functional properties of the biomass of three microalgal species. *J. Food Eng.* 65: 511–517.

Hannon, M., Gimpel, J., Tran, M., Rasala, B., and Mayfield, S. 2010. Biofuels from algae: Challenges and potential. *Biofuels.* 1: 763–784.

Harun, R., Danquah, M.K. and Forde, G.M. 2010. Microalgal biomass as a fermentation feedstock for bioethanol production. *J. Chem. Technol. Biotechnol.* 85: 199–203.

Harun, R., Davidson, M., Doyle, M., Gopiraj, R., Danquah, M., and Forde, G. 2011. Techno-economic analysis of an integrated microalgae photobioreactor, biodiesel and biogas production facility. *Biomass Bioenerg.* 35: 741–747.

Haugan, J.A. and Liaaen-Jensen, S. 1994. Algal carotenoids 54. Carotenoids of brown algae (Phaeophyceae). *Biochem. Syst. Ecol.* 22: 31–41.

Hellingwerf, K.J. and Teixeira de Mattos, M.J. 2009. Alternative routes to biofuels: Light-driven biofuel formation from CO_2 and water based on the 'photanol' approach. *J. Biotechnol.* 142: 87–90.

Hernandez, J.P., De-Bashan, L.E., and Bashan, Y. 2006. Starvation enhances phosphorus removal from wastewater by the microalga *Chlorella*, spp. co-immobilized with *Azospirillum brasilense. Enzyme Microb. Technol.* 38: 190–198.

Holdt, S. and Kraan, S. 2011. Bioactive compounds in seaweed: Functional food applications and legislation. *J. Appl. Phycol.* 23: 543–597.

Huang, H.L. and Wang, B.G. 2004. Antioxidant capacity and lipophilic content of seaweeds collected from the Qingdao coastline. *J. Agric. Food Chem.* 52: 4993–4997.

Janssen, M., Slenders, P., Tramper, J., Mur, L.R., and Wijffels, R.H. 2001. Photosynthetic efficiency of *Dunaliella tertiolecta* under short light/dark cycles. *Enzyme Microb. Technol.* 29: 298–305.

Jones, C.S. and Mayfield, S.P. 2012. Algae biofuels: Versatility for the future of bioenergy. *Curr. Opin. Biotechnol.* 23: 346–351.

Kamiya, A. and Kowallik, W. 1987. The inhibitory effect of light on proton-coupled hexose uptake in *Chlorella*. *Plant Cell. Physiol.* 28: 621–625.

Kaplan, D., Richmond, A.E., Dubinsky, Z., and Aaronson, S. 1986. Algal nutrition. In: Richmond, A. (Ed.), *Handbook for microalgal mass culture.* Boca Raton, FL: CRC Press, pp. 147–198.

Kong, W.-B., Song, H., Hua, S.-F., Yang, H., Yang, Q., and Xia, C.-G. 2012. Enhancement of biomass and hydrocarbon productivities of *Botryococcus braunii* by mixotrophic cultivation and its application in brewery wastewater treatment. *Afr. J. Microbiol. Res.* 61489–61496.

Lee, K. and Lee, C.G. 2002. Nitrogen removal from wastewaters by microalgae without consuming organic carbon sources. *J. Microbiol. Biotechnol.* 12: 979–985.

Leppälä, P. 2000. Bioactive peptides derived from bovine whey proteins: Opioid and ace-inhibitory peptides. *Trends Food Sci. Technol.* 11: 347–356.

Li, Y., Horsman, M., Wu, N., Lan, C.Q., and Calero, N.D. 2008. Biofuels from microalgae. *Biotechnol. Prog.* 24: 815–820.

Liu, B. and Benning, C. 2012. Lipid metabolism in microalgae distinguishes itself. *Curr. Opin. Biotechnol.* 24: 300–309.

Liu, X., Sheng, J., and Curtiss, R. 2011. Fatty acid production in genetically modified cyanobacteria. *Proc. Natl. Acad. Sci. U. S. A.* 108: 6899–6904.

Lobban, C.S. and Harrison, P.J. 1994. *Seaweed ecology and physiology.* New York, NY: Cambridge University Press.

Machado, I.M. and Atsumi, S. 2012. Cyanobacterial biofuel production. *J. Biotechnol.* 162: 50–56.

Mallick, N. 2002. Biotechnological potential of immobilized algae for wastewater N, P and metal removal: A review. *BioMetals.* 15: 377–390.

Mata, T.M., Martins, A.A., and Caetano, N.S. 2010. Microalgae for biodiesel production and other applications: A review. *Renew. Sustain. Energ. Rev.* 14: 217–232.

Mohan, D., Pittman, C.U., Jr., and Steele, P.H. 2006. Pyrolysis of wood/biomass for bio-oil: A critical review. *Energ. Fuels.* 20: 848–889.

Mohanakrishna, G. and Venkata Mohan, S. 2013. Multiple process integrations for broad perspective analysis of fermentative H_2 production from wastewater treatment: Technical and environmental considerations. *Appl. Energ.* 107: 244–254.

Morrissey, J., Kraan, S., and Guiry, M.D. 2001. *A guide to commercially important seaweeds on the Irish coast.* Dublin: Bord Iascaigh Mhara/Irish Sea Fisheries Board.

Munoz, R. and Guieysse, B. 2006. Algal-bacterial processes for the treatment of hazardous contaminants: A review. *Water Res.* 40: 2799–2815.

Murata, N. and Siegenthaler, P.A. 2004. *Lipids in photosynthesis: Structure, function and genetics.* Advances in Photosynthesis and Respiration. Vol. 6. Dordrecht: Kluwer Academic Publishers, pp. 1–20.

Mussgnug, J.H., Klassen, V., Schluter, A., and Kruse, O. 2010. Microalgae as substrates for fermentative biogas production in a combined biorefinery concept. *J. Biotechnol.* 150: 51–56.

Neilson, A.H. and Lewin, R.A. 1974. The uptake and utilization of organic carbon by algae: An essay in comparative biochemistry. *Phycologia.* 13: 227–264.

Nelson, J.A., Savereide, P.B., and Lefebvre, P.A. 1994. The CRY1 gene in *Chlamydomonas reinhardtii*: Structure and use as a dominant selectable marker for nuclear transformation. *Mol. Cell Biol.* 14: 4011–4019.

Ogawa, T. and Aiba, S. 1981. Bioenergetic analysis of mixotrophic growth in *Chlorella vulgaris* and *Scenedesmus acutus. Biotechnol. Bioeng.* 23: 1121–1132.

Ohlroggeav, J. and Browseb, J. 1995. Lipid biosynthesis. *Plant Cell.* 7: 957–970.

Oliver, J.W.K., Machado, I.M.P., Yoneda, H., and Atsumi, S. 2013. Cyanobacterial conversion of carbon dioxide to 2,3-butanediol. *Proc. Natl. Acad. Sci. U. S. A.* 110: 1249–1254.

Perez-Garcia, R.O., Bashan, Y., and Puente, M.E. 2011. Organic carbon supplementation of municipal wastewater is essential for heterotrophic growth and ammonium removing by the microalgae *Chlorella vulgaris*. *J. Phycol.* 47: 190–199.

Radakovits, R., Jinkerson, R.E., Darzins, A., and Posewitz, M.C. 2010. Genetic engineering of algae for enhanced biofuel production. *Eukaryot. Cell.* 9: 486–501.

Rasmussen, R.S. and Morrissey, M.T. 2007. Marine biotechnology for production of food ingredients. In: Taylor, S.L. (Ed.), *Advances in food and nutrition research*. Boston, MA: Academic Press, p. 52.

Rosenberg, J.N., Oyler, G.A., Wilkinson, L., and Betenbaugh, M.J. 2008. A green light for engineered algae: Redirecting metabolism to fuel a biotechnology revolution. *Curr. Opin. Biotechnol.* 19: 430–436.

Ruffing, A.M. 2011. Engineered cyanobacteria: Teaching an old bug new tricks. *Bioeng Bugs.* 2: 136–149.

Samarakoon, K. and Jeon, Y.J. 2012. Bio-functionalities of proteins derived from marine algae: A review. *Food Res. Int.* 48: 948–960.

Sarkar, O., Agarwal, M., Kumar, A.N., and Venkata Mohan, S. 2015. Retrofitting hetrotrophically cultivated algae biomass as pyrolytic feedstock for biogas, bio-char and bio-oil production encompassing biorefinery. *Bioresour. Technol.* 178: 132–138.

Schwender, J. and Ohlrogge, J.B. 2002. Probing in vivo metabolism by stable isotope labeling of storage lipids and proteins in developing *Brassica napus* embryos. *Plant Physiol.* 130: 347–361.

Seibert, M., King, P., Posewitz, M.C., Melis, A., and Ghirardi, M.L. 2008. Photosynthetic water-splitting for hydrogen production. In: Wall, J., Harwood, C., and Demain, A. (Eds.), *Bioenergy*. Washington, DC: ASM Press, pp. 273–291.

Sekar, S. and Chandramohan, M. 2008. Phycobiliproteins as a commodity: Trends in applied research, patents and commercialization. *J. Appl. Phycol.* 20: 113–136.

Shariati, M. and Hadi, M.R. 2011. Microalgal biotechnology and bioenergy in *Dunaliella*. In: Carpi, A. (Ed.), *Progress in molecular and environmental bioengineering—From analysis and modeling to technology applications*. Angelo Corpi, Intech Chapters, Croatia, pp. 483–506.

Sheehan, J., Dunahay, T., Benemann, J., and Roessler, P. 1998. *A look back at the U.S. Department of Energy's Aquatic Species Program—Biodiesel from algae*. Golden, CO: National Renewable Energy Laboratory.

Somerville, C., Browse, J., Jaworski, J.G., and Ohlrogge, J.B. 2000. Lipids. In: Buchanan, B., Gruissem, W., and Jones, R. (Eds.), *Biochemistry and molecular biology of plants*. Rockville, MD: American Society of Plant Physiology, pp. 465–527.

Strik, D.P., Terlouw, H., Hamelers, H.V., and Buisman, C.J. 2008. Renewable sustainable biocatalyzed electricity production in a photosynthetic algal microbial fuel cell (PAMFC). *Appl. Microbiol. Biotechnol.* 81: 659–668.

Tamer, E., Amin, M.A., Ossama, E.T., Bo, M., and Benoit, G. 2006. Biological treatment of industrial wastes in a photobioreactor. *Water Sci. Technol.* 53: 117–125.

Tanner, W. 1969. Light-driven active uptake of 3-Omethylglucose via an inducible hexose uptake system of *Chlorella*. *Biochem. Biophys. Res. Commun.* 36: 278–283.

Tredici, M.R. 2010. Photobiology of microalgae mass cultures: Understanding the tools for the next green revolution. *Biofuels.* 1: 143–162.

Tseng, C.K. 2001. Algal biotechnology industries and research activities in China. *J. Appl. Phycol.* 13: 375–380.

Venkata Mohan, S. and Devi, M.P. 2012. Fatty acid rich effluent from acidogenic biohydrogen reactor as substrate for lipid accumulation in heterotrophic microalgae with simultaneous treatment. *Bioresour. Technol.* 123: 627–635.

Venkata Mohan, S. and Devi, M.P. 2014. Salinity stress induced lipid synthesis to harness biodiesel during dual mode cultivation of mixotrophic microalgae. *Bioresour. Technol.* 165: 288–294.

Venkata Mohan, S., Devi, M.P., Mohanakrishna, G., Amarnath, N., Lenin Babu, M., and Sarma, P.N. 2011. Potential of mixed microalgae to harness biodiesel from ecological water-bodies with simultaneous treatment. *Bioresour. Technol.* 102: 1109–1117.

Venkata Mohan, S., Devi, M.P., Subhash, G.V., and Chandra, R. 2014. Algae oils as fuels. In: Pandey, A., Lee, D.J., Chisti, Y., and School, C.J.L. (Eds.), *Biofuels from algae*. Elsevier, NY, pp. 155–187.

Venkata Mohan, S., Reddy, M.V., Chandra, R., Subhash, G.V., Devi, M.P., and Srikanth, S. 2013. Bacteria for bioenergy: A sustainable approach towardstoward renewability. In: Gaspard, S. and Ncibi, M. (Eds.), *Biomass for sustainable applications: Pollution, remediation and energy*. Cambridge: RSC Publishers, pp. 249–287.

Venkata Mohan, S., Rohit, M.V., Chiranjeevi, P., Chandra, R., Navaneeth, B. 2014. Heterotrophic microalgae cultivation to synergize biodiesel production with waste remediation: Progress and perspectives. *Bioresour. Technol.* 14: S0960-8524. doi: 10.1016/j.biortech.2014.10.056.

Venkata Mohan, S., Srikanth, S., Raghavulu, S.V., Mohanakrishna, G., Kiran Kumar, A., and Sarma, P.N. 2009. Evaluation of the potential of various aquatic ecosystems in harnessing bioelectricity through benthic fuel cell: Effect of electrode assembly and water characteristics. *Bioresour. Technol.* 100: 2240–2246.

Venkata Subhash, G., Chandra, R., and Venkata Mohan, S. 2013. Microalgae mediated bio-electrocatalytic fuel cell facilitates bioelectricity generation through oxygenic photomixotrophic mechanism. *Bioresour. Technol.* 136: 644–653.

Venkata Subhash, G., and Mohan, S.V. 2014. Deoiled algal cake as feedstock for dark fermentative biohy-drogen production: An integrated biorefinery approach. *International journal of hydrogen energy.* 39: 9573–9579.

Xu, H., Miao, X., and Wu, Q. 2006. High quality biodiesel production from a microalga *Chlorella* protothecoides by heterotrophic growth in fermenters. *J. Biotechnol.* 126: 499–507.

Yan, Y., Chuda, M., Suzuki, M., and Nagata, T. 1999. Fucoxanthin as the major antioxidant in Hijikiafusiformis, a common edible seaweed. *Biosci. Biotechnol. Biochem.* 63: 605–607.

Yang, Z., Guo, R., Xu, X., Fan, X., and Luo, S. 2011. Fermentative hydrogen production from lipid-extracted microalgal biomass residues. *Appl. Energ.* 88: 3468–3472.

Yoo, C., Jun, S.Y., Lee, J.Y., Ahn, C.Y., and Oh, H.M. 2010. Selection of microalgae for lipid production under high level of carbon dioxide. *Bioresour. Technol.* 101: 71–74.

Yu, W.L., Ansari, W., Schoepp, N.G., Hannon, M.J., Mayfield, S.P., and Burkart, M.D. 2011. Modifications of the metabolic pathways of lipid and triacylglycerol production in microalgae. *Microb. Cell Fact.* 10: 91.

PART III

Ecological Economics and Policy

CHAPTER 12

Process Systems Engineering Approach to Biofuel Plant Design

Mariano Martín[1] and Ignacio E. Grossmann[2]
[1]Departamento de Ingeniería Química, Universidad de Salamanca, Salamanca, Spain
[2]Department of Chemical Engineering, Carnegie Mellon University, Pittsburgh, PA, USA

CONTENTS

12.1 CONTRIBUTION OF BIOFUELS TO THE ENERGY MARKET

12.1.1 Energy Demand and Emissions

Mankind has been running on renewable sources of energy until very recently. From a recent energy outlook prepared by ExxonMobil (Figure 12.1), it is possible to see that in the 1800s, biomass supplied more than 90% of the energy and fuel needs. From the late 1800s to the early 1900s, fossil fuels became the preferred energy resource because of their easy availability and low cost. Since then, energy consumption has been increasing due to industrialization and population growth. By the 1920s, coal and petroleum had largely replaced biomass sources in industrialized countries, although wood for home heating and hydroelectric power generation remained in wide use. By the end of the twentieth century, nearly 90% of the commercial energy supply was from fossil fuels; only very recently have other renewable and hydro resources been increasing their share (USEIA 2001; Klass 2004; BP 2012; IEA 2012; ExxonMobil 2013).

Emissions are another important concern that parallels energy production and usage. Carbon dioxide (CO_2) concentrations in the atmosphere have been increasing over the past century—from the almost steady level in the preindustrial era (280 ppm) to a rate of 1.9 ppm/yr in the last decade alone and recently surpassing 400 ppm. Thus, over the last 40 years, the emissions of CO_2 from combustion have doubled, reaching 30 $GtCO_2$ in 2012 (IEA 2012) (IEA 2004). In terms of sectors, transportation contributes more than 22% and is second only to electricity production, which reached more than 40%; just beneath transportation is industry, representing 20% (IEA 2004; Cole 2007).

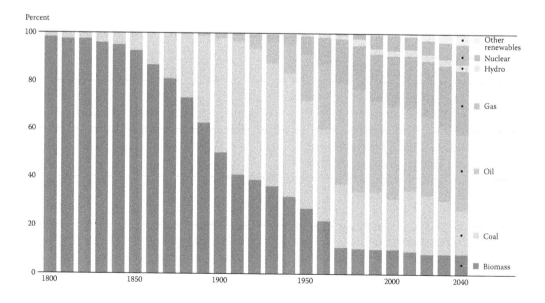

Figure 12.1   The outlook for energy, a view to 2040. (From ExxonMobil, *Outlook for energy 2013–2040*, 2013.)

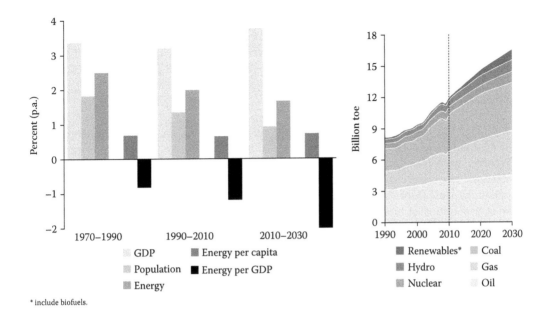

Figure 12.2   Energy outlook, 2012–2030. (From BP. 2012. *Statistical Review of World Energy*. Available at: http://www.bp.com/content/dam/bp/pdf/Statistical-Review-2012/statistical_review_of_world_ energy_2012.pdf)

The recent financial crisis has not helped efforts toward emissions reduction (IEA 2004). Significant increases have occurred in the levels of methane (CH_4) and nitrous oxide (N_2O) (IEA 2004). Emission mitigation is gaining attention through production control, and a market of emissions has been developed to exchanges emission rights based on several commitments, such as the Kyoto Protocol of the United Nations or the newest The 17th Conference of the Parties (COP17). Although industry and electricity emissions can be mitigated at a cost because the production of CO_2 is localized, the transportation section again represents a major challenge. Thus, the use of biofuels to produce high density energy carriers with close to neutral emissions has become an important alternative in the short and medium term.

Figure 12.2 shows the trend in energy consumption. Given the expected growth in consumption over the next decades, together with the growth in population (IEA 2004), there is a need to develop alternative energy sources that are competitive, renewable, and that have no significant impact on the environment, including emissions, in order to address simultaneously the availability of sources and the emission targets. The renewable sources available are solar, wind, thermal, hydroelectric, and biomass (Saxena et al. 2009; BP 2012). The most challenging sector is that of transportation for which biomass is currently the main renewable source. Presently, biomass contributes 10%–14% of the world's energy supply (Hoekman 2009). The worldwide raw biomass energy potential for 2050 has been estimated to be between 150 and 450 EJ/year, or 25×10^9 to 76×10^9 barrel of oil equivalent (BOE) (Putun et al. 2001; IEA 2004).

12.1.2 Biofuels: Challenges and Advantages

In this section, we address the main general challenges related to biofuel production including the competition with agricultural land and natural resources. In each of the topics we subdivide the discussion into two parts. The first one is devoted to ethanol production, but it can also be extended to any biofuel produced from the same raw materials as bioethanol. The second subsection is focused on biodiesel; in this case, we consider algae and seeds as sources for oil. The four main topics we discuss here are: the use of land and the yield from the land of the various raw materials, the consumption of water either for irrigation or as process water, the energy balance for the production of different biofuels and, finally, the expected advantage in emission mitigation as a result of the use of biofuels.

12.1.2.1 Land Usage

Land usage has been the main reason for our current effort to abandon the development of first-generation biofuels. Biofuel production capacity is limited by the availability of raw material. Whether we look at corn, wheat, seeds, or we move to second-generation biofuels from lignocellulosic raw materials or algae, there are two main concerns related to biomass production and supply: the yield from the field to the crop—and ultimately to the biofuel (DG Energy 2010) and the transportation of biomass. The second concern is due to the low density of the biomass. The transportation costs are high, and the production capacity of the plants is typically limited by the economical amount of biomass produced within a certain distance. Although densification techniques such as pelletization are used, increasing the bulk density by an order of magnitude, the production of pellets is not cheap (Sokhansanj and Felton 2006). Therefore, the collection radius and the yield of a particular land determine the processable biomass and the plant size, especially for grains and crop and forest residues (Wooley et al. 1999). As a result, the process capacity hinges on the yield of biomass from the field. Considering to the yield of different crops or algae and how the politics of land usage for biofuels may affect the food market, there are only two ways to expand the biofuel production—land extensification and/or land intensification. We will start with the production of bioethanol and move to biodiesel and other fuels—or the evolution from the first to the second generation of biofuels. In fact, there is no agreement as to whether biofuel targets can be met with minor adjustments on the use of land or not at all (Dale et al. 2010).

First-generation bioethanol has two major drawbacks in terms of land. On the one hand, the yield from the ground of corn or wheat is smaller than that of other energy crops, resulting in the need for large crop growing area—even though this mostly depends on the region (Bauen et al. 2004; Hammond et al. 2007; Keeney 2009). On the other hand, the land used to produce corn devoted to bioethanol production is somehow withdrawn from the production of grain as food. This last statement presents a series of ethical and economic implications. From the ethical standpoint, using food for fuel is something we should not allow as a society. From the economical standpoint, the use of crops for fuel increases food prices and results in the lower availability of biomass and land for food production and also in financial speculation (Ajanovic 2011). It is difficult to establish the exact amount of ethanol from corn that can be produced, but good estimates of the reasonable increment in its production can be presented based on data on expected yield from the literature for different focused areas such as the United States, the European Union (EU), and Brazil.

The U.S. Department of Agriculture (USDA) has been releasing a series of annual reports on productivity and land usage for the eight most important crops. Based on historical data, the yield of corn in the United States has increased 5% per year over the last 40 years (Shapouri et al. 2004). Furthermore, according to the latest report (Westcott 2009), an improvement in the production of corn is expected based on future yield increases (from 151.1 Bu/acre in 2007 to 175 bu/acre in 2018). However, the planted and harvested land will remain constant or even decrease over the same time period (around 91 and 83 millions of acres, respectively, from a total of 250 million acres), maintaining the amount of corn dedicated to food. Thus, production is expected to increase from 13,074 MMBu to 14,580 MMBu. Therefore, in order to annually increase the production of ethanol within these limits (i.e., it is not expected that more land is going to be used to produce energy crops), the focus will be on improving the yield of the land by improving the production process.

In the case of the second world producer of ethanol, Brazil, land productivity for sugarcane has increased 3.77% per year in the last 29 years (Goldemberg 2008). A recent study has also reported that improvements in cattle industry pasture management combined with the introduction of new sugarcane varieties (with annual yield increases of 1.6%) and new ethanol technologies, will make it possible for Brazil to replace 5% of world's gasoline demand with ethanol by 2025, even with reducing the land requirements by 29%–38% (Leite et al. 2009).

The EU has also evaluated the feasibility of its biofuel policies, requiring 10% replacement of petrol-based fuels by 2020. According to EU studies (EU 2007), the impact on land use in the EU-27 is relatively modest. The substitution of 10% of petrol usage represents around 14.2 billion liters. For that, about 15% of arable land would be used. The total land used for first- and second-generation biofuel production would then be 17.5 million ha in 2020. Thus, the impact of the new legislation on land use for biofuel production is relatively modest. Only about an additional 5–7 million ha would be used, depending on the share of contribution by second-generation fuels. Therefore, the additional land-use requirements would not overly draw on the land resources of the EU-27. Moreover, the more even distribution of production capacities over the EU assures that an overconcentration of biofuel feedstock production in only a few regions could be avoided.

However, there are also dissenting opinions that claimed a few years ago that it would not be feasible to meet the EU policy in the United Kingdom. It is fair to say that they did not consider the expected increase in the yield (Hammond et al. 2007).

So far we have been commenting on future expectations based on yield increases. It is clear that first-generation fuel from grain may reach its limit at some point. Thus, in order to increase the productivity of biofuels, expectations have been placed on the so-called second-generation biofuels. First of all, together with the production of grain (corn or wheat), we produce as much as another 50% by mass of biomass, such as corn stover or straw. Therefore, we have another source of biomass already available. We must bear in mind that it is necessary to leave at least 30% of these residues on the ground to protect the soil from erosion and to protect nutrients to maintain the soil's productivity (Atchison and Hettenhaus 2003; Nielsen 2009). On the other hand, we can also discuss energy crops that do not use agricultural land as well as nonland-based sources (such as animal waste and slaughtering residuals) (EU 2007; Parker et al. 2010). Forest residues include naturally grown trees that may be of poor quality or too small to be used commercially, the thinning of timberland, residues left in the forest after commercial logging, residues from clearing rotten trees that could cause forest fires, and residues from processing mills. Currently, these residues, which are sustainable and plentiful, are burned, left in the forest to decay, or are sent to landfills. However, collecting residues from within a forest can be difficult because efficient equipment has not been developed, and most commercial logging operations are not set up to handle residues. Furthermore, the use of biomass from forest residues, as well as from crops, requires the consideration of a number of issues including nutrient cycling and soil productivity, maintenance of biodiversity, water quality, and wildlife habitat. These factors, and the resulting constraints on forest operations, are generally very site-specific. The cost of logging residues is estimated to be $30/dry ton. Municipal solid waste (MSW) is a heterogeneous source of biomass. There is a fractional amount (5%) of urban tree and landscape residue (UTR) within MSW, which includes wood chips, logs, tops and brush, mixed wood, whole stumps; when combined together with leaves collected during seasonal leaf collection as well as grass clippings, this accounts for up to 13% of MSW. In the United States, more than 200 million cubic yards of UTR is generated annually, with 88% potentially available for fuel use. Furthermore, unrecycled paper, which represents 16% of MSW, can also be partially recovered. The costs for separating the different fractions of MSW ranges from $27 t^{-1} dry basis to $33 t^{-1} dry basis for the paper to wood fractions, and $11 t^{-1} dry basis for the yard/green waste fractions, which are source separated at the source (Dorr and Mizroch 2008; Parker et al. 2010). Among the energy crops, it is interesting to highlight those with the highest yield to ethanol—such as switchgrass (3100–7000 L/ha), which almost doubles the current yield of corn (3100–4000 L/ha) and comes close to sugarcane (6800–8000 L/ha). A large part of the United States is capable of producing switchgrass with high yields.

The use of these resources requires new production processes, but we should be able to produce the same amount of energy with less land or less arable land devoted to biofuels. Table 12.1 shows the

Table 12.1 Calculations of Net Yield for Three Crops

Crop	Yield Grain (bu/ac)	Dry Grain (t/ha)	Straw/ Grain Ratio	Gross Yield (t/ha)	Max Fraction Removed for Soil Fertility	Fraction Machine Can Remove	Estimate of Losses from Harvest to Biorefinery	Net Yield (t/ha)
Wheat straw	60	3.5	1.3	4.6	0.5	0.75	0.20	1.822
Corn stover	150	8.1	1.0	8.1	0.7	0.75	0.35	3.677
Switchgrass	NA	NA	NA	10.0	0.8	0.75	0.10	6.750

data provided by Sokhansanj and Fenton (2006). We can see that the net yield of switchgrass per ha is twice that of corn stover or wheat straw; compared to corn, the yield is three times larger (Yusuf 2007).

Thus, taking into account the higher yield of switchgrass compared to corn (two to three times), together with the increment in the productivity of the production processes and if we also consider the increase in the yield of the land due to improvements in the agricultural practices, with no increase in the land used but just substituting the current extension occupied for corn by energy crops and assuming that the yield of the land will increase by 3–5% a year, in 20 years' time (by 2030), the production of ethanol could be around five times larger than it is currently. Therefore, based on this approximation, by 2030 it could be possible to substitute 30% of U.S. gasoline; and, in Europe, up to a 20% substitution could be obtained.

So far, we have discussed land used for production of raw materials devoted mainly to ethanol, although lignocelluloses can also be used to obtain Fischer-Tropsch type fuels. However, the crops used for biodiesel production are different. The demand for biodiesel has turned into an increasing demand for lipid feedstocks such as soybean, canola, rapeseed, sunflower, palm, and coconut oils. However, according to the American Soybean Association (ASA 2007), 96% of the production of soybean oil goes to the food industry. Thus, new lipid sources are needed in order to avoid using food-dedicated oil as fuel. The advantage in favor of selecting a different raw material is that using soybeans for fuel production is expensive (Haas and Foglia 2005). The same can be said about sunflowers, according to the NSA (1993).

The EU's directorate general for agriculture and rural development issued a report in 2007 that claimed that it is feasible to produce biodiesel to partially replace the consumption of diesel up to the levels projected by the new regulations (10% biofuel by 2020), but the land required will be within the availability, therefore it will not affect its use for other crops (EU 2007). A study by the USDA presents improvements in crop production yields (Westcott 2009). Particularly for soybeans, the area used is to be kept constant at around 70 million acres, and the yield is expected to increase from 41.7 to 46.5 bushels per acre. This should result in an increase in production from 2676 to 3260 million bushels. Therefore, there are only two means to increase the production of biodiesel beyond this point:

1. Using wastes as raw material: The first advantage of second-generation biodiesel is that it is based on residues that do not need additional land to be produced. However, the amount of biodiesel produced will depend on the capability of extracting oil from the residues or of producing them directly (such as in case of waste frying/cooking oil). The second advantage is their low cost as raw materials (Mondala et al. 2009). However, the production of biodiesel will have an upper limit related to the production of wastes.
2. Incremental increases in yields either by using a different feedstock—algae presents a yield an order of magnitude larger than jatropha, rapeseed, or palm oil (Yusuf 2007)—or by improving the agricultural method, which will provide an extra 3%–5% annual increment in the yield (Westcott 2009).

The use of microalgae as feedstock for the production of biodiesel was first sponsored by the U.S. Department of Energy (USDOE) in the 1970s. The availability of crude and its low prices closed the program in the 1990s (Sheehan et al. 1998). The yield from ground to fuel is the highest compared to any other crop in more than an order of magnitude, including bioethanol production from corn, lingo, or biodiesel production from different seeds. Typically, values of 5000 gal·acre^{-1}·yr^{-1} are reported (Yusuf 2007). Furthermore, it is reported that microalgae can be produced using saltwater (eliminating the problem of water usage). Using 1–3 million acres, or about 2%–5% of the cropland currently in use in the United States, it would be possible to meet U.S. diesel needs. Furthermore, it is possible to produce diesel from algae far less expensively than the current cost of diesel (Pimentel and Patzek 2005). The main problem with producing biodiesel from algae is making it commercially viable, including the need for a plentiful supply of a concentrated CO_2 source.

Other alternatives are nonedible crops such as *Jatropha curcas* (Foidl et al. 1996), animal fats such as lard (Lee et al. 2002) and beef tallow (Nelson and Schrock 2006), waste cooking oil (Zhang et al. 2003), or municipal sludge (Mondala et al. 2009). Not all of these affect land usage in the same way. Although residues are expected to have almost no effect because they are already a product of the current system, the production capacity of fuels made from them is limited by the main activity that generates them, as we mentioned earlier. Moreover, the absence of market today leads to low prices, but if the biofuel market from them is normalized, the benefits may decrease. The use of nonedible crops, such as jatropha may avoid the problem of direct competition with the food market, providing a higher yield than soybean or rapeseed (Yusuf 2007; Makkar and Becker 2009). But it may also compete with land allocated for food, even though in principle marginal soils could also be used and jatropha can be adapted to different climatic conditions (Makkar and Becker 2009).

Wastewater treatment facilities annually produce approximately 6.2 million metric tons (dry mass) of sludge in the United States (WEF 2002). Due to urbanization and industrialization, an increase in this value is expected. The disposal of sludge is a big problem because either incineration or land filling results in emissions of dioxins and heavy metals. Thus, its use for the production of biodiesel is a good treatment process. The lipid fraction of municipal wastewater sludge is reported to contain up to 38.5% of polar fraction where fatty acids and steroids are found (Jardé et al. 2005). The fatty acids were predominantly in the range of C10 to C18, which are excellent for the production of methyl esters. Boocock et al. (1992) obtained 17–18 wt.% lipids from raw sewage sludge by boiling solvent extraction. Approximately 65 wt.% of the extracts were found to be free fatty acids and 7 wt.% were glyceride fatty acids. On the other hand, Shen and Zhang (2003) investigated the pyrolysis of secondary sewage sludge under inert conditions. They obtained a maximum of 30 wt.% oil from dry sludge.

Finally, waste cooking oils are currently a cheap raw material for the production of biodiesel. In the past, they were sold as animal food. However, the EU has banned this practice because of the harmful chemical properties formed during frying as a result of thermolytic, oxidative, and hydrolytic reactions (Kulkarni and Dalai 2006), which eventually would enter the human food chain. The disposal of this residue is a problem, and its use a raw material for the production of biodiesel can be a solution. For example, 700,000–1,000,000 tonnes/yr are produced in the EU (Supplea et al. 2002). In the United States, it is reported that an average of nine pounds of yellow grease per person are produced annually (Wiltsee 1998); in Canada, 120,000 tonnes/yr of yellow grease are produced (Zhang et al. 2003). Using these numbers and with production yields of 88% from the cooking oil to biodiesel (Phan and Phan 2008), it would be possible to produce 220 billion gallons per year in Europe, 320 billion gallons in the United States, and 31 billion gallons in Canada. In Europe, the share of diesel cars in the automotive market is much more important than in the United States, and although these values will increase the current production of biodiesel by 50% (García and García 2006), they do not mean a solution for the 70 million tons a year consumed by light duty vehicles (EBB 2008). In the United States, this will cover the consumption of diesel, which is around 55 billion gallons a year (Butzen 2006) so that the problem related to the area available for biodiesel production can be partially solved.

12.1.2.2 *Water Usage*

Water is a priceless resource; 74% of its withdrawal is devoted to agriculture, 18% to industry, and 8% to municipalities. Unfortunately, it is expected that by 2025, two-thirds of the world's

Little or no water scarcity

Physical water scarcity

Approaching physical water scarcity

Economic water scarcity

Not estimated

Figure 12.3   (See color insert.) Water availability. (From IWMI, *Insights from the comprehensive assessment of water management in agriculture*, World Water Week, Stockholm, 2006.)

population will suffer water stress (IWMI 2006); see Figure 12.3. Water usage is another growing concern in the production of biofuels. There are two main water uses for the production of biofuels: the water that is used in the production of the biomass, mainly due to irrigation (Elcock 2008), and the process water to transform the biomass into fuels.

If we start by considering growing crops, there is a range of values of water consumed in the production of fuel. In general, it is reported that, on average, roughly 2500 L of crop evapotranspiration and 820 L of irrigation water is withdrawn to produce one liter of biofuel (De Fraiture et al. 2008). In particular, between 785 and 832 gallons water per gallon of ethanol are necessary to produce irrigated corn, which represents 13.3% of the total harvested corn (Elcock 2008; Jacobson 2009). For sugarcane, values ranging between 927 and 1391 gallons of water per gallon of ethanol are presented (SIU 2007). However, even though the values are large, 80% of the crop production is rain fed and only 20% is irrigated (IWMI 2006). Irrigation depends heavily on the region. That is the case in the production of sugarcane in Brazil, where climate conditions, a 365-day growing season, and ample rainfall at the right times, allow sugarcane production at high yields with minimal or no irrigation (Leite et al. 2009). In the United States, between 7 and 14 gal of irrigation water per gallon of corn ethanol are used in favorable regions. But, for regions not appropriate, this could reach 320 gal per gal, which justifies plantation allocation as one of the most important decisions of corn production (Wu et al. 2009). In terms of biodiesel from soy, per gallon of soybean oil we need from 175 to 1000 gal with a mean value of 666 gal per gal (USDOE 2006a). Furthermore, different feedstocks may reduce the water needed. For example, if corn stover is used together with corn grain, because both grow together, the water consumption is cut by almost half. Another lignocellulosic raw material, switchgrass, is a perennial warm season grass that has grown for decades on marginal lands not well suited for conventional crop production. One of the characteristics of switchgrass (as well as other perennial grasses) is that they are deep-rooted and efficient in their use of nutrients and water, and therefore are relatively drought tolerant. If switchgrass grows in its native habitat, we can obtain yields from 4.5 to 8 dry tons per acre without irrigation and, although irrigation could increase yield, it may not be sufficient to offset the additional cost (e.g., for water, pumping, and energy). However, if switchgrass is grown in nonnative regions, irrigation is needed (Wu et al. 2009). For the case of algae, it is estimated that we need from 3.15 to 3650 gal of water per gal of algae fuel. However, the use of wastewater or saline water has also been reported, in which case, the consumption of freshwater in the production of oil from algae should be almost negligible (NSC 2009).

Different studies have been carried out to analyze the impact of the current policies in favor of the production of biofuels (USDOE 2006b; De Fraiture et al. 2008; Pate et al. 2011). De Fraiture et al. (2008) reported that the effect of biofuel policies on irrigation water can be important in certain regions such as South Africa and the United States, reaching a share of more than 20% of all irrigation water used, but on the whole, it will affect less than 5%. In contrast, if the amount of biofuel to be produced is far larger in order to fully replace the current consumption of gasoline and diesel, the production regions must be carefully selected not to have a big impact on food production and water availability.

In terms of process water, the consumption depends on both the raw material used and the processing technology. In the literature, we can find a wide range of values for water consumption but actual data from real plants is only available for first-generation biofuels. Thus, most data related to second-generation biofuels are based on process design studies. The most reliable data come again from the corn-ethanol industry. Minnesota Technical Assistance Program (MTAP) produced a report in 2008 where data about the water consumed at old and new plants in the region was presented. The average consumption of water for old plants is 4.6 gal per gal; for the new ones, this value went down to 3.4 gal per gal. This fact is important because further development of the processes is expected to provide improved results. Delta T claimed values of 1.5 gal per gal. In terms of lignocellulosic raw materials, water consumption ranges from 1 to 9.8 gal per gal (Aden 2007). In actuality, these values must be divided between biochemical processing, for which values of 6–10 gal/gal are typically reported, and thermochemical or thermobiochemical processes, in which case 2 gal per gal commonly are reported (Wu et al. 2009). Coskata (2008) reported a value of 1 gal per gal (Coskata 2008). Finally, in terms of biodiesel production processes, Pate (2008) reported a value from 0.32 to 1 gal per gal. Some optimization studies at the level of conceptual design have not only validated the claims by companies such as Coskata but also have proven that further improvements are expected for better process design (Martín et al. 2011a; Martín and Grossmann 2012a). We present the data in Section 12.3, following the description of the design techniques used.

12.1.2.3 Energy Balance

There has been an important debate on the net energy balance of ethanol and, in particular, of first-generation bioethanol from corn. Different feedstocks require various technologies to process the biomass into ethanol and, as a result, the energy consumption varies from one to another. Although most of the raw materials show positive net balances, the controversy has focused on the energy balance of corn ethanol.

Figure 12.4 presents the data for the net energy balance of corn ethanol reported over the years by different authors. It can be seen that only the first studies, which were conducted in the 1980s by Chambers et al. (1979), Weinblatt et al. (1982), and Ho (1989), showed negative net energy balances. However, as the time line advances, it becomes clear that the net energy balance of the production of ethanol was actually positive. Most of the detailed studies, in 1995 and later in 1999, from the Department of Energy or the USDA, are among the first ones to provide positive values. From that moment on, there was a debate between the USDA and those who kept reporting negative net energy balances. The positive values also came from different national labs in the United States—such as the work by Delucchi (2003) and Wang et al. (1999, 2007) at Argonne National Lab, as well as some updates from Shapouri et al.'s work in 2002 and 2004. The policy of changing from first to second generation reduced the debate. However, it is important to highlight that much of the debate generated was focused on the energy needed for raising the crop (energy for the production of fertilizer, seed corn, etc.) and much less attention has been placed upon the production process itself.

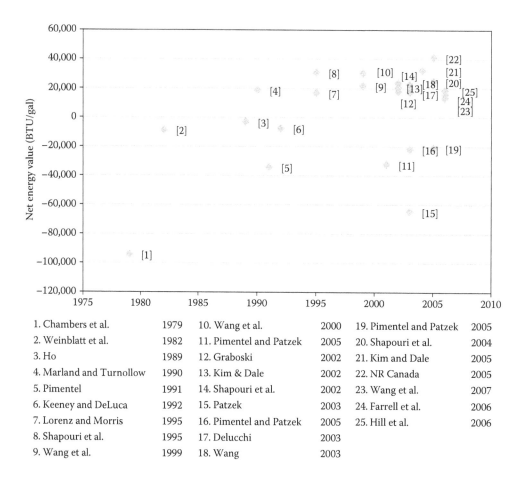

1. Chambers et al.	1979	10. Wang et al.	2000	19. Pimentel and Patzek	2005
2. Weinblatt et al.	1982	11. Pimentel and Patzek	2005	20. Shapouri et al.	2004
3. Ho	1989	12. Graboski	2002	21. Kim and Dale	2005
4. Marland and Turnollow	1990	13. Kim & Dale	2002	22. NR Canada	2005
5. Pimentel	1991	14. Shapouri et al.	2002	23. Wang et al.	2007
6. Keeney and DeLuca	1992	15. Patzek	2003	24. Farrell et al.	2006
7. Lorenz and Morris	1995	16. Pimentel and Patzek	2005	25. Hill et al.	2006
8. Shapouri et al.	1995	17. Delucchi	2003		
9. Wang et al.	1999	18. Wang	2003		

Figure 12.4   Corn-based net energy balance.

Regarding the production process, Figure 12.5 shows the energy needed for the production of a gallon of ethanol. In fact, there were few studies that actually developed a production process to obtain the energy values needed for the transformation. The studies from the USDA, Shapouri´s papers, modeled the process using a commercial software, ASPEN plus. Apart from this group, Taylor et al. (2000), McAloon et al. (2000), and Kwiatkowski et al. (2006) presented several reports and papers evaluating the process (the last one using SuperPro designer). Franceschin et al. (2008) also included pinch analysis for energy integration; Quintero et al. (2008) also showed simulations of the process to determine the usage of raw materials, but the energy consumed was not explicitly reported. However, the process can be improved based on the use of a synthesis approach. Karuppiah et al. (2008) presented a superstructure optimization approach for the design of the optimal process in terms of energy followed by the design of the optimal heat exchanger network. Very promising results were obtained as the technology improved. In addition, the use of process design techniques, process simulation, and systematic approach are incorporated in the design of corn-based bioethanol processes. Although these are data from conceptual studies, in 2008, MTAP presented a report that gathers information from actual plants. The old ones, built in the 1990s, are represented by single letters—the new ones, built in the 2000s, with double letters. We can see that energy consumption is greatly improved in the new plants and that values reach the best predictions of the optimization studies, see Figure 12.6.

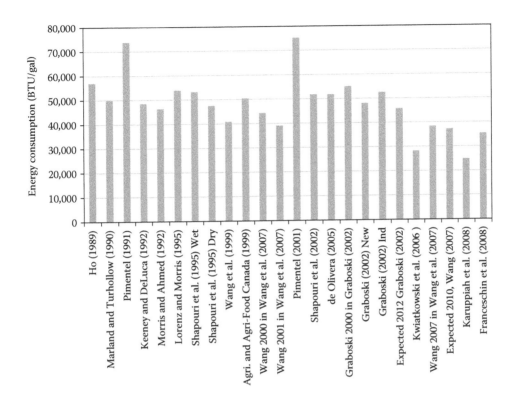

Figure 12.5 Energy consumption in biomass process from corn to ethanol.

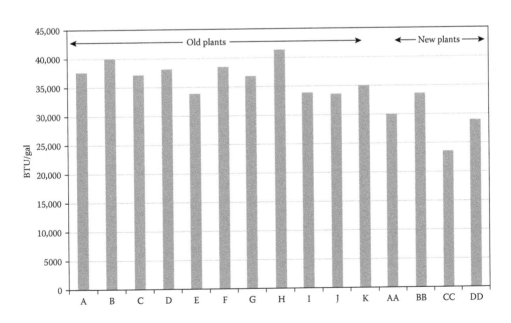

Figure 12.6 Actual energy consumption in corn-ethanol plants. (From MTAP, *Ethanol benchmarking and best practices. The production process and potential for improvement*. Minnesota Technical Assistance Program, University of Minnesota, Minneapolis, MN, 2008.)

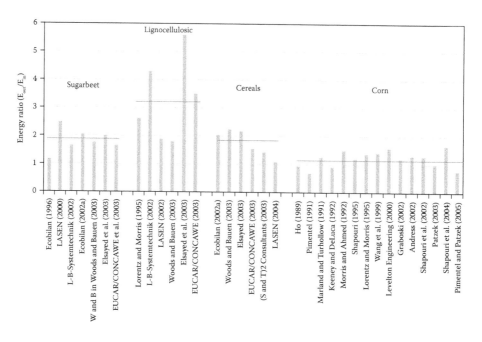

Figure 12.7   Energy net balance for different feedstocks.

Since most of the controversy was focused on corn, other raw materials have drawn less attention. But for corn (see Figure 12.7), all the studies regarding other raw materials such as cereals, lignocellulosics, and sugarbeet report a ratio of energy output to energy input larger than 1. With regard to corn, only the older studies report energy ratios below 1. However, second-generation bioethanol from lignocellulosic materials are still at a low level of development, and a number of alternatives in terms of production routes are available.

If we move now to biodiesel, there has never been a debate regarding the energy balance in biodiesel production due to the small energy consumption when using oil directly as raw material. There are a number of studies reporting the ratio between the energy produced from biodiesel and that consumed in its production for different raw materials and countries. We can find values of 2.5 for rapeseed, 3 for soybeans, 5–6 for waste oils, and 9 for palm oil (Worldwatch Institute 2006). In Greece, sunflower seed presents an energy ratio of 4.5:1 (Kallivroussis et al. 2002). In Germany, studies reveal values between 1.7 and 3 (da Costa et al. 2006). For Poland, using rapeseed oil as raw material, values from 1.51 to 2.75 are reported; for the Netherlands, the values are within the range 2.15–2.93. On average, for Europe, the output/input ratio of energy ranges from 1.24 to 2.34 (Firrisa 2011). However, in Colombia and Brazil, the reported values for the energy ratio are 6 and 8 to 1, respectively (da Costa et al. 2006). In Peru, values in the range of 6–6.5 to 1 have been reported (Rojas and Gutierrez 2008). Donato et al. (2008) reported values from 1.94, without considering by-products, and up to 6.48 taking by-products into account. The USDOE and USDA completed the first comprehensive life cycle analysis for biodiesel production back in 1998. That study found a 3.2-to-1 energy balance. The energy inventory for this study was updated in 2009 using 2002 data, resulting in an improved ratio of 4.56:1, and recently an even better value of 5.54:1 has been reported (Pradhan et al. 2011). Apart from the fact that over the years the energy ratio has improved, a particular trend worth mentioning is that the more industrialized the country, the lower the positive return; but it is still positive. The use of algae is still under consideration due to the energy required for algae harvesting. This topic is still under research, and different technologies are being evaluated to avoid the common process that involves flotation, centrifugation, and drying.

12.1.2.4 Emissions

One of the main reasons for using biofuels is their expected carbon neutrality. Biomass grows by capturing CO_2 from the atmosphere and transforming it into carbohydrates in photosynthesis. However, emissions are a very complex topic with two points of view. On the one hand, we consider the effect of the use of land and the production of biomass. On the other hand, we need to consider the effect of using biofuels instead of fossil-based fuels. This also raises a question regarding the time horizon considered in the computation. When land changes, there is an effect on the CO_2 emitted. However, in the long run, we still keep reducing the emissions by using biofuels instead of fossil fuels.

We start by focusing on the first topic, land usage, to discuss the effect of the conversion of rain forests, peatlands, savannas, or grasslands into agricultural use for the production of biofuels. This first-time conversion of natural or native land is considered to generate anywhere from 17 to 420 times more carbon dioxide than one those fuels would provide from displacing fossil fuels (Dale et al. 2010) and resulting in carbon debts (the time needed to counterbalance the CO_2 emissions) for many years (Fargione et al. 2008). Furthermore, indirect land-use change effects associated with an increased use of corn-based ethanol could potentially double greenhouse gas (GHG) emissions in the next 30 years. Biofuels produced from switchgrass, if grown on areas formerly used for the production of corn, could potentially increase emissions by 50% (Searchinger et al. 2008). The information presented here is based on the assumption that crop yields and biofuel demand result in a continued increase in agricultural area. The expansion of agricultural area is assumed to come at the expense of other land uses in the same proportion as seen in historical land-use conversion patterns. On the other hand, plant growing each year also represents CO_2 capture and sequestration, which should be equivalent to the emissions. Therefore, it comes as a natural conclusion that the best practice is to use appropriate land for energy crops that does not compete with the already allocated land for food or other usage and not to displace the already established usage. Thus, social and environmental sustainability are also enhanced (Dale et al. 2010). It is important to bear in mind that any land not native to a particular crop may provide low yield or may require a certain consumption of water and/or fertilizers.

Furthermore, the actual production of the crops requires processing and transportation. The emissions from biofuel production and processing have been studied with a classic life cycle approach and show that, except for corn ethanol grown in energy intensive agrosystems in the United States, most of the crops have net GHG substitution savings between 20% and 90% (Thow and Warhurst 2007).

With regard to actual emissions when using biofuels, we again divide the study into bioethanol and biodiesel. Ethanol burns cleaner than regular gasoline due to its higher heat of vaporization producing less carbon monoxide, hydrocarbons, and nitrogen oxides. However, it increases aldehyde emissions, which play an important role in formation of photochemical smog (Agarwal 2007). The advantage is that new catalyst systems can be used due to the absence of sulfur. Furthermore, it has also been reported that biofuels produce more ozone than gasoline (Ginnebaugh et al. 2010). If we focus on the first generation, Menichetti and Otto (2008) provided a review of GHG improvement for different biomass sources and countries for bioethanol and biodiesel. In case of bioethanol, most of the studies gathered reported huge advantages in terms of GHG emission reduction. The most efficient raw material is sugarcane, which is the typical one used in Brazil, while corn is again the worst in comparison. No effect in the change on land is included in the results of this (Figure 12.8).

The first generation of biofuels seems to have created some skepticism in society as well as among scientists due to its interference with the food supply chain. Therefore, the expectations for the second generation of bioethanol are high. However, the real potential of the so-called second generation of bioethanol is not clear yet. The possibilities in terms of feedstocks for lignocellulosic

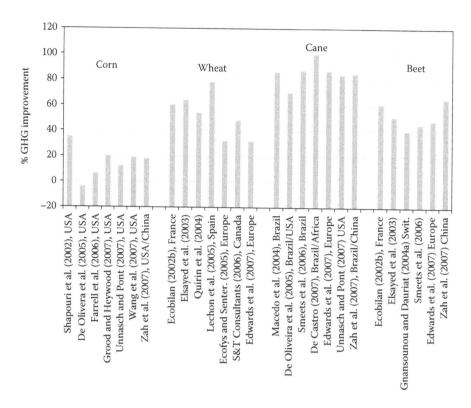

Figure 12.8   Greenhouse gas improvements due to first-generation ethanol.

ethanol are much broader, such as agricultural residues that do not require the use of land because they are produced together with the food-dedicated crops (e.g., corn stover, crop straws, sugarcane bagasse), herbaceous crops (e.g., alfalfa, switchgrass), forestry wastes and wood (hardwoods, softwoods), wastepaper, and other wastes such as municipal waste (Huang 2008; Menichetti and Otto 2008). Cellulosic ethanol can be more effective and promising as an alternative renewable biofuel than corn ethanol in the long run because it could greatly reduce the net GHG emissions as well as provide higher net fossil fuel displacement potential (Kszos 2006; Senternovem 2006); see Figure 12.9.

Tables 12.2 and 12.3 show the expected evolution in GHG emissions depending on the raw material and the production process for second-generation bioethanol. The expectations toward obtaining ethanol by gasification and fermentation of biomass are quite high (SenterNovem 2006). However, it will take time to develop and optimize the technology.

Finally, we compare the effect of using mixtures of ethanol, from different biomass types and processes, with gasoline when it is used in cars. Wang (2005) reported impressive advantages related to the use of lignocellulosic-based ethanol compared to corn-based and also reported that the dry-grind process was better than the wet counterpart. Finally, there is almost the same advantage when we use E10 mixtures (10% ethanol in gasoline) or E85, see Figure 12.10.

Because biodiesel is free from sulfur, less sulfate emissions are generated. Furthermore, biodiesel is an oxygenated fuel (providing a more complete combustion) and causes less particulate formation, CO emissions, and hydrocarbons with reductions up to 50%, 50%, and 70%, respectively, when only biodiesel is burned, depending on the percentage of biodiesel in the fuel mixture. However, there are several reported results of a slight increase in NOx emissions due to improved combustion that increases the temperature in the combustion chamber up to 10% if biodiesel

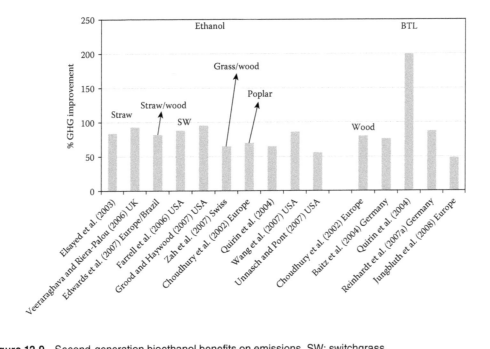

Figure 12.9 Second-generation bioethanol benefits on emissions. SW: switchgrass.

Table 12.2 Expected Decrease in Emissions for Second-Generation Biofuels

Feedstock (Process)	GHG% Emission Reduction When Used Biofuels Compared to Petrol		
	2006	2010	2020
Sugar beets	40	60	60
Grains	20	40	40
Potatoes	20	40	40
Residual starch	50	50	60
Lignocellulosic (Enzymatic hydrolysis)		80	85
Lignocellulosic (Gasification and synthesis)			90
Lignocellulosic (Gasification and fermentation)			85

Source: SenterNovem, Bioethanol in Europe Overview and comparison of production processes Rapport 2GA VE0601, 2006, www.senternovem.nl.

Table 12.3 Emission Mitigation Costs

Feedstock (Process)	GHG% Emission Reduction When Used Biofuels Compared to Petrol		
	2006	2010	2020
Sugar beets	370–450	250–300	190–250
Grains	900–1070	450–530	370–450
Potatoes	1880–2050	940–1020	860–940
Residual starch	190–450	190–450	110–330
Lignocellulosic (Enzymatic hydrolysis)		180–390	20–170
Lignocellulosic (Gasification and synthesis)			20–260
Lignocellulosic (Gasification and fermentation)			240–510

Source: SenterNovem, Bioethanol in Europe Overview and comparison of production processes Rapport 2GA VE0601, 2006, www.senternovem.nl.

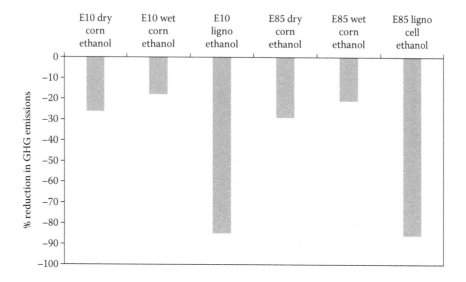

Figure 12.10   Reduction of greenhouse gas emissions for ethanol–gasoline mixtures.

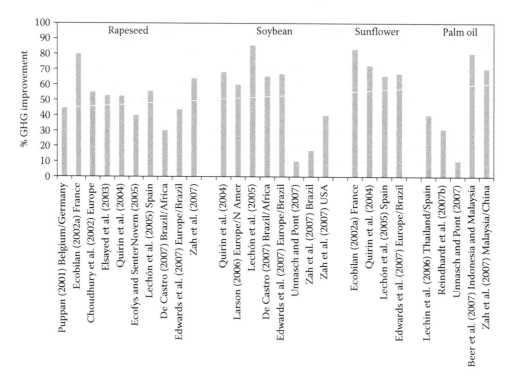

Figure 12.11   Greenhouse gas improvement when using biodiesel.

is burned alone (EPA 2002). However, biodiesel's lower sulfur content allows the use of NOx control technologies that otherwise cannot be used with conventional diesel. Biodiesel use also shows a reduction in polycyclic aromatic hydrocarbons (Klass 1998; Agarwal 2007).

Figure 12.11 represents the improvement in GHG emissions due to the use of biodiesel from different raw materials. There is no agreement as to which raw material provides better improvement in emissions generated when substituting mineral diesel. It is highly region dependent. Only the use of sunflower is quite stable within the range of values reported. The use of wastes to generate biodiesel will reduce emissions because currently these wastes are already generated.

12.1.3 Process Synthesis: Challenges in Biofuel Production Facilities

Bioprocesses present some characteristics that make them different from traditional processes in the chemical industry. In general, we can identify a number of challenges, as discussed in Martín and Grossmann (2013a, 2013b).

First, there are a large number of alternative technologies for processing raw materials into products that must be screened before detailed process design is attempted. Furthermore, any new process involves complex and not well understood chemical reactions, comprising a large number of chemicals and raw material pretreatments, the yield of which depends on the actual raw material used and the operating conditions. Furthermore, product separations include complex multiphase liquid–liquid or vapor–liquid equilibria. Additionally, the presence of biological material in the processes requires controlling the operating temperatures and pressures in the reactors to avoid loss of activity. This is also a source of uncertainty, and much experimental work is being carried out to understand hydrolysis and fermentation of a range of raw materials using different enzymes. A lot of experimental and modeling work is being carried out to unveil the principles of the processes and the effect of the operating variables and the products. Typically, plant design is being carried out based on limited data including experimental results at lab scale (Martín and Grossmann 2012a), and thus the models present a certain degree of uncertainty.

Another important challenge is that most of the processes rely on fermentations that are mostly exothermic but that operate at low temperatures (Karuppiah et al. 2008; Martín and Grossmann 2012b). This results in the need for a large amount of cooling water. Part of the water is lost by evaporation when cooled down in cooling towers, representing the major water consumption of the production facilities (Ahmetovic et al. 2010).

In a number of processes, including those for ethanol, the main product obtained is in a dilute solution of water. We need to separate the desired product from the stream by removing a large amount of water. The energy consumed in this stage, mainly by means of distillation followed by pervaporation or molecular sieves, represents more than half the energy consumption in a bio-ethanol production facility. The distillation has two main products, the concentrated biofuel and a stream that is mostly water. In order to save water, this stream can be treated to be reused within the process. However, before treatment, we have to cool it down, requiring cooling water that must be also processed through the cooling tower.

Most main products are produced in chemical reactions that are in fact in equilibrium. The main example is biodiesel where the transesterification reaction yield depends on a number of variables including catalyst type and load, temperature, pressure, and feed composition (such as excess of alcohol). We need to recover that excess, but it requires energy; thus, the evaluation of the operating conditions at the reactor must be carried out simultaneously with the separation and purification stages (Martín and Grossmann 2012b; Severson et al. 2013). As well as biodiesel, the production of methanol from glycerol or syngas, ethers from glycerol (Martín and Grossmann 2013b), are also complex equilibria where the conversion at the reactor and the separation stages must be evaluated simultaneously.

Therefore, the energy and water optimization problem requires a different approach because there is commonly no source of energy at a high temperature in the reactors (as in most petrochemical processes), which implies that the heat recovery within the process only has a modest impact. Moreover, the low price of freshwater makes its optimization as part of the total cost to have very little effect because the economic benefit of reducing the freshwater consumption versus other utilities is currently still marginal at best.

12.2 MATHEMATICAL PROGRAMING TECHNIQUES FOR BIOFUEL PRODUCTION PROCESSES

12.2.1 Superstructure Optimization

There are a large number of alternative technologies being developed for each stage or transformation from pretreatments for the biomass (i.e., dilute acid, AFEX), gasification processes (i.e., direct gasification, indirect gasification), hydrocarbon removal (i.e., partial oxidation, steam reforming, autoreforming), carbon capture (i.e., absorption in ethanol amines, Pressure Swing Adsoprtion (PSA), synthesis (i.e., sugar fermentation, syngas fermentation, FT type synthesis), and product purification (i.e., distillation, molecular sieves, pervaporation). In order to select the technologies and build the optimal flow sheet, a systematic procedure is the most adequate approach, and superstructure optimization is a powerful tool for determining an optimal topology and operating conditions at the level of conceptual design. For this purpose, we need to model each of the technologies in a way that the models accurately reproduce the operation of the unit, but are simple enough to be integrated in the bigger framework that the superstructure represents. As we mentioned in the previous section, modeling new processes is one of the challenges and, on top of that, modeling processes involving biomass is a bigger challenge. Currently, process simulator companies are working to develop rigorous models for some of the most common units, but units such as gasification of biomass, reforming, fermentations (either of syngas or of sugars), solids pretreatment, saccharification, hydrocracking, and transesterification are especially challenging ones because the physicochemical processes involved are complex due to nonideal behavior, including the need for consideration of toxic concentrations, product inhibition, temperature bounds, bioreactor or pretreatment conversions, generation of chars, and so forth. Commercial process simulators do not have most of these units in their libraries, or they have problems when dealing with solid residues or polymers. The advantage, however, is that some of the units are common to several bioprocesses. There are a number of techniques that can be used to develop algebraic models for bioprocess benchmarking that we describe here. The use of one or some of them largely depends on the data available.

12.2.1.1 Modeling Approaches

We present the main approaches for developing algebraic models for different units. The general advantage of all of them is that we can easily integrate new experimental data as soon as they are available.

Shortcut models are based on mass and energy balances and on thermodynamic principles that are widely used as a first approach for distillation columns, heat exchangers, phase separations, and absorption columns (Biegler et al. 1997). The main advantage is that they have clear physical meaning and they are based on simple first principles. However, we must bear in mind that these models have difficulties in predicting nonideal behavior.

Mechanistic models are more detailed but yet tractable models that can be obtained for units the operation of which can be explained based on chemical kinetics or chemical equilibrium. These models typically require more information in the form of equilibrium or kinetics constants as a function of the temperature and pressure that may not available for any process, and complex parameter estimation problems need to be solved off-line. The advantage is that the understanding of the process is deeper when the proposed mechanism is capable of accurately predicting the behavior of the unit.

Rules of thumb-based models apply when simple first principles alone cannot capture the behavior of a particular unit well enough. Examples include not only distillation columns or cooling towers but adsorbent beds, gas treatment, or solid separations. Rules of thumb are a helpful way to predict the performance of such units. They are listed in common chemical engineering literature

(Wallas 1990; Sinnott 1999; Branan 2000; GPSA 2004) and can be used to propose black box models to the operations based on experience. These models are limited to the typical range of operating variables.

Dimensionless correlations, such as Reynolds, Sherwood, Prandtl, Nusselt, and so forth have been used for many years to design units whose operation was governed by mass or heat transfer limitations, including heat exchangers, cooling towers, absorption towers, or spray driers. The pi theorem provided a basis to correlate a particular coefficient as a function of variables that are grouped in these dimensionless entities (Buckingham 1914). This approach is capable of addressing scale-up as well as involving physical meaning (Martín et al. 2007). However, in order to obtain a representative correlation, we must capture the right variables—although for optimization purposes, the model presents a large number of convex X^Y terms.

Factorial design of experiments is another option for dealing with equipment whose performance depends on a number of variables such as temperature, feed composition, pressure, catalyst load, and so on. This technique (Montgomery 2001) allows developing black box type models that predict the outcome of a process as a function of a number of input variables. It is a purely statistical method that can only predict within the range of variables used in its development. Furthermore, the adjustable parameters depend on the physical system from which the model has been obtained; from a mathematical standpoint, it introduces a large number of XY terms requiring tight bounds for the variables to obtain good solutions. The main advantage is that this method allows the systematic study of the effect of a large number of variables on the output of a process.

In some cases, authors develop *experimentally based correlations*. Sometimes experimental studies are complete enough to present profiles of the output variables as function of a few input operating variables. In some cases, raw experimental data are available in the literature. In other cases, parameter estimation problems can be formulated to develop correlations that predict the outcome of a unit as long as the independent variables are only a few. These correlations involve experimental validation and physical meaning, but sometimes it is not easy to provide a correlation between the independent variables and the outcome of a process, and the mathematical form of such models can also be complex.

Kriging interpolation is another statistical method based on the approximation of a function by sampling the space of variables so that the predicted value is calculated by optimal interpolation. It can be easily used to obtain black box type of models with better accuracy compared to the ones provided by factorial design. However, the method typically may require a large number of points to represent the geometry of the function that is being approximated (Caballero and Grossmann 2008).

Artificial neural networks are also a statistical based modeling approach for correlating the outcome of a process as a function of a number of independent variables. As in previous black box statistical models, the correlations are valid only for the range of the variables studied; thus, the model is subjected to scale-up problems. More importantly, they need a large number of data points to provide accurate predictions (Lemoine et al. 2008; Henao and Maravelias 2011).

12.2.1.2 Superstructure Generation

Once the models for the different units are developed, the next stage is to put them together to formulate a superstructure of alternatives for the systematic design of chemical processes. We actually need to link the different models to build the flowsheet that embeds the options. In Figure 12.12, we present an example of the method. The solution of such a superstructure is a function of the objective to be considered, whether we are looking for minimum energy consumption, environmental impact, heat and water consumption, or a combination of several objectives.

In Figure 12.12, we present an example of the main modeling methods used for the formulation of the superstructure used in the optimal conceptual design of the optimal production of bioethanol from switchgrass.

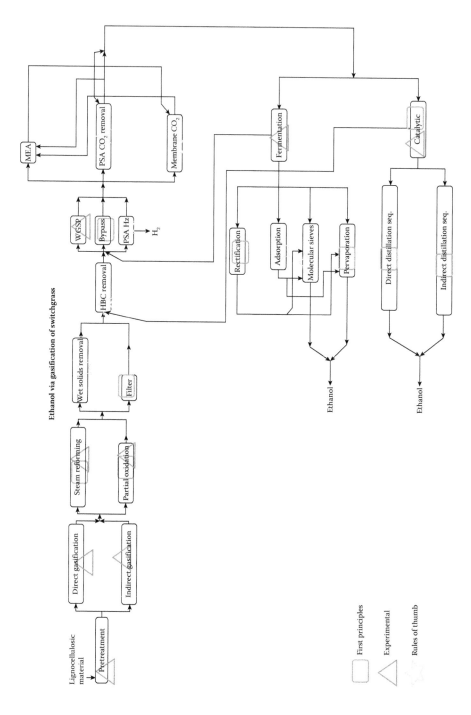

Figure 12.12 Superstructure generation and modeling strategies. (From Martín, M., and Grossmann, I. E., *AIChE J.*, 57, 3408–3428, 2011.)

12.2.2 Energy and Water Integration

Although the superstructure generated includes all the options in terms of processing units, the integration of energy and water can be carried out simultaneously or in sequence. We first present the sequential approach, which stems from the traditional rigorous study of heat and water treatment networks. Next, we comment on simultaneous optimization and heat and/or water integration efforts.

12.2.2.1 Heat Exchanger Networks

Traditionally, the methods for heat exchanger network design can be divided into heuristic based and mathematical optimization formulations. In both cases, the first approach was to use a sequential or step-wise solution procedure (Gundersen and Naess 1988). If we start with the heuristic-based methods, it is important to mention the set of rules proposed by Masso and Rudd (1969) and the ones by Ponton and Donaldson (1974) to match the hottest stream with the cold stream with the highest target temperature. The thermodynamic approach, while intuitively appealing, has been largely replaced by mathematical optimization, but it is the first rigorous method to establish the minimum utility target ahead of design (Hohmann and Lockhart 1976). However, the most important method of this class is based on the discovery of the pinch as a bottleneck for energy transfer. Linnhoff and Hindmarsh (1983) proposed this method, which uses the cost as its objective function in order to minimize energy consumption. In this way, it establishes the minimum heat and cooling needs and the location of the pinch. We use the pinch to divide the network into subproblems because there is no energy transfer across the pinch. Now, for each partition, we look for the subnetwork using heuristics and guidelines that meet targets for minimum utility consumption and minimum number of units. Although these methods are widely used, the highly combinatorial problem results in difficulties when applying it to a real case study.

On the other hand, we have the mathematical programing based methods. Initially, these methods were also based on a multistep procedure. Floudas et al. (1986) proposed a three-stage algorithm: MAGNETS. The heat exchanger network (HEN) design problem is decomposed into three stages. The first two steps involve the solution of the LP and MILP transshipment model of Papoulias and Grossmann (1983) to determine the targets in terms of energy and cooling needs and number of units. Thus, for a particular heat recovery approach temperature (HRAT), the LP model determines the minimum utility requirement for the network. With the utility consumption fixed at the LP solution, the MILP model is solved to determine the minimum number of matches and their corresponding heat demands. Finally, in the third step, heat loads and matches are fixed and the area cost is minimized by the solution of an NLP model (Floudas et al. 1986) to determine the optimal network configuration. The limitation of any sequential synthesis method is that the trade-offs among different costs cannot be analyzed simultaneously. The selection of the HRAT and the partition of the problem have serious implications in the number of units and the configuration of the network. Typically, sequential approaches lead to suboptimal networks that are near optimal.

After the work by Yee and Grossmann (1988), Floudas and Ciric (1989), and Dolan et al. (1987, 1989), in 1990, Yee and Grossmann presented SYNHEAT. Here, a superstructure optimization approach is used that accounts for all costs (piping layout, fix and variable heat exchanger costs, and utility cost); but, still assuming isothermal stream mixing, see Figure 12.13. The solution defines the network by providing the following input:

- Utilities required
- Stream matches and the number of units
- Heat loads and operating temperatures of each exchanger

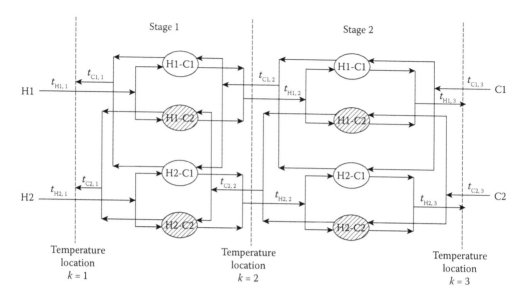

Figure 12.13 Two-stage superstructure. (From Yee, T. F., and Grossmann, I. E., *Comput. Chem. Eng.*, 14, 1165–1184, 1990. With permission.)

- Network configuration and flows for all branches
- Area of each exchanger

Furthermore, constraints on stream matches, stream splits, and number of units can also be specified. In the proposed method, no parameters are required to be fixed—that is, level of energy recovery (HBAT), exchanger minimum approach temperature (EMAT), number of units, and matches. Also, there is no need to perform partitioning into subnetworks, and the pinch point location(s) are not predetermined but rather optimized simultaneously. For a detailed review of heat exchanger methodologies, see Furman and Sahinidis (2002).

12.2.2.2 Heat Integrated Distillation Columns and Column Sequencing

12.2.2.2.1 Same Task: Multieffect Columns

Multieffect columns are a particular integration technique used in the petrochemical industry that has an interesting application in biofuel production processes, in particular those dealing with fermentations. In general, distillation columns are very energy intense equipment. During the ethanol production process, either first or second generation, where glucose, glucose and xylose, or syngas are fermented, a dilute mixture of ethanol is obtained. The dehydration of this ethanol contributes to almost 50% of the energy consumption of the process. The multieffect column arrangement consists of splitting the feed at different pressures so that the boiler of the lower pressure column acts as a condenser for the higher pressure column. The variables are the operating pressures of the columns as well as the fraction of the feed (α_i) that goes to each one of them (see Figure 12.14). Using multieffect columns, we not only reduce the energy consumption by nearly one-half, but the cooling needs are also reduced (Ahmetovic et al. 2010).

12.2.2.2.2 Distillation Sequences and Heat Integration in Different Tasks

Most of the time, when a series of products are obtained at a reactor, the use of a sequence of distillation columns is the appropriate separation technology. We find this is the case when mixed

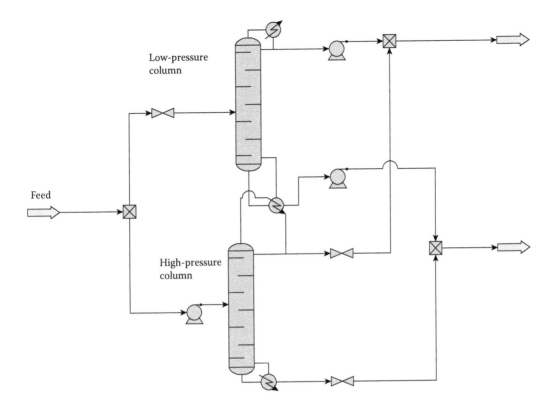

Figure 12.14 Multieffect column.

alcohol synthesis is used to produce ethanol from syngas. Typically methanol, ethanol, propanol, and small amounts of butanol and even pentanol are obtained. Thus, a distillation sequence is selected in order to recycle the methanol to the reactor, obtain ethanol, and to separate the higher alcohols. The problem of distillation sequences has been addressed in the literature formulating MILP problems to select the sequence (Andrecovic and Westerberg 1985; Eliceche and Sargent 1986). Furthermore, on top of that model, we can consider the integration of energy from the condenser and reboilers of the distillation columns of that sequence (Floudas and Paules 1988).

12.2.2.3 Simultaneous Optimization and Heat Integration

In the synthesis of process flowsheets, the flow rates and temperatures of the process streams are unknown because they must be determined so as to define an optimal processing scheme. Both have an impact on the energy recovery. Therefore, heat integration should be taken into account together with the synthesis problem. In particular, when the conversion in the reactor is low and/or is governed by chemical equilibrium, the operating conditions determine the separation stages, the energy, and the cooling requirements. In the biofuel industry, biodiesel production is an example of this particular case. Biodiesel production is based on the transesterification of oil with an excess of alcohol. The conversion of this reaction not only depends on the excess of alcohol but also on the operating pressure and temperature, the catalyst load, and composition. Furthermore, the energy consumption in the process relies on the recovery of the excess of alcohol. Thus, simultaneous optimization and heat integration is the appropriate approach for this example. To make this simultaneous approach possible, the heat integration problem should be formulated so as to allow for variable flow rates and temperatures of the process streams.

Duran and Grossmann (1986) proposed a procedure for solving nonlinear optimization and synthesis problems of chemical processes simultaneously with the minimum utility target for heat recovery networks. It is based on the use of a pinch point location method that can be formulated for embedding the minimum utility target within the process optimization. Because no temperature intervals are required in the proposed procedure, variable flow rates and temperatures of the process streams can be handled, and thus process models can be treated explicitly.

In order to have optimization based software for process design, Kravanja and Grossmann (1990) made use of several concepts in order to develop PROSYM, currently renamed as MYPSYN (Kravanja 2010). This software is an implementation of the modeling and decomposition (MID) strategy and the outer approximation and equality relaxation algorithm (OA/ER) of Kocis and Grossmann (1987, 1989). The main characteristic is that it enables automated execution of simultaneous topology and parameter optimization of processes, including simultaneous optimization and heat integration. Specifically it has,

- The capability of defining NLP subproblems at each step of the strategy where the structure of constraints, variables, and objective function changes. It is only assumed that the model of the overall superstructure is provided by the user (see Figure 12.15).
- The capability of defining linear approximations of the nonlinear process models, with the flexibility to modify the linearizations for the master problem.
- Flexibility for the initialization of the first NLP subproblems.
- The capability to provide the user with good control and supervision of the calculations, with the option of automatically performing most of the calculations.
- The capability to interface various NLP and MILP packages (see Figure 12.16).

Optimization of each NLP subproblem is performed only on the existing units rather than on the entire superstructure, which substantially reduces the size of the NLP subproblems.

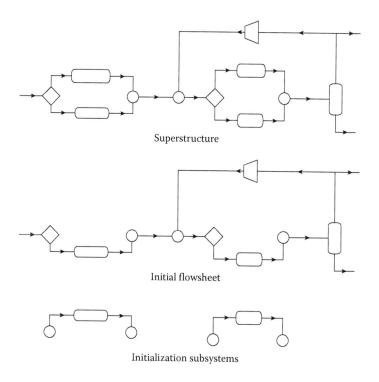

Superstructure

Initial flowsheet

Initialization subsystems

Figure 12.15   Decomposition strategy for a simple example.

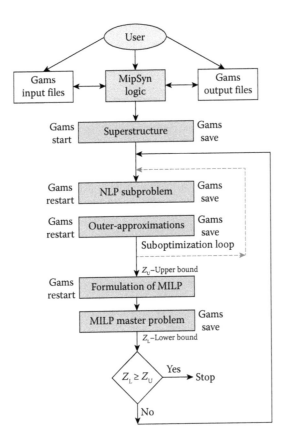

Figure 12.16 Solution algorithm for MIPSYN. (From Kravanja, Z., *Comput. Chem. Eng.*, 34, 1831–1848, 2010. With permission.)

12.2.2.4 Water Networks

Efficient water usage has been addressed since the 1980s in order to reuse the large amounts of water involved in many processes as a means of reducing the freshwater intake. In general, wastewater treatment plants are divided into three or four main steps, including physical treatment, soluble removal (secondary treatment), tertiary treatment (biological removal), and sometimes quaternary treatment to finish polishing the effluent before the discharge (Bagajewicz 2000). Although this is the basic structure, the decision on the actual treatments may vary depending on the main contaminants; several procedures have been proposed to design the most economical treatment. In the beginning, only rules of thumb were applied providing mainly end-of-pipe non-distributed solutions. A good review can be found in Belhateche (1995). In general, we have two approaches for optimal water/wastewater treatment and allocation: (a) the conceptual design and (b) mathematical programing. Apart from the initial works performed by the industry (Carnes et al. 1973; Hospondarec and Thompson 1974; Skylov and Stenzel 1974; Mishra et al. 1975; Anderson 1977; Sane and Atkins 1977), the seminal paper on water networks was written in 1980 by Takama et al. who proposed a mathematical programing approach to define a structure involving all water, using operation and treatment processes. However, the difficulties found in the implementation and solution procedure for the problem led to the use of conceptual design for many years. In 1989 and 1990, El-Halwagi and Manousiouthakis proposed a targeting graphical method to plot the cumulative exchanged mass versus the composition for a set of contaminated streams. Later in 1994,

Wang and Smith (1994a) proposed the so-called "water pinch," based on the same principles as the "temperature pinch." The method is based on the following assumptions:

- Several streams available for cleaning can be split and sent to different treatment operations. Thus, no merging of these streams is assumed.
- The flow rate of water through the processes is constant.
- The treatment units have fixed pollutant removal ratios.
- Cost of treatment is assumed proportional to the flow rate of the stream to clean up. A concentration–load diagram discussion justifies this simplifying assumption.

Once the problem has been expressed in the framework of concentration–load diagrams, a composite curve representing all wastewater streams can be constructed. A minimum treatment flow rate is then obtained by assuming a fixed removal ratio. This is accomplished by rotating a treatment flow rate line around the origin.

It is still an end-of-pipe approach, but most importantly, it is useful for one contaminant only, and the network configuration is built based on rules of thumb. When several contaminants are involved, a network for each contaminant is designed, and after that, a merging procedure results in the complete network. Some modifications were later introduced, including the use of a hierarchical design approach but this method cannot guarantee optimality.

It took several years before the work by Takama et al. (1980) was revisited in 1998. The basic idea holds that a superstructure of all possible reuse and recycle opportunities is formulated to come up with the optimal water network. The difficulty lies in the solution of the MINLP formulations. The main complexity in the nonlinear model is due to the bilinear terms in the mass balance equations (flow rate times concentration) and the concave cost terms in the objective function with which the solution for the total water network is not guaranteed to be the global optimum. Furthermore, it is worth pointing out that most of the published articles do not consider all possible interconnections, multiple sources of water of different quality, pretreatment of the water, and mass transfer and nonmass transfer water-using operations. Moreover, in many articles, the cost of water pumping through pipes and the investment costs for pipes are not included in the objective function; typically, the total water network is decomposed into two parts (network with water-using operations and wastewater treatment network) that are solved separately.

Over the years, in order to obtain a solution, some simplifications have been made. The most important was considering only the water allocation problem or the water treatment. Apart from this major simplification, a few common considerations to solve the problems include the lineal relaxation of the problem for obtaining a lower bound (Quesada and Grossmann 1995) or the use of some considerations over the network including water losses, constant removal ratio, and bounds on the flow rates with no optimality guaranteed (Alva-Argáez et al. 1998a, 1998b). Some authors also use linear models to initialize the original problem (Doyle and Smith 1997); others use heuristic search procedures based on the successive solution of a lineal relaxation and the original nonconvex model (Galan and Grossmann 1998) or the definition of the necessary optimality conditions (maximum outlet concentrations from water-using units and concentration monotonicy) to eliminate the nonlinearities arising from the mass balances (bilinear terms concentration times flow rate) (Savelski and Bagajewicz 2000, 2003). Gunaratnam et al. (2005) also presented a two-stage optimization approach involving an MILP in the first stage to initialize the problem. In the second stage, the design is fine-tuned using an MINLP formulation. In addition to this, the network complexity is controlled by specifying the minimum allowed interconnections in the network. In spite of the effort, it does not necessarily yield the global optimum. Karuppiah and Grossmann (2008) proposed a spatial branch and contract algorithm for the rigorous global optimization of the nonlinear program of the integrated water system design. Li and Chang (2007) developed an efficient initialization strategy to solve the NLP and MINLP for water network applications. In the MINLP model, they formulated structural constraints to manipulate structural complexity, but global optimality is not guaranteed.

They also reported that the optimum solution obtained by the initialization strategy is at least as good as the results reported in the literature but with less computation time to achieve convergence. In the same year, Alva-Argáez et al. (2007 a and b) proposed a methodology based on the water-pinch decomposition making use of the water-pinch insights to define successive projections in the solution space and also considered the solution of a sequence of MILP that replaces the original problem. A similar approach was used by Faria and Bagajewicz (2010). During the same period, Castro et al. (2009) proposed a two-stage solution strategy where the first stage employs a decomposition that replaces the nonlinear program by a succession of linear programs, one for each treatment unit. In the second stage, the resulting network is used as a starting point for the solution of the nonlinear model with a local optimization solver. Karuppiah and Grossmann (2006) proposed the first rigorous global optimization approach for the optimal synthesis of water networks for which they proposed a spatial branch and bound method coupled with a novel cut for strengthening the lower bound of the cost. Recently, Ahmetovic and Grossmann (2011) proposed a superstructure that consists of multiple sources of water, water-using processes (involving mass transfer or not), water treatment technologies, and all possible connections in the network. The model includes cost of piping, which represents a concave function, the cost of water pumping (a linear function of the flow rate) so that it was possible to account for the trade-offs between processing and piping (see Figure 12.17).

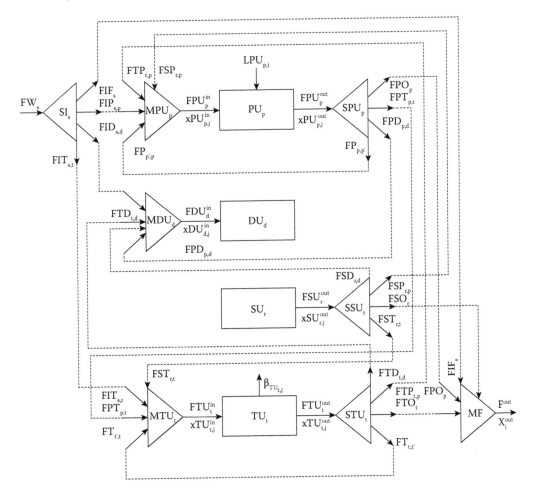

Figure 12.17 Water network superstructure. (From Ahmetovic, E., and Grossmann, I.E., *AIChE. J.*, 57, 434–457, 2011. With permission.)

Current efforts are focused on the simultaneous optimization of heat and water integration. The energy integration by itself reduces the cooling needs and, as a result, the freshwater. Grossmann and Martín (2010) presented a two-stage procedure where, based on this principle, they optimized the energy consumption of the plant based on superstructure optimization and the design of the optimal HEN. On top of that, they developed the water network for the optimal consumption of freshwater. This procedure cannot prove global optimality for the entire process water consumption. Recently, Yang and Grossmann developed a targeting procedure to simultaneously account for water and energy integration (2013). As we have presented, the simultaneous solution of the water network together with the flowsheet optimization is a much harder problem. Yang and Grossmann proposed a two-step procedure. An LP targeting model is used to address the minimum water consumption for the water network (WN) with only water-using process units based on the superstructure proposed by Ahmetovic and Grossmann (2011). Thus, in a first step, the economics of the flowsheet as well as the cost of HEN and WN targets subject to process constraints are simultaneously optimized. The first step then fixes the operating conditions of the flowsheet. The second step consists of determining the detailed HEN and WN structures and corresponding capital and utility costs using the fixed heat capacity flow rate, inlet and outlet temperatures, and water-using process unit flow rates. The targets are not used in the network synthesis problem so that the trade-off of the capital costs of the HEN and WN are now considered.

12.2.3 Environmental Impact Metrics

The optimization of energy and water presents a number of trade-offs difficult to resolve. In general, energy is expensive, and its production contributes to the generation of CO_2 and GHG emissions. On the other hand, freshwater is cheap, even though it also shares a great impact with its extensive use. The use of life cycle assessments (LCAs) is becoming another tool to address the optimization of the plants. It evaluates the environmental impacts resulting from any activity, product, or process (IRAM-ISO-14040 2006). The method can be divided in three steps. First, we set the starting and final phases for the LCA. Next, we define the objective of the analysis and identify the environmental metrics. Finally, we need to quantify the materials and energy used in the process as well as the waste produced. With this information, we compute the environmental impact that can be used as an objective function for any design problem (Gebreslassie et al. 2013).

12.3 SYSTEMATIC DESIGN OF BIOFUEL PRODUCTION PROCESSES

In this section, we present the use of the previous methods for addressing the design of biofuel production processes. We comment on the challenges by type of process, the approach to each particular problem, and finally, provide a summary of the techno-economical data.

12.3.1 Biochemical Processes

12.3.1.1 Particular Challenges

Biochemical processes are characterized by the chemical reaction that produces the desired product. In this way, if we have fermentation, the optimal operating temperatures vary from 32°C to 38°C, and they are mostly exothermic (Jacques et al. 1999; Kazi et al. 2010). This results in the need for cooling water as a cooling agent and the loss of this energy as waste heat because it is very difficult to use the energy generated in the reaction. Furthermore, fermentations involve a large amount of water that accompanies the products, namely ethanol but also acetic acid, lactic acid, and so forth (Jacques et al. 1999). In this case, we need to remove a large amount of water, which requires a large amount of energy. The amount of water that accompanies the product depends on the biomass and the bacteria that are used for the fermentation. For instance, 15% ethanol can be obtained in first-generation bioethanol (Jacques et al. 1999), although for second-generation bioethanol, initially no more than 6% could be obtained (Piccolo and Bezzo 2009). It is true that the development of the processes and the bacteria improve the results and thus reduce the energy required to remove the water (Dimian and Sorin 2008). Finally, the production of ethanol from the raw material depends on how we can breakdown the structure of the biomass (i.e., corn, lignocellulosic). Starchy biomass follows a cooking process capable of breaking the cellulose into glucose using first steam and later saccharification and liquefaction (Jacques et al. 1999). However, lignocellulosic raw materials are stronger in the sense that more energy is needed to break down the biomass into its constituents such as cellulose, hemicelluloses, and from them hexoses and pentoses. The limiting parameter for the yield of the pretreatment is the amount of lignin. A number of pretreatments are available, to reduce the size and crystallinity of the raw material. Following grinding, physicochemical pretreatment (i.e steam explosion, ammonia fiber explosion), chemical pretreatment (i.e. ozone, acids, alkali) and biological pretreatment are the better known (Sun and Cheng 2002).

On the other hand, we have algae growing and oil transesterification. Transesterification can only be considered completely a biochemical process if enzymes are used. A number of different technologies can be used, the typical ones based on alkali or acid catalyst, the newest ones on heterogeneous catalysts including supported enzymes or under supercritical conditions. The transesterification is an equilibrium that is driven toward biodiesel either by the operating conditions (i.e., temperature, catalyst load) or by using an excess of one of the reactants, typically the alcohol (Kulkarni and Dalai 2006). Methanol has been extensively used for a long time based on its quicker reaction rate and lower costs. However, in a biorefinery complex, ethanol is available and can be used as transesterifying agent. In this way, biodiesel production shares some similarities in the sense that algae growth requires the removal of a large amount of water and the product, biodiesel, comes together with alcohol that is to be removed. The trade-off we find is the excess of alcohol necessary to increase the conversion and the energy consumed to recover the unreacted alcohol. The synthesis reaction is endothermic, but the low operating temperature, unless supercritical conditions are used, allows the use of the energy available within the process.

12.3.1.2 Cases of Study

12.3.1.2.1 Ethanol

First generation: Using grain, corn, or wheat, as raw material, the main problem is deciding on the dehydration stage. The basic process can be found in Jacques et al. (1999). Karuppiah et al. (2008) presented a sequential approach for the optimal production of bioethanol from corn where a superstructure of options was formulated so that biomass breakdown using steam was followed by liquefaction and saccharification to obtain the sugars, mainly glucose, and next glucose is fermented into ethanol. They proposed the separation of the solids before or after the beer column and three alternatives for the final dehydration including rectification, whose maximum ethanol purity is the azeotrope, adsorption in corn grits, or molecular sieves. The three alternatives can work in parallel or in sequence but only molecular sieves can discharge ethanol with fuel quality. After the optimization of the superstructure, the system selected the elimination of the solids before the beer column. It is also better to reduce the flow treated in the rectification, due to the high energy consumption, by using adsorption in corn grits. Finally, molecular sieves are used for fuel quality ethanol. Next, the beer column was substituted by a multieffect column to reduce the energy consumption and cooling needs, and finally an HEN was designed based on Yee and Grossmann (1990). The results revealed that, from a base case scenario where no optimization is performed, the energy consumption is reduced by half and the cooling needs by one-third (Karuppiah et al. 2008). Finally, using Ahmetovic ant Grossmann's (2011) model, the WN is designed with promising results because on the one hand, the results proved what Delta T company claimed, 1.5 gal water/gal ethanol, and that ethanol is not as water demanding as it was considered before (Ahmetovic et al. 2010).

Second generation: Using corn stover or switchgrass as raw materials requires decisions at two levels: the pretreatment to be selected and, finally, the dehydration technologies. The procedure followed was similar to the case of the first-generation bioethanol. First, a superstructure of options is considered involving dilute acid and Ammonia fiber explosion (AFEX) as pretreatment technologies; the best two are proved, followed by hydrolysis and sugar (glucose and xylose) fermentation. The dilute mixture is treated in a multieffect distillation column and four options were considered for ethanol: dehydration, rectification adsorption in corn grits, molecular sieves, and pervaporation. These four can work in parallel and/or in sequence but only pervaporation and molecular sieves can provide fuel quality ethanol. The energy optimization of the superstructure results in the selection of dilute acid as pretreatment and molecular sieves as final dehydration technology. Multieffect distillation column is also included together with heat integration by designing a HEN based on the Yee and Grossmann (1990) model. The process in principle requires energy. However, by using lignin to produce energy, the net energy balance of the process is positive (Martín and Grossmann 2012a). The consumption of water is computed in a third stage, based on the design of the optimal WN. It turns out that because the processes are more energy intense to break down a stronger structure, the water consumption is higher than for first-generation bioethanol (Martín et al. 2011a).

12.3.1.2.2 Biodiesel

The process consists of pretreatment, if required by the catalysis used, followed by transesterification, alcohol recovery, catalyst removal, and product (biodiesel and glycerol) purification (Zhang et al. 2003; West et al. 2009). There are structural decisions as well as process conditions to be considered. On the one hand, we have several transesterification reactions classified by the presence of a catalyst or not, and whether it is homogeneous or heterogeneous. Under supercritical conditions, we need to operate the reactor at really high pressures and temperatures, but the separation stages are reduced because there is no catalyst. In catalyzed processes, we can distinguish between homogeneous or heterogeneous ones. The homogeneous catalysts, alkali or acid, require washing and neutralization, while heterogeneous catalysts allow simple purification stages because the catalysts are separated by filtration (Martín and Grossmann 2012b). The use of an alkali catalyst requires pretreatment to avoid the formation of soaps but so far this catalysis presents the highest yields. In terms of the process conditions, we have to avoid product decomposition and operate the equilibrium of the transesterification. The first constraint is related to the operating temperature at the distillation columns—for glycerol, below 150°C, although the biodiesel should be kept below 250°C and the oil below 350°C (Zhang et al. 2003). This fact determines the operating pressure of the columns. In terms of the equilibrium, we need to drive it to biodiesel with the proper pressure, temperature, catalyst load, and alcohol excess. One key parameter is the excess of alcohol used because we need to recover the excess, and the energy consumed in the separation steps is responsible for the biofuel cost. Apart from rigorous kinetic models, which present difficulties in predicting the effect of the number of variables on the yield, surface response models are suitable for providing reasonable insight with reduced computational effort in order to determine the operating conditions of the reactor. This example matches perfectly the Duran and Grossmann (1986) model because simultaneous optimization and heat integration provides the optimal conditions at the reactor together with the adjustment

Table 12.4   Summary of Techno-Economic Details for Biodiesel

	Biodiesel (Ethanol)	Biodiesel (Methanol)	
	(Algae)	(Cooking)	(Algae)
Investment ($MM)	112	17	110
Capacity (MMgal/yr)	68	72	69
Biofuel yield (kg/kg$_{wet}$)	0.47	0.96	0.48
Production cost ($/gal)	0.54	0.66	0.42
Water consumption (gal/gal)	0.35	0.33	0.60
Energy consump. (MJ/gal)	1.93	1.94	1.94
ROI (%)	84.70	565.4	91.27
Pay out (yr)	0.56	0.09	0.52
By-product	Glycerol	Glycerol	Glycerol Fertilizer

of the temperatures across the flowsheet (Martín and Grossmann 2012b). Finally, there is the option of using ethanol or methanol. Because the number of alternatives is small, it is simple to solve each alternative as a case of study. For each one, simultaneous optimization and heat integration is performed. Next, an HEN is designed using the results in the previous optimization, and finally a WN is developed. The results show that the operating conditions at the reactor are different than the ones reported in the literature when only the reactor is considered (Martín and Grossmann 2012b, Severson et al. 2013) because the optimization adjusts the alcohol excess used. Furthermore, in the case of the use of methanol, the most robust process is the one that uses a heterogeneous catalyst because it can be used for cooking oil or algae with promising values for energy and water consumption, see Table 12.4. In the case of using ethanol, there are two promising catalyst alternatives, alkali (if no impurities accompany the oil) or enzymes. In fact, the enzymatic based process requires lower energy and freshwater, but the enzyme cost is still high resulting in a processing cost slightly higher than the one using alkali. Finally, the comparison between the use of methanol (Martín and Grossmann 2012b) and the use of ethanol (Severson et al. 2013) is based on availability and cost. In an integrated refinery, ethanol is competitive (Martín and Grossmann 2013b) but its use as biofuel may limit this option.

12.3.2 Thermal Processes

12.3.2.1 Particular Challenges

These processes are fairly similar to petrochemical ones. The high operating temperatures in the process allow easier energy integration, together with the fact that the exothermic synthesis is a source of energy at high temperature that can be easily used. In terms of alternatives, we can find a large number of different technologies for each of the steps—from gasification (direct or indirect), reforming (steam reforming, partial oxidation, autoreforming, CO_2 reforming), solids removal (cold, hot), sour gases removal (absorption, adsorption, membranes), and synthetic paths, either synthesis gas fermentation or catalytic mixed alcohol synthesis. In this case, reactor conversions may be low, but in general the stream from the reactor contains a range of products that must be separated. A sequence of distillation columns is a common technology to address this issue, or a fractionation column such as the one in crude refining. For the fermentation case, we end up with a dilute mixture of ethanol and water, more dilute that the ones obtained in the case of biochemical processes—typically 5% (Martín and Grossmann 2011a)—that is to be dehydrated as before. Furthermore, we also have the problem of dealing with a reaction that is exothermic and that operates at relatively

Table 12.5 Summary of Techno-Economic Details Using Switchgrass

	Ethanol				
	Hydrolysis	**Gasification and Catalysis**	**Gasification and Fermentation**	**FT Diesel**	**H$_2$**
Investment ($MM)	169	335	260	216	148
Capacity (MMgal/yr)	60	60	60	60	60
Biofuel yield (kg/kg$_{wet}$)	0.28	0.20	0.33	0.24	0.11
Production cost ($/gal)	0.80	0.41	0.81	0.72	0.68
Water consumption (gal/gal)	1.66	0.36	1.59	0.15	—
Energy consump. (MJ/gal)	−10.2	−9.5	27.2	−62.0	−3.84
ROI (%)	44.91	26.15	29.08	36.25	16.86
Pay out (yr)	1.02	1.66	1.51	1.24	2.40
By-product	Energy CO$_2$	Hydrogen Mix alcohols Energy CO$_2$	Hydrogen CO$_2$	Green Gasoline Energy CO$_2$	Energy CO$_2$

low temperatures so that the generated heat cannot be easily reused as in the biochemical process. In the case of using a catalytic path, we obtain a distribution of products. The purification of this stream can be performed using a sequence of distillation columns. The advantage is that the synthesis is exothermic, but it operates at high temperatures, in which case energy integration is easier. Although the high temperatures of operation may be of concern in terms of water consumption, good heat integration can actually help in reducing cooling needs drastically (see Table 12.5).

12.3.2.2 Cases of Study

12.3.2.2.1 Ethanol

Figure 12.12 presents the superstructure for the conceptual design of a second-generation bioethanol plant based on gasification. In this case, the biggest challenge is to come up with a process out of the different technologies traditionally used for carbon processing. More specifically, we need to evaluate trade-offs at each step. Direct gasification operates at high pressure and requires the use of pure oxygen while producing a gas with a lower amount of hydrocarbons. Indirect gasification can use air; because the combustion of the char actually takes place in a second piece of equipment, we have to pay for two units, requiring lower injection of steam and operating at low pressure but generating more hydrocarbons (Phillips et al. 2007; Dutta and Phillips 2009). Once we have obtained the gas, we start by reducing the hydrocarbons. The two extreme approaches are the use of steam reforming, which is endothermic, reducing the energy available within the system but generating larger amount of hydrogen, or the use of partial oxidation, which is exothermic but generating a larger concentration of hydrogen. In between, we can find autothermal reforming—by combining the previous two, or dry reforming—by using CO$_2$, but this alternative is under development. Next, we remove the solids, and typically, hot cleaning by means of the use of filters or cold cleaning, using a scrubber, can be considered. Subsequently, the gas composition may need to be adjusted so that the syngas has the proper composition for the production of ethanol in this case. A ratio H$_2$:CO of 1 is required. We may need to generate more hydrogen for which water gas shift reaction can be used, consuming steam, or removing part of the hydrogen in excess, where PSA/membrane systems can be employed. At this point, we must remove sour gases such as CO$_2$ and H$_2$S. Carbon capture technologies are used for the step. We have chemical absorption—using different solutions such as ethanol amines, and physical absorption—including processes such as Rectisol, Selexol, PSA, or membranes. Although physical absorption typically requires high CO_2 pressure, the use of any of the two absorption technologies requires a large energy consumption to recover the solvent. PSA consists of the adsorption of the gases on zeolites. CO_2 can easily be adsorbed, and new beds are being developed to be able to adsorb H_2S. Finally, porous membranes can be interesting, too, but there is a need for the use of a carrier for the CO_2 that is eliminated, and thus solvent recovery is also expensive. Once the syngas is prepared, we can send it to synthesis. Mainly two paths can be used. On the one hand is syngas fermentation, in which case we need a large amount of water producing a dilute mixture, typically below 5% ethanol. The reaction has been carried up to industrial scale by Coskata (2008), but it has not received much support. In this case, we need to dehydrate the mixture and multieffect columns to operate because beer columns are an interesting technology. Finally, the last purification step is similar to the biochemical case involving rectification, adsorption in corn grits, molecular sieves, and pervaporation. Although in general the process has energy available, the high consumption of steam to dehydrate such a dilute mixture results in an actual negative energy balance (Martín and Grossmann 2011a). On the other hand, we have the mixture of alcohol synthesis of ethanol. In this case, the conversion and the yield are lower obtaining a number of compounds such as propanol—butanol that can only be sold at a low price. The separation of the alcohol mixture can make use of a distillation column sequence as presented here; direct or indirect sequences are proposed.

After solving the superstructure for minimum energy consumption, the same procedure used for other cases is applied. We design the HEN including multieffect columns and next the optimal WN. Table 12.5 shows the results for the optimal processes using each of the synthetic paths. The production of an excess of hydrogen is an advantage for the direct gasification. In terms of reforming, the high price of hydrogen leads us to select the steam reforming. Next, for sour gases removal, the optimization suggests the use of PSA first to adsorb CO_2 so that the Monoethanolamine (MEA) system is used to remove a small amount of CO_2 plus the H_2S. Due to the smaller volume of sour gases treated in this second stage, the MEA recovery does not consume that much energy. Finally, the best process corresponds to the use of mix alcohol synthesis due to the large energy consumed in the dehydration of the ethanol produced via syngas fermentation.

12.3.2.2.2 Hydrogen and FT Fuels

The initial steps of these cases are similar to the previous ones including gasification, reforming, solids removal, and composition adjustment.

In the case of the production of hydrogen, the problem is to adjust the operating conditions of the gasifiers in order to improve the yield to hydrogen. Basically, what we are doing is (1) breaking down biomass into gas and modifying the composition to obtain CO and H_2, and (2) we also are breaking down steam to generate H_2 out of it. Therefore, the trade-off to be considered is the consumption of steam and the production of CO_2 from a raw material that should be more than a source of hydrogen alone but a source of carbon, too. Finally, hydrogen must be purified. Hydrogen is a small molecule that can actually diffuse through metals. Lately, membrane reactors have been used for the production of hydrogen (Ji et al. 2009; Martín and Grossmann 2011b). The advantage of this system is that typically the production of hydrogen is through an equilibrium, so that if we remove one of the products we drive the equilibrium to produce more products. Using the same design procedure as before, the optimal configuration involves indirect gasification, due to lower steam costs, and steam reforming, since increases the yield to hydrogen.

Another set of fuels that can be produced using gasification of biomass are the so-called FT fuels (Swanson et al. 2010). As before, the building block is the syngas. In this particular case, the ratio H_2:CO depends entirely on the fraction to be obtained. Therefore, apart from the technology selection, resulting in the use of indirect gasification and steam reforming, we need to determine the proper composition of the gas and the operating conditions for the production of

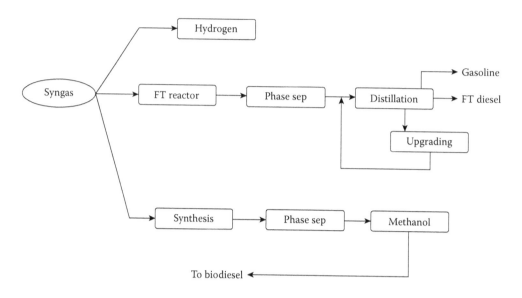

Figure 12.18 Products from syngas for integration purposes.

the fraction of choice. In this particular case, the modeling of the synthesis reactor must be more detailed so that we can adjust the operating conditions to optimize the amount of the desired fraction. Typically, the mechanism of reaction for the FT synthesis is based on a polymerization reactor where the chain of the hydrocarbon molecules grows until the chosen length. Because ethanol is a substitute for gasoline, FT synthesis from biomass can be well considered to be used for the production of diesel substitutes. Apart from the reactor conditions, we need to separate the products obtained. Gasoline, diesel, and heavy products are commonly obtained, and thus product upgrading is also an option to enhance the production of one particular fraction. Among them, hydrocracking of the heavy products is one of the most commonly used. Therefore, in the process for the production of diesel from lignocellulosics, we need to decide on the appropriate technologies for gas production, the adjustment of the composition of the syngas, and the operating conditions at the reactor as well as those in the upgrading (see Figure 12.18). It turns out that optimization chooses the use of indirect gasification followed by steam reforming for obtaining a H_2:CO ratio of 1.7, reactor temperature of around 200°C, 30 bar so that the diesel fraction is maximized (Martín and Grossmann 2011c). Energy and water are generated by the synthesis process and, if properly purified, the water can be reused within the system. Table 12.5 summarizes the major techno-economic results.

12.3.3 Summary of Plant Operation

We divide the presentation of the results into two tables, Tables 12.4 and 12.5 (Martín and Grossmann 2013a). The first one reports the data on biofuels produced from cooking oil or algae oil and the second one uses for lignocellulosics as raw material for biofuels. We can see a number of trade-offs. The process with the lowest production cost, the production of ethanol via gasification using mix alcohol synthesis, is the one that requires higher investment. The production of hydrogen, in spite of the low yield to fuel, has the lowest investment cost due to the smaller number of stages in the process using switchgrass. However, the production of biodiesel requires even lower investment and is competitive—even more so if there is oil such as waste or cooking oil. Biodiesel is competitive compared to any other fuel—and more importantly, compared to FT diesel—with the advantage of lower investment. However, FT diesel is a process that generates energy, while biodiesel requires a certain amount of it.

12.4 INTEGRATION OF PROCESSES: THE BIOREFINERY COMPLEX

In this section, we comment on the integration opportunities within the individual processes presented here based on the technical characteristics.

12.4.1 Integrating First- and Second-Generation Ethanol

Here, we discuss the possibility of using the entire corn plant, the grain, and the stover, for the integrated production of corn and ethanol. In the previous section, we have seen that, on the one hand biochemical production of ethanol from corn operates mainly at low temperatures and requires energy, in spite of the exothermic fermentation. On the other hand, the gasification based process and, in particular, if mixed alcohol synthesis is considered, operates at high temperatures so that it is as interesting feature for energy integration. Furthermore, the CO_2 generated in the fermentation can be captured in the sour gases removal system already included in the second-generation bioethanol process. Another technology integration opportunity is the dehydration stage. In case the syngas fermentation is considered, the purification steps are similar for the product from fermenting glucose and when fermenting syngas. Therefore, using the entire plant of corn or wheat, we can exploit the integration of the raw material, the energy, and the technologies. Cucek et al. (2011) presented the optimal integration of such a plant using MIPSYN to compare the use of thermochemical second-generation bioethanol with thermobiochemical in an integrated facility for the simultaneous production of ethanol and corn, evaluating the economics of the plant (see Figure 12.19). Table 12.6 presents the main techno-economical results.

The lowest cost integrated process uses the thermochemical path for transforming the lignocellulosic material into ethanol especially due to good heat integration in spite of a lower yield toward ethanol ($0.28 \text{ kg}_{ethanol}/\text{kg}_{biomass}$ vs. $0.30 \text{ kg}_{ethanol}/\text{kg}_{biomass}$). However, the highest profit flowsheet consists of the dry-grind process and thermobiochemical route based on the higher yield. Furthermore, the more lignocellulosic material is processed, either because corn is sold as food or because we use other lignocellulosic residues, the lower the production cost. The reason is the generation of hydrogen, a valuable by-product, and the excess of energy available in the system. Finally, the cost of the raw materials determines the distribution of products. If the grain is expensive, it is more interesting to sell it as food rather than use it to produce ethanol.

12.4.2 Integrated Production of Ethanol and Biodiesel from Algae

Algae are a versatile raw material. Although typically in the biofuel industry they have been used for the production of oil, the composition of the dry biomass also includes protein and carbohydrates (Mata et al. 2010). Therefore, algae can be used for the production of other biofuels and high value products. In this case, we comment on the use of algae for the simultaneous production of ethanol and biodiesel (FAEE). Severson et al. (2013) demonstrated the technical and economical feasibility of the use of ethanol instead of methanol for transesterification of the oil. The ethanol becomes attractive for prices below $0.75/gal. Thus, because it is possible to produce ethanol from the carbohydrates, the next step would be to integrate both processes. Martín and Grossmann (2013b) proposed the optimal simultaneous production of bioethanol and biodiesel from algae, see Figure 12.20. This process starts with the oil extraction. Once the oil is available, it follows the typical process for biodiesel production presented earlier. For this case, only the two most promising transesterification technologies identified by Severson et al. (2013) were considered—namely, alkali or enzymatic catalyst. From the biomass remaining after oil extraction, the process is similar to the dry grind used for corn-based ethanol because the carbohydrates in the algae are mostly starch (Mata et al. 2010). The main challenge is the energy consumption in the beer column, and thus a

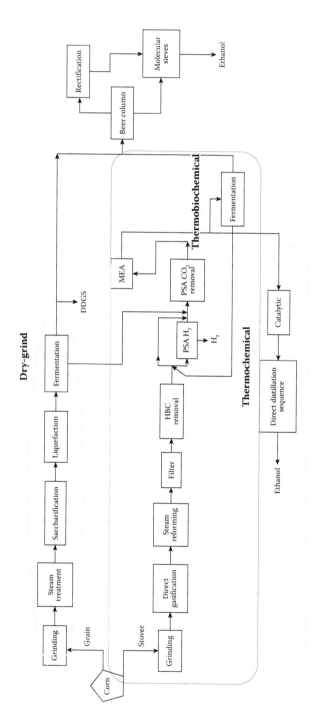

Figure 12.19   Superstructure of alternatives.

Table 12.6   Summary of Integrated Processes

Process Path	Steam (MW)	Electricity (MW)	Cooling (MW)	Cost ($/kg)	Ethanol (kt/yr)	Profit (M$/yr)
Thermobiochemical	63	−1.83	60	0.43	266.7	35.5
Thermochemical	17	0.62	50	0.41	248.5	28

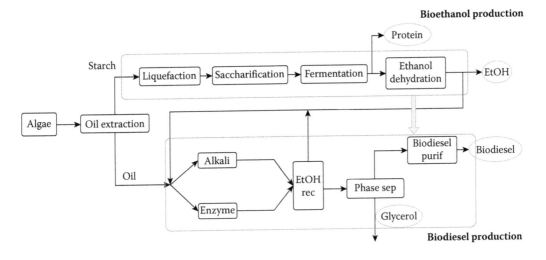

Figure 12.20   Superstructure for the simultaneous production of ethanol and biodiesel.

multieffect column is included in the formulation of the superstructure. Apart from determining the optimal operating conditions at the transesterification reactor, at the biodiesel distillation columns, and at the beer multieffect column, there is another interesting variable, the algae composition. The composition of the algae depends on the growth procedure and the particular species (Mata et al. 2010). In fact, typically the interest has been placed in increasing the lipid composition, but we can tune the composition. In the integrated process, the optimal algae composition for the simultaneous optimization and heat integration of the production of bioethanol and biodiesel (FAEE) are variables to be determined. The results are promising. On the one hand, the operating conditions at the reactor change again, but only in the enzymatic case. Because we have different streams from the bioethanol production section, we are interested in operating the reactor at low temperature so that we can provide that energy within the process. Other than that, the optimal conditions reported an algae composition of 60% lipids, 30% starch, and 10% protein. They are typical achievable values that allow the production of the bioethanol needed within the plant, as well as an excess that can be sold together with biodiesel. The lower energy demand in the biodiesel section is responsible for the 60% in lipids. The production costs of the biofuel are promising compared to the ones presented in Table 12.7. The energy consumption is twice that of the stand-alone plant, but the production capacity increases by 20%.

12.4.3 Hybrid Fuels

Currently, there is high demand for liquid fuels, and biomass is not at a stage of development that allows the substitution of current petrol based fuels; needless to say, the availability of biomass is limited by the harvesting area (as we discussed at the beginning of this chapter). In past years, hybrid processes have been evaluated to consider the simultaneous use of coal, natural gas, and biomass to produce liquid fuels based on FT technologies. Baliban et al. (2010) and

Table 12.7   Summary of Techno-Economic
Results for Simultaneous Production
of Ethanol and Biodiesel

	Alkali Cat	Enzymatic
\$/gal$_{biofuel}$	0.32	0.35
Energy (MJ/gal$_{biofuel}$)	6.72	4.00
Water (gal/gal$_{biofuel}$)	0.77	0.59
Investment (MM\$)	175	180
Capacity (Mgal/yr)	91	90

Elia et al. (2010) proposed the first study on the topic based on the optimization of a superstructure of options considering limited operating conditions for the different units involved. In another work based on the same topic (Baliban et al. 2013), they also included a water treatment in the formulation using a rigorous global optimization branch-and-bound strategy. The boom of shale gas in the United States also supported this kind of strategy; similarly, Martín and Grossmann (2013c) focused on the evaluation of the operation of a hybrid plant based on shale gas and biomass as a function of the price of the raw materials. Biomass price becomes critical when it surpasses \$100/t, although shale gas is attractive at current prices. In which case, as long as we meet the demand for liquid fuels, the excess can be used to produce hydrogen because it is a very valuable product. When the prices of shale gas exceeded \$10/MMBTU, the shale gas used is only to increase the yield to liquid fuels.

12.4.4 By-Product Integration

Although some by-products from the biofuels industry are interesting to provide extra credit, such as the Dried Distillers Grains with Solubles (DDGS) in the first-generation bioethanol production facilities, the fertilizers in biodiesel facilities based on alkali catalytic technology, or the hydrogen in second-generation bioethanol via gasification, there are a couple of interesting by-products for further applications. On the one hand, in the second generation of bioethanol using the biochemical pathway, we can obtain lignin. A number of products can be obtained by breaking up lignin following pyrolysis to obtain bio-oil, or by gasification, in which case we can have the same products as from lignocellulosic raw materials as presented earlier. Furthermore, we can use lignin to produce macromolecules such as carbon fibers or resins, or it can also be a source for aromatics, Benzene, Toluene and Xylenes (BTX) fraction, or phenol (Holladay et al. 2007). It is a broad future area that can be explored. Apart from lignin, we also have glycerol from the biodiesel production process. We can get about 1 kg glycerol per 10 kg of biodiesel. For some time, its use in the cosmetic and food industries has commanded a high price, which contributed to the economics of the bio-diesel production cost (Pagliaro and Rossi 2010). However, the expected increase in the production capacities of biodiesel may saturate the market. Because biodiesel production requires energy, we can further produce hydrogen and/or FT fuels out of glycerol (Martín and Grossmann 2014). The first stage is the reforming of glycerol, where steam reforming or autoreforming are the best known technologies (although we can add aqueous phase reforming if the desired product is hydrogen). By glycerol reforming, we obtain a syngas that can follow any synthetic pathway described earlier for switchgrass (see Figure 12.18). However, the products are not competitive with the ones produced from lignocellulosic biomass unless the cost of glycerol is reduced below \$0.05/lb.

Alternatively, with the syngas, we can produce methanol that can be used within the production of biodiesel to reduce the dependency on fossil fuel based methanol. Martín and Grossmann (2013d) consider this option by developing a superstructure optimization approach for the simultaneous optimization and heat integration of the production of methanol from glycerol and its use in the adjacent biodiesel process. It was possible to reduce the methanol demand to 70%, but the investment increases by 15% with respect to the stand-alone production of biodiesel from algae, and the energy and water consumption is doubled (Martín and Grossmann 2013d). Therefore, we have an interesting trade-off because in order to reduce the environmental impact of the use of fossil-based methanol in the production of biodiesel, we need to pay $0.15/gal higher for the price of ethanol. There are other options under evaluation including ethanol production.

ACKNOWLEDGMENT

The authors gratefully acknowledge National Science Foundation Grant CBET0966524 and the Center for Advanced Process Decision-Making at Carnegie Mellon University.

REFERENCES

Aden, A. (2007). Water usage for current and future ethanol production. *Southwest Hydrology*, September/ October 2007, pp. 22–23.

Agarwal, A.K. (2007). Biofuels (alcohols and biodiesel) applications as fuels for internal combustion engines. *Prog. Energy Combust. Sci.* 33, 233–271.

Agriculture and Agri-Food Canada. (1999). *Assessment of net emissions of greenhouse gases from ethanol–Gasoline blends in Southern Ontario*. Levelton Engineering Ltd, Richmond, B.C.

Agriculture and Agri-Food Canada. (2002). Canadian wheat classes. *Bi-Weekly Bulletin*. 15(7).

Ahmetovic, E. and Grossmann, I.E. (2011). Global superstructure optimization for the design of integrated process water networks. *AIChE. J.* 57(2), 434–457.

Ahmetovic, E., Martín, M., and Grossmann, I.E. (2010). Optimization of water consumption in process industry: Corn—Based ethanol case study. *Ind. Eng. Chem. Res.* 49(17), 7972–7982.

Andres, D. (2002). *Ethanol energy balances*. David Andres and Associates, California.

Ajanovic, A. (2011). Biofuels versus food production: Does biofuels production increase food prices? *Energy*. 36(4), 2070–2076.

Alva-Argáez, A., Kokossis, A.C., and Smith, R. (1998a). Automated design of industrial water networks. *American Institute of Chemical Engineering Annual Meeting*, Paper 13f, Miami, FL.

Alva-Argáez, A., Kokossis, A.C., and Smith, R. (1998b). Wastewater minimisation of industrial systems using an integrated approach. *Comput. Chem. Eng.* 22(Suppl.), S741–S744.

Alva-Argáez, A., Kokossis, A.C., and Smith, R. (2007a). A conceptual decomposition of MINLP models for the design of water-using systems. *Int. J. Environ. Pollut.* 29, 177–205.

Alva-Argáez, A., Kokossis, A.C., and Smith, R. (2007b). The design of water-using systems in petroleum refining using a water-pinch decomposition. *Chem. Eng. J.* 128, 33–46.

Anderson, D. (1977). Practical aspects of industrial reuse. *Int. Chem. Eng. Symp. Ser.* 52, 2–11.

Andrecovic, M.J. and Westerberg, A.W. (1985). An MILP formulation for heat integrated distillation sequence synthesis. *AIChE J.* 31, 1461.

ASA (American Soybean Association). (2007). Soy stats 2007.

Atchison, J. and Hettenhaus, J. (2003). *Innovative methods for corn stover collecting, handling, storage, and transporting*. Report No. ACO-1-31042-01. Golden, CO: National Renewable Energy Laboratory.

Bagajewicz. M. (2000). A review of recent design procedures for water networks in refineries and process plants. *Comput. Chem. Eng.* 24, 2093–2113.

Baitz, M., Binder, M., Degen, W., Deimling, S., Krinke, S., and Rudloff, M. (2004). *Comparative life cycle assessment for SunDiesel (Choren Process) and conventional diesel fuel, Executive Summary*. Leinfelden-Echterdingen, Germany: PE-Europe GmbH.

Baliban, R.C., Elia, J.A., and Floudas, C.A. (2010). Toward novel hybrid biomass, coal, and natural gas processes for satisfying current transportation fuel demands, 1: Process alternatives, gasification modeling, process simulation, and economic analysis. *Ind. Eng. Chem. Res.* 49(16), 7343–7370.

Baliban, R.C., Elia, J.A., and Floudas, C.A. (2013). Biomass to liquid transportation fuels (BTL) systems: Process synthesis and global optimization framework. *Energy Environ. Sci.* 6(1), 267–287.

Bauen, A., Woods, J., and Hailes, R. (2004). *BioPowerswitch! A biomass blueprint to meet 15% of OECD electricity demand by 2020*. Brussels, Belgium: WWF International.

Belhateche, D.H. (1995). Choose appropriate wastewater treatment technologies. *Chemical Engineering Progress*, August, p. 32

Beer, T., Grant, T., and Campbell, P.K. (2007). *The greenhouse and air quality emissions of biodiesel blends in Australia*. Clayton South: CSIRO.

Biegler, L.T., Grossmann, I.E., and Westerberg, A.W. (1997). *Systematic methods of chemical process design*. Upper Saddle River, NJ: Prentice Hall.

Boocock, D.G.B., Konar, S.K., Leung, A., and Ly, L.D. (1992). Fuels and chemicals from sewage sludge: 1. The solvent extraction and composition of a lipid from raw sewage sludge. *Fuel*. 71(11), 1283–1289.

BP. (2012). Statistical Review of World Energy. http://www.bp.com/content/dam/bp/pdf/Statistical-Review-2012/statistical_review_of_world_energy_2012.pdf

Branan, C.R. (2000). *Rules of thumb for chemical engineers*. New York: McGraw Hill.

Buckingham, E. (1914). On physically similar systems; illustrations of the use of dimensional analysis. *Phys. Rev.* 4, 345–376.

Butzen S. (2006). Biodiesel Production in the US Crop Insights, 16, 1–4.

Caballero, J.A. and Grossmann, I.E. (2008). An algorithm for the use of surrogate models in modular flowsheet optimization. *AIChE J.* 54(10), 2633–2650.

Carnes, B.A., Ford, D.L., and Brady, S.O. (1973). Treatment of refinery wastewaters for reuse. *Proceedings of the National Conference on Complete Water Reuse*, Washington, DC.

Castro, P., Teles, J.P., and Novais, A. (2009). Linear program-based algorithm for the optimal design of wastewater treatment systems. *Clean Technol. Environ. Policy*. 11, 83–93.

Chambers, R.S., Herendeen, R.A., Joyce, J.J., and Penner, P.S. (1979). Gasohol: Does it or doesn't it produce positive net energy? *Science*. 206, 790–795.

Choudhury, R., Weber, T., Schindler, J., Weindorf, W., and Wurster, R. (2002). *GM study: Well to wheel analysis of energy use and greenhouse gas emissions of advanced fuel/vehicle systems—A European study*. Ottobrunn, Germany: LB Systemtechnik GmbH.

Cole, D.E. (2007). Issues facing the auto industry: Alternative fuels, technologies, and policies. *ACP Meeting*, Eagle Crest Conference Center, 20 June 2007.

Coskata. (2008). Advantages of the Coskata process. http://www.coskata.com/process/?source=D5E7FB22-6034-405B-898B-58DE4651645D

Cucek, L., Martín, M., Kravanja, Z., and Grossmann, I.E. (2011). Integration of process technologies for the simultaneous production of fuel ethanol and food from corn grain and stover. *Comput. Chem. Eng.* 35(8), 1547–1557.

da Costa, R.E., Lora, E.E.S., Angarita, E.Y., and Torres, E. (2006). The energy balance in the production of palm oil biodiesel two case studies: Brazil and Colombia. *Proceedings, World Bioenergy 2006. Conference and Exhibition on Biomass for Energy*, 30 May–1 June 2006, Jönköping, Sweden, The Swedish Bioenergy Association, Stockholm. http://www.galeon.com/densidadaceite/balancenergia.pdf

Dale, V.H., Kline, K.L., Wiens, J., and Fargione, J. (2010). *Biofuels: Implications for land use and biodiversity*. Biofuels and Sustainability Reports. Washington, DC: Ecological Society of America.

De Castro, J.F.M. (2007). *Biofuels: An overview*. Final report. The Hague, Netherlands: Environmental Infrastructure and Impact Division, Environment and Water Department, Directorate-General for International Cooperation (DGIS).

De Fraiture, C., Giordano, M., and Liao, Y. (2008). Biofuels and implications for agricultural water use: Blue impacts of green energy. *Water Policy*. 10, 67–81.

Delucchi, M.A. (2003). *A Lifecycle Emissions Model (LEM): Lifecycle Emissions from Transportation Fuels, Motor Vehicles, Transportation Modes, Electricity Use, Heating and Cooking Fuels, and Materials; MAIN REPORT*. Institute of Transportation Studies, University of California, avis, Research Report UCD-ITS-RR-03-17-MAIN.

De Oliveira, M.E.D., Vaughan, B.E., and Rykeil, E.J., Jr. (2005). Ethanol as fuel: Energy, carbon dioxide balances, and ecological footprint. *BioScience*. 55(7), 593–602.

DG Energy. (2010). *Impact of land use change on green house gas emissions from biofuels and bioliquids*. DG Energy.

Dimian, A.C. and Sorin, C. (2008). *Chemical process design*. Computer-Aided Case Studies. Weinheim: Wiley-VCH.

Dolan, W.B., Cummings, P.T., and LeVan, M.D. (1987). Heat exchanger network design by simulated annealing. *Proceedings of the First International Conference on Foundations of Computer Aided Process Operations*, July 5–10, 1987, Park City, UT.

Dolan, W.B., Cummings, P.T., and LeVan, M.D. (1989). Process optimization via simulated annealing: Application to network design. *AIChE J.* 35, 725–736.

Donato, L.B., Huerga, I.R., and Hilbert, J.A. (2008). *Energy balance of soybean-based biodiesel production in Argentina.* No. Doc IIR-BC-INF-10-08, Buenos Aires Argentina.

Dorr, T.C. and Mizroch, J., Eds. (2008). *The economics of biomass feedstocks in the United States. A review of the literature.* Occasional Paper No. 1. Biomass Research and Development Board. http://www.biomassboard.gov/pdfs/feedstocks_literature_review.pdf

Doyle, S.J. and Smith, R. (1997). Targeting water reuse with multiple contaminants. *Process Saf. Environ. Protect.* 75, 181–189.

Duran, M.A. and Grossmann, I.E. (1986). Simultaneous optimization and heat integration of chemical processes. *AIChE J.* 32, 123–138.

Dutta, A. and Phillips, S.D. (2009). *Thermochemical ethanol via direct gasification and mixed alcohol synthesis of lignocellulosic biomass.* NREL/TP-510-45913. Golden, CO: National Renewable Energy Laboratory.

EBB. (2008). An Economic and Security of Supply analysis of the widening EU Diesel Deficit How Biodiesel can provide a solution Brussels, October 1, 2008. Available at www.ebb-eu.org.

Ecobilan. (1996). *Ecobilan de l'ETBE de betterave.* France: Ecobilan.

Ecobilan. (2002a). *Energy balances and GHG of motor biofuels production pathways in France.* Neuilly-sur-Seine, France: Ecobilan PWC.

Ecobilan. (2002b). *Bilans énergétiques et gaz à effet de serre des filières de production de biocarburants.* Technical report, Final version. Neuilly-sur-Seine, France: Ecobilan PwC.

Edwards, R., Larivé, J.F., Mahieu, V., and Rouveirolles, P. (2007). *Well-to-wheels analysis of future automotive fuels and power trains in the European context.* v.2c. WTW report 010307. JRC-IES/EUCAR/CONCAWE. http://iet.jrc.ec.europa.eu/about-jec/sites/iet.jrc.ec.europa.eu.about-jec/files/documents/wtw3_wtw_report_eurformat.pdf

Elcock, D. (2008). *Baseline and projected water demand data for energy and competing water use sectors.* ANL/EVS/TM/08-8, Argonne Nationlal Lab. Chicago, IL.

El-Halwagi, M.M. and Manousiouthakis, V. (1989). Synthesis of mass exchange networks. *AIChE J.* 35, 1233–1244.

El-Halwagi, M.M. and Manousiouthakis, V. (1990). Automatic synthesis of mass-exchange networks with single-component targets. *Chem. Eng. Sci.* 45, 2813–2831.

Elia, J.A., Baliban, R.C., and Floudas, C.A. (2010). Toward novel hybrid biomass, coal, and natural gas processes for satisfying current transportation fuel demands, 2: Simultaneous heat and power integration. *Ind. Eng. Chem. Res.* 49(16), 7371–7388.

Eliceche, A.M. and Sargent, R.W.H. (1986). Synthesis and design of distillation of sequences. *IChemE. Symp. Ser.* 61, 1–22.

Elsayed, M.A., Matthews, R., and Mortimer, N.D. (2003). *Carbon and energy balances for a range of biofuel options.* Project no. B/B6/00784/REP URN 03/836. Sheffield, UK: DTI Sustainable Energy Programmes by Resources Research Unit, Sheffield Hallam University.

EPA. (2002). *A comprehensive analysis of biodiesel impacts on exhaust emissions draft.* Technical report EPA420-P-02-001. EPA, USA.

EU. (2007). *The impact of a minimum 10% obligation for biofuel use in the EU-27 in 2020 on agricultural markets.* Directorate G. Economic analysis, perspectives and evaluations G.2.

Economic analysis of EU agriculture Brussels, 30 April 2007 AGRI G-2/WM D(2007). Directorate-General for Agriculture and Rural Development. http://ec.europa.eu/agriculture/analysis/markets/biofuel/impact042007/text_en.pdf

EUCAR/CONCAWE. (2003). *Well-to-wheels analysis of future automotive fuels and powertrains in the European context.* EUCAR, Luxembourg.

ExxonMobil. (2013). *Outlook for energy 2013–2040.* http://corporate.exxonmobil.com/en/energy/energy-outlook

Faria, D.C. and Bagajewicz, M.J. (2010). On the appropriate modeling of process plant water systems. *AIChE J.* 56, 668–689.

Fargione, J., Hill, J., Tilman, D., Polasky, S., and Hawthorne, P. (2008). Land clearing and the biofuel carbon debt. *Science*. 319(5867), 1235–1238.

Farrell, A.E., Plevin, R.J., Turner, B.T., Jones, A.D., O'Hara, M., and Kammen, D.M. (2006). Ethanol can contribute to energy and environmental goals. *Science*. 311, 506–508.

Firrisa, M.T. (2011). *Energy efficiency of rapeseed biofuels production in different agro-ecological systems.* MSc Thesis. http://www.itc.nl/library/papers_2011/msc/gem/firrisa.pdf

Floudas, C.A. and Ciric, A.R. (1989). Strategies for overcoming uncertainties in heat exchanger network synthesis. *Comput. Chem. Eng.* 13, 1133–1152.

Floudas, C.A., Ciric, A.R., and Grossmann, I.E. (1986). Automatic synthesis of optimum heat exchanger network configurations. *AIChE J.* 32, 276–290.

Floudas, C.A. and Paules, G.E., IV. (1988). A mixed integer nonlinear programming formulation for the synthesis of heat—Integrated distillation sequences. *Comput. Chem. Eng.* 12, 531–546.

Foidl, N., Foidl, G., Sanchez, M., Mittelbach, M., and Hackel, S. (1996). *Jatropha curcas* L. as a source for the production of biofuel in Nicaragua. *Bioresour. Technol.* 58, 77.

Franceschin, G., Zamboni, A., Bezzo, F., and Bertucco, A. (2008). Ethanol from corn: A technical and economical assessment based on different scenarios. *Chem. Eng. Res. Des.* 86, 488–498.

Furman, K.C. and Sahinidis, N.V. (2002). A critical review and annotated bibliography for heat exchanger network synthesis in the 20th century. *Ind. Eng. Chem. Res.* 41, 2335–2370.

Galan, B. and Grossmann, I.E. (1998). Optimal design of distributed wastewater treatment networks. *Ind. Eng. Chem. Res.* 37, 4036–4048.

García-Camús, J.M. and Garcia-Laborda J.A. (2006). Biocarburantes líquidos: biodiésel y bioetanol. Fundación para el conocimiento madri+d, CEIM.

Gebreslassie, B.H., Slivinsky, M., Wang, B., and You, F. (2013). Life cycle optimization for sustainable design and operations of hydrocarbon biorefinery via fast pyrolysis, hydrotreating and hydrocracking. *Comput. Chem. Eng.* 50, 71–91.

Giampietro, M., Ulgiati, S., and Pimentel, D. (1997). Feasibility of large-scale biofuel production. *BioScience*. 47(9), 587–600.

Ginnebaugh, D.L., Liang, J., and Jacobson, M.Z. (2010). Examining the temperature dependence of ethanol (E85) versus gasoline emissions on air pollution with a largely-explicit chemical mechanism. *Atmos. Environ.* 44, 1192–1199.

Gnansounou, E. and Dauriat, A. (2004a). *Energy balance of bioethanol: A synthesis*. Lausanne, Switzerland: Lasen, Ecole Polytechnique Fédérale de Lausanne.

Gnansounou, E. and Dauriat, A. (2004b). *Etude comparative de carburants par analyse de leur cycle de vie*. Final report. Lausanne, Switzerland: Lasen, Ecole Polytechnique Fédérale de Lausanne.

Goldemberg, J. (2008). The Brazilian biofuels industry. *Biotechnol. Biofuels*. 1(6), 4096–4117.

Hammond, G.P., Kallu, S., and McManus, M.C. (2007). Development of biofuels for the UK automotive market. *Appl. Energ.* 85(6), 506–515.

GPSA. (2004). Engineering_Data_Book. FPS VERSION 21–10.

Graboski, M. (2002). *Fossil energy use in the manufacture of corn ethanol*. Colorado School of Mines, Denver, CO.

Groode, T.A. and Heywood, J.B. (2007). *Ethanol: A look ahead*. Cambridge, MA: MIT Publication.

Grossmann, I.E. and Martin, M. (2010). Energy and water optimization in biofuel plants. *Chin. J. Chem. Eng.* 18(6), 914–922.

Gunaratnam, M., Alva-Aragez, A., Kokossis, A0, Kim, J.K., and Smith, R. (2005). Automated design of total water systems. *Ind. Eng. Chem. Res.* 44, 588–599.

Gundersen, T. and Naess, L. (1988). The synthesis of cost optimal heat exchanger network synthesis—An industrial review of the state of the art. *Comput. Chem. Eng.* 12, 503–530.

Hass, M.J. and Foglia, T.A. (2005). Alternate feedstocks and technologies for biodiesel production. In: Knothe, G., Gerpen, J., and Krahl, J. (eds). The biodiesel handbook. *AOCS Press*. Champaign, pp. 42–61.

Henao, C.A. and Maravelias, C. (2011). Surrogate-based superstructure optimization framework. *AIChE J.* 57(5), 1216–1232.

Hill, J., Nelson, E., Tilman, D., Polasky, S., and Tiffany, D. (2006). Environmental, economic, and energetic costs and benefits of biodiesel and ethanol biofuels. *Proc. Natl. Acad. Sci. U. S. A.* 103(30), 11206–11210.

Ho, S.P. (1989). Global warming impact of ethanol versus gasoline. *Presented at the 1989 National Conference "Clean air Issues and America's Motor Fuel Business,"* October 3–5, Washington, DC.

Hodson, P. (2008). GHG methodology in the EU renewable energy directive. *Presented at GBEP 2nd TF Meeting on GHG Methodologies*, Washington, 6–7 March 2008.

Hoekman, S.K. (2009). Biofuels in the U.S.—Challenges and opportunities. *Renew. Energ.* 34, 14–22.

Hohmann, E.C. and Lockhart, F.J. (1976). Optimum heat exchanger network synthesis. *AIChE Meeting*, Atlantic City, NJ, 29 August–1 September 1976.

Holladay, J.E., Bozell, J.J., White, J.F., and Johnson, D. (2007). *Top value-added chemicals from biomass volume II—Results of screening for potential candidates from biorefinery lignin.* PNNL-16983. http://www1.eere.energy.gov/biomass/pdfs/pnnl-16983.pdf

Hospondarec, R.W. and Thompson, S.J. (1974). Oil-steam system for water reuse. *Proceedings of the American Institute of Chemical Engineering Workshop*, Vol. 7, AIChE, NY.

Huang, H.J., Ramaswamya, S., Tschirner, U.W., and Ramarao, B.V. (2008). A review of separation technologies in current and future biorefineries. *Separ. Purif. Technol.* 62, 1–21.

IEA (International Energy Agency). (2004). *Biofuels for transport: An international perspective.* Paris, France: International Energy Agency.

IEA (International Energy Agency). (2012). CO_2 emissions from fuel combustion. Highlights, Paris, France.

IRAM-ISO-14040. (2006). *Environmental management-life cycle assessment principles and frame work.* International Standard: ISO Standards.

IWMI (International Water Management Institute). (2006). *Insights from the comprehensive assessment of water management in agriculture.* Stockholm: World Water Week.

Jacobson, M.Z. (2009). Review of solutions to global warming, air pollution, and energy security. *Energy Environ. Sci.* 2, 148–173.

Jacques, K., Lyons, T.P., and Kelsall, D.R. (1999). *The alcohol textbook.* 3rd ed. Nottingham, UK: Nottingham University Press.

Jardé, E., Mansuy, L., and Faure, P. (2005). Organic markers in the lipidic fraction of sewage sludges. *Water Res.* 39, 1215–1232.

Ji, P., Feng, W., and Chen, B. (2009). Production of ultrapure hydrogen from biomass gasification with air. *Chem. Eng. Sci.* 64, 582–592.

Kallivroussis, L., Natsis, A., and Papadakis, G. (2002). The energy balance of sunflower production for biodiesel in Greece. *Biosyst. Eng.* 81(3), 347–354.

Karuppiah, R. and Grossmann, I.E. (2006). Global Optimization for the Synthesis of Integrated Water Systems in Chemical Processes. *Comp. Chem. Eng.* 30, 650–673.

Karuppiah, R. and Grossmann, I.E. (2008). Global optimization of multiscenario mixed integer nonlinear programing models arising in the synthesis of integrated water networks under uncertainty. *Comput. Chem. Eng.* 32, 145–160.

Karuppiah, R., Peschel, A., Grossmann, I.E., Martín, M., Martinson, W., and Zullo, L. (2008). Energy optimization of an ethanol plant. *AIChE J.* 54(6), 1499–1525.

Kazi, F.K., Fortman, J.A., Anex, R.P., Hsu, D.D., Aden, A., Dutta, A., and Kothandaraman, G. (2010). Techno-economic comparison of process technologies for biochemical ethanol production from corn stover. *Fuel.* 89(1), S20–S28.

Keeney, D. (2009). Ethanol USA. *Environ. Sci. Technol.* 43(1), 8–11.

Keeney, D.R. and DeLuca, T.H. (1992). Biomass as an energy source for the mid-western US. *Am. J. Alternative Agr.* 7, 137–143.

Kim, S. and Dale, B.E. (2002). Allocation procedure in ethanol production system from corn grain. *Int. J. Life Cycle Ass.* 7, 237–243.

Kim, S. and Dale, B.E. (2005). Environmental aspects of ethanol derived from no-tilled corn grain: Nonrenewable energy consumption and greenhouse gas emissions. *Biomass Bioenergy.* 28, 475–489.

Klass, D.L. (1998). *Biomass for renewable energy, fuels and chemicals.* San Diego, CA: Academic Press.

Klass, D.L. (2004). *In encyclopedia of energy.* Vol. 1. pp. 193–212, Cleveland, C.J., Ed. London: Elsevier.

Kocis, G.R. and Grossmann, I.E. (1987). Relaxation strategy for the structural optimization of process flowsheets. *Ind. Eng. Chem. Res.* 26, 1869–1880.

Kocis, G.R. and Grossmann, I.E. (1989). A modelling and decomposition strategy for the MINLP optimization of process flowsheets. *Comput. Chem. Eng.* 13, 797–819.

Kravanja, Z. (2010). Challenges in sustainable integrated process synthesis and the capabilities of an MINLP process synthesizer MipSyn. *Comput. Chem. Eng.* 34, 1831–1848.

Kravanja, Z. and Grossmann, I.E. (1990). PROSYN—An MINLP process synthesizer, *Comput. Chem. Eng.* 14, 1363–1378.

Kszos, L.A. (2006). Bioenergy from switchgrass: Reducing production costs by improving yield and optimizing crop management. http://www.ornl.gov/~webworks/cppr/y2001/pres/114121.pdf

Kulkarni, M.G. and Dalai, A.K. (2006). Waste cooking oils. An economical source for biodiesel: A review. *Ind. Eng. Chem. Res.* 45, 2901–2913.

Kwiatkowski, J.R., McAloon, A.J., Taylor, F., and Johnston, D.B. (2006). Modeling the process and costs of fuel ethanol production by the corn dry-grind process. *Ind. Crop. Prod.* 23, 288–296.

Larson, E.D. (2006). A review of LCA studies on liquid biofuels systems for the transport sector. *Energ. Sust. Dev.* 10, 109–126.

LASEN. (2000). *Caractérisation de filières de production de bioéthanol dans le contexte helvétique.* Swiss Federal Office of Energy, CH.

LASEN. (2002). *Etude comparative de carburants par analyse de leur cycle de vie.* Alcosuisse, CH.

LASEN. (2004). *Etude comparative de carburants par analyse de leur cycle de vie.* Alcosuisse, CH.

L-B-Systemtechnik. (2002). *Well-to-wheel analysis of energy use and greenhouse gas emissions of advanced fuel/vehicle systems: A European study.* L-B-Systemtechnik.

Lechón, Y., Cabal, H., de la Rúa, C., Lago, C., Izquierdo, L., and Sáez. R. (2007). Life cycle environmental aspects of biofuel goals in Spain. Scenarios 2010. *15th European Biomass Conference and Exhibition—From Research to Market Deployment,* 7–11 May 2007, Berlin.

Lechón, Y., Cabal, H., de la Rúa, C., Lago, C., Izquierdo, L., Sáez, R., and Fernández, M. (2006). *Análisis del ciclo de vida de combustibles alternativos para el transporte. Fase II. Análisis de ciclo de vida comparativo del biodiésel y del diésel. Energía y Cambio Climático.* Ed. Centro de Publicaciones Secretaría General Técnica Ministerio de Medio Ambiente, Madrid.

Lechón, Y., Cabal, H., Lago, C., de la Rua, C., Sáez, R., and Fernández, M. (2005). *Análisis del ciclo de vida de combustibles alternativos para el transporte. Fase I. Análisis de ciclo de vida comparativo del etanol de cereales y de la gasolina. Energía y Cambio Climático.* Ed. Centro de Publicaciones Secretaría General Técnica Ministerio de Medio Ambiente, Madrid.

Lee, K.T., Foglia, T.A., and Chang, K.S. (2002). Production of alkyl esters as biodiesel fuel from fractionated lard and restaurant grease. *J. Am. Oil Chem. Soc.* 79, 191–195.

Leite, R.C.C., Leal, M.R.L.V., Cortez, L.A.B., Griffin, W.M., and Scandiffio, M.I.G. (2009). Can Brazil replace 5% of the 2025 gasoline world demand with ethanol? *Energy.* 34, 655–661.

Lemoine, R., Behkish, A., Sehabiague, L., Heintz, Y.J., Oukaci, R., and Morsi, B.I. (2008). An algorithm for predicting the hydrodynamic and mass transfer parameters in bubble column and slurry bubble column reactors. *Fuel Proc. Technol.* 89, 322–343.

Levelton Engineering Ltd. (2000a). *Assessment of net emissions of greenhouse gases from ethanol-gasoline blends in Canada: Lignocellulosic feedstocks.* Levelton Engineering Ltd.

Levelton Engineering Ltd. (2000b). *Assessment of net emissions of greenhouse gases from ethanol-gasoline blends in Southern Ontario.* Levelton Engineering Ltd.

Li, B.H. and Chang, C.T. (2007). A simple and efficient initialization strategy for optimizing water-using network designs. *Ind. Eng. Chem. Res.* 46, 8781–8786.

Linnhoff, B. and Hindmarsh, E. (1983). The pinch design method for heat exchanger networks. *Chem. Eng. Sci.* 38, 745–763.

Lorenz, D. and Morris, D. (1995). *How much energy does it take to make a gallon of ethanol? Updated and revised.* Institute for Local Self-Reliance.

Macedo, I., Lima, M.R., Leal, V., and da Silva, J.E.A.R. (2004). *Assessment of greenhouse gas emissions in the production and use of fuel ethanol in Brazil.* São Paulo, Brazil: Government of the State of São Paulo.

Makkar, H.P.S. and Becker, K. (2009). Jatropha curcas, a promising crop for the generation of biodiesel and value-added coproducts. *Eur. J. Lipid. Sci. Technol.* 111, 773–787.

Marland, G. and Turhollow, A. (1990). CO2 emissions from the production and combustion of fuel ethanol from corn. *Energy.* 16, 1307–1316.

Martín, M., Ahmetovic, E., and Grossmann, I.E. (2011a). Optimization of water consumption in second generation bio-ethanol plants. *IandECR.* 50(7), 3705–3721.

Martín, M., Galan, M.A., Cerro, R.L., and Montes, F.J. (2011b). Bubble oscillations: Hydrodynamics and mass transfer. A review. *Bubble Sci. Eng. Technol.* 3(2), 48–63.

Martín, M. and Grossmann, I.E. (2011a). Energy optimization of lignocellulosic bioethanol production via gasification. *AIChE J.* 57(12), 3408–3428.

Martín, M. and Grossmann, I.E. (2011b). Energy optimization of hydrogen production from biomass. *Comput. Chem. Eng.* 35(9), 1798–1806.

Martín, M. and Grossmann, I.E. (2011c). Process optimization of FT-diesel production from biomass. *Ind. Eng. Chem Res.* 50(23), 13485–13499.

Martín, M. and Grossmann, I.E. (2012a). Energy optimization of lignocellulosic bioethanol production via hydrolysis of switchgrass. *AIChE J.* 58(5), 1538–1549.

Martín, M. and Grossmann, I.E. (2012b). Simultaneous optimization and heat integration for biodiesel production from cooking oil and algae. *Ind. Eng. Chem Res.* 51(23), 7998–8014.

Martín, M. and Grossmann, I.E. (2013a). On the systematic synthesis of sustainable biorefineries. *Ind. Eng. Chem. Res.* 52(9), 3044–3064.

Martín, M. and Grossmann, I.E. (2013b). Optimal engineered algae composition for the integrated simultaneous production of bioethanol and biodiesel. *AIChE J.* 59, 2872–2883.

Martín, M. and Grossmann, I.E. (2013c). Optimal use of hybrid feedstock, switchgrass and shale gas, for the simultaneous production of hydrogen and liquid fuels. *Energy.* 55, 378–391.

Martín, M. and Grossmann, I.E. (2014). Optimal Simultaneous Production of Hydrogen and Liquid Fuels from Glycerol: Integrating the Use of Biodiesel Byproducts. *Ind. Eng. Chem. Res.* 53(18), 7730–7745.

Martín, M., Montes, F.J., and Galán, M.A. (2007). Oxygen transfer from growing bubbles: Effect of the physical properties of the liquid. *Chem. Eng. J.* 128, 21–32.

Masso, A.H. and Rudd, D.F. (1969). The synthesis of systems designs II. Heuristic structuring. *AIChE. J.* 15, 10–17.

Mata, T.M., Martíns, A.A., and Caetano, N.S. (2010). Microalgae for biodiesel production and other applications: A review. *Renew. Sust. Energ. Rev.* 14, 217–232.

McAloon, A., Taylor, F., and Yee, W. (2000). *Determining the cost of producing ethanol from corn starch and lignocellulosic feedstocks.* NREL report TP-580-28893. Golden, CO: National Renewable Energy Laboratory.

Menichetti, E. and Otto, M. (2008). Energy balance and greenhouse gas emissions of biofuel from a life cycle perspective. http://new.unep.org/bioenergy/Portals/48107/publications/LCA%20Study.pdf

Mishra, P.N., Fan, L.T., and Erickson, L.E. (1975). Application of mathematical optimization techniques in computer aided design of wastewater treatment systems. Water (II). *Am. Inst. Chem. Eng. Symp. Ser.* 71, 145.

Mondala, A., Liang, K., Toghiani, H., Hernandez, R., and French, T. (2009). Biodiesel production by in situ transesterification of municipal primary and secondary sludges. *Bioresour. Technol.* 100, 1203–1210.

Montgomery, D.C. (2001). *Design and analysis of experiments.* 5th ed, John Wiley & Sons, Inc., New York.

Morris, D. and Ahmed, I. (1992). *How much energy does it take to make a gallon of ethanol?* Institute for Local Self-Reliance, Minneapolis, MN.

MTAP (Minnesota Technical Assistance Program). (2008). *Ethanol benchmarking and best practices. The production process and potential for improvement.* Minneapolis, MN: Minnesota Technical Assistance Program, University of Minnesota.

Natural Resources Canada. (2005). *Ethanol GHG emissions using GHGenius: An update.* Ottawa, Canada: (SandT)2 Consulting, Inc.

Nelson, R.G. and Schrock, M.D. (2006). Energetic and economic feasibility associated with the production, processing, and conversion of beef tallow to a substitute diesel fuel. *Biomass Bioenergy.* 30, 584–591.

Nielsen. (2009). Questions Relative to Harvesting & Storing Corn Stover. http://www.agry.purdue.edu/ext/corn/pubs/agry9509.htm

NSA (National Sunflower Association). (1993). *U.S. Sunflower Crop Quality report.* Bismarck, ND: NSA.

NSC. (2009). *Sustainable development of algal biofuels in the United States.* Washington, DC: The National Academies Press.

Pagliaro, M. and Rossi, M. (2010). *Future of glycerol.* 2nd ed. Cambridge: The Royal Society of Chemistry.

Papoulias, S.A. and Grossmann, I.E. (1983). A structural optimization approach in process synthesis-II. Heat recovery networks. *Comput. Chem. Eng.* 7, 707–721.

Parker, N., Tittmann, P., Hart, Q., Nelson, R., Skog, K., Schmidt, A., Gray, E., and Jenkins, B. (2010). Development of a biorefinery optimized biofuel supply curve for the western United States. *Biomass Bioenergy.* 34, 1597–1607.

Pate, R. (2008). Biofuels and the energy—Water nexus. *Presented at AAAS/SWARM,* Albuquerque, NM, 11 April 2008.

Pate, R., Kilse, G., and Wu, B. (2011). Resource demand implications for US algae biofuels production scale-up. *Appl. Energy.* 88, 3377–3388.

Patzek, T. (2003). Ethanol from corn: Clean renewable fuel for the future, or drain on our resources and pockets? *Nat. Resour. Res.* 1–9.

Phan, A.N. and Phan, T.M. (2008). Biodiesel production from waste cooking oils. *Fuel.* 87, 3490–3496.

Phillips, S., Aden, A., Jechura, J., Dayton, D., and Eggeman, T. (2007). *Thermochemical ethanol via indirect gasification and mixed alcohol synthesis of lignocellulosic biomass.* NREL/TP-510-41168. Golden, CO: National Renewable Energy Laboratory.

Piccolo, C. and Bezzo, F. (2009). A techno-economic comparison between two technologies for bioethanol production from lignocelluloses. *Biomass Bioenergy.* 33, 478–491.

Pimentel, D. (2003). Ethanol fuels: Energy security, economics, and the environment. *J. Agri. Environ. Ethics.* 4, 1–13.

Pimentel, D. and Patzek, T. (2005). Ethanol production using corn, switchgrass, and wood; biodiesel production using soybean and sunflower. *Nat. Resour. Res.* 14(1), 65–76.

Ponton, J.W. and Donaldson, R.A.B. (1974). A fast method for the synthesis of multipass heat exchanger networks. *Chem. Eng. Sci.* 29, 2375–2377.

Pradhan, A., Shrestha, D.S., McAloon, A., Yee, W., Haas, M., and Duffield, J.A. (2011). Energy life-cycle assessment of soybean biodiesel revisited. *Trans. ASABE.* 54(3), 1031–1039.

Puppán, D. (2001). Environmental evaluation of biofuels. *Periodica Polytechnica Ser. Soc. Man. Sci.* 10(1), 95–116.

Putun, A.E., Ozcan, A., and Gercel, H.F. (2001). Production of biocrudes from biomass in a fixed bed tubular reactor; product yields and compositions. *Fuel.* 80, 1371–1378.

Quesada, I. and Grossmann, I.E. (1995). Global optimization of bilinear process networks with multicomponent flows. *Comput. Chem. Eng.* 19, 1219–1242.

Quintero, J.A., Montoya, M.I., Sáncheza, O.J., Giraldo, O.H., and Cardona, C.A. (2008). Fuel ethanol production from sugarcane and corn: Comparative analysis for a Colombian case. *Energy.* 33, 385–399.

Quirin, M., Gärtner, S.O., Pehnt, M., and Reinhardt, G.A. (2004). *CO2-neutrale Wege zukünftiger Mobilität durch Biokraftstoffe: Eine Bestandsaufnahme.* Final report. Frankfurt, Germany.

Reinhardt, G., Rettenmaier, N., Gärtner, S., and Pastowski, A. (2007a). *Rain forest for biodiesel? Ecological effects of using palm oil as a source of energy.* Frankfurt am Main, Germany: WWF.

Reinhardt, G.A., Gärtner, S., Rettenmaier, N., Münch, J., and Falkenstein, E.V. (2007b). *Screening life cycle assessment of jatropha biodiesel.* Final report. Heidelberg, Germany: IFEU.

Rojas, M. and Gutierrez, M. (2008). Palm biodiesel production in the Peruvian amazon: Energy balance and carbon emissions. *Paper presented at the Annual Meeting of the International Congress for Conservation Biology*, Convention Center, Chattanooga, TN, 23 May 2009. http://www.allacademic.com/meta/p244221_index.html

SandT Consultants. (2005). *Ethanol GHG emissions using GHGenius—An update.* Delta, BC, Canada: SandT Consultants.

SandT Consultants. (2006). *Sensitivity analysis of GHG emissions from biofuels in Canada.* Delta, BC, Canada: SandT Consultants.

(SandT)2 Consultants Inc. (2003). *The addition of ethanol from wheat to GHGenius.* (SandT)2 Consultants Inc.

Sane, M. and Atkins, U.S. (1977). Industrial water management. *Int. Chem. Eng. Symp. Ser.* 52, 1–9.

Savelski, M. and Bagajewicz, M. (2003). On the necessary conditions of optimality of water utilization systems in process plants with multiple contaminants. *Chem. Eng. Sci.* 58, 5349–5362.

Savelski, M.J. and Bagajewicz, M.J. (2000). On the optimality conditions of water utilization systems in process plants with single contaminants. *Chem. Eng. Sci.* 55, 5035–5048.

Saxena, R.C., Adhikari, D.K., and Goyal, H.B. (2009). Biomass-based energy fuel through biochemical routes: A review. *Renew. Sust. Energ. Rev.* 13, 167–178.

Searchinger, T., Heimlich, R., Houghton, R.A., Dong, F.X., Elobeid, A., Fabiosa, J., Tokgoz, S., Hayes, D., and Yu, T.H. (2008). Use of U.S. croplands for biofuels increases greenhouse gases through emissions from land use change. *Science.* 319, 1238–1240.

SenterNovem. (2006). *Bioethanol in Europe overview and comparison of production processes.* Rapport 2GAVE0601. http://aoatools.aua.gr/pilotec/files/bibliography/Bioethanol%20in%20Europe,%20overview%20and%20comparison%28SenterNovem%29-0720881920/Bioethanol%20in%20Europe,%20overview%20and%20comparison%28SenterNovem%29.pdf

Severson, K., Martín, M., and Grossmann, I.E. (2013). Process optimization biodiesel production using bioethanol. *AIChE J.* 59(3), 834–844.

Shapouri, H., Duffield, J., and Graboski, M.S. (1995). *Estimating the net energy balance of corn ethanol.* Agricultural Economic Report No. 721. U.S. Department of Agriculture (USDA), Economic Research Service.

Shapouri, H., Duffield, J., McAloon, A., and Wang, M. (2004). *The 2001 net energy balance of corn-ethanol (preliminary).* Washington, DC: US Department of Agriculture.

Shapouri, H., Duffield, J., and Wang, M. (2002). *The energy balance of corn ethanol: An update.* Agricultural Economic Report No. 813. Washington, DC: US Department of Agriculture, Office of the Chief Economist, Office of Energy Policy and New Uses.

Sheehan, J., Dunahay, T., Benemann, J., and Roessler, P. (1998). *A look back at the U.S. Department of Energy's aquatic species program—Biodiesel from Algae.* NREL/TP-580-24190. http://www.nrel.gov/docs/legosti/fy98/24190.pdf

Shen, L. and Zhang, D.K. (2003). An experimental study of oil recovery from sewage sludge by low-temperature pyrolysis in a fluidised-bed. *Fuel.* 82, 465–472.

Sinnott, R.K. (1999). *Coulson and Richardson.* 3rd ed. Vol. 6. Singapore: Butterworth Heinemann.

SIU. (2007). *Sugarcane industry union.* Sao Paulo, Brazil: UNICA.

Skylov, V. and Stenzel, R.A. (1974). Reuse of wastewaters—Possibilities and problems. *Proceedings of the American Institute of Chemical Engineering Workshop*, Vol. 7.

Smeets, E., Junginger, M., Faaij, A., Walter, A., and Dolzan, P. (2006). *Sustainability of Brazilian bioethanol.* Utrecht, Netherlands: Copernicus Institute, University of Utrecht.

Sokhansanj, S. and Fenton, J. (2006). *Cost benefit of biomass supply and pre-processing.* A BIOCAP research integration program synthesis paper. http://www.cesarnet.ca/biocap-archive/rif/report/Sokhansanj_S.pdf

Sun, Y. and Cheng, J. (2002). Hydrolysis of lignocellulosic materials for ethanol production: A review. *Bioresour. Technol.* 83, 1–11.

Supplea, B., Holward-Hildige, R., Gonzalez-Gomez, E., and Leahy, J.J. (2002). The effect of steam treating waste cooking oil on the yield of methyl ester. *J. Am. Oil Chem. Soc.* 79(2), 175–178.

Swanson, R.M., Platon, A., Satrio, J.A., and Brown, R.C. (2010). Techno-economic analysis of biomass-to-liquids production based on gasification. *Fuel.* 89(Suppl. 1), S11–S19.

Takama, N., Kuriyama, T., Shiroko, K., Umeda, T. (1980). Optimal water allocation in a petroleum refinery, *Comp. Chem. Eng.* 4, 251–258.

Taylor, F., Kurantz, M.J., Goldberg, N., McAloon, M.J., and Craig, J.C., Jr. (2000). Dry grind process for fuel ethanol by continuous fermentation and stripping. *Biotechnol. Prog.* 16, 541–547.

Thow, A. and Warhurst, A. (2007). *Biofuels and Sustainable Development.* Maplecroft, Wales.

Unnasch, S. and Pont, J. (2007). *Full fuel cycle assessment: Well to wheels energy inputs, emissions, and water impacts.* Cupertino CA: California Energy Commission.

USDOE (United States Department of Energy). (2006a). Energy demands on water resources. http://www.sandia.gov/energy-water/docs/121-RptToCongress-EWwEIAcomments-FINAL.pdf

USDOE (United States Department of Energy). (2006b). *Report to congress on the interdependency of energy and water.* Washington, DC: United States Department of Energy.

USEIA (U.S. Energy Information Administration). (2001). *Annual energy review 2000.* Washington, DC: U.S. Department of Energy.

Veeraraghava, S. and Riera-Palou, X. (2006). *Well-to-wheel performance of Logen lignocellulosic ethanol.* Shell Global Solutions International.

Wallas, S.M. (1990). *Chemical process equipment.* Elsevier, Oxford, UK.

Wang, M. (2005). Updated energy and greenhouse gas emission results of fuel ethanol. *The 15th International Symposium on Alcohol Fuels*, 26–28 September 2005, San Diego, CA.

Wang, M. and Haq, Z. (2008). Letter to science in response to Searchinger et al. (2008), Sciencexpress, February 7, 2008.

Wang, M., Saricks, C., and Santini, D. (1999). *Effects of fuel ethanol use on fuel-cycle energy and greenhouse gas emissions.* Argonne, IL: Center for Transportation Research, USDOE Argonne National Laboratory.

Wang, M., We, M., and Huo, H. (2007). Life cycle energy and greenhouse gas emission impacts of different corn ethanol plant types. *Environ. Res. Lett.* 2, 024001.

Wang, Y.P. and Smith, R. (1994a). Wastewater minimization. *Chem. Eng. Sci.* 49, 981–1006.

Wang, Y.P. and Smith, R. (1994b). Design of distributed effluent treatment systems. *Chem. Eng. Sci.* 49, 3127–3145.

Wang, Y.P. and Smith, R. (2000). Wastewater minimization with flow rate constraints. *Chem. Eng. Res. Des.* 73, 889–904.

WEF (Water Environment Federation). (2002). *Activated sludge MOP OM-9.* 2nd ed. Alexandria, VA: Water Environment Federation.

Weinblatt, H., Reddy, T.S., and Turhollow, A. (1982). *Energy and precious fuels requirements of fuel alcohol production.* Washington, DC: U.S. Department of Energy.

West, A.H., Posarac, D., and Ellis, N. (2009). Assessment of four biodiesel production processes using HYSYS plant. *Bioresour. Technol.* 99, 6587–6601.

Westcott, P. (2009). USDA agricultural projections to 2018 (Outlook No. OCE-2009-1). 106 pp, February 2009. http://www.ers.usda.gov/publications/oce-usda-agricultural-projections/oce-2009-1.aspx

Wiltsee, G. (1998). Waste grease resource in 30 US Metropolitan areas. *The Proceedings of Bioenergy '98 Conference,* Madison, WI, November 1998, pp. 956–963.

Woods, J. and Bauen A. (2003). *Technology status review and carbon abatement potential of renewable transport fuels in the UK.* London, UK: Centre for Energy Policy and Technology, Imperial College.

Wooley, R., Ruth, M., Sheehan, J., and Ibsen, K. (1999). *Lignocellulosic biomass to ethanol process design and economics utilizing co-current dilute acid prehydrolysis and enzymatic hydrolysis current and futuristic scenarios.* NREL report TP-580-26157.

Worldwatch Institute. (2006). *Biofuels for transport. Global potential and implications for sustainable agriculture and energy in the 21st century.* Washington, DC: Worldwatch Institute.

Wu, M., Mintz, M., Wang, M., and Arora, S. (2009). *Consumptive water used in the production of ethanol and petroleum gasoline.* ANL/ESD/09-1. Argonne National Lab, Chicago, IL.

Yang, L. and Grossmann, I.E. (2013). Water targeting models for simultaneous flowsheet optimization. *Ind. Eng. Chem. Res.* 52, 3209–3224.

Yee, T.F. and Grossmann, I.E. (1988). A screening and optimization approach for the retrofit of heat exchanger networks. *Annual AIChE Meeting,* November 1998, Washington, DC.

Yee, T.F. and Grossmann, I.E. (1990). Simultaneous optimization models for heat integration. II. Heat exchanger network synthesis. *Comput. Chem. Eng.* 14, 1165–1184.

Yee, T.F., Grossmann, I.E., and Kravanja, Z. (1990a). Simultaneous optimization models for heat integration. I. Area and energy targeting and modeling of multi-stream exchangers. *Comput. Chem. Eng.* 14, 1151–1164.

Yee, T.F., Grossmann, I.E., and Kravanja, Z. (1990b). Simultaneous optimization models for heat integration—III. Process and heat exchanger network optimization. *Comput. Chem. Eng.* 14, 1185–1200.

Yusuf, C. (2007). Biodiesel from microalgae. *Biotechnol. Adv.* 25, 294–306.

Zah, R., Böni, H., Gauch, M., Hischier, R., Lehmann, M., and Wäger, P. (2007). *Ökobilanz von Energieprodukten: Ökologische Bewertung von Biotreibstoffen.* St. Gallen, Switzerland: EMPA.

Zhang, Y., Dube, M.A., McLean, D.D., and Kates, M. (2003). Biodiesel production from waste cooking oil: 1. Process design and technological assessment. *Bioresour. Technol.* 89, 1–16.

CHAPTER 13

Impact of Allocation Procedures on the Greenhouse Gas Intensity of Wood-Based Cellulosic Ethanol

Puneet Dwivedi
Daniel B. Warnell School of Forestry and Natural Resources
University of Georgia, Athens, GA, USA

CONTENTS

13.1 INTRODUCTION

The Energy and Independence Security Act of 2007 enacted by the U.S. Congress has set a target of producing 60.5 billion liters of cellulosic biofuels by 2022 (US Congress 2007). It is generally believed that the use of cellulosic biofuels will reduce greenhouse gas (GHG) emissions from the transportation sector of the economy and decrease the nation's dependency on foreign energy sources (Somerville et al. 2010). It is estimated that, apart from agriculture-based feedstocks such as crop residues and perennial grasses, woody feedstocks obtained from the nation's forestlands would provide up to 30% of the total demand of cellulosic feedstocks for bioenergy development at the national level (ORNL 2011).

Several studies have analyzed relative savings in GHG emissions when ethanol derived from woody feedstocks is used in place of gasoline. Dwivedi et al. (2012) reported that the use of ethanol derived from woody feedstocks could save 76% of GHG emissions relative to gasoline over the average life span of a small passenger car in the United States. Eriksson and Kjellström (2010) found that the GHG intensity of ethanol produced at a combined heat and power plant was 25% less than the ethanol produced from other locally available options. Bright and Strømman (2009) reported that the use of ethanol derived from boreal forests of Scandinavia could save GHG emissions ranging from 46% to 68% per unit of gasoline depending upon the adopted conversion technology. Kemppainen and Shonnard (2005) found that the aggregated life cycle environmental impact of ethanol derived from virgin timber and recycled newsprint feedstocks was the same but these feedstocks differed in consumption of fossil energy use because ethanol derived from virgin timber generated excess electricity. Other studies have evaluated the potential of different existing conversion technologies (Galbe and Zacchi 2002; Sun and Cheng 2002; Dwivedi et al. 2009) and the economics of utilizing woody feedstocks for ethanol production (Kaylen et al. 2000; Wingren et al. 2003; Frederick et al. 2008; Hu et al. 2008; Huang et al. 2009).

A review of existing literature suggests that present studies compare GHG intensity and unit cost of ethanol derived from different wood-based feedstocks using the same technology, else they focus on comparing GHG intensity and unit cost of ethanol derived from different technologies using the same feedstock. No study to date has assessed the impact of different allocation procedures on the GHG intensity of produced ethanol derived from woody feedstocks. This information gap is critical because based on an adopted allocation procedure, the GHG intensity of ethanol derived from woody feedstocks could change significantly. For example, Luo et al. (2009) found that the GHG intensity of ethanol derived from corn stover was smaller when energy and material inputs were allocated on the mass rather than the economic basis. Similarly, Wang et al. (2011) found that the relative savings in GHG emissions could range from 25% to 60% for ethanol derived from corn depending upon adopted allocation criteria. Other studies (Slade et al. 2009; Singh et al. 2010) also deliberate upon the importance of allocation procedures on the GHG intensity of produced ethanol. However, no study, to the best of author's knowledge, has determined the impact of allocation procedures on the GHG intensity of ethanol derived from woody feedstocks in the context of the southern United States, a major roundwood producing region of the country.

In order to fill a critical gap in our understanding about the overall potential of cellulosic ethanol derived from woody feedstocks in saving GHG emissions, this study analyzes a total of 186 different scenarios: two forest management choices (intensive and nonintensive), three feedstocks (logging residues, pulpwood, and pulpwood and logging residues combined), and 31 harvest ages (10–40 years in steps of 1 year). For each scenario, GHG emissions related to plantation management were allocated by the percentage weight and revenue contributed by selected feedstocks. The GHG intensity for each scenario was separately calculated for both allocation procedures. Estimated GHG intensity was compared with the GHG intensity of gasoline to ascertain relative savings in GHG emissions. This study uses standard guidelines of life cycle assessment to estimate the GHG intensity of produced ethanol for both allocation procedures. The life cycle assessment is a technique to assess the environmental aspects and potential impacts associated with a product, process, or service, by: compiling an inventory of relevant energy and material inputs and environmental releases; evaluating the potential environmental impacts associated with identified inputs and releases; and interpreting the results to help make a more informed decision (EPA 2012).

The following five steps were considered as a part of ethanol production supply chain: (a) production of woody feedstocks; (b) chipping of feedstocks; (c) transportation of chipped feedstocks; (d) ethanol production; and (e) transportation of ethanol. The use of produced ethanol in an automobile was considered as carbon neutral. The functional unit of analysis was a megajoule (MJ) of energy derived from produced ethanol. Slash pine (*Pinus elliottii*) was selected as a representative species. This species covered an area of about 52,609 km^2 in 2009, mostly in states of Florida and Georgia (Smith et al. 2009). This study focuses on the southern region of the United States because this area contributed to about 62% of total timber harvested nationwide in 2006 (Smith et al. 2009). Under intensive forest management choice, herbicides (at the plantation establishment year) and fertilizers (at the second and twelfth year of plantation) were applied. These inputs were not used under nonintensive forest management choice.

13.2 METHODS

A popular growth and yield model of slash pine was used to estimate availability of three timber products (sawtimber, chip-n-saw, and pulpwood) from a hectare of slash pine plantation under intensive and nonintensive forest management choices (Yin et al. 1998). The availability of logging residues at a harvest age was calculated as the difference between total biomass present in stems and biomass available in sawtimber, chip-n-saw, and pulpwood at the stand level plus 20% of all biomass present in sawtimber, chip-n-saw, and pulpwood at the same harvest age (Jenkins et al. 2003). The additional 20% biomass was added as a proxy for biomass available in branches and treetops (Jenkins et al. 2003).

Total GHG emission related to plantation management under intensive forest management choice was 4803 kg CO_2e ha^{-1} when the harvest age was equal or greater than 12 years. It was 2431 kg CO_2e ha^{-1} when the harvest age was 10 and 11 years (Dwivedi et al. 2011). For nonintensive forest management choice, total GHG emission was 2200 kg CO_2e ha^{-1} for the selected range of harvest ages (Dwivedi et al. 2011). An updated value of nitrous oxide emissions was used based on the latest downloadable version of the Greenhouse Gases, Regulated Emissions, and Energy Use in Transportation (GREET) life cycle model (Wang 2001). GHG emissions related to plantation management were allocated to available timber products by the percentage weight contributed by each timber product toward the combined weight of timber products available at any harvest age. Similarly, these GHG emissions were allocated to available timber products by the percentage revenue contributed by each timber product toward the combined revenue of timber products available at any harvest age. For ascertaining total revenue, prices of sawtimber, chip-n-saw, and pulpwood were taken as US\$27.10 Mg^{-1}, US\$16.50 Mg^{-1}, and US\$10.50 Mg^{-1}, respectively (TMS 2013). The price of logging residues was taken as half of current pulpwood prices (US\$5.30 Mg^{-1}) because markets for logging residues are still under development. No GHG emissions were allocated to logging residues when pulpwood was used as the only feedstock under the assumption that logging residues were left in field when not used as a feedstock.

The GHG emission factor related to chipping of feedstocks was taken as 4 kg CO_2e Mg^{-1} of feedstock chipped (PRé-Consultants 2013a). This GHG emission factor was multiplied with the quantity of feedstock (Mg ha^{-1}) to determine GHG emissions related to chipping of feedstocks. A similar procedure was followed for ascertaining GHG emissions related to other steps present in the supply chain (Table 13.1). GHG emissions related to individual steps were summed to determine overall GHG emissions (g CO_2e ha^{-1}).

Total quantities of ethanol produced (MJ ha^{-1}) were estimated using Equation 13.1:

$$Q_{EtOH} = Yield_{EtOH} * MC_{feedstock} * Q_{feedstock} * CV_{EtOH} \qquad (13.1)$$

where $Yield_{EtOH}$ represents ethanol yield (377.2 L/dry metric ton of feedstock) (Wang 2001). Variables $MC_{feedstock}$ and $Q_{feedstock}$ represent percentage moisture content (50%) and quantities of incoming green feedstock (Mg ha^{-1}), respectively. The CV_{EtOH} represents the calorific value of ethanol (21.3 MJ/L) (Wang 2001). The value of $Q_{feedstock}$ is dependent on the type of feedstocks considered for ethanol production (logging residues, pulpwood, or logging residues and pulpwood combined), harvest age, and choice of forest management. Estimated quantity of overall GHG emission (g CO_2e ha^{-1}) was divided by total quantities of ethanol produced (MJ ha^{-1}) to determine the GHG intensity of produced ethanol (g CO_2e MJ^{-1}). This GHG intensity was compared with the GHG intensity of gasoline (94 g CO_2e MJ^{-1}) to determine relative savings in GHG emissions (Wang 2001).

Table 13.1 Parameters Used for Determining GHG Emissions of Steps Present in Supply Chain

Details	GHG Factor	Distance (km)	References
Chipping of feedstocks	4 kg CO_2e Mg^{-1} of feedstock		PRé-Consultants 2013a
Transport of chipped feedstocks	0.133 kg Mg^{-1} km^{-1}	100	PRé-Consultants 2013b
Ethanol production	7.32 g CO_2e MJ^{-1}		Wang 2001
Transportation of produced ethanol	0.133 kg Mg^{-1} km^{-1}	50	PRé-Consultants 2013b

Note: No GHG emissions are attributed to the use of ethanol in an automobile.

13.3 RESULTS

The availability of logging residues, chip-n-saw, and sawtimber increased with a rise in the plantation age, whereas availability of pulpwood decreased with an increase in plantation age (Figure 13.1). The combined availability of logging residues and pulpwood was always greater under intensive than nonintensive forest management choice ranging from 68 Mg ha^{-1} to 4 Mg ha^{-1} at harvest ages 9 and 50 years, respectively. Similarly, the availability of logging residues was always greater under intensive than nonintensive forest management. The highest availability of logging residues under intensive and nonintensive forest management choices occurred at harvest ages 33 years (84 Mg ha^{-1}) and 40 years (72 Mg ha^{-1}), respectively. The pulpwood availability was higher under intensive than nonintensive forest management choice until the fifteenth year of plantation only.

The percentage weight of the logging residues under both forest management choices stabilized at about 17% of total weight of timber products (Figure 13.2). The percentage weight of pulpwood increased first and then started to decrease with a rise in the plantation age. This can be attributed to the decreased availability of pulpwood at mature plantation ages. A similar case was observed for the combined availability of pulpwood and logging residues. Figure 13.3 shows percentage revenue contributed by selected feedstocks out of the total revenue of timber products. The percentage revenue from logging residues under intensive and nonintensive forest management choices stabilized at about 5% of total revenue. Appendix (at the end of the chapter) summarizes percentage contribution when only pulpwood was used as feedstock. Percentage contribution of selected feedstocks when allocated by weight was typically higher relative to the situation when selected feedstocks were allocated by revenue for the same harvest age.

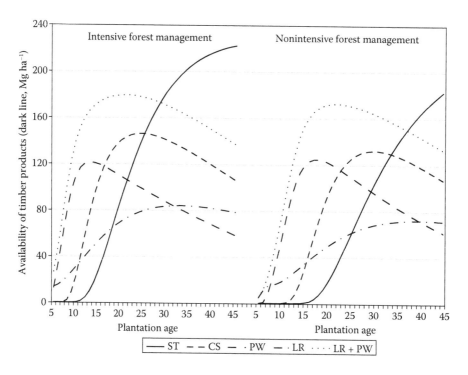

Figure 13.1   Availability of timber products with respect to plantation age. Site index is 21.4 m at twenty-fifth year of plantation. Initial plantation density is 1236 seedlings per hectare. ST: sawtimber, CS: chip-n-saw, PW: pulpwood, LR: logging residues. Combined availability of pulpwood and logging residues is shown separately.

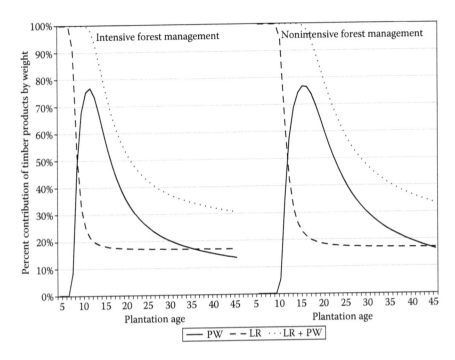

Figure 13.2 Weight of selected feedstocks relative to the weight of all timber products combined. Combined percentage weight contributed by pulpwood and logging residues is shown separately. PW: pulpwood; LR: logging residues.

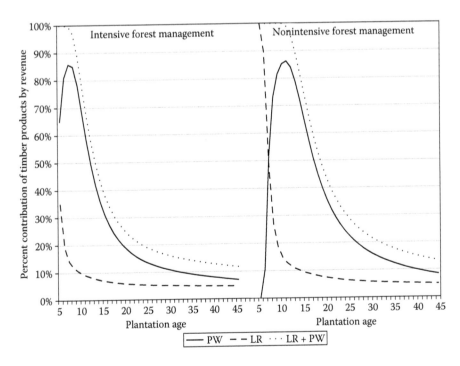

Figure 13.3 Revenue from selected feedstocks relative to revenue from all timber products combined. Combined percentage revenue contributed by pulpwood and logging residues is shown separately. PW: pulpwood; LR: logging residues.

Figure 13.4 shows distribution of allocated plantation management related GHG emissions to selected feedstocks. The allocated GHG emissions were higher under intensive than nonintensive forest management especially after the twelfth year of plantation because of application of fertilizers at that age. The percentage GHG emissions allocated by weight were typically higher than when GHG emissions were allocated by revenue for a given harvest age equal to or greater than 12 years.

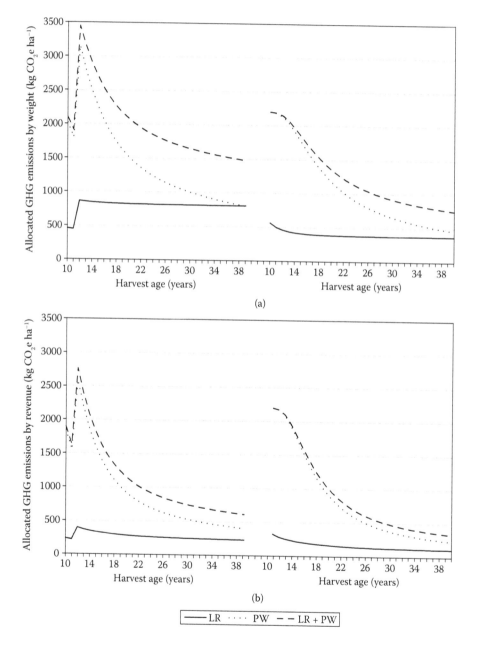

Figure 13.4 Allocation of GHG emissions related to plantation management to selected feedstocks: (a) shows distribution when GHG emissions related to plantation management are allocated by weight and (b) shows distribution when GHG emissions related to plantation management are allocated by revenue. PW: pulpwood, LR: logging residues.

Ethanol availability was directly proportional to the availability of selected feedstocks (Figure 13.5). Figures 13.6 and 13.7 show the contribution of different steps present in the system boundary towards overall GHG emissions when GHG emissions related with the stage of plantation management were allocated by weight and revenue basis, respectively. The contribution of GHG emissions related to step ethanol production and transportation to a nearby pump ranged between 34% and 57% of total GHG emissions followed by GHG emissions related to transportation of biomass from a harvested site to ethanol mill (between 15% and 25%). The contribution of GHG emissions related to chipping of feedstocks was approximately 8%. The contribution of GHG emissions related to plantation management toward overall GHG emissions varied from 10% to 45%. The variability in total GHG emissions across harvest ages was only due to GHG emissions related to plantation management. The overall GHG intensity varied from 13 to 22 g CO_2e MJ^{-1} for selected scenarios for both GHG allocation methods.

Relative savings in GHG emissions across scenarios (Figure 13.8) ranged between 77% and 88% relative to gasoline indicating that the use of ethanol would save considerable GHG emissions irrespective of allocation method initially used for allocating GHG emissions incurred at the stage of plantation management to selected feedstocks. Relative GHG savings were highest when only logging residues were used as a feedstock followed by when both pulpwood and logging residues were used as a feedstock. Figure 13.9 shows the difference between relative GHG savings when GHG emissions linked to plantation management were allocated by weight and revenue. GHG savings were higher when they were allocated by revenue rather then by weight ranging from 0% to 3%.

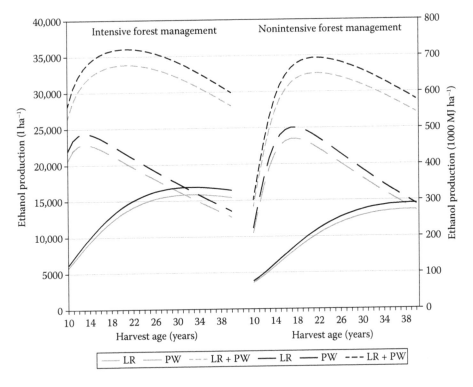

Figure 13.5 Quantities of ethanol produced. Gray lines show ethanol production on volume basis (L ha⁻¹), whereas black lines show ethanol production on energy basis (1000 MJ ha⁻¹). PW: pulpwood, LR: logging residues.

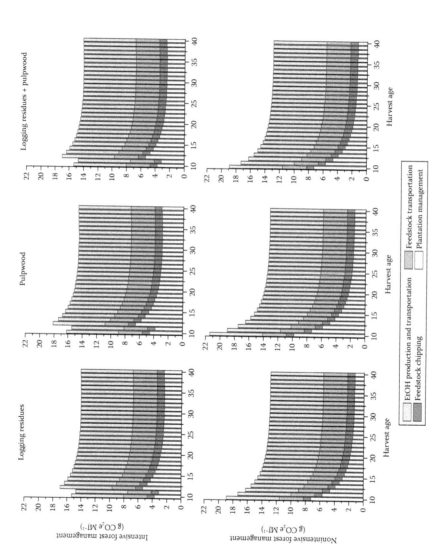

Figure 13.6 Contribution of steps toward overall GHG emission when plantation management related GHG emissions are allocated by weight.

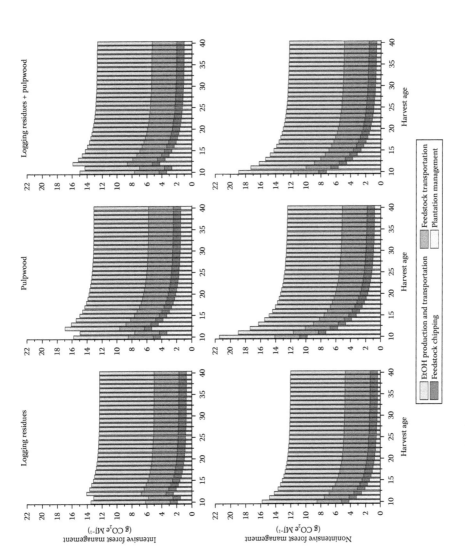

Figure 13.7 Contribution of steps toward overall GHG emission when plantation management related GHG emissions are allocated by revenue.

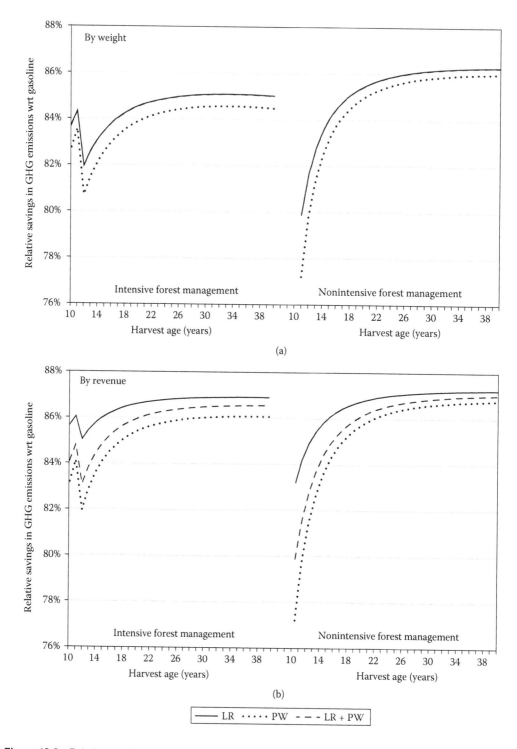

Figure 13.8 Relative savings in GHG emissions: (a) shows distribution when GHG emissions related to plantation management are allocated by weight and (b) shows distribution when GHG emissions related to plantation management are allocated by revenue. PW: pulpwood, LR: logging residues.

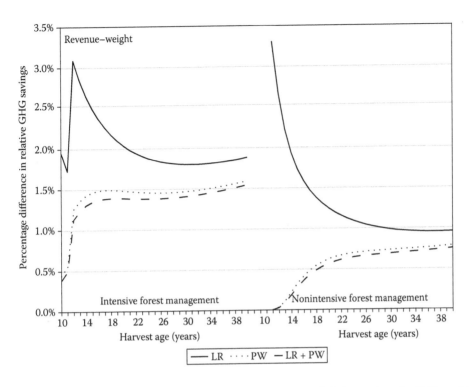

Figure 13.9 Difference between relative savings in GHG emissions. Relative savings in GHG emissions reported in (a) are subtracted from relative savings in GHG emissions reported in (b). PW: pulpwood, LR: logging residues.

13.4 DISCUSSION AND CONCLUSION

This study assesses the impact of two different allocation procedures on relative savings in GHG emissions associated with the use of ethanol instead of gasoline. GHG emissions related to plantation management were allocated by weight to different feedstocks under the first allocation method. In the second allocation method, these GHG emissions were allocated by revenue to different feedstocks.

Results indicate that allocating GHG emissions related to plantation management by revenue increases overall relative GHG savings of produced ethanol compared to when GHG emissions were allocated by weight. This finding contrasts the finding of Luo et al. (2009). Identifying an exact reason for this difference is difficult because Luo et al. (2009) use corn stover as a feedstock whereas this study uses woody feedstocks. Magnitude and trajectory of reduction in GHG emissions varied across feedstocks. Difference in relative savings when only logging residues were used as a feedstock for ethanol production were much more sensitive to the selected allocation procedure compared to the situation when only pulpwood or both pulpwood and logging residues were used as feedstocks. Results of this study suggest that the method of allocating GHG emissions related to plantation management does not significantly impact overall relative GHG saving of ethanol relative to gasoline as the overall relative saving in GHG emissions under both allocation methods was between 77% and 88% across scenarios. This result is similar to Wang et al. (2011), which also found that relative savings in GHG emissions did not differ much for ethanol derived from switchgrass when GHG emissions related to material and energy inputs were allocated based on energy and market basis.

This study clearly indicates that the use of cellulosic ethanol derived from woody feedstocks instead of gasoline is a viable mitigation strategy from a carbon perspective especially when forestland owners maintain the same rotation age. However, this study does not analyze economic viability of the same. Additionally, this study only focuses on only one species for biomass procurement. Silvicultural activities differ across regions and species. Annual biomass availability differs from species to species as well. Therefore, it will be prudent to analyze other forestry species that potentially could be used as a feedstock for bioenergy development at the national and regional levels using the framework developed in this study. Moreover, this study does not compare GHG intensity of ethanol obtained from other potential cellulosic feedstocks such as miscanthus and switchgrass. It is expected that future studies will help fill remaining knowledge gaps.

APPENDIX

Percentage contribution of pulpwood toward GHG emissions related to plantation management when pulpwood is used as an only feedstock.

Allocation Method	Intensive Forest Management		Nonintensive Forest Management	
Harvest Age	By Weight	By Revenue	By Weight	By Revenue
5	1.00	0.91	1.00	0.91
6	1.00	0.91	1.00	0.91
7	1.00	0.90	1.00	0.91
8	0.98	0.87	1.00	0.91
9	0.92	0.79	1.00	0.91
10	0.83	0.68	1.00	0.91
11	0.74	0.58	0.99	0.90
12	0.66	0.49	0.97	0.86
13	0.58	0.41	0.92	0.80
14	0.52	0.35	0.86	0.72
15	0.47	0.30	0.80	0.64
16	0.43	0.27	0.73	0.57
17	0.40	0.24	0.67	0.50
18	0.37	0.21	0.61	0.44
19	0.34	0.19	0.56	0.39
20	0.32	0.18	0.52	0.35
21	0.30	0.16	0.48	0.31
22	0.29	0.15	0.45	0.28
23	0.27	0.14	0.42	0.26
24	0.26	0.13	0.40	0.24
25	0.25	0.13	0.37	0.22
26	0.24	0.12	0.35	0.20
27	0.23	0.11	0.33	0.19
28	0.23	0.11	0.32	0.17
29	0.22	0.11	0.30	0.16
30	0.21	0.10	0.29	0.15
31	0.21	0.10	0.28	0.14
32	0.20	0.09	0.27	0.14
33	0.20	0.09	0.26	0.13
34	0.19	0.09	0.25	0.12

Continued

Allocation Method	Intensive Forest Management		Nonintensive Forest Management	
Harvest Age	By Weight	By Revenue	By Weight	By Revenue
35	0.19	0.09	0.24	0.12
36	0.18	0.08	0.23	0.11
37	0.18	0.08	0.22	0.11
38	0.17	0.08	0.22	0.10
39	0.17	0.08	0.21	0.10
40	0.17	0.08	0.20	0.10
41	0.16	0.07	0.20	0.09
42	0.16	0.07	0.19	0.09
43	0.16	0.07	0.19	0.09
44	0.15	0.07	0.18	0.08
45	0.15	0.07	0.18	0.08

REFERENCES

Bright RM, Strømman AH. 2009. Life cycle assessment of second generation bioethanols produced from Scandinavian boreal forest resources. *J. Ind. Ecol.* 13, 514–31.

Dwivedi P, Alavalapati JRR, Lal P. 2009. Cellulosic ethanol production in the United States: Conversion technologies, current production status, economics, and emerging developments. *Energy Sustain. Dev.* 13, 174–82.

Dwivedi P, Bailis R, Alavalapati JRR, Nesbit T. 2012. Global warming impact of E85 fuel derived from forest biomass: A case study from Southern USA. *BioEnergy Res.* 5, 470–80.

Dwivedi P, Bailis R, Bush TG, Marinescu M. 2011. Quantifying GWI of wood pellet production in the southern United States. *BioEnergy Res.* 4, 180–92.

EPA. 2012. *Life Cycle Assessment (LCA)*. Washington, DC: United States Environmental Protection Agency.

Eriksson G, Kjellström B. 2010. Assessment of combined heat and power (CHP) integrated with wood-based ethanol production. *Appl. Energy.* 87, 3632–41.

Frederick WJ, Lien SJ, Courchene CE, DeMartini NA, Ragauskas AJ, Iisa K. 2008. Production of ethanol from carbohydrates from loblolly pine: A technical and economic assessment. *Bioresour. Technol.* 99, 5051–7.

Galbe M, Zacchi G. 2002. A review of the production of ethanol from softwood. *Appl. Microbiol. Biotechnol.* 59, 618–28.

Hu G, Heitmann JA, Rojas OJ. 2008. Feedstock pretreatment strategies for producing ethanol from wood, bark, and forest residues. *Bioresources.* 3, 270–94.

Huang H-J, Ramaswamy S, Al-Dajani W, Tschirner U, Cairncross RA. 2009. Effect of biomass species and plant size on cellulosic ethanol: A comparative process and economic analysis. *Biomass Bioenergy.* 33, 234–46.

Jenkins J, Chojnacky D, Heath L, Birdsey R. 2003. National scale biomass estimators for United States tree species. *For. Sci.* 49, 12–35.

Kaylen M, Van Dyne DL, Choi Y-S, Blase M. 2000. Economic feasibility of producing ethanol from lignocellulosic feedstocks. *Bioresour. Technol.* 72, 19–32.

Kempainen AJ, Shonnard DR. 2005. Comparative life cycle assessments for biomass-to-ethanol production from different regional feedstocks. *Biotechnol. Prog.* 21, 1075–84.

Luo L, Voet E, Huppes G, Udo de Haes HA. 2009. Allocation issues in LCA methodology: A case study of corn stover-based fuel ethanol. *Int. J. Life Cycle Assess.* 14, 529–39.

ORNL. 2011. *U.S. Billion-Ton Update: Biomass Supply for a Bioenergy and Bioproducts Industry*. Oak Ridge, TN: Oak Ridge National Laboratory.

PRé-Consultants. 2013a. *Ecoinvent Database Simapro LCA Software*. Amersfoort, The Netherlands: PRé-Consultants.

PRé-Consultants. 2013b. *US LCI Database Simapro LCA Software*. Amersfoort, The Netherlands: PRé-Consultants.

Singh A, Pant D, Korres NE, Nizami A-S, Prasad S, Murphy JD. 2010. Key issues in life cycle assessment of ethanol production from lignocellulosic biomass: Challenges and perspectives. *Bioresour. Technol.* 101, 5003–12.

Slade R, Bauen A, Shah N. 2009. The greenhouse gas emissions performance of cellulosic ethanol supply chains in Europe. *Biotechnol. Biofuels.* 2, 15.

Smith W, Miles P, Perry C, Pugh S. 2009. *Forest Resources of the United States, 2007: A Technical Document Supporting the Forest Service 2010 RPA Assessment.* Washington, DC: United States Forest Service.

Somerville C, Youngs H, Taylor C, Davis SC, Long SP. 2010. Feedstocks for lignocellulosic biofuels. *Science.* 329, 790–2.

Sun Y, Cheng J. 2002. Hydrolysis of lignocellulosic materials for ethanol production: A review. *Bioresour. Technol.* 83, 1–11.

TMS. 2013. *Southeastern Average Stumpage Prices Timber Mart South, Warnell School of Foresrt and Natural Resources at the University of Georgia.* Athens, GA: TMS.

US Congress. 2007. The Energy Independence and Security Act (EISA) of 2007.

Wang M. 2001. *Development and Use of GREET 1.6 Fuel-Cycle Model for Transportation Fuels and Vehicle Technologies.* Lemont, IL: Argonne National Laboratory.

Wang M, Huo H, Arora S. 2011. Methods of dealing with co-products of biofuels in life cycle analysis and consequent results within the U.S. context. *Energy Policy.* 39, 5726–36.

Wingren A, Galbe M, Zacchi G. 2003. Techno-economic evaluation of producing ethanol from softwood: Comparison of SSF and SHF and identification of bottlenecks. *Biotechnol. Prog.* 19, 1109–17.

Yin R, Pienaar L, Aronow M. 1998. The productivity and profitability of fiber farming. *J. For.* 96, 13–18.

CHAPTER 14

Policy Mechanisms to Implement and Support Biomass and Biofuel Projects in the United States

Pankaj Lal,[1] **Pralhad Burli,**[1] **and Janaki Alavalapati**[2]

[1]Department of Earth and Environmental Studies, Montclair State University, Montclair, NJ, USA

[2]Department of Forest Resources and Environmental Conservation,
 Virginia Polytechnic Institute and State University, Blacksburg, VA, USA

CONTENTS

14.1 INTRODUCTION

In recent years, bioenergy—energy produced using plants, trees, and other organic matter—has aroused interest, given its potential to bolster a country's energy security, manufacturing, job creation, and the possibility of reducing environmental impacts through greenhouse gas (GHG) reductions (Childs and Bradley 2007). Forest and agricultural biomass-based energy sources are being considered as an immediately feasible alternative for reducing GHG, especially carbon dioxide emissions from fossil fuel burning and land-use change (Abbasi and Abbasi 2010; U.S. Department of Energy [USDOE] 2011a; Gan and Cashore 2013). Government and private sector investment in biomass-based energy sources has surged in many areas of the world (Lal et al. 2013).

Governments worldwide are emphasizing development of alternative energy sources, and bioenergy has emerged as one of the options. Public policies such as renewable portfolio standards, subsidies, and import protection coupled with volatile oil prices are key factors that contribute toward this expansion (Taheripour and Tyner 2010). Furthermore, the physical and chemical properties of cellulosic biofuels require relatively limited modifications to engine technology and fueling infrastructure (Rajagopal et al. 2007). There are several provisions, incentives, and mandates in the United States—such as farm bills, 2005 Energy Policy, and the 2007 Energy Independence and Security Act (EISA)—that support and encourage the bioenergy industry. Some of the salient U.S. policy interventions include minimum renewable fuel usage requirements, tax credits on production and blending, import tariffs, loans, and research grants (Schnepf and Yacobucci 2013). The expansion of the bioenergy industry can be gauged by the fact that between 2000 and 2011, ethanol production in the United States rose from around 105,500 barrels per day to more than 908,600 barrels per day (Energy Information Administration [EIA], International Energy Statistics 2014a), an average annual growth rate in excess of 22% (see Figure 14.1). Meanwhile, biodiesel production rose from around 500 barrels per day in 2001 to more than 63,000 barrels per day in 2011. The wood pellet industry is also forecast to maintain an upward export trend (see Figure 14.2).

Public policy interventions target multiple beneficiaries, including farmers, biofuel producers, petroleum suppliers, small rural businesses, and fuel marketers (Yacobucci 2012). Besides producing and using electricity and thermal heating from biomass sources, there is an increasing trend of exporting densified wood fiber in the form of pellets and wood briquettes to Europe and emerging Asian economies. Projected European biomass demand over the next 20 years, due to a 20% renewable portfolio goal of European Union (EU) nations, ranges from 35 to 315 million tons, with imports estimated to account for 16–60 million tons of this volume (Joudrey et al. 2012).

The development of bioenergy has not been without controversy, and there are viewpoints that argue that bioenergy policies cause price distortion and favor the bioenergy industry at a cost to traditional industries facing higher biomass prices. Studies also suggest that rapid expansion of the bioenergy industry in the United States can have wide ranging effects on agricultural land-use changes, forest landscapes, and ecological systems (Searchinger et al. 2008; Kim et al. 2009;

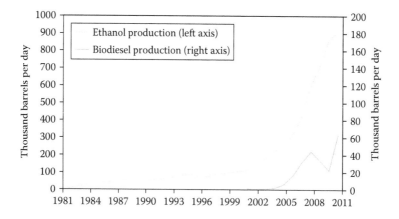

Figure 14.1 Biofuel production in the United States. [From EIA, International Energy Statistics. (2014). Retrieved from http://www.eia.gov/cfapps/ipdbproject/IEDIndex3.cfm?tid = 79&pid = 79&aid = 1 (Accessed February 15, 2014).]

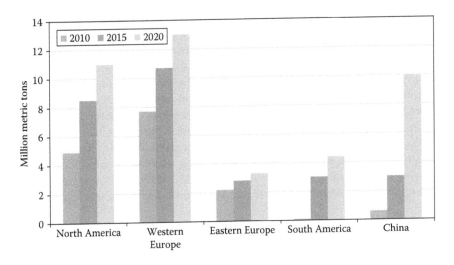

Figure 14.2   Global wood pellet production outlook. (Modified from Site Selection Magazine, July 2013 World Reports. Available at http://www.siteselection.com/issues/2013/jul/world-reports.cfm.)

Fargione et al. 2010; Havlík et al. 2011). First-generation biofuels could potentially result in food shortages arising from diversion of food crops to biofuel production (Pimentel 2009). In addition, the competition for land between food crops and biofuel crops poses a host of challenges (Doornbosch and Steenblik 2007). Fargione et al. (2008) contend that whether biofuels will result in net carbon savings depends largely on the methods and means adopted in production.

Against this backdrop, the focus has shifted to the development of cellulosic biofuels—biofuels manufactured from nonfood organic matter, including forest biomass, woody crops, corn stover, and energy grasses, among others. The organic inputs required for the production of cellulosic biofuels are potentially low-cost, diverse, and relatively abundant (Bracmort et al. 2011). They are considered less land and water intensive and could make a significant contribution to the renewable energy supply (Carriquiry et al. 2011).

In this chapter, we showcase leading government programs and analyze them based on intensive literature review. The subsequent sections of this chapter are organized as follows. First, we provide a historical perspective of U.S. bioenergy policies. We then turn to discussing government policy interventions over the past decade with a particular focus on production targets, financial assistance to biomass producers, import tariffs, fiscal incentives, and research and development support, both in the United States and globally. In the next section, we discuss voluntary mechanisms to ensure sustainable bioenergy production through certifications and best management practices (BMPs). Next, we discuss a few case studies that highlight some of the successes and shortcomings of bioenergy development. Lastly, we discuss and summarize observations and provide perspectives on the future policy mechanisms pertaining to prospective development of bioenergy markets.

14.2 BIOENERGY POLICIES—PAST, PRESENT, AND FUTURE

As per the EIA (2014b), in 1862, the U.S. Congress imposed a $2 per gallon excise tax on ethanol to pay for the Civil War. Ethanol, the major illuminating oil in the years prior to the Civil War, was too expensive to be used this way thereafter. In 1906, more than 50 years after the imposition, Congress repealed the excise tax, making ethanol an alternative transportation fuel. An upturn in wartime demand for fuel saw increased production of ethanol in the early 1940s, but it waned soon after. One of the early instances of federal intervention to promote ethanol as a fuel came as early as

1974 in the form of the Solar Energy Research, Development, and Demonstration Act. Thus began the research and development support to convert cellulosic and organic inputs into energy fuels.

The Energy Tax Act of 1978 introduced "gasohol," a blend of gasoline with at least 10% alcohol, excluding alcohol made from petroleum, natural gas, or coal. This bill resulted in a $0.40 per gallon subsidy for every gallon of ethanol blended into gasoline and led to the use of bioethanol as the blended fuel of choice. Since then, numerous incentives and mandates have been targeted toward biofuel producers and blenders in order to promote the growth of this industry. The Energy Security Act of 1980 offered loans of up to $1 million to small ethanol producers (less than one million gallons per year) to cover up to 90% of construction costs for ethanol production plants. By 1984, the ethanol subsidy had increased to $0.60 per gallon and has gradually fallen since. Meanwhile, technology has advanced, vehicle fleets capable of operating with an ethanol blend of 85% (E85) have risen, and fueling stations have been set up across the country—all with the aim of increasing production and consumption of biofuels.

14.2.1 Bioenergy Policies—The Last Decade

In order to provide a shot in the arm to bioenergy markets, federal and state governments in the United States have used diverse policy prescriptions. Sometimes mandates are more likely than incentives to achieve desired results. In other cases, financial assistance tends to influence behavior and encourages large-scale adoption of technology. From a trade perspective, tariffs and quotas can be used to accomplish desired results, but one has to account for ensuing distributional impacts. The next few sections look at the different public policy mechanisms in the United States and around the world.

14.2.1.1 Production Targets

In recent years, the key federal policy pertaining to bioenergy has been the promulgation of renewable portfolio standards, under which the government stipulates that a minimum portion of energy should come from renewable sources such as biomass, wind, solar, and geothermal (Lal et al. 2013). Similarly, renewable fuel standards (RFSs) pertain to mandates concerning renewable transportation fuel volume. The U.S. Congress introduced the RFS for the first time with the enactment of the Energy Policy Act of 2005. Referred to as RFS1, it mandated a minimum of four billion gallons of biofuel production in 2006, increasing to 7.5 billion gallons in 2012. However, in 2007, the EISA expanded the requirements under the revised RFS to 36 billion gallons by 2022. The proportion of corn-based ethanol was capped at 15 billion gallons, with cellulosic ethanol and other advanced biofuels making up the remaining 21 billion gallons (see Figure 14.3).

RFS2 also introduced minimum targets for lifecycle GHG emission reductions, and land-use restrictions were outlined. Sorda et al. (2010) contend that the RFS policy not only incentivized the production of second-generation biofuels, but also emphasized that they are produced in a sustainable manner.

The U.S. Environmental Protection Agency (USEPA) revises its biofuel production mandates based on waiver requests, bearing in mind the practicality and the industry potential to meet those requirements. The underlying guiding principle is to reduce the country's dependence on imported petroleum and to ensure a steady path for the development of the biofuels industry in the United States (EPA 2010). Revisions to the RFS also recognized challenges arising from blending limitations often termed as the "blend wall." The "blend wall," in effect, sets a limit on the consumption of ethanol due to restrictions arising from demand for transportation fuel as E10 (i.e., gasoline containing 10% ethanol). If actual gasoline demand fell short of the initial 2007 projections, the targets for ethanol consumption were revised. The most significant revision was made to the cellulosic biofuel standard, wherein the limit was lowered drastically in November 2013 from 1.75 billion gallons to 17 million gallons (EPA 2013).

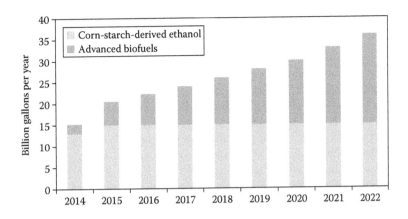

Figure 14.3 RFS2 renewable fuel requirements. [From EPA. (2010). Regulatory announcement #EPA-420-F-10-007 issued by the Office of Transport and Air Quality. Retrieved from http://www.epa.gov/otaq/renewablefuels/420f10007.pdf (Accessed February 20, 2014); and EPA. (2013). Regulatory announcement #EPA-420-F-13-048 issued by the Office of Transport and Air Quality. Retrieved from http://www.epa.gov/otaq/fuels/renewablefuels/documents/420f13048.pdf (Accessed February 20, 2014).]

PRODUCTION MANDATES FOR BIOFUELS BEYOND THE UNITED STATES

The EU has a 5.75% mandate for renewable fuels. Although this mandate was expected to increase to 10% by 2020, the European Parliament voted to raise the cap to just 6%. It is anticipated that around 70% of this mandate would be met through biofuels. Brazil has one of the highest mandates for ethanol content, with a minimum blend of 18%. The Brazilian government has lowered its mandate from 25% owing to rising global prices of sugar. It is projected that biofuel demand in Brazil could reach eight billion gallons by 2022. China is targeting a 10% biofuels mandate by 2020, primarily through ethanol blending. This mandate would translate into an estimated biofuel demand of approximately 2–3 billion gallons. India has an E5 mandate that translates to a 20% blending target for both biodiesel and bioethanol by 2017. It is highly unlikely that this target will be achieved, given current production levels and capacity.

Source: Biofuels Digest. (2011a). Biofuels mandates around the world. Retrieved from http://www.biofuelsdigest.com/bdigest/2011/07/21/biofuels-mandates-around-the-world/ (Accessed March 10, 2014); *Biofuels Digest.* (2013). Biofuels mandates around the world 2014. Retrieved from http://www.biofuelsdigest.com/bdigest/2013/12/31/biofuels-mandates-around-the-world-2014/ (Accessed March 10, 2014).

14.2.1.2 Assistance to Biomass Producers

In the Food, Conservation, and Energy Act of 2008, the U.S. government announced the Biomass Crop Assistance Program (BCAP) with the goal of improving the country's energy security while supporting the agrarian economy and spurring job growth. According to the U.S. Department of Agriculture (USDA), under the BCAP the federal government "provides financial assistance to owners and operators of agricultural and non-industrial private forest land who wish to establish, produce, and deliver biomass feedstocks." By reducing uncertainty and associated financial risks, the BCAP supports landowners who decide to grow unconventional energy crops, including switchgrass, miscanthus, fast-growing woody poplar, algae, and energy cane, among others.

THE BIOMASS CROP ASSISTANCE PROGRAM IS NOT ALL GOOD!

The Wood Fiber Coalition (2012) in its statement to the House Agriculture Subcommittee on Conservation, Energy, and Forestry noted that the BCAP program negatively impacted several longstanding industries. In particular, the clause related to matching payments resulted in the diversion of wood by-products including woodchips, bark, and sawdust from traditional markets and diverted the biomass toward bioenergy industry. They highlighted that "there are at least 14 federal definitions of renewable biomass for renewable energy programs, most of which classify the raw materials used to manufacture our products as waste materials" indicating that a common definition must be agreed upon for the benefit of all industry participants. Subsequently, the farm bill was amended and clarified such that "any woody eligible material collected or harvested outside contract acreage that would otherwise be used for existing market products" was deemed ineligible for BCAP subsidies.

Selected crop producers are eligible for reimbursements of up to 75% of the cost of establishing a perennial bioenergy crop. Producers can receive up to five years of annual payments for nonwoody crops and up to 15 years of annual payments for woody crops. In addition, the BCAP provides a matching payment of up to $45 per ton to producers for collection, harvest, storage, and transportation of their crops to qualified conversion facilities. These facilities are not restricted to biofuel production but also include conversion facilities producing heat, power, and other biobased products. Furthermore, producers who enter into contracts with the Commodity Credit Corporation to produce eligible biomass crops could be eligible to receive annual payments for cultivation on contract acres within BCAP project areas.

It was anticipated that the establishment of the BCAP program would support the country's energy and environmental strategy as well as bring rich dividends to the economy. The USDA Farm Service Agency (2013) report states that under the BCAP, county offices have enrolled more than 50,000 acres dedicated to the production of nonfood bioenergy crops. In his testimony to the U.S. Senate Committee on Agriculture on behalf of the National Farmers Union (NFU 2011), Steve Flick, the founder of Show Me Energy Cooperative, highlighted that under the Show Me Energy BCAP project, 39 counties spread across western Missouri and eastern Kansas would grow native grasses and polycultures, leading to the creation of hundreds of direct jobs and thousands of indirect jobs.

14.2.1.3 Import Tariffs

In the early 1980s, the U.S. government imposed an import tariff on foreign-produced ethanol, primarily to restrict competition from the Brazilian ethanol industry. Elliott (2013) states that many countries impose trade restrictions on feedstocks and biofuel, often to protect domestic producers but also to offer preferential tariffs or quotas to some countries. Ethanol producers in the United States enjoyed such protectionist policies for more than three decades. The $0.54 per gallon tax on imported ethanol was scrapped in December 2011, opening up the market to overseas producers. The sugarcane industry association in Brazil welcomed the move, claiming that it was the first time ever that the world's top two ethanol producers were not imposing an import tariff. Most Brazilian ethanol is produced from sugarcane. Subsequent to the tariff repeal, ethanol imports from Brazil rose sharply in 2012 (see Figure 14.4).

Import tariffs as public policy interventions have been used in order to protect the domestic biofuels industry in many countries (Sorda et al. 2010). The EU tariff on denatured ethanol (ethanol unsuitable for human consumption) is 102 Euros per thousand liters; the tariff is 192 Euros for

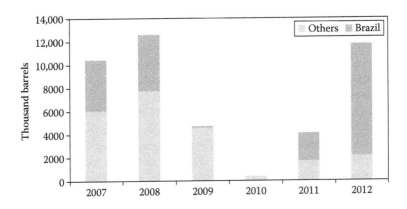

Figure 14.4 U.S. fuel ethanol imports. [From EIA, International Energy Statistics. (2014c). Retrieved from http://www.eia.gov/dnav/pet/pet_move_impcus_a2_nus_epooxe_im0_mbbl_a.htm (Accessed February 15, 2014).]

IS BIOFUEL PRODUCTION CAPACITY HIGHER IN U.S. STATES WITH MORE POLICY INCENTIVES?

There are approximately 139 policy incentives targeted toward the biodiesel industry, and the combined production capacity across all the states is estimated at 2236 million barrels per day. However, when we estimate correlation between number of policies providing some kind of impetus to bioenergy production and operation capacity, the correlation coefficient is a paltry 0.24. Similarly, the combined ethanol production capacity across all U.S. states is about 14,875 million barrels per day, and there are 136 policies supporting the ethanol industry. In this case, too, the estimated correlation coefficient is just 0.21. Furthermore, both these correlation coefficients are statistically insignificant, suggesting that biofuel production capacity appears to be unrelated to the number of policy incentives being offered.

Note: Data for performing this analysis was collated from Database of State Incentives for Renewables & Efficiency retrieved from http://www.dsireusa.org/ (Accessed February 17, 2014).

undenatured ethanol. For biodiesel, the EU imposes a tariff of 6.5% on imports and an additional antidumping or countervailing duty on some U.S. companies and most Canadian ones (Flach et al. 2013). In Brazil, a zero import tariff is applicable to all ethanol containing less than 1% water, a policy that is likely to prevail until December 2015 (Barros 2013). On the other hand, the import tariff on biodiesel is set at 14%. China imposes an import tariff on denatured ethanol at 5% and on undenatured ethanol at 40%. In addition, a 17% value-added tax and a 5% consumption tax are applied to both denatured and undenatured ethanol (Scott and Junyang 2012). India imposes a 7.55% tariff on denatured ethyl alcohol and spirits including ethanol. However, ethanol imports from Brazil are levied at a preferential tariff rate of 6% (Aradhey 2013).

Reduction in trade barriers tends to increase competition, leading to higher efficiency and lower prices to consumers. However, in an increasingly globalized and interconnected economy, policy interventions in one part of the world can have adverse impacts elsewhere. Kojima et al. (2007) observed that trade barriers may enhance overall welfare in cases where biofuels are produced at the cost of destroying rain forests. On the other hand, biofuel policies in the EU are sometimes blamed for deforestation in developing countries such as Indonesia (Blonz et al. 2008).

14.2.1.4 Fiscal Incentives

Since the early 1980s, the U.S. government has adopted a slew of policies that encouraged and incentivized the development of the biofuels industry. Under the Volumetric Ethanol Excise Tax Credit (VEETC), the U.S. government provided tax incentives to eligible ethanol blenders that would apply against their federal excise tax obligation. At its peak, this subsidy amounted to $0.60 per gallon in 1984, but it was subsequently reduced to $0.54 in the 1990 Omnibus Reconciliation Act and further to $0.45 per gallon under the 2008 Farm Bill. Despite the reductions in per gallon tax credit, the U.S. ethanol industry expanded sharply during this period. Even after several decades of government support, it was noted that in May 2010, the plant-gate ethanol prices in Iowa were nearly $0.50 lower than Brazilian ethanol prices. The VEETC was discontinued in December 2011. According to Miller and Coble (2011), the VEETC had relatively small impact on the biofuel industry until 2005, when the price of petroleum-based fuels increased along with the increased use of ethanol as a gasoline oxygenate. However, by the time the VEETC was repealed, the U.S. ethanol industry—corn-based ethanol in particular—had matured and was expected to grow further, albeit at a slightly lower rate (Natural Resources Defense Council 2010).

Along with the VEETC, the Small Ethanol Producer Tax Credit and the Small Agri-Biodiesel Producer Tax Credit, providing $0.10 per gallon tax incentives for the first 15 million gallons of fuel produced during a tax year, were also discontinued on December 31, 2011. More recently, the Alternative Fuel Mixture Excise Tax Credit, a $0.50 per gallon tax incentive, and the Biodiesel Mixture Excise Tax Credit, a $1.00 per gallon tax incentive, expired on December 31, 2013. Table 14.1 provides an aggregated regional level snapshot of a number of state incentives and regulations and the production capacity of biodiesel and ethanol. Although aggregating, we do not assess effectiveness of the policy incentives or regulations at the microlevel; rather, the focus is on the number of public policy options that have been promulgated and on associated production capacity. The table shows that the number of policy incentives and regulations do not show any discernible pattern across regions.

Other countries have used fiscal incentives to spur their renewable energy portfolios as well. The Brazilian government operates a preferential tax regime for flex-fuel vehicles, provides tax exemptions for sugarcane supplied for ethanol production, and exempts biodiesel and ethanol sales from the industrialized products tax (Barros 2013). The government also provides loans at low interest rates and extends credit for ethanol storage. Following the success of the ethanol industry, the government is focusing its attention on biodiesel production through credit extension and regional tax exemptions. China, on the other hand, does not provide any direct subsidies for biodiesel production but exempts 5% consumption tax on biodiesel produced from waste cooking oil (Scott and Junyang 2012). The Indian government uses a number of fiscal instruments to incentivize biofuel production (Raju et al. 2012). These include providing loans at subsidized rates—up to a maximum of 40% of project

Table 14.1   Aggregate Regional Level Incentives and Regulations Along with the Production Capacity for Biodiesel and Ethanol

Region	Biodiesel Production Capacity (MGY)	Ethanol Production Capacity (MGY)	Number of Biodiesel Incentives	Number of Biodiesel Regulations	Number of Ethanol Incentives	Number of Ethanol Regulations
Northeast	125	274	13	30	11	27
Midwest	1055	12,832	42	56	48	59
South	853	836	54	64	52	59
West	202	659	30	65	25	58

Source: EIA (2014d); Government of Nebraska. (2014). Retrieved from http://www.neo.ne.gov/statshtml/121.htm (Accessed February 15, 2014); and Alternative Fuels Data Center. (2014). Retrieved from http://www.afdc. energy. gov/laws/matrix/incentive (Accessed February 15, 2014).

costs—to sugar mills setting up ethanol production facilities. Aradhey (2013) suggests that a proposal to categorize biofuels under the category of "declared goods" is under active consideration in order to facilitate unrestricted interstate movement and exemption from all central government taxes and duties, barring a concessional excise tax of 16% on bioethanol. Another policy incentive provided is a minimum support price (MSP) provided to nonfood oil seeds to support biofuel feedstock producers. The MSP as the price guarantee for producers is revised periodically by the government.

14.2.1.5 Research and Development Support

The U.S. government provides funds to institutions of higher learning, national laboratories, federal research agencies, private sector organizations, and nonprofits to develop technologies in the biomass-based industry and to increase understanding of the socioeconomic and environmental implications of biofuel use. The Biomass Research and Development Initiative (BRDI) is one such program run jointly by the USDA's National Institute of Food and Agriculture and the USDOE Office of Biomass Programs. The BRDI program focuses on addressing the practical and technological aspects of bioenergy development. The Biorefinery Assistance Program gives loan guarantees for the development, construction, and modernization of biorefineries producing advanced biofuels. Under this program, a loan guarantee up to $250 million and grant funding up to 50% of project costs is permissible. In addition, monies from the Advanced Research Projects Agency and the Surface Transportation Research, Development, and Deployment Program support projects that develop technologies and innovative transportation infrastructure.

Countries elsewhere are emphasizing bioenergy research and development as well. China is investing in research facilities to develop technology that will help produce ethanol from sweet sorghum and nongrain feedstocks. Research priorities include development of procedures for collection, handling, and transportation of feedstocks with a focus on developing feedstock production facilities on marginal land (Scott and Junyang 2012). The Indian government provides grants for the development of research facilities and advanced technologies for conversion processes (Aradhey 2013).

14.3 CERTIFICATION AND BMPs

Buyers in Europe, the EU, and regulators in importing countries are increasingly demanding sustainability and emissions accounting of sourced biomass used for pellet production. With this backdrop, voluntary efforts such as certification programs, guidelines for biomass harvest, and BMPs are gaining traction as instruments for ensuring sustainability of bioenergy products and processes. This subsection briefly outlines sustainability initiatives and associated complexities pertaining to bioenergy policy development.

Regulatory approaches that include procedural rules, legislatively prescribed practices, reporting, monitoring, compliance, and enforcement are useful tools in ensuring sustainability (Ellefson et al. 2004; Lal et al. 2013). Equally important in this context are nonregulatory schemes such as extension education, information sharing, technical assistance, and other financial incentives. One such example in the woody biomass context is the SHARP Logger program in the state of Virginia, which utilizes continuing education courses from Virginia Tech to train loggers in chipping and woody biomass harvesting techniques. The result is vastly improved practices for biomass harvesting in integrated logging operations in the region (Barrett et al. 2012). Most states in the United States rely on a framework of agriculture, water, and forestry guidelines and BMPs focused mainly on a variety of issues (e.g., water quality, pest management) that are bound together by voluntary programs focused on outreach to landowners, land managers, farmers, and loggers. State level laws are often supplemented by local ordinances that offer a further degree of control over feedstock cultivation activities (Kittler et al. 2012).

TRANSACTION COSTS OF ENSURING SUSTAINABILITY: FORESTRY BEST MANAGEMENT PRACTICES AND BIOMASS HARVESTING GUIDELINES

Loggers, often operating on tight margins, have expressed negative perceptions of BMPs and biomass harvesting guidelines, mostly due to what they consider unnecessary cost burdens (Fielding 2011). A study of harvests in Georgia, Florida, and Alabama found the aggregate marginal cost of implementing state BMPs on approximately 4000 acres of forestland to be around $50,000 or 2.9% of gross harvest revenue (Lickwar et al. 1992). The study estimated the cost of BMPs implementation ranges between $4.50 and $25 per acre with a mean value of $12.45 per acre. Cubbage (2004), in a field study, estimated a much higher cost of BMP implementation, ranging from $8 to $29 per acre with a mean of $19 per acre. Voluntary biomass harvesting guidelines implementation costs are not yet available because they have been developed very recently (Fielding 2011).

One of the key voluntary approaches for sustainability compliance is a certification system. Certification, a system of standards, rules, and procedures for assessing conformity with specific requirements, generally contains feedstock management certification standards (e.g., principles, criteria, and indicators, or similar nomenclature) and outlines the procedures by which a third party evaluates performance against the system's biomass management standards. This requires an accreditation process through which third-party auditors are deemed competent to carry out certification audits and procedures to enforce rules for entities making claims about their environmental performance relative to a particular standard (Kittler et al. 2012).

The existing programs deal with certifying feedstock management and production systems (e.g., forest certification program for sustainable forest management and forest products) rather than biomass-based energy products. For example, the Forest Stewardship Council (FSC) management standards for the United States do not specifically discuss biomass harvests as a particular type of removal because the FSC feels that the key environmental considerations associated with biomass are already addressed by their standard. For example, principle 6 addresses the environmental impacts of harvesting operations, and indicator 6.3.f requires that "management maintains, enhances, or restores habitat components and associated stand structures, in abundance and distribution that could be expected from naturally occurring processes," which includes "live trees with decay or declining health, snags, and well-distributed coarse down and dead woody material" (Evans et al. 2010, p. 21). On the other hand, existing certification standards can be challenging for intensively managed forests geared toward biomass production following a "grow and harvest principle." Certification can be a costly proposition for small landowners or land managers. Identifying this constraint, certifiers such as the FSC do allow for the grouping of several small parcels that can help spread the costs of certification and auditing across more acres, potentially increasing access to certified fiber (Bowyer et al. 2011).

There are a number of certification programs and guidelines that exist in the United States and around the globe; however, there is wide variation in terms of what is acceptable and what is not. For example, the FSC bans the use of genetically modified organisms, but their use is allowed under other forest certification standards. Although forest and agricultural biomass can be sourced from the same forest or farm, the standards for farm product and forest management systems might not necessarily suffice for bioenergy production (RSB 2014). The impacts of bioenergy feedstock production depend on several factors, including previous land-use and agricultural practices employed. A credible assurance mechanism of sustainable sourcing of agricultural and wood fiber can act as an important first step in the biomass and bioenergy policy sphere.

14.4 CASE STUDIES

In this section, we discuss selected case studies highlighting successes and failures in the U.S. bioenergy sector. We have chosen cases that encompass starch and cellulosic ethanol and biodiesel production, as well as biomass pellet production for heat and electricity generation.

14.4.1 Case Study 1: Corn Ethanol and Biodiesel Production in Iowa

As of December 2013, the state of Iowa was the largest producer of ethanol and the second largest producer of biodiesel in the United States. Iowa produced more than 3.6 billion gallons of ethanol in 2013, almost 27% of the country's ethanol output. According to a recent report entitled "Contribution of the Renewable Fuels Industry to the Economy of Iowa," published by Agriculture and Biofuels Consulting LLP (Urbanchuk 2014), the renewable fuels industry accounted for nearly $5.6 billion, about 4% of Iowa's gross domestic product (GDP), and supported more than 62,000 jobs. The state government has initiated several incentives to promote the biofuels industry, such as an "Ethanol Promotion Tax Credit" up to $0.08 per gallon for pure ethanol blended into gasoline. The state grants tax credits to retail stations dispensing fuel blends of ethanol and biodiesel blends. According to the Iowa Department of Revenue, E85 sales set a new record in 2013, growing by nearly 20% from 2012 levels. The state also has provisions for infrastructure cost-share grants under the Renewable Fuel Infrastructure Program. According to this program, 70% of the total cost of the project, not exceeding $50,000, can be provided to install new E85 or biodiesel infrastructure. A similar grant is also available to upgrade or replace existing infrastructure.

Urbanchuk (2014) highlighted the fact that the state was able to produce commercial-scale, second-generation feedstock ethanol for the first time in 2013. He contends that the state benefited from the installation of additional production capacity, higher corn output, and a consequent decline in feedstock prices, and several policy interventions. Identifying benefits of public policy supports, the Iowa Renewable Fuels Association (IRFA) has been a strong advocate of maintaining high targets for the RFS, often suggesting that the EPA should not reverse course on the initial level targets. An IRFA press release dated February 25, 2014, claims that the USDA Preliminary 2012 Farm Census shows that the RFS has revitalized Iowa's rural economy (IRFA 2014). The press release highlights the fact that Iowa has witnessed an increase in the value of farm products and crop values to the tune of almost 51% and 67%, respectively. Meanwhile, the state has announced a new program called "Fueling Our Future" whereby federal funds will be directed to improve the state's renewable fuel infrastructure and to establish more blender pumps that supply E30 and biodiesel. Thus far, the policy interventions and their focus on developing Iowa's biofuel industry have yielded rich dividends in the form of growth in the agricultural and allied sectors.

14.4.2 Case Study 2: Biomass Pellets in Georgia

Crop residues are the largest source of biomass in the United States; however, the availability of crop-related biomass in Georgia is below the national average (Noonan 2012). On the other hand, Georgia ranks first and second in the biomass categories of primary mill and forest residues. According to Milbrandt (2005), the state had an estimated 7.23 million dry tons in primary mill residues and 3.62 million dry tons in forest residues. According to Biomass Magazine (2014), the state biomass pellet plants have a current operating capacity of 1.67 million tons (see Figure 14.5).

Additionally, a pellet plant with a capacity of 0.55 million tons is under construction and several other plants with a combined capacity of 1.71 million tons have been proposed. Noonan (2012) claimed that state economic goals have placed a huge emphasis on the development of biomass production facilities. Moreover, according to a press release by Noonan (2012), "the state has successfully attracted new bioenergy industry despite a sluggish economic environment."

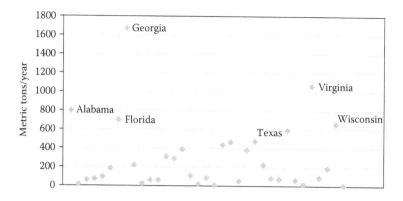

Figure 14.5 Pellet plant capacity in different states of the United States. [From Biomass Magazine. (2014). Location of US Pellet plants and production capacity. Retrieved from http://biomassmagazine. com/plants/listplants/pellet/US/Operational/page:1/sort:capacity/direction:desc (Accessed March 22, 2014).]

The state of Georgia has been able to provide an enabling environment for the biomass industry by encouraging research at universities and by addressing the needs of feedstock producers and corporations. This was evidenced by the fact that, in April 2006, the state government enacted a legislation exempting biomass materials from the state's sales and use taxes. According to the legislation, Georgia Code Title 48. Revenue And Taxation, Chapter 8. Sales And Use Taxes, Article 1 State Sales And Use Tax, Part 1. General Provisions O.C.G.A. § 48-8-3, "Biomass material [would include] organic matter, excluding fossil fuels, including agricultural crops, plants, trees, wood, wood wastes and residues, sawmill waste, sawdust, wood chips, bark chips, and forest thinning, harvesting, or clearing residues; wood waste from pallets or other wood demolition debris; peanut shells; pecan shells; cotton plants; corn stalks; and plant matter, including aquatic plants, grasses, stalks, vegetation, and residues, including hulls, shells, or cellulose-containing fibers." Georgia is one of the very few states that has successfully developed policy instruments uniquely targeting woody biomass for energy (Aguilar and Saunders 2010).

14.4.3 Case Study 3: Cellulosic Ethanol Production in Florida

In July 2013, Florida became the first state in the United States to produce commercial-scale cellulosic ethanol (USDOE 2013). Developed through a joint venture between INEOS Bio and New Planet Energy, the Indian River County Bio Energy Center is expected to produce an annual output of eight million gallons of cellulosic ethanol using a combination of vegetative wastes, agricultural wastes, and municipal solid wastes. However, the cellulosic ethanol industry operates on a very limited scale because there are several challenges that must be overcome before wider commercialization and use is achieved. Bracmort et al. (2011) highlighted three focus areas: feedstock availability at volumes needed by refineries; advances in conversion technology to make costs competitive; and improvements in distribution and vehicle infrastructure in order to absorb increases in ethanol output. The state of Florida provides an income tax credit of 75% of all capital, operation, maintenance, and research and development costs incurred in connection with an investment in the production, storage, and distribution of biodiesel, ethanol, or other renewable fuel in the state, up to predefined limits. These tax credits can be availed on costs incurred between July 1, 2012, and June 30, 2016. Additionally, if the credit is not fully used in any one year, the unused amount may be carried forward through December 31, 2018 (USDOE 2014). With production of corn-based ethanol at near peak levels and the mandated cap at 15 billion gallons by 2015, development of cellulosic ethanol is expected to attract a disproportionate amount of attention, incentives, and policy initiatives in the years to come.

14.4.4 Case Study 4: Range Fuels Ethanol Facility in Georgia

Range Fuels' plant was expected to be the first commercial-scale biorefinery in the United States. It was anticipated that the biorefinery would capitalize on the abundant supply of woody biomass in the vicinity of Soperton, Georgia, to produce cellulosic ethanol. The refinery's production capacity was pegged at around 20 million gallons of biofuels each year, and it was expected to create around 70 direct full-time jobs and around 250 jobs during the construction phase (USDOE 2011b). Range Fuels received a construction loan to the tune of $80 million from AgSouth Farm Credit Bank and also availed itself of the first ever biofuel loan guarantee from the USDA, a total of $64 million in addition to a $76 million USDOE grant (Biofuels Digest 2011b). The biorefinery was expected to produce ethanol starting in 2008, but it was not until 2010 that the USEPA announced that the plant would finally produce some fuel. Ironically, the output was a mere four million gallons of methanol, a fuel that is not categorized as renewable biofuel. Eventually, Range Fuels defaulted on its loan in January 2011 and was foreclosed on, leading to much criticism and debate on whether taxpayers' money was wasted and whether the government should engage in funding high-risk businesses. The Soperton facility was sold to LanzaTech in 2012, and the company is trying to produce ethanol using a different technology.

14.5 DISCUSSION AND CONCLUSIONS

These four case studies provide insights into the development of the bioenergy industry in the United States. It is clear that this industry, especially that of advanced biofuels, is still at a very nascent stage of development, and there are several challenges that must be mitigated before large-scale commercialization is possible. The process is likely to be slow given the inherent risks associated with working with new technologies. It is extremely difficult to conceive a one-size-fits-all policy strategy, given that experiences differ depending upon location, nature of technology used, sourcing of feedstock inputs, and so forth. Thus, state governments will likely have to tailor their policy interventions with respect to regional resource endowments, technological capacity, infrastructure, and economic priorities. It is evident that when these policies have been well structured, local economies have benefited through job creation and income generation. Technological advancements are critical for this industry to become competitive, as is government support though an increased focus on research and development.

Government policies and growing demand for alternative energy are expected to result in dramatic increases in bioenergy production in the United States. Although there is a substantive body of research assessing the potential impacts of bioenergy-driven land-use change on issues such as GHG and regional water quality, surprisingly little attention has been given to the market distortions associated with public policy interventions that favor large-scale bioenergy production. Incorporating these public policy implications into future bioenergy programs and projects would help ensure the sustainability of biomass sourcing with minimal adverse impacts. It is hoped that some uncertainty will be resolved once the European Commission (EC) finalizes recommendations for EU-level sustainability criteria and the ISO standards for bioenergy sustainability certification are established. Because many states in the United States are promoting wood and agricultural biomass-based energy, new cofiring and combined heat and power plants as well as ethanol biorefineries are likely to be established in the future. Expansion of this sector—more agriculture or woody biomass-based energy plants or expansion of existing facilities to achieve economies of scale—will be associated with an increase in the demand for energy fiber.

The BMPs and harvesting guidelines developed especially for bioenergy could reduce harvesting impacts through minimum tillage and reduced applications of fertilizers and pesticides; protect wildlife corridors, riparian zones, and other sensitive areas; adopt wildlife habitat enhancement measures such as leaving patches of undisturbed areas, promoting certain species mixtures, and crop rotations; and retaining some harvest residue, litter, deadwood, snags, and den trees (Fletcher et al. 2009).

Potential impacts on agriculture and forest ecosystems at local and regional levels are most likely to challenge the bioenergy feedstock producing community to update existing certification systems with guidelines on how, when, and where woody biomass removals should be conducted, as has been propounded by the following authors:

- Dale et al. (2012) focus on invasiveness, water use, and social acceptance to address the concerns for sustainability. Specifically, they discuss the opportunities and constraints of sustainable production of grass, eucalyptus, and other feedstocks in the United States by focusing on the biofuel life cycle's impacts on sustainability.
- Janowiak and Webster (2010) provide a framework that includes adapting management to site conditions, increasing forested land where feasible, using biomass harvests as a restoration tool, evaluating the possibility of fertilization and wood ash recycling, and retaining deadwood and structural heterogeneity for biodiversity.
- Lal et al. (2011) report a set of nine criteria that are necessary to the pursuit of sustainable woody biomass extraction: (i) reforestation and productive capacity, (ii) land-use change, (iii) biodiversity conservation, (iv) soil quality and erosion prevention, (v) hydrologic processes, (vi) profitability, (vii) community benefits, (viii) stakeholder participation, and (ix) community and human rights.

Certification programs and biomass harvesting guidelines are intended to improve the sustainability of biomass procurement in ways that are economically viable to bioenergy developers as well as accessible to many mid- and small-scale forest owners (Kittler et al. 2012). U.S. foresters have almost 20 years of experience with sustainable forest management (SFM) certifications, which are widely looked to as a potential basis for the emerging bioenergy certification programs in the United States (Gan and Cashore 2013). However, SFM programs do not yet have a stand-alone certification for bioenergy production. Because the main export market in the EU is expected to demand more stringent regulations in terms of addressing sustainability concerns regarding biomass feedstock sources for pellet or chip production, there is currently a clear incentive for producers to work with a broad range of stakeholders to develop certification programs that are protective of natural resources and biodiversity. Realizing this opportunity, multistakeholder efforts such as the Roundtable on Sustainable Biomaterials (RSB) and the Global Bioenergy Partnership are already underway. Both are in the process of developing principles and criteria for a set of global, science-based indicators coupled with field examples and best practices (including benchmarks for GHG reduction potential) to ensure bioenergy sustainability.

In addition to the overall scale of biomass production, the location and methods of feedstock production and biomass harvest can affect the health, vitality, and ecological functions of forests and agricultural landscapes. Modified certification programs can play a pivotal role in safeguarding the ecological integrity of these farmlands and forests. Certification can act as an important communication tool to educate customers about a supplier's commitment to sustainable coexistence through a process of verification and auditing; and it can offer some assurance that a supplier has adopted a reasonable level of best practices in environmental management and that their growth pursuits are not at the cost of the environment. Existing certification systems such as the FSC, American Tree Farm System, Sustainable Forestry Initiative, and chain-of-custody have criteria and indicators to safeguard site productivity, water quality, and biodiversity, but some additional indicators may be required for agricultural and woody biomass harvests. For example, an indicator might be needed to address forest harvest residues left on site to maintain habitat for small mammals, insects, reptiles, and amphibians.

14.5.1 The Road Ahead

Bioenergy markets are envisaged as an integral part of the solution toward a renewable energy future for the United States. The sharp bioenergy production increase observed during the last decade has been accompanied by a simultaneous increase in government mandates and other policy instruments. Coexistence of many policies makes discerning individual policy impacts extremely difficult. Yet it is increasingly evident that bioenergy production is impacting farm incomes and boosting rural economies through higher economic activity. Bioenergy markets tend to differ not only according to the type of bioenergy feedstock but also according to cultivation, harvesting, conversion, and distribution pathways. This diversity suggests that uniform public policy toward bioenergy development may not necessarily yield desired outcomes. Moreover, some policy initiatives have resulted in unintended consequences, often distorting markets. For example, incentives and protectionist policies have enabled the corn-based ethanol industry in the United States to mature but have not been very successful in the case of cellulosic biofuels.

Public and private investments are needed to establish and strengthen the infrastructure required for a successful bioenergy industry, including distribution channels and other logistical infrastructure and vehicular fleets. The policy interventions should follow a portfolio approach emphasizing different feedstocks and technologies. It is imperative that existing certification programs be modified to incorporate sustainability concerns arising due to the expansion of agriculture- and forestry-based bioenergy. It would also be beneficial to develop bioenergy-specific certification schemes that evaluate adherence to standards in the supply chain of growing, harvest, transportation, conversion, and waste disposal. So far, initiatives by governments or international bodies, such as the RSB, have developed sustainability standards that are primarily aimed at limiting GHG emissions, regulating biofuel production in biodiverse areas, and promoting environmental management, food security, and social justice.

As bioenergy production grows and trade increases, public policies should not only focus on its expansion but also prioritize sustainability standards along the bioenergy supply chain. The bioenergy industry and feedstock producers can strive to ensure that bioenergy products developed in the country as well as in other parts of the world do not have an irreversible negative impact on ecological functions or on regional economies. Going forward, the priority areas should emphasize activities that make bioenergy commercially viable and ecologically sustainable.

ACKNOWLEDGMENT

This work was partially supported by funding from the U.S. Department of Energy's Office of Energy Efficiency and Renewable Energy, Bioenergy Technologies Office, and sponsored by the U.S. Department of Energy's International Affairs under award number, DE-PI0000031 and from the U.S. Department of Agriculture National Institute of Food and Agriculture, award number 2012-67009-19742.

REFERENCES

Abbasi, T. and Abbasi, S. (2010). Biomass energy and the environmental impacts associated with its production and utilization. *Renewable and Sustainable Energy Reviews*, 14(3), 919–937.

Aguilar, F. X. and Saunders, A. (2010). Policy instruments promoting wood-to-energy uses in the continental United States. *Journal of Forestry*, 108(3), 132–140.

Alternative Fuels Data Center. (2014). All Laws and Incentives Sorted by Type. Retrieved from http://www.afdc.energy.gov/laws/matrix/incentive (Accessed February 15, 2014).

Aradhey, A. (2013). India: Biofuels Annual, GAIN Report No. IN3073, USDA Foreign Agricultural Service. Available at http://www.google.co.in/url?sa=t&rct=j&q=&esrc=s&source=web&cd=2&ved=0CC8QFjAB&url=http%3A%2F%2Fgain.fas.usda.gov%2FRecent%2520GAIN%2520Publications%2FBiofuels%2520Annual_New%2520Delhi_India_8-13-2013.pdf&ei=JvB9UvL0J4GRrQek0oDQAw&usg=AFQjCNH1dqH-2EPe1sa4Re9J-XrJe9SqAg&bvm=bv.56146854,d.bmk&cad=rja (Accessed November 4, 2013).

Barrett, S. M., Bolding, M. C., and Munsell, J. F. (2012). Evaluating continuing education needs and program effectiveness using a survey of Virginia's SHARP logger program participants. *Journal of Extension*, 50, 1RIB8.

Barros, S. (2013). *Brazil—Biofuels Annual—Annual Report 2013*. USDA Foreign Agricultural Service, GAIN report BR13005, December 2013. http://gain.fas.usda.gov/Recent%20GAIN%20Publications/Biofuels%20Annual_Sao%20Paulo%20ATO_Brazil_9-12-2013.pdf

Biofuels Digest. (2011a). Biofuels mandates around the world. Retrieved from http://www.biofuelsdigest.com/bdigest/2011/07/21/biofuels-mandates-around-the-world/ (Accessed March 10, 2014).

Biofuels Digest. (2011b). The Range Fuels failure. Retrieved from http://www.biofuelsdigest.com/bdigest/2011/12/05/the-range-fuels-failure/ (Accessed March 22, 2014).

Biofuels Digest. (2013). Biofuels mandates around the world 2014. Retrieved from http://www.biofuelsdigest.com/bdigest/2013/12/31/biofuels-mandates-around-the-world-2014/ (Accessed March 10, 2014).

Biomass Magazine. (2014). Location of US Pellet plants and production capacity. Retrieved from http://biomassmagazine.com/plants/listplants/pellet/US/Operational/page:1/sort:capacity/direction:desc (Accessed March 22, 2014).

Blonz, J. A., Vajjhala, S. P., and Safirova, E. (2008). *Growing complexities: A cross-sector review of US biofuels policies and their interactions*. Washington, DC: Resources for the Future.

Bowyer, J., Stai, S., Bratkovich, S., and Howe, J. (2011). *Differences between the Forest Stewardship Council (FSC) and Sustainable Forestry Initiative (SFI) certification standards for forest management*. Dovetail Partners. Retrieved from http://www.sfiprogram.org/files/pdf/dovetailfscsficomparison32811pdf/ (Accessed March 17, 2014).

Bracmort, K., Schnepf, R., Stubbs, M., and Yacobucci, B. D. (2011). *Cellulosic biofuels: Analysis of policy issues for congress*. Congressional Research Service, 7-5700, The Library of Congress 101 Independence Ave, SE Washington, DC.

Carriquiry, M. A., Du, X., and Timilsina, G. R. (2011). Second generation biofuels: Economics and policies. *Energy Policy*, 39(7), 4222–4234.

Childs, B. and Bradley, R. (2007). *Plants at the pump: Biofuels, climate change, and sustainability*. Washington, DC: World Resources Institute.

Cubbage, F. W. (2004). Costs of forestry best management practices in the south: A review. *Water, Air, & Soil Pollution: Focus*, 4, 131–142.

Dale, V. H., Langholtz, M. H., Wesh, B. M., and Eaton, L. M. (2012). Environmental and socioeconomic indicators for bioenergy sustainability as applied to *Eucalyptus*. *International Journal of Forestry Research*, 2013, 215276.

Doornbosch, R. and Steenblik, R. (2007). *Biofuels: Is the cure worse than the disease?* Report SG/SD/RT(2007)3. Prepared for Round Table on Sustainable Development. Paris: Organization for Economic Cooperation and Development (OECD).

EPA. (2010). Regulatory announcement #EPA-420-F-10-007 issued by the Office of Transport and Air Quality. Retrieved from http://www.epa.gov/otaq/renewablefuels/420f10007.pdf (Accessed February 20, 2014).

EPA. (2013). Regulatory announcement #EPA-420-F-13-048 issued by the Office of Transport and Air Quality. Retrieved from http://www.epa.gov/otaq/fuels/renewablefuels/documents/420f13048.pdf (Accessed February 20, 2014).

Ellefson, P. V., Kilgore, M. A., Hibbard, C. M., and Granskog, J. E. (2004). *Regulation of forestry practices on private land in the United States: Assessment of state agency responsibilities and program effectiveness*. St. Paul, MN: Department of Forest Resources, University of Minnesota.

Elliott, K. (2013). *Subsidizing farmers and biofuels in rich countries: An incoherent agenda for food security*. CGD Policy Paper 032. Washington, DC: Center for Global Development.

Evans, Z., Perschel, B., and Kittler, B. (2010). *Revised assessment of biomass harvesting and retention guidelines*. Santa Fe, NM: The Forest Guild, pp. 1–21.

Fargione, J., Hill, J., Tilman, D., Polasky, S., and Hawthorne, P. (2008). Land clearing and biofuel carbon debt. *Science*, 319(29), 1235–1238.

Fargione, J. E., Plevin, R. J., and Hill, J. D. (2010). The ecological impact of biofuels. *Annual Review of Ecology, Evolution, and Systematics*, 41, 351–377.

Fielding, D. (2011). *Perceptions of biomass harvesting guidelines in North Carolina: A qualitative analysis of forest managers, loggers and landowners.* M.S. Thesis, North Carolina State University, Raleigh, NC.

Flach, R., Bendz, K., Krautgartner, R., and Lieberz, S. (2013). *EU biofuels annual report 2013.* Hague: USDA Foreign Agricultural Service.

Fletcher, R. J., Jr., Alavalapati, J., Evans, J., and Jao, R. (2009). *WHPRP final report: Impacts of bioenergy production on the conservation of wildlife habitat.* Washington, DC: National Council for Science and the Environment.

Gan, J. and Cashore, B. (2013). Opportunities and challenges for integrating bioenergy into sustainable forest management certification programs. *Journal of Forestry*, 111(1), 11–16.

Government of Nebraska. (2014). Ethanol Facilities Nameplate Capacity and Production Capacity Ranked by State. Retrieved from http://www.neo.ne.gov/statshtml/121.htm (Accessed February 15, 2014).

Havlík, P., Schneider, U. A., Schmid, E., Böttcher, H., Fritz, S., Skalský, R., De Cara, S., et al. (2011). Global land-use implications of first and second generation biofuel targets. *Energy Policy*, 39(10), 5690–5702.

IRFA (Iowa Renewable Fuels Association). (2014). USDA preliminary 2012 farm census data shows RFS is revitalizing rural Iowa. Retrieved from http://www.iowarfa.org/RFSRevitalizesRuralIowa.php (Accessed March 22, 2014).

Janowiak, M. K. and Webster, C. R. (2010). Promoting ecological sustainability in woody biomass harvesting. *Journal of Forestry*, 108(1), 16–23.

Joudrey, J., McDow, W., Smith, T., and Larson, B. (2012). *European power from U.S. forests: How evolving EU policy is shaping the transatlantic trade in wood biomass.* Environmental Defense Fund. Retrieved from http:www.edf.org/sites/default/files/europeanPowerFrom USForests.pdf (Accessed March 18, 2014).

Kim, H., Kim, S., and Dale, B. E. (2009). Biofuels, land use change, and greenhouse gas emissions: Some unexplored variables. *Environmental Science & Technology*, 43(3), 961–967.

Kittler, B., Price, W., McDow, W., and Larson, B. (2012). *Pathways to sustainability: An evaluation of forestry programs to meet European biomass supply chain requirements.* EDF and Pinchot Institute for Conservation. Retrieved from http://www.edf.org/energy/sustainable-bioenergy (Accessed March 20, 2014).

Kojima, M., Mitchell, D., and Ward, W. A. (2007). *Considering trade policies for liquid biofuels.* Washington, DC: Energy Sector Management Assistance Program.

Lal, P., Alavalapati, J., Marinescu, M., Matta, J. R., Dwivedi P., and Susaeta, A. (2011). Developing sustainability indicators for woody biomass harvesting in the United States. *Journal of Sustainable Forestry*, 30, 736–755.

Lal, P., Upadhyay, T., and Alavlapati, J. R. R. (2013). Woody biomass for bioenergy: A policy overview. In: Evans, J. M., Fletcher, R. J., Jr., Alavalapati, J. R. R., Smith, A. L., Geller, D., Lal, P., Vasudev, D., Acevedo, M., Calabria, J., and Upadhyay, T. (eds.), *Forestry bioenergy in the Southeast United States: Implications for wildlife habitat and biodiversity.* Merrifield, VA: National Wildlife Federation. Retrieved from http://www.nwf.org/~/media/PDFs/Wildlife/Conservation/NWF_Biomass_Biodiversity_Final.ashx (Accessed March 21, 2014).

Lickwar, P., Hickman, C., and Cubbage, F. C. (1992). Costs of protecting water quality during harvesting on private forestlands in the southeast. *Southern Journal of Applied Forestry*, 16, 13–20.

Milbrandt, A. (2005). *A geographic perspective on the current biomass resource availability in the United States.* Golden, CO: National Renewable Energy Laboratory.

Miller, J. C. and Coble, K. H. (2011). Incentives matter: Assessing biofuels policies in the South. *Journal of Agricultural and Applied Economics*, 43, 413–421.

Natural Resources Defense Council. (2010). Let the VEETC expire: Moving beyond corn ethanol means less waste, less pollution and more jobs.

NFU (National Farmers Union). (2011). Testimony to the US Senate Committee on Agriculture. Retrieved from http://www.ag.senate.gov/download/?id = eb614af7-ebb5-4279-aafb-168caa4fb9ad (Accessed February 21, 2014).

Noonan, D. (2012). *Biomass Policy and Georgia. Georgia Public Policy Foundation.* Retrieved from http://www.georgiapolicy.org/biomass-policy-and-georgia/ (Accessed 20 February 2014).

Pimentel, D., Marklein, A., Toth, M. A., Karpoff, M. N., Paul, G. S., McCormack, R., Kyriazis, R., and Krueger, T. (2009). Food versus biofuels: Environmental and economic costs. *Human Ecology*, 37(1), 1–12.

Rajagopal, D., Sexton, S. E., Roland-Holst, D., and Zilberman, D. (2007). Challenge of biofuel: Filling the tank without emptying the stomach? *Environmental Research Letters*, 2(4), 044004.

Raju, S., Parappurathu, S., Chand, R., Joshi, P., Kumar, P., and Msangi, S. (2012). *Biofuels in India: Potential, policy and emerging paradigms*. Policy Paper 27. New Delhi, India: National Centre for Agricultural Economics and Policy Research.

RSB (Roundtable on Sustainable Biomaterials). (2014). RSB sustainability standards. Retrieved from http://rsb.org/sustainability/rsb-sustainability-standards/ (Accessed March 21, 2014).

Schnepf, R. and Yacobucci, B. D. (2013). *Renewable Fuel Standard (RFS): Overview and issues*. Washington, DC: Congressional Research Service.

Scott, R. and Junyang, J. (2012). China annual report #12044. Global Agricultural Information Network (GAIN), USDA Foreign Agricultural Service. Retrieved from http://gain.fas.usda.gov/Recent%20GAIN%20Publications/Biofuels%20Annual_Beijing_China%20-%20Peoples%20Republic%20of_7-9-2012.pdf (Accessed January 17, 2015).

Searchinger, T., Heimlich, R., Houghton, R. A., Dong, F., Elobeid, A., Fabiosa, J., Tokgoz, S., Hayes, D., and Yu, T. (2008). Use of U.S. croplands for biofuels increases greenhouse gases through emissions from land-use change. *Science*, 319(29), 1238–1240.

Sorda, G., Banse, M., and Kemfert, C. (2010). An overview of biofuel policies across the world. *Energy Policy*, 38(11), 6977–6988.

Taheripour, F. and Tyner, W. (2010). Biofuels, policy options, and their implications: Analyses using partial and general equilibrium approaches. In: Khanna, M., Scheffran, J., and Zilberman, D. (eds.), *Handbook of bioenergy economics and policy*. Vol. 33. Series of Natural Resource Management and Policy, pp. 365–383, Springer Science+Business Media LLC, NY.

Urbanchuk, J. M. (2014). *Contribution of the renewable fuels industry to the economy of Iowa*. Agriculture and Biofuel Consulting LLP, PA.

U.S. Cong. Compilation of the Energy Security Act. (1980). 1980 Amendments to the Defense Production Act of 1950. Cong. Bill. Washington: U.S. G.P.O., 1980. Print. Retrieved from http://www.govtrack.us/congress/bills/96/s932/text. (Accessed February 15, 2014).

U.S. Cong. Energy Tax Act. (1978). Conference Report to Accompany H.R. 5263. Cong. Bill. Washington: U.S. Govt. Print. Off., 1978. Print. Retrieved from https://www.govtrack.us/congress/bills/95/hr5263/text. (Accessed February 15, 2014).

U.S. Cong. Energy Policy Act. (2005). Cong. Bill. Washington, D.C.: U.S. G.P.O., 2005. Print. Retrieved from http://energy.gov/eere/femp/downloads/energy-policy-act-epact-2005. (Accessed February 15, 2014).

U.S. Cong. Food, Conservation, and Energy Act (2008). Conference Report to Accompany H.R. 2419. Cong. Rept. Washington: U.S. G.P.O., 2008. Print. Retrieved from https://www.govtrack.us/congress/bills/110/hr2419/text. (Accessed February 15, 2014).

USDA Farm Service Agency. (2013). BCAP: Biomass Crop Assistance Program, Energy Feedstocks from Farmers & Foresters. Retrieved from http://fsa.usda.gov/Internet/FSA_File/bcap_documentation.pdf (Accessed 15 January 2015).

USDA Farm Service Agency. (2014). Biomass Crop Assistance Program. Retrieved from http://www.fsa.usda.gov/FSA/webapp?area=home&subject=ener&topic=landing (Accessed February 18, 2014).

USDOE (U.S. Department of Energy). (2011a). U.S. Billion-Ton Update: Biomass Supply for a Bioenergy and Bioproducts Industry. In: Perlack, R. D. and Stokes, B. J. (eds), ORNL/TM-2011/224. Oak Ridge National Laboratory, Oak Ridge, TN. 227p.

USDOE (U.S. Department of Energy). (2011b). Range biofuels commercial scale biorefinery. Retrieved from http://www1.eere.energy.gov/biomass/pdfs/ibr_commercial_rangefuels.pdf (Accessed March 22, 2014).

USDOE (U.S. Department of Energy). (2013). Florida project produces nation's first cellulosic ethanol at commercial-scale. Retrieved from http://energy.gov/articles/florida-project-produces-nation-s-first-cellulosic-ethanol-commercial-scale-0 (Accessed March 22, 2014).

USDOE (U.S. Department of Energy). (2014). Biofuels investment tax credit. Retrieved from http://www.afdc.energy.gov/laws/law/FL/6074 (Accessed March 22, 2014).

U.S. Energy Information Administration, International Energy Statistics. Biofuels Production (2014a). Retrieved from http://www.eia.gov/cfapps/ipdbproject/EDIndex3.cfm?tid = 79&pid = 79&aid = 1 (Accessed February 15, 2014).

U.S. Energy Information Administration, International Energy Statistics. (2014b). Energy timelines ethanol. Retrieved from http://www.eia.gov/kids/energy.cfm?page = tl_ethanol (Accessed February 15, 2014).

U.S. Energy Information Administration, International Energy Statistics. (2014c). U.S. Imports by Country of Origin. Retrieved from http://www.eia.gov/dnav/pet/pet_move_impcus_a2_nus_epooxe_im0_mbbl_a.htm (Accessed February 15, 2014).

U.S. Energy Information Administration, International Energy Statistics. (2014d). U.S. Biodiesel production capacity and production. Retrieved from http://www.eia.gov/renewable/data.cfm#biomass (Accessed February 15, 2014).

Wood Fiber Coalition. (2012). Testimony to the subcommittee on conservation, energy and forestry. Retrieved from http://www.compositepanel.org/userfiles/filemanager/878/ (Accessed February 21, 2014); United States. Cong. House. Conference Report 113-333 To Accompany HR 2642. Agricultural Act of 2014 (p. 291). 113 Cong., 2nd sess. Retrieved from http://www.gpo.gov/fdsys/pkg/CRPT-113hrpt333/pdf/CRPT-113hrpt333.pdf on 21 February 2014.

Yacobucci, B. D. (2012). *Biofuels incentives: A summary of federal programs*. Congressional Research Service, The Library of Congress 101 Independence Ave, SE Washington, DC.

PART IV

Case Studies

CHAPTER 15

Cellulosic Biofuel in the United States
Targets, Achievements, Bottlenecks, and a Case Study of Three Advanced Biofuel Facilities

Sougata Bardhan, Shibu Jose, and Larry Godsey
The Center for Agroforestry, University of Missouri, Columbia, MO, USA

CONTENTS

15.1 INTRODUCTION

Fossil fuels supply an estimated 80% of the primary energy consumed globally, of which 58% is utilized in the transportation sector as liquid fuel (Escobar et al. 2007). According to current predictions, worldwide demand for oil will be around 110–120 million barrels per day during the period from 2030 to 2040 (CERA Report 2009; AEO 2013). Worldwide, for every four barrels of oil that are consumed, only one new barrel of oil is found (Aleklett and Campbell 2003). It is also envisioned that, with all the current reserves and new oil discoveries along with tar sand reserves, there still will be a large gap in demand and supply, which could be tens of millions of barrels per day. Liquid drop-in biofuels are renewable sources of transportation fuel that could become critical in order to bridge this gap around the world and at the same time reduce U.S. dependence on foreign oil. Conversion of biomass and its utilization as a renewable source of energy is essential for transformation toward a more sustainable planet (Tran et al. 2010). Besides, renewable energy is viewed as one of the principal ways of reducing GHG emissions and of creating a sustainable, environmentally friendly energy economy that could provide new employment opportunities in rural communities (Goldemberg 2000; Goldemberg et al. 2004; Jefferson 2006).

Although biofuel can be produced from any kind of biomass, recent focus has been on production of second-generation or advanced biofuels from nonfood sources such as cellulosic feedstock, algae, and bacteria. Advanced biofuels are not only beneficial in terms of geopolitical implications, but they are also much more sustainable in comparison with fossil fuel and even in comparison with first-generation biofuels. Biofuel, as a source of renewable energy, started to receive attention toward the later part of the last century in the United States, although the push came only in spurts and sustained efforts to develop biofuels were never maintained. Mostly, renewable energy produced from biomass involved ethanol produced from corn and sugarcane and biodiesel from soybean. However, these crops have demand as food commodities, and diversion of these crops from food to fuel has been attributed to higher food prices and food shortages around the world. As a result, in the recent past, we have seen grain commodity prices become highly unstable (Zhang et al. 2009, 2010; Mueller et al. 2011). It is estimated that ethanol production in the United States consumes about 40% of the domestic corn crop (based on 2012 figures) and has been the reason for serious debate over the importance of food versus fuel.

Although prices have stabilized somewhat since that time, there is always the potential for a future impact on the economy from continued diversion of food crops for biofuel production. It is thus obvious that there is a certain limitation that creates a bottleneck for production of biofuels from first-generation feedstock that could easily threaten food security and biodiversity (Doornbusch and Steenblik 2007; Fargione et al. 2008). However, development of second-generation biofuels (such as cellulosic biofuel) could create an environment that will mitigate and reduce the negative impact of food crop–based biofuel production. (Gonzalez-Garcia et al. 2009; Mueller et al. 2011).

There has been a tremendous increase in the production of biofuels around the world in the past few years. During the first decade of this century, fuel ethanol production worldwide has increased from 16.9 to 72.0 billion liters, while bio diesel production grew from 0.8 to 14.7 billion liters (Licht and Brown 2009). At present, production of biofuel is unprofitable and state support is necessary—through tax exemptions, subsidies, and other financial incentives (Rajagopal and Zilberman 2007)—in order to achieve several essential objectives, such as to reduce dependence on fossil fuel, to mitigate environmental pollution and global warming, and to create a sustainable renewable energy economy. To address the issue of sustainable biofuel production, several legislative tools targeted at promoting the biofuel industry have been approved by governments around the world. At present, there are 62 countries that have some sort of mandate or directive for production and integration of biofuels in their energy scheme. Among the important mandates, 27 countries from the European Union and 13 countries in the Americas have established a mandate along with 12 from Asia Pacific and 10 from Africa. These directives include blending of biofuel with mineral fuel at rates ranging from 5% to 25% in different countries and within variable time frames.

Lignocellulosic biofuels produced from materials such as corn stover, wood chips, or dedicated feedstocks developed specifically for the production of advanced biofuels have been suggested as a sustainable alternative (Nigam and Singh 2011). Although the technology for converting any organic material to ethanol has been in existence for many years, the economic feasibility of producing large-scale biofuel refineries has been inefficient. Therefore, there remain several genuine challenges toward achieving that goal and toward the creation of a sustainable advanced biofuel economy (Stevens et al. 2004).

In this chapter, we will discuss some of the issues related to increasing advanced biofuel production in the United States (the current state of affairs, challenges, and a blueprint for success, we will explore three different biofuel companies that are in operation making use of current technologies to produce advanced biofuels.

15.2 GOALS, ACHIEVEMENTS, AND BOTTLENECKS

Historically, there have been few discrete policies on biofuels and renewable energy in the United States, and these have been focused mostly on ethanol. For example, the Energy Tax Act of 1978 provided tax credits for ethanol blenders. The first comprehensive policy on biofuels by the U.S. government was initiated late in the last century under the Energy Policy Act of 1992. The primary goal of this act was to draft a reasonable goal for clean energy use and improved overall energy efficiency in the country. The next major piece of biofuel legislation was called the EP Act of 2005. Along with several other clean energy initiatives, this legislation mandated four billion gallons of biofuels be mixed with gasoline by 2006, which was to be further increased to 6.1 billion gallons by 2009, and up to 7.5 billion gallons by 2012. In 2007, the Energy Independence and Security Act (EISA 2007) was signed into law by President George W. Bush, and it significantly increased the ante of reducing dependence on fossil fuel through the "20 in 10" challenge to reduce gasoline consumption by 20% in 10 years. The Renewable Fuel Standard 2 (RFS2) that was introduced in July 2010 mandates the use of biofuels to increase to 36 billion gallons per year (BGY) by 2022, with 21 BGY coming from advanced biofuels. RFS2 also requires that producers of advanced biofuels achieve at least a 50% reduction in lifecycle GHG emissions along with a 20% reduction for first-generation biofuel producers.

In the original RFS (RFS1), the requirement for total biofuel use was 4 billion gallons in 2006, which was to increase to 7.5 billion gallons by 2012. The RFS2 mandated inclusion of 100 million gallons of cellulosic biofuels to be included and blended in the national transportation fuel supply in 2010, which would increase to 1.75 billion gallons by 2014. However, the first registered production of cellulosic biofuels in the United States was only achieved in 2012. Figure 15.1 explains the disparity between the RFS2 and actual production in the United States.

Several factors were responsible for the disparity between RFS2 and actual production and use of cellulosic biofuels. RFS2 was a very highly ambitious mandate that did not take into account the possible pitfalls and other setbacks that could slow the development of cellulosic biofuel technology. Second, to qualify for the RFS, the lifecycle GHG emissions had to be reduced drastically by 20%–60% for first-generation to cellulosic biofuels, respectively. Moreover, these reductions in GHG emissions should consider land-use change as a factor. Other important issues that delayed

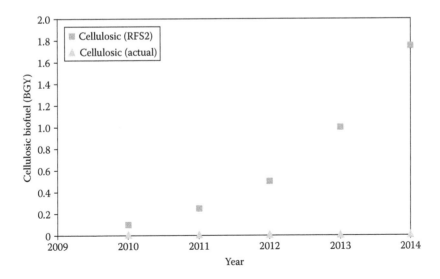

Figure 15.1 Disparity between RFS2 and actual production of cellulosic biofuel.

the progress of cellulosic biofuel production in the United States include lack of technological breakthrough and setbacks in the conversion technologies necessary to bring down the cost of production for commercial purposes. Also, minimal financial support was also provided, which was not sufficient to strengthen the biofuels economy.

A reliable supply of biomass feedstock is crucial for the success of a biorefinery. In many instances, there was no guaranteed year-round supply of adequate biomass. Based on the billion-ton report (US DOE and USDA, 2005), approximately 1.3 billion tons of cellulosic biomass could be available each year by 2050. Assuming suitable and cost-effective technologies are available by the time, the United States will be able to produce between 65 and 130 billion gallons of cellulosic ethanol at a conversion rate of 50–100 gallons per ton of dry biomass. However, the billion-ton report is based on aggressive harvesting of biomass, which may be harmful for the overall sustainability of the soil ecosystem. Therefore, dedicated energy crops should be established to support the biorefineries and to provide a continuous year-round supply of feedstock.

15.3 CASE STUDY OF BIOREFINERIES

Although steady progress has been made in technical and economical aspects of biofuel production from second-generation feedstocks, commercial success still faces an uphill climb in the long run. However a cost-effective second-generation biofuel production technology is in its early stages and requires further investigation and investment. Second-generation biofuels can be produced either through biochemical (ethanol and butanol) or thermochemical processes (Fischer–Tropsch liquid or semirefined crude oil) (Farias et al. 2007). For the production of ethanol, lignocellulosic biomass—primarily cellulose and hemicellulose—is converted into sugars by hydrolysis, and then the sugars are fermented into ethanol using existing technology. Another approach of converting lignocellulosic biomass into biofuels involves pyrolysis and gasification of feedstock into "syngas," which is composed of carbon monoxide, hydrogen, and other compounds. The syngas can then be converted into various types of fuels and chemicals using the Fischer–Tropsch synthesis. The thermochemical process results in higher yields of end products in comparison with the hydrolysis method as lignin, which constitutes a bulk of the feedstock, is also processed to produce biofuels.

Although significant progress has been made in terms of deployment of advanced biofuel production facilities around the country (Figure 15.2), most of the cellulosic biofuel production facilities are in the demonstration stage (Table 15.1) and the cost of production is still relatively high. However, with new technological breakthroughs and further development in current technologies, it is expected that in the long run biofuels will be able to compete with other sources of fuel, especially in a carbon constrained environment. In this chapter, we will look at three different biofuel companies from three U.S. regions that are producing three different types of biofuels (Table 15.2). The three companies are KiOR (biocrude), POET-DSM (bio-butanol), and ZeaChem (bioethanol). Among these companies, only KiOR had developed a commercial scale biorefinery (in Columbus, MS), and it has plans for two more plants.

15.3.1 Abengoa

Abengoa, S.A., was founded in 1941 by two Spanish entrepreneurs and was initially involved in telecommunication and transportation. Subsequently, they ventured into energy and environment with significant research efforts in sustainable technologies around the world. By 2012, Abengoa and its subsidiaries were operating in 80 countries and employing about 27,000 people. Abengoa is the largest ethanol producer in Europe and one of the major ethanol producers in the world. Worldwide, Abengoa produces about 900 million gallons of ethanol from 15 plants in five countries.

Figure 15.2 Map of the United States showing the different advanced biofuel facilities that are either in operation or under construction. (Modified from Advanced Ethanol Council, the Cellulosic Biofuels Industry Progress Report 2012–2013.)

In the United States, the company has been involved in bioethanol production using their proprietary and partner technology for a long time with one of the oldest dry mill bioethanol facilities at Colwich, KS, which has been operating continuously for the past 25 years. Also, Abengoa has several operational plants in the United States located in the states of Kansas, Nebraska, and Indiana, primarily producing bioethanol from first-generation feedstocks.

Abengoa's model is development of new technologies in biofuel production while employing sustainable sourcing of raw materials and feedstocks. While some of the plants were producing first-generation biofuels from corn or other food sources, Abengoa launched a 2G (second-generation) bioethanol plant in Hugoton, KS, with an annual capacity of 25 million gallons and a daily consumption of 1100 dry tons of biomass feedstock. The construction of the plant began in 2011 and it became fully operational in 2014. This plant will consume approximately 350,000 tons of cellulosic feedstock annually, and the residues of the biofuel process will be further utilized to generate up to 18 megawatts of electricity. The power generated as a result of co-production will make the facility energy efficient and environment friendly. The various types of feedstocks that will be used include a mixture of agricultural waste, nonfeed energy crops, and wood waste.

As a next step, Abengoa plans to offer licenses and assistance to other investors in setting up of biofuel plants around the world, including support with process design, engineering and construction, production, and even downstream marketing of products from the licensed facilities.

Table 15.1 List of Advance Biofuel Establishments[a]

Name	Location	Type	Feedstock	Product	Capacity	Status
Abengoa Bioenergy	Hugoton, KS	Commercial	Agricultural residues, dedicated energy crops, prairie grasses	Cellulosic ethanol, 20 MW renewable electric power	25 MGY	Operational
American Process—Green Power	Alpena, MI	Demo	Mixed hardwood	Cellulosic ethanol, potassium acetate	700,000 GPY	Operational
American Process—AVAP Tech.	Thomaston, GA	Demo	Variety	Cellulosic sugar, ethanol, cellulose	300,000 GPY	Operational
Beta Renewables	Sampson Co, NC	Commercial	Dedicated energy crops	Cellulosic ethanol, biobased chemicals	20 MGY	Under development
Bluefire Renewables	Fulton, MS	Commercial	Forestry residues and cellulosic waste	Cellulosic ethanol, gypsum, lignin, protein cream	19 MGY	Under construction
Enerkem	Pontonoc, MS	Commercial	MSW, wood residue	Syngas, biomethanol, acetates, cellulosic ethanol	10 MGY	Under development
Fiberight	Blairstown, IA	Commercial	MSW, nonfood waste	Cellulosic ethanol, biochemicals	6 MGY	Operational
Fulcrum Bioenergy	McCarran, NV	Commercial	MSW	Advanced ethanol	10 MGY	Under construction
Inbicon Biomass Refinery	Spiritwood, ND	Commercial	Wheat straw	Cellulosic ethanol, renewable power	10 MGY	Engineering/permitting
Ineos Bio	Vero Beach, FL	Commercial	MSW, vegetative and yard waste	Cellulosic ethanol, 6 MW renewable power	8 MGY	Commissioning stage
KiOR	Columbus, MS	Commercial	Forestry residue	Cellulosic gasoline and diesel	13 MGY	Filed for bankruptcy on Nov, 2014
KiOR	Natchez, MS	Commercial	Forestry residue	Cellulosic gasoline and diesel	40 MGY	Filed for bankruptcy on Nov, 2014
Lanzatech	Soperton, GA	Commercial	Forestry waste	Ethanol, chemical, aviation fuel	4 MGY	Under development
Mascoma Corporation	Kinross, MI	Commercial	Wood pulp and chips	Cellulosic ethanol	20 MGY	Final engineering/financing
POET-DSM Advanced Biofuels	Emmetsberg, IA	Commercial	Corn, crop residue	Ethanol, biogas	20 MGY	Operational
ZeaChem Inc.	Boardman, OR	Commercial	Poplar trees, wheat straw	Cellulosic ethanol, biochemical	25 MGY	USDA Loan awarded

Source: Modified from Advanced Ethanol Council. Cellulosic Biofuels Industry Progress Report 2012–2013.
Note: MSW-Municipal solid waste, MGY-Million gallons per year.
[a] Utilizing a variety of feedstocks to produce biofuels to meet the federal renewal fuel standard in the United States.

Table 15. 2   General Information about the Three Biorefineries
Reviewed

	Abengoa, KS	POET-DSM	Zeachem
Established	2014	2014	2002
Feedstock	Cellulosic	Cellulosic	Cellulosic
Product	Bioethanol	Bioethanol	Bioethanol
Stage	Commercial	Commercial	Demo
Radius	50[a]	45	5
Yield	90	—	135
Consumption (ton/day)	1100	770	—
Annual production	25 Mgal	25 Mgal	—

[a] Partial acquisition.

15.3.2 POET-DSM

The first commercially operational cellulosic ethanol plant was inaugurated in September, 2014, in Emmetsburg, IA, and dubbed Project Liberty. The plant is owned by POET-DSM, a joint venture between POET (one of the world's largest ethanol producers, based in Sioux Falls, SD) and Royal-DSM (a global science based company involved in health, nutrition, and material science innovation). Using DSM technology, it is possible to coferment all xylose and arabinose (C6 and C5) sugars in the biomass. After successful operation, POET-DSM intends to license the complete technology and production package to other ventures to promote the production of advanced biofuels globally.

POET is a 25-year-old company with an annual production capacity of more than 1.6 billion gallons of ethanol and 9 billion pounds of high-protein animal feed. The capital cost involved for Project Liberty was approximately $275 million, and it has employment generation potential of 50 direct jobs and 200 indirect jobs in the community. The plant will have a consumption capacity of about 285,000 tons of biomass annually, and procurement will be made from within a 45-mile radius of the plant. The fuel produced at the Project Liberty facility will represent an 85–95% reduction in GHG emissions in comparison with gasoline. The production potential at the onset is estimated to be about 20 MGY (million gallons per year) of cellulosic bioethanol, which will be increased to 25 MGY in the near future.

15.3.3 ZeaChem

ZeaChem, Inc., founded in 2002 and headquartered in Lakewood, CO, has developed a cellulose-based biorefinery platform capable of producing advanced ethanol, fuels, and chemicals. ZeaChem's proprietary patented process generates the highest yield in the most cost-effective way while reducing the carbon footprint to a minimum. ZeaChem carries out its research and development work in a laboratory facility in Menlo Park, CA. In 2010, ZeaChem broke ground for a 250,000-gallon-per year demonstration biorefinery in Boardman, OR, which was completed in 2012. ZeaChem has a long-standing agreement with Greenwood Resources to procure feedstock for its first commercial biorefinery. ZeaChem has also partnered with Procter & Gamble for production of liquid drop-in chemicals and has a strategic partnership with Chrysler Group LLC to promote the use of advanced cellulosic ethanol in the transportation industry. The company also secured a conditional loan guarantee of more than $200 million from the U.S. Department of Agriculture for the commercial scale biorefinery.

ZeaChem use a hybrid biochemical and thermochemical reaction to produce advanced cellulosic ethanol, fuels, and chemicals. ZeaChem also uses biomass feedstocks that contain high levels of cellulose, which is the structural component of all plants and exists in large quantities in nonfood biofuel feedstocks. Like Cobalt, ZeaChem also uses both xylose [C_5] and glucose [C_6] to generate

Table 15.3 Overall Returns of ZeaChem and Other Biofuel Processes

	Farm Yield (Bu/ac/yr or BDT/ac/yr)	Factory Yield (Gal/Bu or Gal/BDT)	Auto Efficiency (mi/Gal EtOH)	Land Productivity (mi/ac/yr)
First-generation: corn ethanol	150	2.7	14	5670
Second-generation: cellulosic ethanol	7.5	90	14	9450
Third-generation: advanced cellulosic ethanol (ZeaChem Inc.)	15	135	14	28,350
Third-generation + auto efficiency	15	135	25	50,625

biofuels and other chemicals through the process of fermentation where an acetogenic process is utilized to ferment the sugars to acetic acid without CO_2 as a by-product (Table 15.3).

In comparison, traditional yeast fermentation creates one molecule of CO_2 for every molecule of ethanol. Thus the carbon efficiency of the ZeaChem fermentation process is nearly 100% vs. 67% for yeast.

ZeaChem takes the lignin residue from the fractionation process and utilizes to create steam and power for the different processes. Obtaining a plant yield of 135 gallons per bone dry ton (gal/BDT), the conversion process is highly efficient as the steam and power generated from the nonhydrogen portion of the syngas stream is used during the process.

15.4 POTENTIAL FOR BIOFUEL ECONOMY

The USDA announced the "Regional Roadmap to Meeting the Biofuel Goals of the Renewable Fuels Standards by 2022," which identified the areas where dedicated energy crops could be grown that will feed the biorefineries. The USDA estimated that, with an average capacity of 40 million gallons a year for a biofuel plant, there will be a need to construct about 500 new biorefineries by 2022 at a cost of approximately $160 billion. According to this roadmap, the Southeast, Hawaii, and the central eastern United States are the most suitable areas to develop a sustainable biofuel industry that can provide up to 94% of the cellulosic biofuel needs.

In the 2008 farm bill, the Biomass Crop Assistance Program (BCAP) was introduced with the intention of jumpstarting the development of advanced biofuels from second-generation feedstocks. As part of this program, individuals or companies will receive payment to plant perennial grasses, harvest agricultural residues, collect wood waste, store biomass, and to transport these materials to bioenergy and biofuel processing centers. The program encountered various implementation problems and, as a result, restrictions were placed on funding this program. In the latest farm bill, the BCAP program has been reauthorized and provided with $25 million of mandatory funding for fiscal years 2014–2018.

Although, the BCAP and other programs such as the Conservation Reserve Program (CRP) are available, it is important to target areas that are not traditionally used for agricultural production. We have already discussed that food versus fuel is a debate that is counterproductive to a sustainable biofuel economy. Therefore, encroaching on lands that are used for growing food crops creates the same problem in a different way. Figure 15.3 shows available land area that could be brought under dedicated energy crops without interfering with prime agricultural land along the Mississippi/Missouri River corridor. This estimate includes land that comes off the CRP and marginal lands where biomass crops can be grown without irrigation and with minimal fertilizer input.

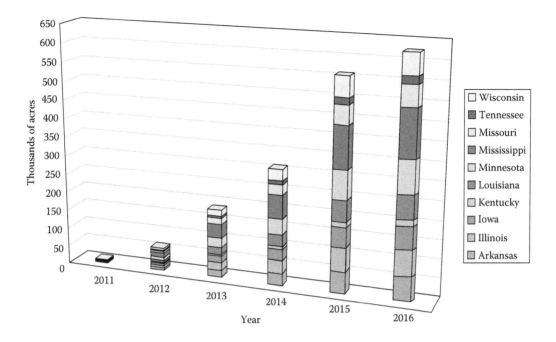

Figure 15.3 Potential land area to establish dedicated energy crops in a ten-state region along the Mississippi/Missouri corridor.

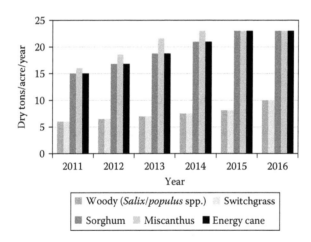

Figure 15.4 Production potential of different feedstock species on marginal lands over time on employing accelerated breeding for crop improvement.

At the same time, we should have research programs that will optimize feedstock species that are suitable for specific soil type or landscape type. Using breeding and selection, it will be possible to increase the productivity of cellulosic feedstocks and thus achieve the goals laid out in the RFS2 (Figure 15.4).

In order to create a truly sustainable biofuel production system, we have to strategically establish biorefineries in local areas that will be fed by feedstock produced in the surrounding neighborhoods. Based on our analyses, we have observed that a 50-mile radius is an ideal footprint for creating such a system where there will be one biorefinery at the center of this area. This hub-and-spoke model, an advanced rural biorefinery (ARBR), can be replicated to scale-up production.

15.5 ARBR MODEL—A SCALE-UP STRATEGY

Under this model a public–private partnership will implement a model for a regional ARBR to sustainably grow, process, and transport native perennial grasses to produce cellulosic biofuel, bio-based products, heat, and power with a commercially viable price structure. Billions of dollars are being invested annually by major private companies, venture capitalists, and the federal government in the development of new technologies to convert woody and nonwoody species into advanced, drop-in biofuels and biobased products. Major companies in the U.S. Midwest are looking to purchase large volumes of advanced biofuels. For example, FedEx has publicly committed to replacing 30% of its annual fuel use with biofuels by 2030. The U.S. Navy, Air Force, and Army have all set ambitious targets for their adoption of renewable, domestically produced biofuels. The Navy has its "Great Green Fleet," a fighting force of ships, submarines, and planes powered entirely by biofuels, with plans for the fleet to be operational by 2016. The Air Force hopes to get half of the fuel it uses for domestic flights from alternative renewable sources by 2016. However, the development of a sustainable, replicable feedstock system and scalable supply chain has not been on pace with the technology development; the result is a bottleneck in which the technology cannot be deployed until the feedstock production, processing, and logistical system is in place.

Theoretically, this model relies on the utilization of 40,000 acres of underutilized land to provide 150,000 tons of biomass to produce combined heat and power (CHP), animal feed, fertilizer, and 10–15 million gallons of cellulosic biofuel per year. In addition to reducing the U.S. demand for fossil fuel, this will create economic development opportunities in rural areas where it is most needed. An average single ARBR can employ 50–65 people directly with an estimated local economic impact of $90–$100 million per year. Overall, the model will result in production and distribution of cellulosic biofuel, biobased products (animal feed and fertilizer), and heat and power with net positive social, environmental, and economic impacts. As advanced biofuel production expands in the region, the ARBR model could be replicated using a franchise model throughout the United States.

15.6 CONCLUSION

At present, the dominant feedstock for biofuel production in the United States is corn, which is used to produce ethanol. However, in the long run, corn-based ethanol is not efficient in comparison to cellulosic ethanol. Therefore, it is desirable to shift the corn-based biofuel production system into more sustainable cellulosic biofuel production systems. At the same time, the conversion technology currently available to process cellulosic plant material to produce biofuel is inefficient and unprofitable. As a result, the cellulosic ethanol targets established in the RFS2 have not been achieved yet in the United States. Interestingly, at this point, several biofuel refineries are transitioning from demonstration scale to commercial scale, and the success of cellulosic ethanol production in the United States will partly depend on the lessons learned from these operations. There is hope that the new cellulosic biorefineries will be able to demonstrate economies of scale. Future research should focus more on aspects that help to establish biomass feedstock production systems for advanced biorefineries and to improve the conversion technologies available to process the produced feedstock to generate renewable biofuels.

ACKNOWLEDGMENT

The authors gratefully acknowledge partial funding from the U.S. Department of Energy's Office of Energy Efficiency and Renewable Energy, Bioenergy Technologies Office and sponsored by the U.S. Department of Energy's International Affairs under award no. DEPI0000031.

REFERENCES

AEO, 2013. Annual Energy outlook 2013. http://www.eia.gov/forecasts/aeo/pdf/0383%282013%29.pdf (URL verified 02/12/2015).

Aleklett K, Campbell CJ. (2003). The peak and decline of world oil and gas production. *Miner Energy*. 18: 35–42.

CERA Report, 2009. Growth in Canadian Oil Sands: Finding the New Balance http://www.circleofblue.org/waternews/wp-content/uploads/2010/08/Cera-oil-sands-report.pdf (URL verified 02/12/2015).

Doornbusch R, Steenblik R. (2007). Biofuels: Is the cure worse than the disease? *Paper prepared for the Round Table on Sustainable Development*, Organisation for Economic Co-operation and Development (OECD), Paris, 11–12 September, SG/SO/RT(2007)3/REV1. http://www.oecd.org/dataoecd/9/3/39411732.pdf

EISA. 2007. The Energy Independence and Security Act. http://www.gpo.gov/fdsys/pkg/BILLS-110hr6enr/pdf/BILLS-110hr6enr.pdf (URL verified 02/12/2015).

Escobar JC, Lora ES, Venturini OJ, Yanez EE, Castillo EF, Almazan O. (2007). Biofuels: Environment, technology and food security. *Renew Sustain Energy Rev*. 13: 1275–1287.

Fargione J, Hill J, Tilman D, Polasky S, Hawthorne P. (2008). Land clearing and the biofuel carbon debt. *Science*. 319: 1235–1240.

Farias FEM, Silva FRC, Cartaxo SJM, Fernandes FAN, Sales FG. (2007). Effect of operating conditions on Fischer–Tropsch liquid products. *Latin Am Appl Res*. 37: 283–287.

Goldemberg J. (2000). *World energy assessment: Energy and the challenge of sustainability*. United Nations. New York: UNDESA and WEC.

Goldemberg J, Coelho ST, Lucon O. (2004). How adequate policies can push renewables. *Energy Policy*. 32(9): 1141–1146.

Gonzalez-Garcia S, Luo L, Moreira M, Feijoo G, Huppes G. (2009). Life cycle assessment of flax shives derived second generation ethanol fueled automobiles in Spain. *Renew Sustain Energ Rev*. 13: 1922–1933.

Jefferson M. (2006). Sustainable energy development: Performance and prospects. *Renew Energ*. 31(5): 571–582.

Licht FO, Brown LR. (2009). *Plan B 4.0: Mobilizing to save civilization*. New York: W. W. Norton.

Mueller SA, Anderson JE, Wallington TJ. (2011). Impact of biofuel production and other supply and demand factors on food price increases in 2008. *Biomass Bioenergy*. 35(5): 1623–1632.

Nigam PS, Singh A. (2011). Production of liquid biofuels from renewable resources. *Prog Energ Combust Sci*. 37(1): 52–68.

Rajagopal D, Zilberman D. (2007). *Review of environmental, economic and policy aspects of biofuels*. Policy Research Working Paper 4341. Washington, DC: The World Bank.

Stevens DJ, Worgetten M, Saddler J. (2004). *Biofuels for transportation: An examination of policy and technical issues*. IEA Bioenergy Task 39, Liquid Biofuels Final Report 2001–2003.

Tran N, Bartlett J, Kannangara G, Milev A, Volk H, Wilson M. (2010). Catalytic upgrading of biorefinery oil from micro-algae. *Fuel*. 89: 265–274.

U.S. Department of Energy, U.S. Department of Agriculture. (2005). Biomass as feedstock for a bioenergy and bioproducts industry: The technical feasibility of a billion-ton annual supply. http://www1.eere.energy.gov/biomass/pdfs/final_billionton_vision_report2.pdf

Zhang Z, Lohr L, Escalante C, Wetzstein M. (2009). Ethanol, corn, and soybean price relations in a volatile vehicle-fuels market. *Energies*. 2: 320–339.

Zhang Z, Lohr L, Escalante C, Wetzstein M. (2010). Food versus fuel: What do prices tell us? *Energy Policy*. 38: 445–451.

CHAPTER 16

Biorefineries for Sustainable Production and Distribution
A Case Study from India

Thallada Bhaskar, Bhavya Balagurumurthy, Rawel Singh, and Azad Kumar
Bio-Fuels Division, CSIR–Indian Institute of Petroleum, Dehradun, India

CONTENTS

16.1 INTRODUCTION

Among the biggest problems that mankind is facing today is the progress toward a sustainable energy supply. Energy scenarios (Global Business Environment Shell International 2001; IEA 2003; European Commission 2006) project that the world's annual energy consumption will increase steeply from the current 500 exajoules (EJ) per annum to 1000–1500 EJ by 2050.

Crude-oil is the world's primary source of energy and chemicals with a current demand of about 12 million tonnes per day (84 million barrels a day) and a projection to 16 million tonnes per day (116 million barrels a day) by 2030 (The Royal Society 2008). There is a general scientific consensus that fossil fuel utilization causes global warming due to emission of greenhouse gases (GHG) such as carbon dioxide (CO_2) and methane (CH_4) (Solmon et al. 2007). Several alternatives are currently being explored that include a range of carbon free and renewable sources (photovoltaic, wind and nuclear power) in an attempt to replace fossil resources in the electricity generation sector. However, as yet, there is no such equivalent in transportation sector because fuel cells, electric/hybrids–based cars are still not mainstream vehicles.

Short- and medium-term alternatives for hydrocarbon resources are needed; ideally, they should reduce the dependence on fossil resources and meet the GHG emissions target. Need for a secure source of transportation fuels and chemicals make it essential to explore biofuels/biobased hydrocarbons as alternatives to mineral oil based hydrocarbons (Luque et al. 2008).

It is necessary for every country to have a national energy policy that strives to create a sustainable energy system, with a long-term vision of obtaining all energy supplies from renewable sources. Elements of this policy must include ensuring a reliable energy supply, increasing the efficiency of energy use, reducing the environmental impacts of energy use, breaking the current dependence on oil, and encouraging cost-effectiveness in energy supply and use.

The refining capacity in India increased to 213.066 metric million tons per annum (MMTPA) as of April 2012 from 187.386 MMTPA as of April 2011 (Basic Statistics on Indian Petroleum and Natural Gas 2009–2010). The production of petroleum products during 2011–2012 was 196.71 MMT (excluding 2.213 MMT of LPG production from natural gas). India exported 60.84 MMT of petroleum products against the imports of 15 MMT. The consumption of petroleum products during 2011–2012 was 147.995 MMT (including sales through private imports), which was 4.93% higher than that of 141.040 million metric tonnes during 2010–2011 (Basic Statistics on Indian Petroleum and Natural Gas 2009–10). Petrochemicals are derived from various chemical compounds, mainly hydrocarbons. These hydrocarbons are derived from crude oil and natural gas. Among the various fractions produced by distillation of crude oil, naphtha, and so forth is the main feedstock for the petrochemical industry. Ethane, propane, and natural gas liquids obtained from natural gas are the other important feedstock used in the petrochemicals industry. Presently there are three naphtha and three gas cracker complexes in operation with a combined ethylene capacity of about 2.9 MMTPA. In addition, there are four aromatic complexes in operation with a combined xylene capacity of about 2.9 MMTPA. The petrochemical industry is mainly comprised of synthetic fiber/yarn, polymers, synthetic rubber (elastomers), synthetic detergent intermediates, performance plastics, and the plastic processing industry (Annual Report 2009–2010). The chemical industry is an important constituent of the Indian economy. Its size is estimated at around US$35 billion, which is equivalent to about 3% of India's gross domestic product (GDP). The total investment in the Indian chemical sector is approximately US$60 billion, and the total employment generated is about one million. The Indian chemical sector accounts for 13%–14% total exports and 8%–9% total imports of the country. In terms of volume, it is the twelfth largest in the world and the third largest in Asia. Currently, per capita consumption of chemical industry products in India is about one-tenth of the world average (Annual Report 2009–2010).

With this background, this chapter addresses the possibility/opportunity for production of hydrocarbons (transportation fuels and chemicals) from Indian lignocellulosic (nonedible) biomass feedstocks, availability, and perspectives for the future. In this chapter, we have aimed to construct a national scale analysis of how biomass could be used efficiently to meet the objectives of producing renewable hydrocarbons for fuels or chemicals and reduction of CO_2 emissions. Our primary focus is on India, but we acknowledge that India exists in a global energy market.

16.2 ENERGY PATTERN IN INDIA

In India, the concept of energy as *Shakti* has been at the focus of philosophic, scientific, and metaphysical thought from time immemorial. India is endowed with rich solar energy resources, with most parts of the country receiving 4–7 Wh of solar radiation/m² per day and 250–300 sunny days in a year. The average intensity of solar radiation received in India is 200 MW/km². With a geographical area of 3.287 million km², this amounts to 657.4 million MW. However, 87.5% of the land is used for agriculture, forests, fallow lands, and so on; 6.7% for housing, industry, and so on; and 5.8% is barren, snowbound, or generally uninhabitable (http://www.indiaenergyportal.org/subthemes.php?text=solar).

The important renewable energy sources include sun, wind, tides, waves, biomass, hydropower (from water), and geothermal energy. The pattern of energy consumption in India shows that 56.5% of total energy is from commercial sources, such as coal, oil, and electricity, and the remaining 43.5% is noncommercial energy. Firewood, charcoal, agricultural residues, vegetable wastes, cow dung, urban and industrial wastes, and forest residues are the main sources of this noncommercial energy (http://www.world-agriculture.com/agricultural-bioenergy/biomass-fuel-history.php).

The rural energy scenario is also pertinent to the biofuel policy context in India. Around 72% of the Indian population still lives in rural areas; 77% of household in rural areas and 23% in urban areas still depend on firewood and dung for fuel (India's Energy Security: Briefing Paper—Dialogue on Globalization, Friedrich Ebert Stiftung, Online source: http://library.fes.de/pdffiles/iez/global/04809.pdf). Biomass sources such as wood, chips, and dung cakes contribute around 30% of the total primary energy consumed in the country (Mission Document in Evolution: National Mission on Decentralized Biomass Energy for Villages and Industries). Village common lands, wastelands, and forestland constitute major source of this biomass.

One kilogram of biomass is capable of producing enough gas to run a generation system to produce one kilowatt hour of electricity. Power is a major issue, and India has an acute shortage of it. According to the Ministry for New and Renewable Energy (MNRE), the power generation potential has been assessed at about 16,000 MW from agro residues. However, as of September 2008, only about 656 MW of grid connected biopower based on agriresidues and plantations could be achieved (http://www.hindu.com/pp/2009/02/14/stories/2009021452830300.htm).

The amount of increased biomass production over the years would depend on forest management intensity (including fertilization), recovery of forest biomass residues (including thinning residues, harvest slash, and stumps), the use of fallow land for biofuel cultivation, the selection of species to be cultivated, the level of wood products manufactured and the recovery of associated by-products, and the recovery of post-consumer wood waste (including demolition residues) (Gustavsson et al. 2007). The most efficient utilization of these resources comes when they are converted to fuels/chemicals by appropriate technologies.

16.3 NECESSITY AND OPPORTUNITY FOR LIGNOCELLULOSIC BIOMASS

Biomass refers to all organic matter that is generated by photosynthesis using solar energy. Researchers feel that energy production from biomass has the advantage of forming smaller amounts of GHGs compared to the conversion of fossil fuels because the CO_2 generated during the energy conversion is consumed during subsequent biomass regrowth.

Growing plants are very advantageous for many reasons including the efficient absorption of carbon dioxide, a harmful GHG. On an average, a tree can absorb 50 lbs of this gas per year. One acre of tree plantations absorbs 27,647 lbs (12,524.1 kg) of carbon dioxide per annum. India emits 479.039 MMT of CO_2 per annum; 479.039 MMT of CO_2 will be absorbed by 38.25 million acres (15.45 million hectares) of tree plantations (www.coloradotrees.org/benefits.htm; http://rainforests.mongabay.com/carbon-emissions/india.html). Sustainably managed biomass systems recycle the carbon that is taken in by photosynthesis and return it to the atmosphere during combustion. The net effect is that solar energy is used to provide energy services, and plants provide energy concentration and store chemical bonds (Gustavsson et al. 2007).

Crop-based fuels denoted as first-generation biofuels—including bioethanol and biodiesel—once emerged as real alternatives to the use of gasoline and conventional diesel in transportation. The serious concern with conventional first-generation biofuels is the fact that their production processes generally involve the use of "food" crops or edible portions (lipid) of food crops; causing food vs. fuel issues.

Second-generation liquid biofuels are generally produced from lignocellulosic biomass, which are either nonedible residues of food crop production, nonedible whole plant biomass (e.g., grasses or trees specifically grown for production of energy), or forest residues. The main advantage of the production of second-generation biofuels from nonedible feedstocks is limiting the direct food versus fuel competition. Feedstock involved in the process can be bred specifically for energy purposes, enabling higher production per unit land area; a greater amount of aboveground plant material can be converted to produce biofuels. As a result, this will further increase land-use efficiency compared to first-generation biofuels.

Lignocellulosic biomass is composed of the following:

- **Cellulose:** Cellulose is the structural component of a plant's cell wall. It is made up of linear polysaccharides in the cell walls of wood fibers, consisting of D-glucose molecules bound together by β-1,4-glycosidic linkages (comprises about 41%).
- **Hemicellulose:** This is an amorphous and heterogeneous group of branched polysaccharides (copolymer of any of the monomers glucose, galactose, mannose, xylose, arabinose, and glucuronic acid). Hemicellulose surrounds the cellulose fibers and is a linkage between cellulose and lignin (about 25%–35% of the dry weight of biomass).
- **Lignin:** Lignin is a polymer that holds cellulose and hemicelluloses together. It is a highly complex three-dimensional polymer of different phenylpropane units bound together by ether and carbon–carbon bounds. Lignin is concentrated among the outer layers of the fibers, leading to structural rigidity and holding the fibers of polysaccharides together (about 27%).

In addition, small amounts of extraneous organic compounds are found in lignocellulosic materials (about 4%) (Stöcker 2008).

As stated by Larson (2008), it is believed that the basic characteristics of feedstocks hold potential for lower costs and significant energy and environmental benefits for the majority of second-generation biofuels.

Third-generation biofuels are obtained from algae (both microalgae and macro algae). In common production units, CO_2 is fed into the algae growth media either from external sources such as power plants (Emma et al. 2000; Hsueh et al. 2007) or in the form of soluble carbonates such as Na_2CO_3 and $NaHCO_3$ (Colman and Rotatore 1995; Suh and Lee 2003). Other inorganic nutrients required for algae production include nitrogen, phosphorus, and silicon (Hu et al. 2008). Mata et al. (2010) suggested that efforts in microalgae production should concentrate on reducing costs in small- and large-scale systems that could be achieved by using cheap sources of CO_2 (flue gas), nutrient-rich wastewaters, inexpensive fertilizers, cheaper design culture systems with automated process control, greenhouses, and heated effluents to increase algal yields. In addition to saving costs of raw materials, these measures would help reduce GHG emissions, waste disposal problems, and feed cost. Availability of microalgae biomass for different applications would increase and contribute to the sustainability and market competitiveness of microalgae industry. Studies regarding the selection of the most efficient algal strains, cultivation of selected strains at best growth rates, designing the metabolic pathways by engineering those reactions that control lipid synthesis to produce algal cells saturated with desirable lipid contents, and optimization of lipid extraction process to standardize an efficient and economical method of oil recovery from the algal cells should be carried out (Nigam and Singh 2011).

16.4 BIOFUEL IMPLEMENTATION STRATEGY

Designing a national strategy should aim at expanding the use of biomass energy involving boundary conditions beyond those linked to techno-economic aspects of energy systems. Policy priorities, short- and long-term goals, available biomass potential, possibilities for technological development, economic situation and identification of niche markets, and opportunities for added value should be included. As the system becomes more complex and the number of factors increases, strategy development becomes tough, but regional and international perspectives are important when developing national strategies because additional opportunities may arise and implementation priorities may shift as the geographic scale of analysis becomes larger. Usage of biomass energy as inexpensively as possible will facilitate an expansion of its use, and using it as efficiently as possible will increase the impact of its use on achieving policy objectives. This would be possible if biomass is used in place of fossil fuels and in applications where it most effectively serves society's objectives. The choice of biomass energy systems and the parameters chosen for comparing them would vary according to the objectives of the analysis. As an example, GHG benefits of biomass use can be optimized with respect to any of several limiting factors, including per ton of biomass feedstock, per hectare of land, per unit of monetary resources spent for carbon emission reduction, or per unit of bioenergy output that can be absorbed by a specific market or sector (Schlamadinger et al. 2005). Regardless of the factor to be optimized, it has become widely accepted to include the full chain of biomass use within the analytical system boundaries, from primary plant growth to final fuel consumption (Schlamadinger et al. 1997). There has been growing interest in efficient ways to use biomass for the substitution of fossil fuels and nonbiomass materials.

Energy demand is projected to grow by more than 50% by 2025, with much of this increase in demand emerging from several rapidly developing nations. Clearly, increasing demand for finite petroleum resources cannot be a satisfactory policy for the long term. In view of changing world energy needs, a research road map for the biorefinery of the twenty-first century is vital. This biorefinery vision will contribute to sustainability not only by its inherent dependence on sustainable bioresources but also by recycling waste, with the entire process becoming carbon neutral. Complete knowledge of plant genetics, biochemistry, biotechnology, biomass chemistry, separation, and process engineering is required to have a positive impact on the economic, technical, and environmental well-being of society. An integrated biorefinery is an approach that optimizes the use of biomass for the production of biofuels, bioenergy, and biomaterials for both short- and long-term sustainability (Ragauskas et al. 2006).

16.5 BIOFUEL STRATEGIES OF OTHER NATIONS

An emphasis on biofuels as an alternative energy source has been an issue of concern lately. These concerns are related to the impact of biofuel not only on food security (Promotion and Protection of all Human Rights, Civil, Political, Economic, Social and Cultural Rights, Including The Right to Development, Report of the Special Rapporteur on the Right to Food, United Nations General Assembly, Online source http://daccess-ods.un.org/TMP/5810871.html) but also on the environment and on the socioeconomy (An EU Strategy for Biofuel, Communication from the Commission, SEC(2006)142, Online source: ec.europa.eu/agriculture/biomass/biofuel/com2006_34_en.pdf).

- **China:** In an effort to deal with environment pollution and an energy crisis, China has put forward a strategic framework for energy development for the next 20 years (Research Team on the Strategy and Policy for Energy Development of China 2003). In this framework, energy saving is considered to be the top priority, and other strategies include adjustment of energy use structure, use of environmentally friendly energy, and reduction of dependency on imported oil to below 55%. Major pollutants would be attenuated by 46%–60% from their current status. Biomass energy would become the major component in the future sustainable energy system and is expected to account for 40% or more of the total energy consumption by the middle of this century. Energy saving and constantly searching for new energy sources is of paramount importance to China's ongoing economic reform and industrial structure adjustment. Research facilities are being established to breed and select high-quality seed sources targeting these plants, including both natural energy woody plants and common crops (Research Team on the Strategy and Policy for Energy Development of China 2003).

- **European Union:** The European Union (EU) also explicitly discusses energy security as a main concern of its energy policy and aims to address this concern by reducing energy demand, diversifying the sources of energy supply, and increasing reliance on internal sources of supply (Green Paper—Towards a European Strategy for the Security of Energy Supply (COM (2000) 769 Final). European Commission, Brussels). The EU has developed a strategy to increase the use of renewable energy resources (Energy for the Future: Renewable Sources of Energy, White Paper for a Community Strategy and Action Plan (COM (97) 599 final). European Commission, Brussels), including the use of biofuels in the transport sector (Directive 2003/30/EC of the European Parliament and of the Council of 8 May 2003 on the promotion of the use of biofuels or other renewable fuels for transport, IJ L 176, 15.7.2003, European Union, 42–46) and a biomass action plan (Communication from the Commission: Biomass Action Plan (COM (2005) 628 Final). European Commission, Brussels). Recently, the Swedish government has set a target of reducing, by the year 2020, the use of oil in road transport by 40%–50%, in industry by 25%–40%, and in building heating by 100% (Commission on Oil Independence, 2006 Making Sweden an Oil-Free Society. Swedish Government Offices, Stockholm Web-accessible at http://www.sweden.gov.se/sb/d/2031/a/67096S).
- **United States:** The United States has issued a sweeping new set of rules and directives regarding U.S. biofuel policy, including the release of the revised Renewable Fuel Standards from the Environmental Protection Agency (EPA) and a new set of "Lead Agency" assignments to support first-generation biofuels while driving the development and commercialization of advanced fuels, with a focus on drop-in fuels for aviation and ground transportation. The EPA has recalibrated the targets and categories originally established by Congress in 2007 in the Energy Independence and Security Act. The overall target of 36 billion gallons of biofuels by 2022 has been affirmed, along with a goal of 15 billion gallons from first-generation fuels and 21 billion gallons from advanced biofuels, but annual targets have been revised. In the case of modern corn ethanol, there is 21% reduction with international indirect land use change (ILUC) emissions and a 52% reduction without international ILUC change emissions of GHGs is aimed. It was found that ethanol from switchgrass reduces GHG emissions by 110% via the biochemical conversion process and 72% via the thermochemical process. Ethanol from corn stover reduces GHG emissions by 130% (biochemical) and by 93% (thermochemical) (http://www.biofuelsdigest.com/blog2/2010/02/04/obama-administration-reorganizes-us-biofuels-policy-leadership-rules-in-sweeping-change/).
- **Japan:** The Japanese government has officially defined biomass as one of the new energy resources in the "law concerning special measures for promotion of the use of new energy" for the first time (Yoshioka et al. 2005). The targets based on the premise of maximum efforts from the government and the public in the fiscal year 2010 are 340,000 cubic meters of crude oil equivalents with biomass power generation, corresponding to 330 MW capacity of electrical power generation, and 670,000 cubic meters of crude oil equivalent with thermal utilization of biomass (Yoshioka et al. 2005). The utilization of bioenergy is becoming an important issue in Japan. Among various biomass resources, woody biomass particularly attracts a great deal of attention. This is because not only its amount is abundant but the energy utilization of woody biomass is expected to contribute to revitalizing the forest and forest products industries, which have been depressed for a long time, as well as to maintaining the relevant ecological (including biological diversity), economic, and social functions of man-made forests, which are behind in tending. The visions for introduction and diffusion of bioenergy utilization should be adopted at a national level. These visions could consider the quantification of available woody biomass resources for energy, the development of low-cost harvesting and transportation systems, and the conversion processes comprehensively, and these should answer the needs of the social system (Yoshioka et al. 2005).

16.6 INDIAN BIOMASS SCENARIO

Biomass is a material that is very region specific and its composition varies from place to place. This difference is due to the varying soil properties and climatic conditions. Owing to this, the calorific value of biomass also varies as does the cost. The gross calorific value of biomass in different parts of India is shown in Table 16.1. It is proven industrially that the quantity

Table 16.1 Gross Calorific Value of Biomass in Different Parts of India

State	Calorific Value (kcal/kg)
Andhra Pradesh	3275
Haryana	3485
Maharashtra	3611
Madhya Pradesh	3612
Punjab	3368
Rajasthan	3689
Tamil Nadu	3300
Uttar Pradesh	3371
Other States	3467

Table 16.2 Cost of Biomass at Various Parts of India

State	Biomass Price (Rs/MT)
Andhra Pradesh	1301
Haryana	2168
Maharashtra	1801
Madhya Pradesh	1299
Punjab	2092
Rajasthan	1822
Tamil Nadu	1823
Uttar Pradesh	1518
Other States	1797

of moisture content determines the energy required for processing and the resulting product portfolio. This emphasizes the requirement of region specific technology. India sees its biofuel program to be sustainable because of the program's dependence on nonedible feedstock, which is supposed to be derived from wasteland and thus does not entail diversion of farm land or food crops (National Action Plan on Climate Change, Government of India).

Biomass costs for different areas in India are given in Table 16.2 (Gujarat Electricity Regulatory Commission 2010). The price of biomass fuel depends on the price paid to farmers, cost of biomass charged by forest department/state government, cost related to collection, storage, transportation, loading and unloading cost, agent's commission, and so forth. The fuel procurement and transportation of biofuels are handled by an unorganized sector; thus, the prices are influenced by the local factors.

The energy that can be generated using biomass as fuel depends on properties such as moisture content, calorific value, and the noncombustible materials in the biomass. The heat content of the biomass of one crop residue differs from that of another. Biomass used for fuel production could be cost-effective at one location but may prove to be costly at a different location. But the same biomass could be a good source of chemicals in the latter location. The cost also depends on the amount of biomass generated at one place. Thus, economic analysis of biomass-based processes at one location will not be appropriate for another and will follow the normal demand and supply scenario.

Lignocellulosic biomass can be used to make various fuels, chemicals, and petrochemicals that are all high value products when compared to the original feedstock. The exponential increase in the value addition can be achieved by producing high-quality specialty and bulk chemicals, and liquid and gaseous fuels from lignocellulosic biomass.

Under the agro residue category, the amount of biomass generated is 511 MMTPA. Out of this, 18,729 MWe of power is produced, and 145 MMTPA of biomass exists as surplus. Under the forest and wasteland category, the amount of biomass generated is 155 MMTPA. Out of this, 14,567 MWe of power is produced, and 104 MMTPA of biomass exists as surplus (http://lab.cgpl.iisc.ernet.in/ Atlas/Tables/Tables.aspx). This shows us that even after power is produced using biomass, we are left with excess biomass that can be used to make hydrocarbons by the various methods of bio-mass conversion. Increasing the thermal power plant efficiency will help generate more power from lignocellulosic biomass. The data has also been shown as Figures 16.1 and 16.2.

Additional biomass (such as algae etc.) can be produced by using advanced technologies (such as helioculture, etc.) that utilize the anthropogenic CO_2 emitted into the atmosphere, which can add to the existing biomass resources. More than 99% of commercial algae biomass pro-duced worldwide currently is mainly from seaweed farmed near the seashore. India has the opportunity to utilize the algae that can be cultivated in large quantities in the coastal regions due to the tropical climate (http://www.deccanherald.com/content/260639/algae-biofuel-could-offer-solution.html).

Thus we see that the context for the emergence of biofuel policy in India projects a very complex picture. On one hand, India's growing economy is leading to rising oil import bill. On the other hand, the majority of India's rural population subsists on the fuels that are derived from biomass, which in turn comes mainly from forest and nonforest lands. International and national policy scenarios provide an opportunity for initiating programs for alternative energy and regeneration of degraded lands. There is also the problem of addressing the issue of land degradation related to a large livestock population, especially as many people are dependent for their livelihoods on the same degraded lands—the so-called "wastelands." Wasteland is a medieval English term used in a legal sense to refer to land that is unoccupied, undeveloped, or uncultivated. Land resources in India are under tremendous human and livestock pressure as about 16% of the world's people and 20% of the world's livestock are supported by a mere 2.5% of

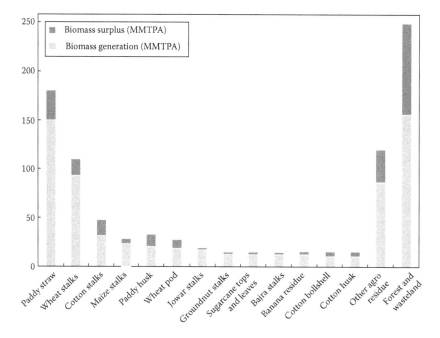

Figure 16.1 Biomass availability in India (absolute).

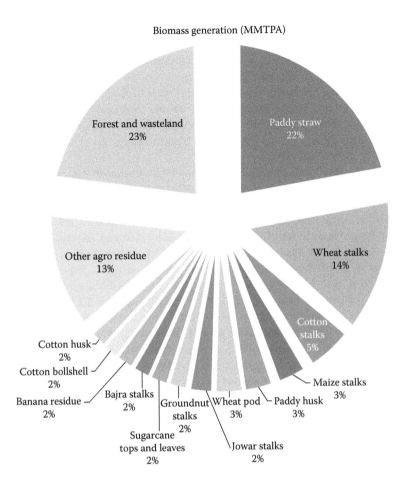

Figure 16.2 Biomass availability in India (percentage).

the world's geographical area (The State of Environment—India 2001). A large part of this land area is also designated as wasteland—around 55 Mha (Nagar, Biofuel policy process in India: context, actors & discourses).

The wasteland can be used for biofuel production in two ways—first, for the generation of feedstock to be used for a national network of small, decentralized biomass power plants and second, for cultivation of lignocellulosic biomass to be used as a feedstock for production of biofuels and value added chemicals (Nagar, Biofuel policy process in India: context, actors & discourses).

There are several factors that determine whether a crop is suitable for energy use. The main material properties of interest during subsequent processing as an energy source relate to a crop's moisture content (intrinsic and extrinsic), calorific value, proportions of fixed carbon and volatiles, ash/residue content, alkali metal content, and cellulose/lignin ratio. The first five properties are significant for dry biomass conversion processes, although the first and last properties are important for wet biomass conversion (McKendry 2002).

16.6.1 Agriculture Crop Residues

The term agricultural residue is used to describe all the organic materials that are produced as the by-products from harvesting and processing of agricultural crops. These residues can be further categorized into primary residues and secondary residues. Agricultural residues, which are generated in the field at the time of harvest, are defined as primary or field based residues (e.g., rice straw and sugarcane tops), whereas those coproduced during processing are called secondary or processing based residues (e.g., rice husk and bagasse).

Availability of primary residues for energy application is usually low because collection is difficult and because they have other uses as fertilizer, animal feed, and so on. However, secondary residues are usually available in relatively large quantities at the processing site and may be used as a captive energy source for the same processing plant involving no or little transportation and handling cost (CES 1995; CMIE 1997).

The current share of biofuels in the consumption of transportation fuels is extremely low and is confined mainly to 5% blending of ethanol in gasoline, which the Indian government had made mandatory in the states of Andhra Pradesh, Goa, Gujarat, Haryana, Karnataka, Maharashtra, Punjab, Tamil Nadu, Uttar Pradesh, and Uttaranchal, and in the union territories of Daman and Diu, Dadra and Nagar Haveli, and Chandigarh. The 2006 demand for ethanol at 5% gasoline doping levels was 0.64 billion liters, although the estimated current demand for 10% blending of the entire amount of gasoline sold in India is about 1.5 billion liters. This demand is projected to be 2.2 billion liters in 2017. The Indian alcohol industry is fairly mature with 295 distilleries operational across the country having an installed annual capacity of 3.2 billion liters. The alcohol produced in India is mainly consumed by the liquor industry with the remaining share going into the chemical industry. According to one survey, the major agroresidues, in terms of volumes generated (in million metric tons, or MMT), are rice straw (112), rice husk (22.4), wheat straw (109.9), sugarcane tops (97.8), and bagasse (101.3) (Pandey et al. 2009). These account for almost 80% of the residue generated by the crops that were studied. Looking at the estimated amounts of residues, it is apparent that the most potent feedstocks (in terms of generation in large quantities) are rice straw, wheat straw, bagasse, and sugarcane tops, followed by other cereal residues. There are also other crop residues—such as those from cotton and chilli cultivation (18.9 and 0.6 MMT, respectively). Residues from processing of forest products such as bamboo and reed may also serve as potent feedstock. Pine needles emerged from the survey as an unexpected feedstock with an estimated annual availability of 1.6 MMT; however, the resource presents a problem with respect to collection and logistics. The physical properties, content of cellulose, and fermentable pentosans in each of these residues are different; accordingly, the processing technologies might have to differ slightly if they are to be used as raw material for ethanol production (Sukumaran et al. 2010).

To explain the spread of biomass across India, a few states have been taken into consideration and data from those places have been provided. Punjab, along with the neighboring state of Haryana, are referred to as the "Grain Bowl of India." Agriculture is the major economic activity, sustaining nearly 70%–80% of the total population. Of the total geographical area, 84% is under agricultural use, 0.18% area is under cultivable wasteland, and 8.94% of the area is not available for cultivation within the state. The cropping intensity in the state is averaged at 188%, which is considered as one of the highest within the country. This is attributed to the significant availability of groundwater for irrigation purposes. It has been estimated that around 55%–60% of the net sown area is irrigated. Making up only 1.6% of India's geographical area, Punjab produces approximately 22% of wheat crops, 10% of rice crops, and 13% of cotton crops of the total production of these crops in the entire country. The total residue generation from all the major and minor crops was reported to be 40.142 Mt y^{-1}. Of this, wheat and paddy in the form of straw and husk alone contributed more than 86%, while the remainder was contributed by residues of cotton (4.12%), fodder (4.11%), sugarcane (2.30%), and so on. The total residue consumption from the agriculture sector in the state is 28.518 Mt y^{-1}, which is about 71.04% of the total generation. Of this, domestic fuel and fodder together consume more than 68%; the rest is used in thatching and in manure form (Chauhan 2012).

16.6.2 West Bengal

Paddy straw, wheat, potato, pulses, and oil seeds are the major crops cultivated in the West Bengal. Tea is also a major crop in North Bengal, especially in the Darjeeling and Jalpaiguri districts. The main crop residue generated here is paddy straw, which is mainly consumed as fodder. Minor crop residues generated in West Bengal are wheat straw, potato stalks, cotton stalks, rapeseed, and mustard stalks. These are used as supplementary domestic fuel (Saud et al. 2011).

16.6.3 Haryana

Wheat, paddy, cotton, jowar (bicolor sorghum), bajra (*Pennisetum glaucum*), pulses, sugarcane, oil seeds, and potato are the major crops grown in the state of Haryana. In light of the good irrigation sources, the cropping intensity in the state is averaged at 170% (http://www.haryana-online.com). There are two major agricultural seasons here, *Rabi* (winter crop) and *Kharif* (summer crop). The major crops grown during the Rabi season are wheat, barley, gram, mustard, cotton, and sugarcane; during the Kharif season, paddy, jowar, bajra, and maize are the major crops. The residues generated from these crops consists of wheat stalk and pod, barley stalk, gram stalk, mustard stalk and husk, cotton stalk, sugarcane top and trash, paddy husk and straw, jowar stalk, bajra stalk and cobs, and maize stalk and cobs. The numerical values are at a lower level and real potential is higher due to nonavailability of statistics.

16.7 VARIOUS BIOMASS CONVERSION TECHNOLOGIES

An abundant raw material—consisting mainly of lignin, cellulose, and hemicellulose—enters the biorefinery. This raw material is converted through a number of different processes into a mixture of products, including biofuels, valuable chemicals, heat, and electricity. However, the logistical problems involved in the running of a modern biorefinery should not be underestimated, especially regarding the gathering of sufficient biomass. This technology requires space and biomass. It is more feasible to set up such a facility in a rural area where biomass procurement is easy. Setting up the plant in a city will entail transportation of biomass and, therefore, higher costs.

Thermochemical methods have been in practice for the production of second-generation biofuels. They involve the cracking or rearrangement of the macromolecules in the presence of heat and/or catalysts into smaller hydrocarbons that can be used as transportation fuels, chemicals, or feedstocks for petrochemical industry. Various thermochemical methods include combustion (exothermic chemical reaction accompanied by large heat generation), gasification (changing solid raw material into fuel gas or chemical feedstock gas, called syngas), pyrolysis/fast pyrolysis (decomposition of *biomass* in the absence of oxygen for production of liquids), carbonization (obtaining charcoal by heating solid biomass in the absence of air or oxygen), and hydrothermal gasification and hydrothermal liquefaction (conversion of solid biomass into gaseous and/or liquid products in the presence of water/steam).

Grass species, trees, and agricultural and industrial residues can be converted via two main pathways: a biochemical or a thermochemical route. The first relies on dedicated enzymes and/or microorganisms that can break down cellulose and/or hemicellulose to reach the sugars contained in the biomass. This pathway yields "cellulosic ethanol." The thermochemical route converts biomass via processes mentioned above.

The *World Energy Outlook 2009* (IEA 2009) projects that biofuels will provide 9% (11.7 EJ) of the total transport fuel demand (126 EJ) in 2030. In the *Blue Map Scenario* of *Energy Technology Perspectives 2008* (IEA 2008), which extends the analysis until 2050, biofuels are expected to provide 26% (29 EJ) of total transportation fuel (112 EJ) in 2050, with second-generation biofuels accounting for roughly 90% of all biofuel. More than half of the second-generation biofuel production in the *Blue Map Scenario* is projected to occur in non-Organization for Economic Cooperation and Development (OECD) countries, with China and India accounting for 19% of the total production.

In many of the other countries (e.g., Cameroon, India, Tanzania, and Thailand), significant investments in technological improvement, new infrastructure, and capacity building are needed to increase the productivity and sustainability of the agricultural sector. This could allow future dedication of agricultural land to second-generation feedstock production.

The constraints related to the availability of additional land suggest that second-generation biofuel industries should focus on currently available feedstock sources in the initial phase of the industry's development. Agricultural and forestry residues form a readily available source of biomass and can provide feedstock from current harvesting activities without the need for additional land cultivation (IEA 2010).

The recent technologies in use for the conversion of lignocellulosic biomass into value added hydrocarbons that have been developed by various groups are summarized here. Dynamotive utilizes fluid-bed technology and has a plant of capacity 100 t/day with waste sawdust as feed. The prepared feedstock (<10% moisture and 1–2 mm particle size) is fed into the bubbling fluid-bed reactor, which is heated to 450°C–500°C in the absence of oxygen. Three products are produced: bio-oil (60%–75% by weight), char (15%–20% by weight), and noncondensable gases (10%–20% by weight) (http://www.dynamotive.com/technology/fast-pyrolysis/; Sandvig et al. 2003; Dynamotive 2007). Aston University, Birmingham, UK, has developed a rotating blade reactor for ablative pyrolysis on a small scale of 3 kg/h and yields up to 80% by weight are obtained (Peacocke and Bridgwater 1995). CNRS (France) has reported bio-oil yields of about 74% using an ablative pyrolysis pilot plant of 250 kg/h (Lede et al. 2007). PYTEC, in Germany, has patented a fast pyrolysis system, based on a reactor, in which the ablative process takes place continuously. At PYTEC's laboratory plant, ablation rates of 4 mm/s have been achieved. PYTEC is planning a 2 ton/h unit in Mecklenburg-Vorpommern (http://www.pytecsite.de/pytec_eng/flash_pyrolyse_2.htm; Meier 2004, 2007; Scholl et al. 2004). TNO has been collaborating with the University of Twente on the development of a reactor, the PyRos, that also separates the products. It is a cyclone reactor with an integrated rotating particle filter in which full pyrolysis and separation can occur rapidly. TNO has a 30 kg/h pilot plant at University of Twente, Netherlands (Bramer et al. 2004; http://www.tno.nl/content.cfm?context=thema&content=markt_product&laag1=895&laag2=190&laag3=224& item_id=350&Taal=2). Enco Enterprises, Inc., has a 200 kg/h fluid-bed system that converts peat moss into liquid fuel (Fransham 2006). Vapo Oil and Fortum Oil have patented a new principle for fluid-bed pyrolysis (Nieminen et al. 2003). Ensyn's Rapid Thermal Processing is a joint venture with UOP and is commercializing the process under the name Envergent Technologies LCC. There are 100 kg/h, 40 kg/h, and 10 kg/h plants in Ottawa and a 20 kg/h plant at VTT, Finland. The weight percent yields from processing dry (~8% moisture) biomass are approximately 65%–80% liquid, with 12%–16% each of char and combustible gas (Graham et al. 1994; Freel 2002; US Pat., 6 485 841, 2002; http://www.ensyn.com/technology/process-yield-product-quality/). Pyrovac International has a 3 ton/h demonstration plant in the city of Saguenay, Quebec, Canada. It first used bark residues and then later crumb rubber. It is a vacuum pyrolysis at reduced pressure of about 20 kPa. The University of West Ontario has developed a circulating fluid bed that gives a bio-oil yield of more than 70 wt.%. BTG has developed a rotating cone reactor and has a 50 t/day plant in Malaysia (Venderbosch and Prins 2010). ABRI-Tech has a 50 ton/day plant at Iowa. The drier design allows large size reduction and moisture removal from biomass and uses much less energy than conventional biomass dryer systems. Also, the rapid heat transfer process provides an efficient method for transferring energy to the reaction. This process utilizes waste heat and energy, limiting the reliance on external power (Badger and Fransham 2003; http://www.advbiorefineryinc.ca/technology/). Forschungszentrum Karlsruhe has developed a fast pyrolysis reactor that uses straw as feedstock (Henrich et al. 2009).

ZeaChem's first commercial biorefinery using hybrid poplar woody biomass is expected to have a capacity of 25 M gallons per year (GPY) and will be located in Boardman, Oregon (http://advancedbiofuelsusa.info/zeachem-signs-binding-feedstock-agreement-with-greenwood-tree-farms-to-supply-its-first-commercial-biorefinery). Mitsubishi Heavy Industries, Ltd. (MHI) has successfully established technology to produce ethanol for automobile fuel, satisfying the Japanese Automotive Standards Organization (JASO), from lignocellulose (soft cellulose) such as rice straw and barley straw. For preprocessing and saccharification, MHI adopted a hydrothermal treatment system that enables continuous processing of feedstocks. In conjunction with the fermentation process, Hakutsuru Sake Brewing, in cooperation with Kobe University, established a technology to convert sugar originating from rice and barley straw into ethanol by using yeast. The yeast is selected from those actually in use or from those bred with nongenetically modified organism (non-GMO) technology.

Kansai Chemical Engineering verified distillation and dehydration technologies for producing bioethanol that satisfy automotive fuel standards. The technology involves use of a new type of distillation column developed by the company that enables longer continuous operation than conventional systems and a liquid phase adsorption type dehydration unit that enables processing with less energy than existing gas phase types.

At the same time, based on the verification results, the bioethanol production cost was estimated for a commercial-scale plant to be built in Hyogo Prefecture. It was confirmed that total running costs from collection and transportation of feedstocks to ethanol production can be achieved below the target of 90 yen per liter (http://advancedbiofuelsusa.info/mhi-establishes-technology-to-produce-biofuel-locally-at-low-cost-from-rice-and-barley-straws). The biobutanol company, Cobalt Technologies, and American Process Inc. (API), a developer of lignocellulosic sugar production technologies, announced an agreement to build the world's first industrial-scale cellulosic biorefinery to produce biobutanol.

GreenPower, using biobutanol technology, selectively converts part of a boiler cellulosic biomass feedstock into renewable biobutanol, an industrial chemical widely used in paints and other coatings, and a platform for production of renewable jet fuel and other valuable compounds.

Under the agreement, Cobalt Technologies and American Process will integrate Cobalt's continuous fermentation and distillation technology into American Process's Alpena Biorefinery, currently under construction in Alpena, Michigan (http://advancedbiofuelsusa.info/cobalt-technologies-american-process-to-build-first-cellulosic-biobutanol-refinery). Cobalt is the first company to produce a drop-in replacement for petroleum and petrochemicals from beetle-affected lodgepole pine. The company's innovative technology offers the potential for converting beetle-killed pine into a low-carbon, sustainable biofuel and chemical (http://advancedbiofuelsusa.info/cobalt-technologies-is-first-to-create-renewable-biobutanol-fuel-from-beetle-killed-pine). HCL CleanTech, a U.S.–Israeli biofuels technology development company, has chosen North Carolina as the site for its administrative headquarters and for its first pilot plant. The company's technology utilizes hydrochloric acid to break down lignocellulose, which releases starches for enzymatic fermentation into fuel alcohol. In the process, HCL CleanTech recovers the hydrochloric acid, which allows maximum efficiency of enzymes while reusing the acid (http://advancedbiofuelsusa.info/hcl-cleantech-establishes-administrative-and-pilot-plant-facilities-in-north-carolina). The existing catalysts and technology are still in the evolutionary stages and not yet applicable for various feedstocks. With this scenario, the development of novel and integrated processes is of immense importance to convert lignocellulosic biomass into value added hydrocarbons. This eventually makes the biofuels directly usable or allows them to be treated easily along with the crude oil distillation products.

The main technologies available to biorefineries can be classified as extraction, biochemical, and thermochemical processes. Currently, these methodologies are usually studied independently of one another, with each jostling for primacy as the biorefinery technology of the future. However, it is essential that the strengths and weaknesses of all the technologies available are recognized to enable the integration and blending of different technologies and feedstocks to best maximize the diversity of applications and products formed. Nonconventional energy sources (microwave, gamma irradiation etc.) may offer advantages in biomass conversion over conventional sources of heating (electrical, gas fired etc.) due to increased decomposition rates, higher heating rates, better controllability, higher energy efficiency and better selectivity in terms of the ratio of solid, can occur at much lower temperatures than previously reported (Guiotoku et al. 2009; Budarin et al. 2010, 2011; Zhang et al. 2010).

The UOP's rapid thermal pyrolysis (RTP) process generates a relatively high yield (i.e., approx 75 weight%) of pourable, liquid bio-oil from residual forestry or agricultural biomass. The process also produces by-product char and noncondensable gas, both of which can be efficiently used to provide process energy, which can then be used in the reheater to maintain the RTP process and/ or in the dryer to condition the biomass. The result is a process that is, for all intents and purposes, self-sustaining. India produces 666 MMTPA (agro, wasteland, and forest residue) of biomass. With a conversion of about 75%, we will obtain 500 MMTPA of liquid hydrocarbons. Liquid hydrocarbons are given the preference because of their high energy density, easy usage, and their wide range of applications. As can be seen, there is an enormous contribution of fossil fuels to the hydrocarbon sector. With the growing fear of fossil fuel unavailability, biomass can be a very good alternative for liquid hydrocarbons (chemicals and fuels). Its various advantages can make it a very good potential hydrocarbon source in a newly industrialized country such as India.

The size necessary for efficient biofuel processes is another key element for evaluation. In fact, it will determine the distance from which it is necessary to transport the biomass to allow the constant feed of the plant (e.g., the penalty in CO_2 emissions due to transport of the feedstock to the plant) and other relevant aspects for sustainability—such as the investment costs, the possibilities of the location, and so on. Flexibility in feed (e.g., the possibility of using multiple types of raw materials) is another key parameter. Using different lignocellulosic sources (agrofood and wood production residues, sorted municipal solid wastes, herbaceous energy crops) makes it possible to limit the distance from which the raw materials are transported, guaranteeing a constant supply of feedstock and efficient utilization of local biomass.

In most of the developed countries, biomass is a limited resource. There are options (such as regional biomass supply) to increase the local added value and transregional supply of biomass and bioenergy. It is preferable from a social perspective for the motivation discussed earlier, but many types of biofuels and biobased chemical production are subject to the system's inherent trade-off between the preferably medium- to large-scale plants with regard to "economy of scale" (and thus a high biomass demand and a relatively small catchment area for the supply of this feedstock demand) (Centi et al. 2008).

Thus there are several parameters determining the optimal choice, and these are significantly dependent on local conditions. An optimum technology for one country may not necessarily be valid in another country. Due to the presence of multiple aspects, advanced analytical techniques such as life cycle analysis (LCA), though necessary, are not sufficient. Each country should develop its own preferable mix of technologies or at least invest significantly in research and development to enable the optimal choices that best suit a country's needs (Centi et al. 2011). The lignocellulosic biomass can thus be a potential substitute for fossil-based resources to meet the future energy demands of a developing country such as India and thereby increase its energy security.

16.8 CONCLUSIONS

It is widely accepted that lignocellulosic biomass has a tremendous potential for reducing concerns about volatility of crude prices and environmental and national energy security. Lignocellulosic biomass is the only sustainable source of carbon that can be used to make renewable fuels and chemicals. It is clear from the available lignocellulosic biomass data that India has the potential to produce valuable hydrocarbons for transportation fuels and chemicals using advanced and integrated conversion technologies. The successful introduction and implementation of biobased fuels depends on various issues and some of them are outlined here.

To begin with, fossil fuels should be replaced with renewable sources of energy wherever it is possible. More plants should be grown because they reduce carbon dioxide in the atmosphere. Biomass has a wide range of functionality, and it can be used for production of value added hydrocarbons (i.e., chemicals—preferably—as well as transportation fuels and other high value materials that can store energy for an extended period of time). When biomass technologies are properly commercialized in small regions, they might solve other problems such as rural unemployment and might help to meet food and hydrocarbon needs of that particular region.

Research and development need to be pursued for fundamental studies in the conversion of lignocellulosic biomass using advanced chemistry, nanotechnology, and engineering aspects. Heterogeneous catalysis made it possible to efficiently convert fossil hydrocarbons to fuels and chemicals, and this has the potential to convert renewable resources to bio-oil and chemicals. Thus, upgrading the bio-oil produced using novel heterogeneous catalysts will yield valuable hydrocarbons useful for transportation fuels, specialty chemicals, and bulk chemicals. Biomass conversion units have the advantage that they can be made having a small capacity depending upon biomass availability. The technology must be region and feed specific, owing to the varying climate and soil conditions. In the area of biotechnology, the development of microorganisms to ferment all kinds of sugars obtained from the hydrolysis of lignocellulosic biomass is required. A net increase in the conversion and energy efficiency of all existing technologies to obtain more bio-oil or other valuable products by producing fewer emissions is the need of the hour.

Research and development is also required in the area of agronomy for developing region-specific seeds that can give high yields when used in the specified soil and climate conditions and also taking into consideration water and fertilizer availability. The immediate task should be to forecast the problems associated with different materials/processes. However, it is still too early to fully assess the potential social, economic, and environmental impacts of large-scale second-generation biofuel production in practice. These research steps are suggested to understand better the potential and impact of second-generation biofuels in developing countries and emerging economies. Once this is successfully done, lignocellulosic biomass will not only be a sustainable resource for hydrocarbons in the future but will also help in a smooth transition from the fossil-based economy to a renewable energy-based economy.

ACKNOWLEDGMENTS

The authors thank The Director, Council of Scientific and Industrial Research (CSIR), Indian Institute of Petroleum, Dehradun, India, for his constant encouragement and support. RS thanks CSIR, New Delhi, for providing senior research fellowship (SRF). The authors thank CSIR for funding in the form of XII Five-Year Plan project (CSC0116/BioEn) and the Ministry of New and Renewable Energy (MNRE), Government of India, for providing financial support.

REFERENCES

An EU Strategy for Biofuel, Communication from the Commission, SEC(2006)142, Online source: ec.europa. eu/agriculture/biomass/biofuel/com2006_34_en.pdf

Annual Report 2009–2010, Department of Chemicals and Petrochemicals, Ministry of Chemicals and Fertilizers, Government of India, New Delhi.

Badger, P.C. and Fransham, P. 2003. Use of mobile fast pyrolysis plants to densify biomass and reduce biomass handling costs. In *IEA Bioenergy Task 31: International Workshop: Impacts on Forest Resources and Utilization of Wood for Energy* (ed. J. Richardson), Elsevier, pp: 321–325, Flagstaff, AZ.

Basic Statistics on Indian Petroleum & Natural Gas, 2009–10, Ministry of Petroleum and Natural Gas, Government of India, New Delhi.

Bramer, E.A., Holthuis, M.R., and Brem, G. 2004. A novel thermogravimetric vortex reactor for the determination of the primary pyrolysis rate of biomass. *Proceedings Conference on Science in Thermal and Chemical Biomass Conversion,* August 30–September 02, 2004, Vancouver.

Budarin, V.L., Clark, J.H., Lanigan, B.A., Shuttleworth, P., and Macquarrie, D.J. 2010. Microwave assisted decomposition of cellulose: A new thermochemical route for biomass exploitation. *Bioresour Technol* 101: 3776–9.

Budarin, V.L., Shuttleworth, P.S., Dodson, J.R., Hunt, A.J., Lanigan, B., Marriott, R., Milkowski, K.J., et al. 2011. Use of green chemical technologies in an integrated biorefinery. *Energy Environ Sci* 4: 471–9.

Centi, G., Lanzafame, P., and Perathoner, S. 2011. Analysis of the alternative routes in the catalytic transformation of lignocellulosic materials. *Catal Today* 167: 14–30.

Centi, G., Fornasiero, P., Kaltschmitt, M., Miertus, S., Müller-Langer, F., Rönsch, S., Sivasamy, A., Thrän, D., Vogel, A., and Zinoviev, S. 2008. *Survey of Future Biofuels and Bio-based Chemicals.* Prepared by a group of ICS-UNIDO experts. Trieste, Italy. Online source: www.ics.trieste.it

CES (Centre for Ecological Sciences). 1995. *Annual Report.* Bangalore: Centre for Ecological Sciences, IISc. Online source: http://lab.cgpl.iisc.ernet.in/Atlas/Atlas.aspx

Chauhan, S. 2012. District wise agriculture biomass resource assessment for power generation: A case study from an Indian state Punjab. *Biomass Bioenerg* 37: 205–12.

CMIE. 1997. *Directory of Indian Agriculture.* Mumbai, India: CMIE.

Colman, B. and Rotatore, C. 1995. Photosynthetic inorganic carbon uptake and accumulation in two marine diatoms. *Plant Cell Environ* 18: 919–24.

Commission on Oil Independence. 2006. *Making Sweden an OIL-FREE Society.* Stockholm: Swedish Government Offices.

Communication from the Commission: Biomass Action Plan (COM (2005) 628 Final). Brussels: European Commission.

European Parliament and Council. *Directive 2003/30/EC of the European Parliament and of the Council of 8 May 2003 on the Promotion of the Use of Biofuels or Other Renewable Fuels for Transport, IJ L 176.* European Union, 2003, pp. 42–46.

Saud, T., Singh, D.P., Mandal, T.K., Ranu, G., Pathak, H., Saxena, M., and Sharma, S.K. 2011. Spatial distribution of biomass consumption as energy in rural areas of the Indo-Gangetic plain. *Biomass and Bioenergy* 35: 932–941.

Dynamotive. 2007. Bio-oil® production at Dynamotive's Guelph plant. *Cellulose based Bio-oil®,* 23: 2.

Emma, H.I., Colman, B., Espie, G.S., and Lubian, L.M. 2000. Active transport of CO_2 by three species of marine microalgae. *J Phycol* 36: 314–20.

Energy for the Future: Renewable Sources of Energy, White Paper for a Community Strategy and Action Plan (COM (97) 599 final). Brussels: European Commission.

European Commission. 2006. *Energy and Transport Trends to 2050-Update 2005.* Brussels: European Commission.

Fransham, P. 2006. Advances in dry distillation technology. *ThermalNet Newsletter* 1, Aston University, Birmingham.

Freel, B. 2002. Ensyn announces commissioning of new biomass facility. *PyNe Newsletter* 14: 1–2.

Freel, B., and Graham R.G. 2002. Bio-oil preservatives. Ensyn Technologies Inc. US Pat., 6485841(B1).

Global Business Environment Shell International. 2001. *Exploring the Future: Energy Needs, Choices and Possibilities—Scenarios to 2050.* Global Business Environment Shell International, London.

Graham, R.G., Freel, B.A., Huffman, D.R., and Bergougnou, M.A. 1994. Commercial-scale rapid thermal processing of biomass. *Biomass Bioenerg* 7: 251–8.

Green Paper—Toward a European Strategy for the Security of Energy Supply (COM (2000) 769 Final). Brussels: European Commission.

Guiotoku, M., Rambo, C.R., Hansel, F.A., Magalhaes, W.L.E., and Hotza, D. 2009. Microwave-assisted hydrothermal carbonization of lignocellulosic materials. *Mater Lett* 63: 2707–9.

Gujarat Electricity Regulatory Commission. 2010. *Determination of Tariff for Procurement of Power by Distribution Licensees from Biomass Based Power Generator and Other Commercial Issues.* Ahmedabad: Gujarat Electricity Regulatory Commission.

Gustavsson, L., Holmberg, J., Dornburg, V., Sathre, R., Eggers, T., Mahapatra, K., and Marland, G. 2007. Using biomass for climate change mitigation and oil use reduction. *Energy Policy* 35: 5671–91.

Hazell, P. and Pachauri, R. K., eds. 2006. *Bioenergy and Agriculture: Promises and Challenges.* 2020 Focus No. 14. International Food Policy Research Institute, Washington, DC.

Henrich, E., Dahmen, N., and Dinjus, E. 2009. Cost estimate for biosynfuel production via biosyncrude gasification. *Biofuels Bioprod Bioref* 3: 28–41.

Hsueh, H.T., Chu, H., and Yu, S.T. 2007. A batch study on the bio-fixation of carbon dioxide in the absorbed solution from a chemical wet scrubber by hot spring and marine algae. *Chemosphere* 66: 878–86.

Hu, Q., Sommerfeld, M., Jarvis, E., Posewitz, M., Seibert, M., and Darzins, A. 2008. Microalgal triacylglycerols as feedstocks for biofuel production: Perspectives and advances. *Plant J* 54: 621–39.

IEA (International Energy Agency). 2003. *Energy to 2050 Scenarios for a Sustainable Future.* Paris: OECD/IEA.

IEA (International Energy Agency). 2008. *Energy Technology Perspectives 2008: Scenarios and Strategies to 2050.* Paris: OECD/IEA.

IEA (International Energy Agency). 2009. *World Energy Outlook 2009.* Paris: OECD/IEA.

IEA (International Energy Agency). 2010. *Sustainable Production of Second-Generation Biofuels.* Paris: OECD/IEA.

India's Energy Security: Briefing Paper—Dialogue on Globalization. Friedrich Ebert Stiftung. Online source: http://library.fes.de/pdffiles/iez/global/04809.pdf

Larson, E.D. 2008. *Biofuel Production Technologies: Status, Prospects and Implications for Trade and Development.* Report No. UNCTAD/DITC/TED/2007/10. New York: United Nations Conference on Trade and Development.

Lede, J., Broust, F., Ndiaye, F.T., and Ferrer, M. 2007. Properties of bio-oils produced by biomass fast pyrolysis in a cyclone reactor. *Fuel* 86: 1800–10.

Luque, R., Herrero-Davila, L., Campelo, J.M., Clark, J.H., Hidalgo, J.M., Luna, D., Marinasa, J.M., and Romero, A.A. 2008. Biofuels: A technological perspective. *Energy Environ Sci* 1: 542–64.

Mata, T.M., Martins, A.A., and Caetano, N.S. 2010. Microalgae for biodiesel production and other applications: A review. *Renew Sustain Energy Rev* 14: 217–32.

McKendry, P. 2002. Energy production from biomass (part 1): Overview of biomass. *Bioresour Technol* 83: 37–46.

Meier, D., Schoell, S., and Klaubert, H. 2004. New ablative pyrolyser in operation in Germany. *PyNe Newsletter* 17: 1–3.

Meier, D. 2007. Aus Holz wird Oel. *Forstmaschinen Profi* 15: 26.

Gokhale, A.M., Gupta, A.K., and Kishwan, J. 2006. National *Mission on Decentralized Biomass Energy for Villages and Industries.*

Mission Document in Evolution: National Mission on Decentralized Biomass Energy for Villages and Industries.

Shailesh, N. 2009. Biofuel policy process in India - Context, Actors and Discourses. In *The Indian Society for Ecological Economics (INSEE) 5th Biennial Conference*; January 21–23, 2009. Available at http://www.ecoinsee.org/fbconf/Sub%20Theme%20C/Shailesh%20Nagar.pdf

Prime Minister's Council. 2008. *National Action Plan on Climate Change.* Government of India, New Delhi.

Nieminem, J-P., Gust, S., Nyrönen, T. 2003. Forestera™ - liquefied wood fuel pilot plant. *PyNe Newsletter* 16: 2–4.

Nigam, P.S. and Singh, A. 2011. Production of liquid biofuels from renewable resources. *Prog Energ Combust Sci* 37: 52–68.

Pandey, A., Biswas, S., Sukumaran, R.K., and Kaushik, N. 2009. *Study on Availability of Indian Biomass Resources for Exploitation: A Report Based on a Nationwide Survey.* New Delhi: TIFAC.

Peacocke, G.V.C. and Bridgwater, A.V. 1995. Ablative plate pyrolysis of biomass for liquids. *Biomass Bioenerg* 7: 147–54.

Promotion and Protection of all Human Rights, Civil, Political, Economic, Social and Cultural Rights, Including The Right to Development. Report of the Special Rapporteur on the Right to Food. United Nations General Assembly. Online source: http://daccess-ods.un.org/TMP/5810871.html

Ragauskas, A.J., Williams, C.K., and Davison, B.H. 2006. The path forward for biofuels and biomaterials *Science* 311: 484–9.

Research Team on the Strategy and Policy for Energy Development of China. 2003. *Basic Framework for Energy Development Strategies of China*. Beijing: People Press, p. 268.

Runge, C.F. and Senauer, B. 2007. How Biofuels Could Starve the Poor. *Foreign Affairs* 3: 86.

Sandvig, E., Walling, G., Brown, R.C., Pletla, R., Radlein, D., and Johnson, W. 2003. *Integrated Pyrolysis Combined Cycle Biomass Power System Concept Definition*. Final Report, Report DE-FS26-01NT41353, USA.

Schlamadinger, B., Apps, M., Bohlin, F., Gustavsson, L., Jungmeier, G., Marland, G., Pingoud, K., and Savolainen, I. 1997. Toward a standard methodology for greenhouse gas balances of bioenergy systems in comparison with fossil energy systems. *Biomass Bioener* 13: 359–75.

Schlamadinger, B., Edwards, R., Byrne, K.A., Cowie, A., Faaij, A., Green, C., Fijan-Parlov, S., et al. 2005. Optimizing the greenhouse gas benefits of bioenergy systems. *Presented at 14th European Biomass Conference*, 17–21 October 2005, Paris, France.

Scholl, S., Klaubert, H., and Meier, D. 2004. Wood liquefaction by flash-pyrolysis with an innovative pyrolysis system. *DGMK Proceedings 2004-1 Contributions to DGMK-Meeting, Energetic Utilization to Biomasses*, April 19–21, Velen, Westf.

Solmon, S., Qin, D., Manning, M., Chen, Z., Marquis, M., Averyt, K.B., Tignor, M., and Miller, H.L., eds. 2007. *Climate Change 2007: The Physical Science Basis*. Contribution of Working Group 1 to the Fourth Assessment Report of the Intergovernmental Panel on Climate Change. Cambridge, UK: Cambridge University Press.

Stöcker, M. 2008. Biofuels and biomass-to-liquid fuels in the biorefinery: Catalytic conversion of lignocellulosic biomass using porous materials. *Angew Chem Int Ed* 47: 9200–11.

Suh, I.S. and Lee, C.G. 2003. Photobioreactor engineering: Design and performance. *Biotechnol Bioprocess Eng* 8: 313–21.

Sukumaran, R.K., Surender, V.J., Sindhu, R., Binod, P., Janu, K.U., Sajna, K.V., Rajasree, K.P., and Pandey, A. 2010. Lignocellulosic ethanol in India: Prospects, challenges and feedstock availability. *Bioresour Technol* 101: 4826–33.

The Royal Society. 2008. *Sustainable Biofuels: Prospects and Challenges*. London: The Royal Society.

United Nations Environment Programme (UNEP). 2001. *The State of Environment*, India. Available at http://www.moef.nic.in/soer/2001/ind_toc.pdf

Venderbosch, R.H. and Prins, W. 2010. Fast pyrolysis technology development. *Biofuels Bioprod Bioref* 4: 178–208.

Yoshioka, T., Hirata, S., Matsumura, Y., and Sakanishi, K. 2005. Woody biomass resources and conversion in Japan: The current situation and projections to 2010 and 2050. *Biomass Bioenerg* 29: 336–46.

Zhang, L., Xu, C., and Champagne, P. 2010. Overview of recent advances in thermo-chemical conversion of biomass. *Energy Convers Manag* 51: 969–82.

CHAPTER 17

Techno-Economic and Environmental Impacts of Biofuel Options in Brazil

Otavio Cavalett, Tassia L. Junqueira, Mateus F. Chagas, Lucas G. Pereira, and Antonio Bonomi
Brazilian Bioethanol Science and Technology Laboratory (CTBE),
Brazilian Center of Research in Energy and Materials (CNPEM),
Campinas, São Paulo, Brazil

CONTENTS

17.1 INTRODUCTION

In Brazil, about half of the total energy supplied comes from renewable sources, mainly hydroelectric power, sugarcane, and wood. Sugarcane bioenergy is very important in this context; in 2012, it accounted for 15.4% (43.6 Mtoe) of the national energy supply—slightly greater than the contribution of hydroelectric power (13.8%) (EPE 2013). In the road transport sector, the share of biofuels has been considerable: 25% in 2012 as anhydrous and hydrated ethanol (19.1 Mm³) and biodiesel (2.2 Mm³) (EPE 2013). The expanding biofuels global market has raised concerns about its effective sustainability (Nogueira and Capaz 2013).

Ethanol is consumed by Brazilian cars as anhydrous ethanol (less than 0.6% of water by mass) blended with gasoline (20%–25% by volume) in conventional gasoline engines and as hydrated ethanol (approximately 6% water), pure or in any blend with Brazilian gasoline, in dedicated engines or flex-fuel engines.

Regarding biodiesel, in 2005, the Brazilian government launched the National Program for Production and Use of Biodiesel (PNPB) (Brazil, 2005). This program was developed to encourage small producers and farmers from the least developed regions of Brazil to become involved with biodiesel production and to set progressive targets for the mandatory use of biodiesel blends in all diesel oil sold in gas stations. This blending obligation started in January 2008 with 2% biodiesel (B2), which then increased to B3 in July 2008 and to B4 in July 2009. According to the original legislation, in January 2013, the biodiesel content should be 5%; but in response to requests from biodiesel producers, the Brazilian authorities decided to move the B5 obligation forward to January 2010 (Nogueira and Capaz 2013). The production of biodiesel has increased exponentially since 2005 and reached 2.71 Mm³ in 2012. The total consumption of biodiesel in that same year was 2.75 Mm³, mainly in road transportation (79%) and in agricultural activities (13%) (EPE 2013).

This study shows a critical review of technical, economic, and environmental impacts of current and future biofuel options in Brazil. Furthermore, trends for future production and use of feedstock and biofuels are presented and discussed in this chapter.

17.2 BIOFUEL PRODUCTION IN BRAZIL

This section presents the availability of various kinds of biomass for biofuels and widely used technologies for production of biofuels in Brazil.

17.2.1 Ethanol from Sugarcane

In Brazil, sugarcane derived ethanol is one of the most successful examples of large-scale biofuel production, distribution, and use. For more than 30 years, since the adoption of the Proalcool program, Brazil has been implementing biofuel policies not only to reduce the country's dependence on fossil fuels but also to benefit from the environmental, economic, and social advantages that go hand to hand with the sustainable production and use of biofuels. In the pursuit of achieving these objectives, Brazil has traveled a long road that demanded not only steady resolve—from the producers and from the government—to invest in the production but also the engagement of research institutions and think tanks in order to provide the best possible technological solutions for both industry and agriculture to increase yield and efficiency in these sectors (Lago et al. 2012). The use of ethanol in transportation and generation of electricity from sugarcane residues accounts for nearly 16% of the total energy supply (EPE 2013), making it the second primary source after oil. In the 2012–2013 harvest season, the national cultivated area was 8.5 Mha, the total sugarcane production was 595 Mt, and the total ethanol production was 23.6 Mm³ (CONAB 2012).

17.2.2 Biodiesel

Although the PNPB was developed to encourage small producers and farmers from the least developed regions of Brazil to become involved with biodiesel production, soybean has traditionally been the feedstock used for biodiesel production in Brazil. With a well-organized supply chain, soybean has been one of the most successful large-scale agribusiness operations and is responsible for approximately 80% of all biodiesel produced in the country (Padula et al. 2012). Additionally, beef tallow, which had been considered an undesirable residue in the past, has become a nonfood feedstock strategy for the success of PNPB, contributing with almost 15% of all national production (ANP 2013). Biodiesel from beef tallow presents advantages if compared to oilseeds, such as higher cetane number and flash point and lower production costs due to the vertical production chain (MAPA 2012a; Padula et al. 2012).

The production of other oilseeds (cotton, castor, rapeseed, palm, and sunflower, among others) has not yet been established (MAPA 2012b, 2012c). The goal of diversifying the feedstock used to produce biodiesel has not been achieved due to a number of difficulties. Oilseeds that do not have structured and organized supply chains, despite having several strengths and enjoying considerable incentives and support from public policies, have yet to overcome the challenges faced in their different stages of production.

For biodiesel production, vegetable oils can be extracted with solvent or mechanically by pressing, which is followed by purification and refining processes (MMA 2006). Vegetal oil has a viscosity 10–17 times higher than petrodiesel, which can cause damage to engines. Therefore, crude vegetable oil is converted into biodiesel through a transesterification or esterification process to have its viscosity reduced (Beltrão and Oliveira 2008). In the transesterification process, reaction occurs between the triglyceride (oil) and an alcohol forming a methyl, ethyl, or another ester as main product and glycerine as by-product. To increase efficiency and reduce reaction time, different catalysts such as acids, bases, or enzymes can be used.

Different types of alcohols can be used in the transesterification process including ethanol, methanol, propanol, butanol, and amyl-alcohol. Currently, the methanol route to biodiesel production is the most commonly used (Bergmann et al. 2013), due to its higher efficiency, higher productivity, and lower temperature (60°C) compared to other alcohols (Freedman et al. 1984). However, because methanol is a toxic substance obtained from a fossil source, its use is expected to be replaced by alternative alcohols. According to Freedman et al. (1984), the transesterification reaction of soybean oil using methanol, ethanol, or butanol with 1% of sulfuric acid, achieved conversions between 96% and 98%, regardless of the alcohol used. Because of the large production of ethanol in Brazil, its use can be envisioned as economically and environmentally advantageous in comparison to methanol for biodiesel production. In spite of that, there are still technical challenges to be addressed on the use of ethanol, such as the difficulty of its separation from the ester formed in the process. Lima (2005) states that studies on the use of ethanol and new catalysts in the production of biodiesel are in progress, but this technology is not yet largely applied in industrial scale.

17.2.3 Second-Generation Ethanol

Current ethanol production is based on a first-generation process, fermenting sugars extracted from sugarcane stalks. However, sugar represents approximately one-third of the energy content of sugarcane. The other two-thirds are composed of straw, which is either burned in the field or left as mulch, and bagasse, the fibrous material left from the juice extraction process, which is mostly used as fuel for process heat and electricity generation at the sugarcane mill. Cellulosic ethanol, also known as second-generation ethanol, can be produced from currently considered agricultural and industrial residues: straw and bagasse.

Traditionally, straw used to be burned prior to the harvesting, but increasing public concern has resulted in a phasing out of this technique throughout the main sugarcane growing regions of the country. Bagasse is readily available at the plant site without collection and transportation costs. However, two trends might put pressure on bagasse availability for cellulosic ethanol production: the declining fiber content of new sugarcane varieties—a target of most crop breeding programs— and the increase in surplus power generation, especially in the new mills that have been investing in efficient, high pressure boilers. On the other hand, the development of commercially available technologies to reduce process steam demand reduces bagasse consumption and increases its availability for cellulosic ethanol production.

There are also potential benefits of leaving crop residues on the field, such as protection against erosion, nutrient cycling, soil carbon sequestration, weed suppression, and soil moisture retention. On the other hand, considering the large quantities of residue generated and their high carbon to nitrogen ratio and fiber content, it is likely that removing part of the straw will still secure most of the environmental and agronomic benefits. Those benefits are site-specific, so the amount of straw that can be sustainably removed should be calculated considering climate, topography, soil, and crop variables. Besides, the task of collecting, transporting, and pretreating this material presents important challenges that need to be overcome before its commercial implementation.

17.2.4 Future Production Targets and Research Directions in Brazil

17.2.4.1 Biodiesel Perspectives

Although biodiesel production has increased over the years in this country, plants have been using approximately 47% of their maximum installed capacity (Ubrabio 2010), and production is still not economically sustainable (Padula et al. 2012). Forecasts from the sector and governmental agencies indicate that, over the next decade, a decrease of 1% per year in the participation of soybean as feedstock for biodiesel production is expected; the total share would be reduced to 70% by 2020. It is foreseen that seeds with higher oil yield than soybeans will have their participation as feedstock increased to 20%–25% content. Beef tallow would also have its share reduced from the current 15% to less than 5% in the same 10-year horizon.

Diversification of crops used for biodiesel production in Brazil is desirable; however, bottlenecks hampering this process will require management activities and practices directly related to the structure and organization of production chains (agriculture, industrialization, distribution, and use). Research leading to improved crop varieties, domestication of more exotic species, information about pathogens, and optimization of crop management practices will likely play a significant role as to which oil crops are to become important biodiesel feedstocks in the future.

According to Ubrabio (2010), given the demand for biodiesel and the mandatory blend regulations, it is likely that production will continue to increase. Considering the figures for current installed capacity of biodiesel production units, B10 demand could be met immediately. However, a B20 scenario would require investments in view of the need for improved industrial units and the development of new feedstock production areas. The Brazilian Oil Company (Petrobras) has been conducting tests using B20 in vehicles since 2010; however, no prediction has been made regarding the practical implementation of this blend and its commercialization.

In spite of optimistic predictions from agencies related to the sector, there is no concrete reason to believe that additional biodiesel demand will be imposed. The regulatory framework for biodiesel established by law does not indicate any compulsory or voluntary blends higher than B5 to be implemented after 2013. The goal of cost efficiency and reduction/elimination of subsidies that currently generally support the bioenergy programs in the world remain key issues in political, technological, and managerial agendas (Northoff 2008; Sorda et al. 2010). The economic feasibility of biodiesel in Brazil can be questioned because it is still strongly supported by tax incentives from the federal government on the production and commercialization stages.

17.2.4.2 Ethanol Perspectives

Recent advances in the cellulosic ethanol technology have directed attention to the use of agricultural residues (e.g., sugarcane bagasse and straw and forest residues) as promising alternatives for liquid biofuels (and also electricity) production. A fundamental advantage for integration of cellulosic ethanol into current first-generation ethanol production from sugarcane in Brazil is the availability of lignocellulosic material (bagasse) at the plant site and the feasible alternative to also use sugarcane crop residues (straw). Regarding this, it is fundamental to consider optimization strategies for first-generation ethanol production, aiming at energy savings, resulting in more surplus lignocellulosic material for second-generation ethanol production. Cellulosic ethanol production in Brazil may also benefit from sharing part of the infrastructure where first-generation ethanol production takes place (for instance, juice concentration, fermentation, distillation, dehydration, storage, and cogeneration facilities). In addition, potential fermentation inhibitors generated in the lignocellulosic material pretreatment may have a minor effect on fermentation yields because the hydrolyzed liquor may be fermented mixed with sugarcane juice, diluting these inhibitors (Dias et al. 2012).

Due to the high potential of biomass for the production of fuels and chemicals, research in Brazil has focused on the hydrolysis of sugarcane bagasse and/or straw for second-generation production. Second-generation ethanol production involves basically four steps: pretreatment, enzymatic hydrolysis, fermentation, and ethanol recovery. In recent years, pilot and demo scale plants have been built in Brazil and worldwide. However, enzymatic technology still faces numerous obstacles and is not yet mature enough for full commercialization scale. Some major challenges faced are the high cost of the second-generation process, as well as the high cost of enzymes due to the fact there are no market and positive learning curve effects for this technology (Dias et al. 2012, 2013a).

17.2.4.3 Alternatives for Harvesting Extension in Sugarcane Biorefineries: Sweet Sorghum and Forest Residues

Sugarcane is a semiperennial crop with a plant crop and successive regrowth crops, known as ratoons. After five or six harvests on average, it is necessary to replant the crop, and there is usually a gap of a few months in which there is a fallow period or there are cover crops, usually legumes. The sweet sorghum (*Sorghum bicolor* L.) production in the sugarcane reforming areas yields an extra harvest without interfering in the sugarcane production cycle. Another advantage that could be pointed out is that agricultural machinery and inputs are similar to those used for sugarcane production. In the same direction, sweet sorghum can be considered a "drop-in feedstock" for the industry because its industrialization requires minor adjustments in the equipment that is already used for sugarcane processing.

Although sweet sorghum is not yet produced on a large scale in Brazil, the benefits of this culture have been discussed in recent years; but uncertainties regarding its benefits remain. At the moment, improvements in the agricultural system focus on development of better varieties, crop management systems, logistics, and suitability of sweet sorghum introduction in the sugarcane chain (May et al. 2012a, 2012b).

Forest residues are defined as the biomass material remaining in harvested forests, which are almost identical in composition to forest thinnings. Forestry residues include logging residues, excess small pole trees, and rough or rotten dead wood. Typically, forest residues are either left in the forest or disposed of via open burning through forest management programs. The interest in the removal of forest residues has been amplified by the possibility of using the biomass for energy purposes, as second-generation ethanol or electricity. It has been estimated that the amount of forest residues (tops and branches) varies from 30 to 100 ton ha^{-1} (Brand et al. 2009; Koopmans and Koppejan 1998) depending on species, age, site type, and wood assortments harvested.

17.3 TECHNO-ECONOMIC AND ENVIRONMENTAL IMPACTS OF BIOFUEL OPTIONS IN BRAZIL

In order to provide a comparison in terms of economic viability, important indicators from economy engineering, such as internal rate of return (IRR), production costs of products, among others, can be calculated considering a set of scenarios related to different biorefinery alternatives.

The principles for the evaluation of economic viability are based on a cash flow projected for each technological scenario to be evaluated, taking into account the investment needed for the project and all expenses and revenues for an expected project lifetime.

In order to evaluate environmental aspects of sugarcane biorefineries, the life cycle assessment (LCA) methodology has been used as a key tool (e.g., Cavalett et al. 2012; Dias et al. 2012, 2013a, 2013b; González-García et al. 2012; Luo et al. 2009, 2010; Melamu and Blottnitz 2011). LCA is a recognized method for quantitatively determining the environmental impacts of a product (or a good or service) during its entire life cycle, from extraction of raw materials through manufacturing, logistics, use, and final disposal or recycling. In LCA, substantially broader environmental aspects can be covered, ranging from GHG emissions and fossil resource depletion to acidification, toxicity, and water and land-use aspects. Hence, it is an appropriate tool for quantifying environmental impacts of a product system. The ISO 14040 series provides a technically rigorous framework for carrying out LCAs (ISO 2006a, 2006b).

17.3.1 First-Generation Ethanol

Ethanol production in annexed plants (producing sugar, ethanol, and in some cases, surplus electricity) presented lower environmental impacts in comparison to autonomous distilleries (producing ethanol and, in some cases, electricity) in most of the environmental impacts categories evaluated in the study from Cavalett et al. (2012) (e.g., global warming potential, abiotic depletion, and human toxicity). However, the advantage presented in the results toward annexed plants in comparison to autonomous plants is small (less than 5% in most of the categories evaluated). Benefits of more flexible scenarios (meaning that sugarcane juice used for ethanol production can vary between 30% and 70% in annexed plants, depending on the relative ethanol and sugar market prices) were also examined. For the more flexible scenarios, capital expenses (CAPEX) proved to be an important issue because CAPEX increases from autonomous to annexed plants and from fixed to flexible plants, having a significant impact on the IRR. Another important observation was that annexed plants present higher IRR for flexible (favoring sugar production) and fixed plants. Although autonomous distillery also presented good results, it is important to take into account that market oscillations can considerably change and flexibility may be decisive for maintaining the sugarcane biorefinery profitability.

Environmental impacts calculated using the LCA methodology indicate that optimization technologies such as high pressure boilers, system integration for decreasing steam consumption, use of molecular sieves for ethanol dehydration, and use of straw for power generation, beyond others have a great potential for a significant decrease on the environmental impacts (around 25%) in comparison to present sugarcane biorefineries (for both autonomous and annexed plants). Results also showed that flexibility capacity in annexed plants produces little effect on the environmental impacts when the entire ethanol production life cycle is considered (Cavalett et al. 2012).

17.3.2 Second-Generation Ethanol

Dias et al. (2012) evaluated different scenarios for integrated and stand-alone second-generation ethanol production from sugarcane bagasse and trash. Five scenarios were selected to demonstrate the economic and environmental impacts of second-generation ethanol production in comparison to an optimized autonomous first-generation ethanol production plant in Brazil. Results showed that the current integrated first- and second-generation ethanol production scenario—characterized by higher investment cost in the second-generation process (due to the fact it will be one of the first plants), higher enzyme cost, lower yield, and lower solids load in the hydrolysis step—presents lower IRR in comparison to the optimized first-generation ethanol production. However, the integrated first- and second-generation ethanol production, considering future scenarios where target parameters are used for second-generation processes and ethanol can be also produced from C5 sugars, is more attractive to the investor than the optimized first-generation ethanol production. Environmental impacts calculated using the LCA methodology showed that current integrated first- and second-generation ethanol production scenario has the potential to decrease the impacts in relation to the optimized first-generation ethanol production. The study identified that the high amount of sodium hydroxide used in the alkaline delignification step considered for future second-generation processes has a strong influence on the environmental results.

Integrated first- and second-generation ethanol production presented lower impacts than first generation for acidification and eutrophication. These impact categories are mainly related to fertilizer use in the agricultural stage. Because more ethanol is produced per unit of area on integrated scenarios, this process presents better results for acidification and eutrophication categories (Dias et al. 2012, 2013b).

Also, regarding biomass pretreatment alternatives, steam explosion pretreatment presents lower life cycle environmental impacts than hydrothermal processing based on liquid hot water, due to higher energy consumption and, consequently, lower ethanol production in comparison to steam explosion (Junqueira et al. 2012).

Regarding cogeneration schemes, systems using 65 bar boilers presented the lower environmental impacts than systems with 22, 42, and 82 bar boilers in several environmental impact categories including global warming potential. This occurs because 65 bar boilers are more efficient than 22 and 42 bar boilers; and higher pressure boilers (82 bar) consume more bagasse than low pressure boilers, resulting in a lower ethanol output (Dias et al. 2013b).

Melamu and Blottnitz (2011) evaluated different second-generation ethanol production scenarios diverting bagasse into second-generation ethanol and replacing it with coal in sugarcane refineries. The work concluded that diverting bagasse into second-generation ethanol in the absence of energy improvements on the sugar mill results in an increase in the greenhouse gases (GHGs) emissions. Similar higher impacts are observed for other impact categories such as nonrenewable energy use, aquatic eutrophication, and terrestrial acidification.

Dias et al. (2013a) evaluated a flexible biorefinery with the possibility of diverting a fraction of the lignocellulosic material (sugarcane bagasse and straw) either for electricity production or as feedstock in second-generation ethanol. Both fixed and flexible plants operating at maximum ethanol production lead to the highest avoided carbon dioxide emissions when usage phase and displacement of equivalent fossil products are considered (avoided emissions due to products). However, biorefineries producing more ethanol presented higher environmental impacts per unit of ethanol produced than the configuration with maximum electricity production due to high impacts of inputs (sodium hydroxide, enzymes, and sulfuric acid, for instance) used in the second-generation process. Therefore, diverting lignocellulosic material to the cogeneration system presents lower environmental impacts than producing second-generation ethanol, even though it has higher emissions in the cogeneration system. Luo et al. (2009) found similar results considering the global warming potential, but the comparison did not include a flexible plant. Summing up, industrial flexibility can be translated by higher investments in equipment that is underutilized, considering the plant maximum annual processing capacity (i.e., part of the plant always would be idle). The fact that more equipment is required in the flexible scenario presented small effect in the environmental impacts considering the life cycle impacts of ethanol production. A similar study from González-Garcia et al. (2012) evaluating ethanol production and power generation systems using *Salix* spp. biomass showed that ethanol production is potentially the best choice from an energy perspective, whereas electricity production seems to be a more suitable alternative when global warming is the decisive factor.

17.3.3 Sweet Sorghum and Forest Residues for Harvesting Extension in the Sugarcane Biorefineries

Cavalett et al. (2013a) presented a technical and economic evaluation of extending the operation of the sugarcane plant using sweet sorghum as feedstock for ethanol production in an autonomous distillery and using only the sugarcane-reforming areas for sweet sorghum production. Different scenarios were constructed to assess potential improvements of the sorghum quality and its processing technology. Obtained results show that sugarcane harvest season extension in an autonomous distillery using sweet sorghum provides better economic returns in some scenarios considering sweet sorghum production with higher sugar content and agricultural productivities. The best case evaluated for sweet sorghum production presented an IRR of 14.7%, higher than that obtained with no harvest extension (13.7%). If it was possible to extend the industrial operation in 60 days, using the best sweet sorghum variety evaluated, the IRR would be increased to 16.2%.

Cavalett et al. (2013b) also evaluated technical, economic, and environmental assessment of different alternatives for harvest season extension in sugarcane biorefineries. The evaluated scenarios included sweet sorghum, concentrated cane juice, stored lignocellulosic material, and use of forest residues for ethanol production during the off-season for sugarcane. Best economic results were obtained when the industrial operation is extended by 60 days, using sweet sorghum. The scenario considering use of forest residues for second-generation ethanol production also presented interesting economic results. Regarding environmental aspects, scenarios using sweet sorghum and concentrated cane juice presented the lowest environmental impacts per liter of ethanol produced in the global warming and use of abiotic resources categories in comparison with other alternatives evaluated for harvesting extension. However, scenarios with second-generation ethanol (stored lignocellulosic material and use of forest residues) presented lower impacts in the categories of acidification, eutrophication, and human toxicity in comparison with other evaluated alternatives.

17.3.4 Biodiesel

Cavalett and Ortega (2010) used a multimethod approach to evaluate environmental impacts of biodiesel production from soybeans in Brazil. The study pointed out that, in spite of a possible contribution to reduce the CO_2 emissions, soybean biodiesel is not a viable alternative taking into consideration materials, energy, and energy assessments. The direct pollution (fertilizers, agrochemicals, and pesticides) and other environmental impacts (soil loss, energy, material, water, and land use), related to the net energy delivered to society as biodiesel, indicate that soybean biodiesel produces a high pressure on the environment. The study concluded that future of biodiesel is very likely to be linked to the ability of clustering biodiesel production with other agro-industrial activities at an appropriate scale and mode of production to take advantage of the potential supply of valuable coproducts.

Borzoni (2011) performed a multiscale integrated assessment of soybean biodiesel in Brazil using a parallel biophysical and economic reading at different scales to shed light on the consequences and sustainability of alternative options to fossil oil. The study showed that energy delivered by soybean biodiesel is higher than the energy invested. However, the resulting net energy is not comparable with the energy source it is intended to substitute. The study concluded that it is becoming increasingly acknowledged that diversification independently from oil sources is urgent. The upshot is that soybean biodiesel production and use in Brazil may be feasible, but its desirability is highly questionable due to its net energy production.

César and Batalha (2010) studied the diesel production from castor oil in Brazil. The study showed that considerable public and private efforts were devoted for a sustainable biodiesel production using castor oil in Brazil. However, biodiesel chain development projects are currently economically inadequate and hugely supported by government incentives. This research work shows that companies are facing substantial problems in implementing contracts with family farmers. In addition to the technological difficulties for the biodiesel production from castor oil to meet the standards required by the National Agency of Petroleum, Natural Gas and Biofuels (ANP), the lack of economic sustainability seems to be the main reason for steering away the castor beans from the social projects that originally focused on biodiesel production for the chemical ricinus segment. The study concluded that biodiesel production based on castor beans is impracticable in Brazil in the short term, and it seems to be a rather distant future possibility.

Nogueira (2011) presented an assessment of biodiesel production comparing four different productive systems. According to this evaluation, soybean and castor are limitedly feasible, whereas tallow and palm oil represent more suitable alternatives. The study also mentioned that to improve the understanding of biofuels, and particularly biodiesel potentials and limits, it is very important to develop more detailed life cycle studies considering existing and future processes, using different feedstock and productive systems. The study concluded that there is a large room for improvement in biodiesel production; however, not all routes present the same perspectives, effects, and costs—therefore development on the matter should start from sound and known basis, such as the results from energy balance studies are able to offer for decision makers.

Kaercher et al. (2013) presented a study for optimization of small-scale biodiesel production considering its environmental impacts. The study concluded that it is possible to develop equipment that allows small-scale biodiesel production for local consumption. The type of alcohol and thermal insulation used during biodiesel production both play significant roles in the context of environmental concerns. Factors that helped reduce the negative impacts of small-scale biodiesel production included the use of renewable feedstocks, the prevention of alcohol vapor loss, and reduction in heat loss. Properly addressing these issues yielded decreases in alcohol emissions and energy consumption. Another important factor to be taken into account for small-scale biodiesel production systems is to assure that the product meets ANP biodiesel standards.

Leoneti et al. (2012) studied several alternatives for the use of unrefined glycerol because it is an important by-product of biodiesel production in Brazil. Exploratory research was carried out to identify these viable alternatives for the use of this by-product. The possibilities included the production of chemical products, fuel additives, production of hydrogen, development of fuel cells, ethanol or methanol production, animal feed, codigestion, and cogasification, among others. The study revealed that there are promising possibilities for the use of unrefined glycerol.

Cunha (2011) emphasizes that the establishment of the sector was supported by the opportunities associated with the consolidated productive chains of soy and cattle slaughtering. Although biodiesel was intended to play an important supporting role in the Brazilian energy matrix as a renewable source of fuel, it has definitely not fulfilled the main objective proposed by the PNPB. Furthermore, participation of small-scale family agriculture in the program, one of the premises—as a creation factor of regional employment and income—has not been significant.

17.4 FINAL REMARKS

The development of the integrated production of first- and second-generation ethanol from sugarcane is an excellent opportunity for efficient integral use of sugarcane, either selling surplus electricity to the grid or increasing ethanol production through optimized integrated second-generation technologies. Improvements in the second-generation processes are expected to improve sustainability impacts of the ethanol production from sugarcane. Regarding biodiesel systems, several studies show that feasibility is limited due to low net energy provided by current oil crops.

REFERENCES

ANP. 2013. *Brazilian Statistical yearbook of petroleum, natural gas and biofuels 2013* (in Portuguese). National Agency of Petroleum, Natural Gas and Biofuels. Available at: http://goo.gl/yHSHsq (Accessed August 8, 2013).

Beltrão NEM, Oliveira MIP. 2008. *Oilseeds and their oils: Advantages and disadvantages for the production of biodiesel* (in Portuguese). Campina Grande: Embrapa.

Bergmann JC. Tupinambá DD, Costa OYA, Almeida JRM, Barreto CC, Quirino BF. 2013. Biodiesel production in Brazil and alternative biomass feedstocks. *Renewable and Sustainable Energy Review* 21: 411–420.

Borzoni M. 2011. Multi-scale integrated assessment of soybean biodiesel in Brazil. *Ecological Economics* 70: 2028–2038.

Brand MA, Furtado TS, Ferreira JC, Costa LEO, Pesco RD, Neves MD. 2009. Residues production in *Pinus* commercial forests for energy production. *XIII World Forestry Congress*, Buenos Aires, Argentina, pp. 18–23. Available at: http://www.solumad.com.br/artigos/201011171824001.pdf (Accessed March 12, 2013).

Cavalett O, Junqueira TL, Dias MOS, Jesus CDF, Mantelatto PE, Cunha MP, Franco HCJ, et al. 2012. Environmental and economic assessment of sugarcane first generation biorefineries in Brazil. *Clean Technologies and Environmental Policy* 14(3): 399–410.

Cavalett O, Cunha MP, Chagas MF, Junqueira TL, Dias MOS, Pavanello LG, Leal MRLV, Rossell CEV, Bonomi A. 2013a. An exploratory economic analysis of sugarcane harvest extension using sweet sorghum in the Brazilian sugarcane industry. *XXVIII ISSCT Congress*, June 24–27, 2013, São Paulo, Brazil, pp. 225–225.

Cavalett O, Junqueira TL, Morais ER, Chagas MF, Pavanello LG, Mantelatto PE, Maciel Filho R, Bonomi A. 2013b. Technical, economic and environmental assessment of different alternatives for harvest season extension in sugarcane biorefineries. *20th International Symposium on Alcohol Fuels—ISAF*, March 25–27, 2013, Consultus Publisher, Stellenbosch, South Africa, pp. 34–35.

Cavalett O, Ortega E. 2010. Integrated environmental assessment of biodiesel production from soybean in Brazil. *Journal of Cleaner Production* 18: 55–70.

César AS, Batalha MO. 2010. Biodiesel production from castor oil in Brazil: A difficult reality. *Energy Policy* 38: 4031–4039.

CONAB. 2012. Companhia Nacional de Abastecimento. Acompanhamento da safra Brasileira. Cana-de-açúcar Safra 2012/2013—Terceiro Levantamento. Available at: http://www.conab.gov.br/OlalaCMS/uploads/arquivos/12_12_12_10_34_43_boletim_cana_portugues_12_2012.pdf (Accessed November 20, 2013).

Cunha MP. 2011. *Socioeconomic and environmental assessment of biodiesel production routes in Brazil, based on input-output analysis* (in Portuguese). PhD Thesis, Mechanical Engineering College, State University of Campinas (Unicamp), Campinas, São Paulo, Brazil, p. 264.

Dias MOS, Junqueira TL, Cavalett O, Cunha MP, Jesus CDF, Mantelatto PE, Rossell CEV, Maciel Filho R, Bonomi A. 2013b. Cogeneration in integrated first and second generation ethanol from sugarcane. *Chemical Engineering Research and Design* 91(8): 1411–1417.

Dias MOS, Junqueira TL, Cavalett O, Cunha MP, Jesus CDF, Rossell CEV, Maciel Filho R, Bonomi A. 2012. Integrated versus stand-alone second generation ethanol production from sugarcane bagasse and trash. *Bioresource Technology* 103: 152–161.

Dias MOS, Junqueira TL, Cavalett O, Pavanello LG, Cunha MP, Jesus CDF, Maciel Filho R, Bonomi A. 2013a. Biorefineries for the production of first and second generation ethanol and electricity from sugarcane. *Applied Energy* 109: 72–78.

EPE. 2013. *Empresa de Pesquisa Energética*. National Energy Balance, Base Year 2013—2012: Final report, p. 283. Available at: http://goo.gl/2dsCd (Accessed December 3, 2013).

Freedman B, Pryde EH, Mounts TL. 1984. Variables affecting the yields of fatty esters from transesterified vegetable oils. *Journal of the American Oil Chemists' Society* 61: 1638–1643.

González-García S, Iribarren D, Susmozas A, Dufour J, Murphy RJ. 2012. Life cycle assessment of two alternative bioenergy systems involving Salix spp. biomass: Bioethanol production and power generation. *Applied Energy* 95: 111–122.

ISO (International Organization for Standardization). 2006a. *ISO 14040—Environmental management—Life cycle assessment—Principles and framework*. London: ISO.

ISO (International Organization for Standardization). 2006b. *ISO 14044—Environmental management—Life cycle assessment—Requirements and guidelines*. London: ISO.

Junqueira TL, Dias MOS, Cavalet O, Jesus CDF, Cunha MP, Rossell CEV, Maciel Filho R, Bonomi A. 2012. Economic and environmental assessment of integrated 1st and 2nd generation sugarcane bioethanol production evaluating different 2nd generation process alternatives. *Computer Aided Chemical Engineering* 30: 177–181.

Kaercher JA, Schneider RCS, Klamt RA, Silva WLT, Schmatz WL, Szarblewski MS, Machado EL. 2013. Optimization of biodiesel production for self-consumption: Considering its environmental impacts. *Journal of Cleaner Production* 46: 74–82.

Koopmans A, Koppejan J. 1998. *Agricultural and forest residues—Generation, utilization and availability.* Regional Consultation on Modern Applications of Biomass Energy. FAO, Kuala Lumpur, Malaysia.

Lago AC, Bonomi A, Cavalett O, Cunha MP, Lima MAP. 2012. Sugarcane as a carbon source: The Brazilian case. *Biomass and Bioenergy* 46: 5–12.

Leoneti AB, Aragão-Leoneti V, Oliveira SVWB. 2012. Glycerol as a by-product of biodiesel production in Brazil: Alternatives for the use of unrefined glycerol. *Renewable Energy* 45: 138–145.

Lima PCR. 2005. *Biodiesel: A new fuel for Brazil* (in Portuguese). Biblioteca Digital da Câmara dos Deputados. Brasília: Consultoria Legislativa.

Luo L, van der Voet E, Huppes G. 2009. Allocation issues in LCA methodology: A case study of corn stover-based fuel ethanol. *The International Journal of Life Cycle Assessment* 14: 529–539.

Luo L, van der Voet E, Huppes G. 2010. Biorefining of lignocellulosic feedstock—Technical, economic and environmental considerations. *Bioresource Technology* 101: 5023–5032.

MAPA. 2012a. *Biodiesel from cattle grease* (in Portuguese). Brazilian Ministry of Agriculture, Cattle Raising and Supply. Available at: http://goo.gl/0rswZn (Accessed June 8, 2013).

MAPA. 2012b. *Sectoral chambers and themes.* Brazilian Ministry of Agriculture, Cattle Raising and Supply. Available at: http://www.agricultura.gov.br/camaras-setoriais-e-tematicas (Accessed July 8, 2013).

MAPA. 2012c. *Plan of agriculture and cattle raising 2011/2012* (in Portuguese). Brazilian Ministry of Agriculture, Cattle Raising and Supply. Available at: http://www.agricultura.gov.br/planoagricola (Accessed August 6, 2013).

May A, Durães FOM, Vasconcellos JH, Parrella RAC, Miranda RA. 2012a. Seminário Temático sobre Sorgo Sacarino. Documentos 137. *Embrapa Milho e Sorgo, Anais do I Seminário Temático sobre Sorgo Sacarino,* September 20–21, 2011, Embrapa Milho e Sorgo, Sete Lagoas, MG, p. 83.

May A, Durães FOM, Pereira Filho IA, Schaffert RE Parrella RAC. 2012b. *Sistema Embrapa de Produção Agroindustrial de Sorgo Sacarino para Bioetanol Sistema BRS1G—Tecnologia Qualidade Embrapa.* Documentos 139. Embrapa Milho e Sorgo, Sete Lagoas, MG, p. 120.

Melamu R, von Blottnitz H. 2011. 2nd Generation biofuels a sure bet? A life cycle assessment of how things could go wrong. *Journal of Cleaner Production* 19: 138–144.

MMA. 2006. *Diagnosis of the production of biodiesel in Brazil* (in Portuguese). Brazilian Ministry of Environment. Available at: http://goo.gl/0oDB6p (Accessed July 8, 2013).

Nogueira LAH. 2011. Does biodiesel make sense? *Energy* 36: 3659–3666.

Nogueira LAH, Capaz RS. 2013. Biofuels in Brazil: Evolution, achievements and perspectives on food security. *Global Food Security* 2: 117–125.

Northoff E. 2008. Reviewing biofuels policies and subsidies. Available at: http://fao.org/newsroom/en/news/2008/1000928/ (Accessed September 8, 2013).

Padula AD, Santos MS, Ferreira L, Borenstein D. 2012. The emergence of the biodiesel industry in Brazil: Current figures and future prospects. *Energy Policy* 44: 395–405.

Sorda G, Banse M, Kenfert C. 2010. An overview of biodiesel policies across the world. *Energy Policy* 38: 6977–6988.

Ubrabio FGV. 2010. *Biodiesel and its contribution to the Brazilian development* (in Portuguese). União Brasileira do Biodiesel and Fundação Getúlio Vargas. Available at: http://www.ubrabio.com.br/sites/1700/1729/00000201.pdf (Accessed August 8, 2013).

CHAPTER 18

Biomass and Biofuels
A Case Study from Nigeria

Muhammad A. Sokoto*

Department of Pure and Applied Chemistry, Usmanu Danfodiyo University, Sokoto, Nigeria

CONTENTS

* Present affiliation: Bio-Fuels Division (BFD), Thermo-Catalytic Processes Area, CSIR-Indian Institute of Petroleum (IIP), Dehradun 248005, India.

18.1 INTRODUCTION

Energy is the mainstay of economical status for many developed nations. The technological operations and industrial processes that are principal components to national development are basically energy dependent. The chains for executing numerous development roles in the duo sectors begin with extraction of primary energy, which is later converted into an energy carrier for various end uses. Thus, energy is vital to all aspects of modern life and indeed the lifeline of industrial production, fuel for transportation, and of electricity generation in thermal power plants (Sambo 2005). In recognition of energy's essentiality, provision of an adequate and affordable supply to the African populace is included in the United Nation's Millennium Developmental Goals (MDGs), in addition to portable water supply, food, health care provisions, shelter, and so on (Ibitoye 2013).

In many developing countries (Nigeria included), several reforms and policies were put in place for the realization of these goals. In Nigeria, for instance, privatization, task force committees, power reform policies, renewable energy policy bills, and a seven-point agenda were introduced, among others. These reforms include the National Electric Power Policy (NEPP) in 2001, Nigeria Integrated Power Project (NIPP) in 2004, the Electric Power Sector Reform Act in 2005, and the Roadmap for Power Sector Reform of August 2010. The Electric Power Sector Reform Act (EPSRA) portrayed how the nation's electrical power supply authority will pave the way to independent power firms for an improved and extended power supply to Nigerians (NPG 2012). However, from its inception until 2010, the available generation capacity was less than the projected output of 4000 MW (NPG 2012). The generation and supply seem to be indifferent prior to 2005 reforms. Also, establishment of the Presidential Task Force on Power (PTFP) in 2010 is thought to be a milestone for stable energy supply in Nigeria. The committee was mandated to implement the reform of Nigeria's power sector and to coordinate the activities of the various agencies connected with power issues as well as monitoring, planning, and executing of various short-term projects in generation, transmission, and distribution of fuel to power. Yet, large percentages of the Nigerian population are experiencing elliptical power supply and total black out in rural areas. The absence of reliable energy supply in Nigeria has left the populace socially backward, deterring effective industrial growth, and leaving much economic potential untapped. Iwayemi (2008) asserted that the persistent energy crisis in Nigeria has weakened industrialization in the country; it has significantly undermined efforts to achieve sustained economic growth and has increased competitiveness of indigenous industries in domestic, regional, and global market as well as in employment opportunities. The present electrical power demand in Nigeria is estimated to be 20,000 MW and the generation stands at about 4500 MW (Nebo 2014).

Many countries across the globe are trying to overcome their energy challenges through diversification of energy resources, especially by increased penetration of renewable resources into their energy mix. Biomass is one of the promising forms of renewable resources that was pervasively harnessed in urban and rural areas. The main focus of this chapter is to highlight the possible contributions of biomass to Nigeria's energy mix toward overcoming its protracted energy crisis. The energy scenario, energy resources, biomass distribution, and availability and modes of utilization for energy generation and its contribution to the energy mix of Nigeria will be discussed.

18.2 ENERGY SCENARIO IN NIGERIA

According to a report by the Energy Information and Administration (EIA 2013), Nigeria has the lowest net per capita electricity generation rate in the world, with electricity generation falling short of demand, resulting in power load shedding, blackouts, and reliance on private generators. Also, World Bank data (2012) showed the electrification rate at about 50% for the country as a whole, leaving approximately 80 million people in Nigeria without access to electricity. Mohammed et al. (2013) reported household accessibility to electrification at about 46% and only 2% for rural households with generation either by rural electrification initiated by the government or self-generation using private generators. As per estimates by EIA (2011), total primary energy consumption in Nigeria was about 4.3 quadrillion British thermal units (Btu). Currently, power generation in Nigeria is less than 4000 MW (Makwe et al. 2012; Ishola et al. 2013). With this scenario, there is a huge gap between the energy demand and its supply, and it will continue to hamper the socioeconomic status of present and future generations. In the transportation sector, fossil fuel is the major source of power for automobile engines. The consumption rate of these fuels is greater than its supply, especially for the refined products, despite the country's large proven reserves and installed refineries. Table 18.1 shows the demand and supply of the refined petroleum product in Nigeria over the years.

At present, the petroleum resource is the major energy base of the country, which also serves the nation as the basic revenue earning. However, the wide gap between the demand and supply chain of these refined petroleum products had to be bridged by import of the refined petroleum products to meet domestic demand. Table 18.1 shows the picture of domestic fuel need and supply over the years. It is an obvious manifestation that some commitments need to be made to improve domestic energy content.

Furthermore, Nigeria has electricity power generation plants of an approximate installed capacity of 10,000 MW, which are mostly the hydro and thermal types. The plants have an estimated available capacity of 6056 MW of which hydrostations account for 10,060 MW (NPG 2012). The plant names and their installed capacities are shown in Table 18.2. At present, there are a number of ongoing power projects in the country with a projected capacity of 5445 MW under the national integrated power projects. Wind farm power, coal-fired, and small and large hydro power plants were envisaged in a strategic plan of the government for enhancing energy security and diversification of energy mix (FMNP 2012). According to the Federal Ministry of Power's annual report in 2011, an installed wind farm power of 10 MW (the first in the country) will be onboard. Although the role of biomass as energy feedstock is recognized, its utilization for power generation at the industrial scale level is still dormant.

Nigeria has a population of 173.62 million and a growth rate of 2.8%, according to Food and Agricultural Organization data (FAOSTAT 2014a). Therefore, with increases in population growth coupled with the present global technological changes and immense concern about the environment, the quest for alternative energy sources could not be ruled out in salvaging the present and future energy needs of the country through exploitation of available resources.

Table 18.1 Nigeria's Daily Demand and Output of Refined Petroleum Products

Refined Product	Demand (1000 b/day)					Output (1000 b/day)				
	2009	2010	2011	2012	2013	2009	2010	2011	2012	2013
Gasoline	168.2	191.5	215.3	241.3	373.9	10.5	23.9	29.7	27.4	30.4
Kerosene	25.3	30.6	43.6	47.7	53.3	7.0	13.7	16.0	12.9	15.8
Distillate	28.2	35.8	43.3	45.1	48.8	10.6	20.6	21.8	18.8	18.4
Residuals	10.9	11.9	8.2	8.6	8.0	12.3	24.1	25.2	18.4	20.7
Other	0.8	0.8	0.9	1.0	1.0	1.1	5.3	7.7	4.8	3.3
Total					485					88.6

Source: OPEC, *Annual statistical bulletin* OPEC, 1–112, 2014. Vienna, Austria. Available at http://www.opec.org

Table 18.2   Power Generation Plants in Nigeria

Plant	Plant Type	Installed Capacity (MW)[a]	Average Availability (MW)[b]
Kinji	Hydro	760	295.38
Jabba	Hydro	578	414.48
Shiroro	Hydro	600	497.46
Afam (I–V)	Thermal	987.2	95.32
Egbin	Thermal	1320	1022.56
Sapele	Thermal	1020	316.78
Ughelli	Thermal	942	N/A
Geregu	Gas	414	274.96

Source: [a]FMNP, Federal Ministry of Power 2011 Annual Report, 2012, Federal Secretariat, pp. 1–137, Abuja; [b]Chukwu, P.U., et al., *Journal of Sustainable Developmental Studies*, 6, 242–259, 2014.

Table 18.3   Nigerian Proven Energy Sources Reserve

Energy Source	Estimated Reserve	Source
Crude oil	37,140,000 (000 bn b)	Xu, and Bell 2013; Jacobs 2014
Natural gas	182 tn cu ft	Xu, and Bell 2013; Jacobs 2014
Tar sands	30 bn b of oil equivalent	Oseni 2012
Coal and lignite	2.7 bn t	NNPC 2013
Hydro power	46,235 MW	NPG 2012
Fuel wood	11 m ha	Naibbi, Al., and Healey, R.G., 2013
Animal waste	61 m t/yr	Oseni 2012
Crop residue	83 m t/yr	Oseni 2012
Solar radiation	3.5–7.5 kWh/m²/day	Oseni 2012
Wind	2–4 m/s at 10 m height	Oseni 2012

18.3 NIGERIA'S ENERGY RESOURCES

Nigeria is rich with versatile and abundant energy resources scattered across the regions of the country. These resources include crude oil, natural gas, bitumen, coal, tar sand, and renewable resources such as wind, hydro, solar, biomass, and so forth. The oil and natural gas resources are primarily located in the Niger Delta region; livestock and cereal crops are mostly produced in the northern part of the country.

Nigeria is the largest oil producer in Africa, the largest holder of natural gas proven reserves in Africa, and the ninth largest holder in the world (EIA 2014; Jacobs 2014). It was also the world's fourth leading exporter of liquefied natural gas (LNG) in 2012 (EIA 2014). Table 18.3 portrays the estimate energy reserves in the country. As of January 2014, Nigerian crude oil proven reserves were estimated at 37,140,000 billion barrels with a daily production of capacity of 2.7 million barrels of crude oil (NNPC 2013). The natural gas proven reserve was 182 trillion cubic feet. Nigeria consumed 270,000 bb/day of petroleum in 2012 and has four refineries (Port Harcourt I and II, Warri, and Kaduna) with a combined crude oil distillation capacity of 445,000 bb/day (OPEC 2014). Despite having a refinery capacity that exceeds domestic demand, the country has to import refined petroleum commodities because the refineries operate below their full capacity.

Other potential energy resources are coal and lignite the quantity of which amounted to 2.7 b t and renewable sources that include hydro, wind, and solar energy. In line with these, it is evident that Nigeria has abundant energy resources that can sustainably accommodate its energy requirements.

18.4 BIOMASS AND BIOFUELS

18.4.1 Biomass

Biomass is a term that describes an organic, nonfossil material of biological origin. It is often called "phytomass," which refers to bioderived resources found in terrestrial and aquatic environments. Organic materials that have been transformed by geological processes into substances such as coal or petroleum are not regarded as biomass materials because they contain carbon that has been "out" of the carbon cycle for a very long time. Their combustion therefore disturbs the carbon dioxide content in the atmosphere.

Biomass has been categorized into primary, secondary, and tertiary (ASABE 2011). Primary biomass is produced directly by photosynthesis and is harvested or collected from the field or forest where it is grown. It comprises grains, perennial grasses, wood crops, crop residues, and residues from logging and forest operations. Secondary biomass consists of residues and by-product streams from food, feed, fiber, wood, and material processing plants (such as sawdust, black liquor, and cheese), and manures from concentrated animal feeding operations. Tertiary biomass sources encompass postconsumer residues and wastes such as fats, greases, oil, wood debris from construction and demolition sites, as well as packaging wastes, municipal solid wastes, and landfill gases.

Biomass is a vital resource for energy generation especially in the developing world. In Nigeria, for instance, the distribution of fuel consumption (Figure 18.1) indicates biomass is the major energy resource for a majority of Nigerians. Traditional biomass and waste (typically consisting of wood, charcoal, manure, and crop residues) accounted for 83% of consumed fuel, usually for heating and cooking needs, mainly in rural areas.

18.4.2 Biofuels

According to Dara (2007), a fuel is any combustible substance that burns in atmospheric air to generate heat that is capable of being economically used for domestic and industrial purposes such as lighting, heating, and generation of power. The term biofuel refers to liquid, gas, or solid fuels predominantly produced from biomass. Varieties of fuels produced from biomass resources include liquid fuels (ethanol, methanol, and biodiesel) and gaseous fuels (such as hydrogen and methane) (Demirbas 2008).

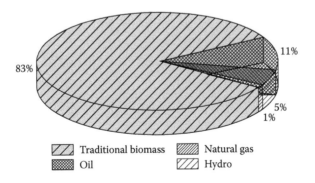

Figure 18.1 Total primary energy consumption in Nigeria in 2011. (From EI A, Independent Statistic Analysis by United State Energy and Information Administration, 2014, http://www.eia.gov./countries/analysisbriefs/Nigeria/nigeria.pdf)

Poonam and Anoop (2011) broadly classified biofuels into primary and secondary biofuels. The primary biofuels are natural and unprocessed forms of biomass mainly used for heating, cooking, or electricity generation in small- and large-scale applications. These fuel types include fuel wood, wood chips and pellets, and so forth. The secondary biofuels are modified forms of primary fuels produced by processing the biomass (e.g., ethanol, biodiesel) and are essentially utilized in vehicles, engines, and lighting materials.

18.5 ASSESSMENT OF BIOMASS RESOURCES IN NIGERIA

Nigeria is endowed with versatile and plentiful biomass resources that are spread across the vegetation zones of the country. Figure 18.2 shows the national biofuel feedstocks in Nigeria. According to Sambo (2009) and Olaoye (2011) in Simonyan and Fasina (2013), the availability of biomass resources in Nigeria follows the nation's vegetation pattern. The rain forest in the south generates the highest quantity of woody biomass while the guinea savannah vegetation of the north central region generates more crop residues than the Sudan and Sahel savannah zones. The biomass resources available in the country include fuel wood, agricultural waste and crop residue, sawdust and wood shavings, animal dung/poultry droppings, sugarcane molasses, and industrial effluents/municipal solid waste. Most of these residues are generated by operations such as thinning, extracting stem wood for pulp and timber, and natural attrition. In sawmills, wood processing generates significant volumes of residues in the form of sawdust, offcuts, bark, and wood-chip rejects. However, crop residue resources are mainly generated through farming activities.

Farming in Nigeria is characterized by peasantship and mechanized form. The peasant farmers are largely dispersed and use either manual tools or animals in their farming activities.

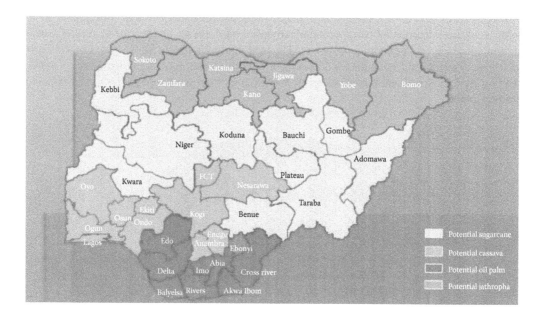

Figure 18.2 Nigeria's biomass feedstock map. (From NNPC, Sustainable biofuels industries in Nigeria—A gate way to economic development. An Address by Group Manger Renewable Industries Division of Nigerian National Petroleum Corporation (NNPC) at the Second Nigerian Alternative Energy Expo, 2013, http://www.slides.net/mathesisslides/sustainable-biofuels-industries-in-nigeria)

They provide most of the major food needs of the country. The mechanized system of farming involves the use of heavy machines in farming operations. Mechanized farming typically takes place on large hectares of land for crop cultivation and livestock production.

Nigeria has a total land area of 924 mha; of this, 91 mha is judged suitable for cultivation (Olanyi 2007). Approximately half of this cultivable land is effectively under permanent arable crop while the rest is covered by forest woodland. As of 1996, a total of 33 mha were cultivated crops, generally out of which 17.7 mha were for staples and 4.9 mha were for industrial crops (Olanyi 2007). The production data for major crops and livestock are shown in Figures 18.3 and 18.4.

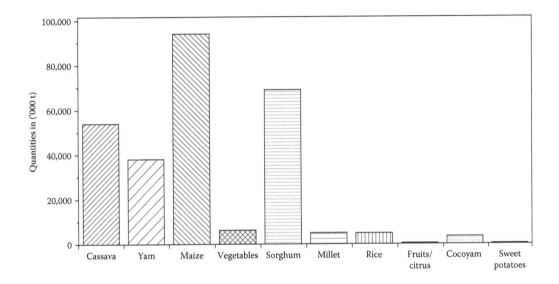

Figure 18.3 Nigeria's top ten commodities production quantities, 2012. (From FAOSTAT. 2014a. *FAO of the UN*. http://faostat.fao.org/site/550/default.aspx#ancor)

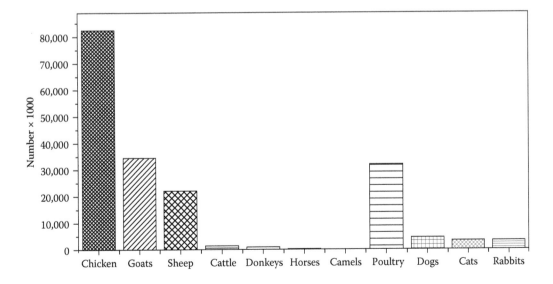

Figure 18.4 Nigeria's livestock and poultry production, 2012. (From FAOSTAT. 2014b. *Food and Agricultural Organization Statistic*. http://Faostat.fao.org/site/567/default.aspx#anchor)

Though the crops are typically meant for food, during harvest and processing, many tons of residues are generated, which were previously burned on the farm to aid in crop production in the next season but currently are harnessed for bioenergy generation. Likewise, the dung, droppings, and feathers from livestock are other feedstocks for biofuels production.

The aforementioned agricultural products and their by-products varied in their quantities, mode of applications, and physicochemical characteristics. The physical and chemical makeup of the feedstocks has an impact on their energy content. According to Dara (2007), the organic components of biomass undergo complete combustion to generate latent heat that is often called heating value. The heat generated can be utilized for heating, cooking, lighting, electricity generation, and so on. The quantum of heat depends on the elemental constituent of the feedstock, especially carbon (Yokomo 2008). Tables 18.4 and 18.5 show the physicochemical properties and elemental composition of some Nigerian agricultural residues.

Moisture content of the biomass feedstock influences its energy content. Substrate with low moisture content generates more latent heat on combustion. The moisture content of the agricultural residue ranges from 3 to 97 wt% (Table 18.3) and heating values from 13.4 to 28.3 MJ/dry kg.

The elemental composition of biomass (Table 18.5) infers that the more the carbon content in the feedstock, the higher the tendency of its generating a high heating value. Pine wood had the highest carbon content with a correspondingly high energy potential. These indicate that feedstocks such as sawdust and other wood derivates are heated energy carriers from which gainful energy can be obtained. Other residues that are in plentiful supply could also contribute megajoules of energy when properly harnessed.

Table 18.4 Physicochemical Properties of Some Agricultural Residues and Coal

Feedstock	Moisture (wt%)	Heating Value (MJ/dry kg)
Cattle manure	20–70	13.4
Sawdust	15–60	20.5
Sorghum	20–70	17.6
Switchgrass	30–70	18.0
Water hyacinth	85–97	16.0
Paper	3–13	17.6
Rice straw	5–15	15.2
Coal	5–10	28.3

Source: Oladeji, J.T., and Oyetunji, O.R., *Journal of Energy Technologies and Policy*, 3, 1–8, 2013; Yokomo, S. 2008. *Japan institute of Energy*. pp. 1–338.

Table 18.5 Elemental Constituents of Selected Agricultural Residues

Element (%)	Cassava Peel[a]	Yam Peel[a]	Cellulose	Pine Wood	Water Hyacinth	Livestock
C	22.08	25.29	44.44	51.8	41.1	35.1
H	13.54	15.19	6.22	6.3	5.29	5.3
O	37.31	15.19	49.34	41.3	28.84	33.2
N	4.4	1.41	–	0.1	1.96	2.5
S	1.84	1.39	–	–	0.34	0.4
HHV	12.765	17.346	17.51	21.24	10.01	13.37

HHV (MJ/dry kg): heating value.
Source: [a]Oladeji, J.T., and Oyetunji, O.R., *Journal of Energy Technologies and Policy*, 3, 1–8, 2013; Yokomo, S. 2008. *Japan institute of Energy*. pp. 1–338.

Table 18.6 Energy Potential of Some Nigerian Agroresidues

Agricultural Crop	Generated Residue	Production Quantity (1000 t)^2	Crop to Residue Ratio	Calculated Generated Residue	Energy Content (MJ/kg)
Maize	Stalk	7303	1.5	10.95	15.48
Rice	Straw	3219	1.5	4829	15.56
Sorghum	Stalk	4784	2.62	12.532	17.00
Wheat	Stalk	34.2	1.5	53.3	19.3
Cocoanut	Shell	170	0.6	102	10.61
Oil palm fruit	Empty fruit bunch	8500	0.25	21.25	15.51
Sugarcane	Bagasse	1414	0.3	424.3	13.88
Cocoa	Husk	428	1.0	428	15.48
Millet	Stalk	4125	3.0	12.375	15.51

Source: FAO, *Statistics of Animal Production, 2009.* FAO, 2010, Rome, Italy.

18.6 ENERGY PROJECTION FROM THE BIOMASS IN NIGERIA

The quantum of energy derivable from biomass residues relies on its availability and magnitude. The magnitude of potential energy per year from biomass components can be quantified from their recoverable quantities and respective heating values. Based on a Nigerian Power Guidelines NPG (2012) estimate, the amount of biomass resources in Nigeria is 144 mt/yr. The energy content of these resources can be estimated from the product of biomass component and its corresponding heat values. Table 18.6 shows some selected agroresidues and their calculated potent energy. The total gainful energy of 697.15 TJ can be generated from the crop residue (32,879,559 t). However, shrubs and forage grasses are other biomass resources that have biofuel potential. An estimate of more than 200 million tons of these resources can be obtained in Nigerian territory annually; an estimate of 2.28×10^6 MJ can be generated from these biomasses (Sambo 2005). Table 18.6 showed an estimated potent energy that can be generated from some livestock excretions. A total gainful energy of 697.15 TJ can be generated from the forestry and crop residues, while about 857,031 MJ/t/day can be generated daily from livestock manure.

18.7 BIOMASS FEEDSTOCKS

18.7.1 Dung for Fuel

Dung (a concentrated solid excrete from livestock) is widely used as fuel either by direct burning or as a feedstock for biogas production. On the basis of livestock production (Figure 18.4), millions of tons of this substrate could be generated for numerous purposes that entail generation of heat. The estimated quantity of livestock dung generated in 1985 was 227,500 tons (Edirin and Nosa 2012) and 61 mt/yr in 2004 (Iloeje 2004). In the context of the current production trend, the population of livestock in Nigeria is continuously increasing, likewise the potential of the manure for energy utilization (Table 18.7). In many countries, animal dung is a valued fuel for cooking and heating. It represents the major fuel for household use by millions of farmers in Asia and Africa and in parts of the Near East and Latin America.

18.7.2 Wood Fuel and Charcoal

Fuel wood is a form of primary biomass used for cooking and heating by low-income earners residing in rural and urban settlements. In developing countries, the greatest number of

Table 18.7   Energy Potential of Some Nigerian Livestock and Poultry Manure

Species	Number (10³)	Average Animal Weight (kg)	Generated Manure/ Dead/Day (kg)	Quantity of Manure (T) per Day	Heating Value (MJ/Dry kg)[a]	Energy Potential (MJ/T/day)	Source
Chickens	82,400		0.1	8240	13.53	111,487.2	(Collins et al. 1999; Williams, et al. 1999)
Sheep	22,100	45	1.0	22,100	17.82	293,822	(Kaith and Bhardwaj 2009)
Cattle	1390	272	14	19,460	15.73	306,105.8	(Kaith and Bhardwaj 2009)
Horses	200	634	25	5000			(Kaith and Bhardwaj 2009)
Other poultry[b] (turkey)	31,900		0.255	8135	13.49	109,741.15	(Collins et al. 1999; Williams et al. 1999)
Guinea pig	500	90	25	12,500	2.87	35,875	(Kaith and Bhardwaj 2009)

Source: Bourn, D., et al., *Nigerian Livestock Survey. A Report submitted to Federal Government of Nigeria by Resource Inventory Limited,* 1992, http://fao.org/livestock/agap/Frg/FEDback/war/t1300b/t1300bog. htm#potential%20application%20elsewhere

[a] Echiegu, E.A., et al., *Scientific and Engineering Research,* 4, 999–1004, 2013.
[b] Includes pigeons, ducks, guinea fowl, and turkeys.

the population relied on fuel wood and other forest products as their primary energy source (see Table 18.8). Available data on wood fuel consumption in Nigeria are based largely on estimates because the greater part of wood fuel production and usage occurs outside commercial channels and thus goes unrecorded. Over the years from 1980 to 1999, the consumption of fuel wood in Nigeria was estimated at more than 1.4 b t, with corresponding energy content of 143,488 PJ (Figure 18.5). However, the total charcoal consumption over the period 1989–1996 was 21,477 (000 mt). The figures have succinctly showed that biomass has played a pivotal role in meeting Nigerian energy needs.

The total area projected for fuel wood production was 17,279.5 ha, transmission poles 21.384, match splint 0.4585, pine wood 10.155, and so forth; yet, the total production revealed there is an undisturbed forest area of 121,14 km² or about 1.3% of the total country land area. This infers that there is a lot of biomass that can be exploited to augment the energy needs of Nigerians.

18.7.3 Domestic and Industrial Waste

Agricultural, human, and industrial activities are generally perceived to generate waste materials in routine operations. Agricultural residues are generated in the course of crop growth and harvest. Similarly, livestock farming produces a large volume of by-products in terms of urine and dung. Human activities also discharge waste components in the form of garbage, sewage, and pebbles from construction sites in addition to other gaseous waste components. The combination of these wastes creates nuisance in the environment if not properly managed. Blockage of drains and roads, and foul odors are prominent issues generated by heaps of waste from dump sites in most of the major cities in Nigeria. Daily waste generation per person in Nigeria is estimated at 0.58 kg/person/day with little or no consideration for the rural dwellers that constitute more

Table 18.8 Wood Fuel Consumption in Nigeria and Energy Content

Year	Consumption (000 × m³)	Energy Content (PJ)	Charcoal Consumption (000 × mt)
1980	60,310	663	993
1981	62,030	682	1022
1982	63,810	701	1061
1983	67,660	722	1082
1984	67,580	743	1113
1985	69,590	764	1146
1986	71,590	787	1179
1987	73,700	810	1214
1988	75,880	831	1250
1989	78,150	859	1288
1990	80,510	885	1326
1991	82,970	912	1367
1992	85,570	939	1401
1993	88,140	968	1440
1994	90,820	997	1485
1995	93,540	1028	1540
1996	96,310	1057	1570

Source: FAOSTAT. 2013. http://www.fao.org/docrep/x2740e/x2740e31.pdf

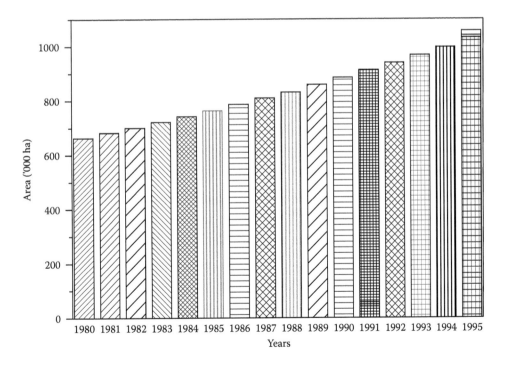

Figure 18.5 Projected plantation area of fuel wood in Nigeria. (From FAOSTAT. 2013. http://www.fao.org/docrep/x2740e/x2740e31.pdf)

than half of the Nigerian population (Ajao and Anurigwe 2002; Abel and Afolabi 2007; Aziegbe 2007). In Nigeria, most of the generated wastes are either dumped in a landfill or burned in the open atmosphere. With time, gaseous products emanate from landfills and open combustion of solid waste build up in the atmosphere because most landfills in Nigeria were not developed to capture the gas (Afolayan et al. 2012). Effective waste management therefore becomes an urgent matter in view of the fact that waste generation is inevitable, and it induces degrading effects into the environment.

In some developed nations, high quantity of waste generation symbolizes a greater opportunity for electricity generation in the form of alternative energy from biowaste resources. Organic fractions of municipal solid waste are anaerobically digested for biogas generation, while inorganic components are pyrolized to bio-oil and char. Thus, establishment of a biofuel plant that utilizes these waste components could be a gateway for creating job opportunities and for safeguarding the environment as well as producing a generation of products that are energy carriers.

18.7.4 Biogas

Biogas refers to a gaseous product produced as a result of biodegradation of organic material under anaerobic conditions by microorganisms. Biomass streams meant for biogas production are usually subjected to anaerobic digestion in which the larger molecules are broken down to monomers by fermentative bacteria via hydrolysis. The generated monomers are converted to organic acids, carbon dioxide, and hydrogen by acetogenic bacteria; finally, methane is generated by the action of methenogens bacteria. Biogas is a mixture of gases—essentially methane (50%–75%), carbon dioxide (25%–45%), 2%–8% water vapor and traces of hydrogen sulfide, hydrogen, nitrogen, ammonia, and oxygen (Fachagentur 2009). The composition and yield of biogas depends upon feed material, the plant design, fermentation temperature, retention time, and so forth. Biogas can be used in similar ways to natural gas in gas stoves, lamps, or as fuel for engines. In comparison with natural gas, which contains 80% to 90% methane, the energy content of the gas depends mainly on its methane content. The average methane content per meter cubic volatile solids of different substrates is shown in Table 18.9. The average calorific value of biogas is about 21–23.5 MJ/m^3. Thus, 1 m^3 of biogas corresponds to 0.5–0.6 l diesel fuel or about 6 kWh (Fachagentur 2009).

In the Nigerian context, biogas technology has been domesticated and a number of pilot plants have been built. Most of the installed systems or plants are in government institutions for pilot scheme and are mainly for thermal applications, such as heating and cooking purposes.

Table 18.9 Estimated Quantity of Methane from Some Biomass Substrates

Substrate	Approximate CH$_4$ m^3/VS kg
Food waste	400–600
Fruit and vegetable waste	200–500
Manure from cattle, pigs, chickens	100–300
Slaughterhouse waste	700
Cereals	300–400
Silage grasses	350–390
Straw	100–320
Municipal sludge	160–350
Distillation waste	300–400

Source: Schnurer, A., and Jarvis, A., *Microbiology Handbook for Biogas Plants*. Swedish Waste Management U2009:03, Swedish gas report. pp. 1–134, 2010.

Conceptualization of biogas by the wider population is very minimal and basically limited to thermal usage. The application of biogas has gone beyond the thermal usage; rather it is currently devised as a source of electric power and vehicular fuels. The use of biogas for energy generation is well-developed in many countries. In Germany, for instance, biogas plants utilize maize silage and cow manure as feedstocks to yield electrical power of 2 kWel (48 kWhel per day).

Therefore, the use of modern biogas technologies in Nigeria will make a biogas plant a viable source of energy beyond heating and lighting to electric power, especially in rural communities where electricity demand is minimal.

18.7.5 Biodiesel

Biodiesel is a fuel composed of mono-alkyl esters of long chain fatty acids derived from lipid feedstock, such as vegetables, animal fats, and waste oils. It is also a nonpetroleum-based fuel consisting of alkyl esters derived from either the transesterification of triglycerides (TGs) or esterification of free fatty acids (FFAs) with low molecular weight alcohols. Biodiesel is a liquid form of fuel that is miscible with conventional transportation fuel in all proportions. Currently, a blend of biodiesel (B20) is used in automobile ignition engines without modification. Its usage can reduce the magnitude of greenhouse emissions from automobile exhaust. There are vast numbers of oil seeds crops which contain oil in appreciable quantities in their kernel. Jatropha and castor seed oil are the prominent feedstock used for biodiesel production in the country. However, many nonconventional seeds exist that possess substantial extractible oil in their seed kernels. Ajayi (2010) reported the percentage crude lipid content of some nonconventional seeds that include *Adansonia digitata* (baoba) 33%; *Calophyllum inophyllum* linn 49.1%; *Pentaclethra macrophylla Benth* (African bean) 40.29%; *Telfairia occidentalis Hook F.* (fluted pumpkin) 53.63%, *Terminalia catappa* linn 55.11%, and so forth. Also, Hassan and Sani (2010) reported a yield of 39.22% from bottle gourd and 36.7% from *Lagenaria vulgaris* seeds (Sokoto et al. 2013). Proper utilization of these resources can curb the protracted energy crisis in Nigeria and boost the socioeconomic status of its teeming population (Table 18.10).

From the available literature, there is neither established commercial plantation nor an industrial plant for biodiesel production in Nigeria. The biodiesel production is either a projection or pilot scheme. The palm kernel plantation that exists currently in the state of Delta is meant for production of cooking oil. Nonetheless, varieties of nonedible oil crops are evident in the Nigerian domain.

Table 18.10   Typical Yield of Raw Material for Bioethanol and Biodiesel Production in Nigeria

Feedstock	Yield (ton/ha/yr)	Bioethanol/Biodiesel Yield (liters/ton)	Bioethanol/Biodiesel Yield (L/ha/ha/yr)
Sugarcane	50–90	70–90	3500–8000
Sweet sorghum	45–80	60–80	1750–5300
Cassava	10–65	170	1700–11,050
Oil palm	16–35	136	4760
Jatropha	6–10	151.36	1513.6
Castor beans	0.75–1.5	753.6	1130.4

Source: NNPC, Sustainable biofuels industries in Nigeria—A gate way to economic development. An Address by Group Manger Renewable Industries Division of Nigerian National Petroleum Corporation (NNPC) at the Second Nigerian Alternative Energy Expo, 2013, http://www.slides.net/mathesisslides/sustainable-biofuels-industries-in-nigeria

Table 18.11 Ethanol Plants in Nigeria

Company Name	Location Site	Feedstock	Capacity (mL/Yr)
Alconi/Nasak	Lagos	Crude ethanol (imported)	43.8
UNIKEM	Lagos	Crude ethanol (imported)	65.7
Intercontinental distilleries	Ota Idiroko	Crude ethanol (imported)	9.1
Dura clean	Bacita	Molasses/cassava	4.4
Allied Atlantic distilleries	Sango Ota	Cassava	10.9

Source: Edirin, B.A., and Nosa, A.O., *Research Journal in Engineering and Applied Science*, 1, 149–155, 2012.

18.7.6 Bioethanol

Bioethanol is a form of liquid fuel produced from abundant and renewable biomass sources such as agricultural residues, forest residues, and energy crops (Gnansounoun and Dauriat 2010; Wang et al. 2011). Bioethanol is one of the promising biofuels that is gaining world market acceptance especially in transportation sector. An estimated 88.7 million m^3 of fuel ethanol was produced worldwide in 2013 (Luo et al. 2014). The United States and Brazil are the two largest producers of bioethanol, representing 56.8% and 26.8% of the total world's ethanol production in 2013, respectively (USDA 2013). In Africa, bioethanol is not a recent development in a number of countries, such as Zimbabwe, Malawi, and Kenya; they started blending programs in the 1980s (Agboola and Agboola 2011). In Nigeria, bioethanol production is still in the infancy stage and requires a strong commitment from major key players for a smooth acceptance. The use of bioethanol blends in Nigerian transportation fuel (E10) is an effort to minimize carbon emissions, energy insecurity, and to augment a shortfall of energy. The major feedstocks considered as input for Nigerian bioethanol production are cassava, sweet sorghum, and sugarcane bagasse. This may be supported by the fact that Nigeria produces a significant quantity of cassava and has a land mass suitable for crop cultivation. A projected output of 1.27 bn (billion) L/year is the target from bioethanol plants. According to Ohimain (2010), an investment of more than $3.86 billion has been committed to the construction of 19 ethanol biorefineries, 10,000 units of mini-refineries, and feedstock plantations for the production of more than 2.66 billion liters of fuel grade ethanol per annum. Currently, the installed bioethanol plants in the country are Duraclean (located at Bacita), with an output of 4.4 mL/yr, and the AADL plant (at Sango Ota) with an installed capacity of 10.9 mt/yr. At present, neither of the plants is operational due to policy framework and industrial motives which could be due to lack of sustainable bio-fuel policy which regulates sales, use and production (Agboola et al. 2011). Table 18.11 shows the bioethanol plants in Nigeria. As of this writing, the annual demand for ethanol in Nigeria is estimated at 160 mL/yr, and the requisite quantity is met by importation (Nang'ayo et al. 2005).

18.8 BIOMASS PROSPECTS

Critical assessment of Nigeria's energy resources base indicates that the country has a variety and significant number of potent feedstocks that can boost its energy requirement for domestic and industrial purposes. Overdependence on petroleum products as the ultimate energy chain has kept the nation from the benefit of tapping other resources. However, it is evident that, as long as Nigeria hopes to be among the top developed nations by the year 2020 and to compete favorably in the global scene, an integrated energy matrix cannot be compromised. If Nigeria wants to sustain its prominence in the global petroleum market for many years to come, diversification of energy mix through exploration of alternative resources and the use of modern generation techniques

Table 18.12   Prospects for Biofuel Production in Nigeria

Feedstock	Estimated Quantity	Prospective Usage
Population in habitant (millions)	172.294	Strong manpower
Land mass	924,000 sq km	Abundant arable land
Cultivable land Arable crops Forest	91 Mha 33 Mha 58 Mha	Vehicular fuel, lighting, heating, cooking, generator fuel, hydrocarbon source, chemical derivation, biofertilizer, electricity generation
Agricultural residues	2567.1 (1000 t)^2	Vehicular fuel, lighting, heating, cooking, generator fuel, hydrocarbon source, chemical derivation, biofertilizer, electricity generation
Nonedible oil crops	N/A	Vehicular oil blend, biochemical
Manure	27.53 Mt/yr	Biogas, solid fuel

N/A: data not available.

are inevitable. Approaching the adoption of renewable energy sources such as biomass can be an immense opportunity for the country. This approach is supported by the fact that biomass is not exhaustible such as fossil fuel but is ecofriendly (unlike conventional fuel, which is the key precipitator of global climate change), is available in all parts of the country, and the technology for its use is uncomplicated, developed, and is within the capacities of Nigeria as a country (Table 18.12).

It is an obvious fact that Nigeria has strong manpower that can produce substantial energy output from the endowed resources; it also has high potential for biofuel production due to the level of water and arable land available to support both stable food and industrial feedstock.

18.9   OPPORTUNITIES

The production and use of biomass as feedstocks for energy derivation purposes—such as power generation, heating, lighting, and vehicular use—could be the avenues for improving energy security, ensuring economic development, and safeguarding the environment.

18.9.1   Energy Security

Integration of biomass into Nigeria's energy stream will increase the use of domestic fuel and reduce overdependence on fossil fuels and the volume of refined petroleum fuel that has to be imported into the country due to a demand that is beyond the present operational capacities of the refineries. Effective utilization of biomass may salvage a large expense incurred as a result of fuel importation as well as decreasing the vulnerability of petroleum pipes from vandalism.

18.9.2   Economic Development

Biomass power is reliable and cost-effective in the sense that feedstock can be stored and used for power generation at will, unlike other renewable sources (solar, wind). Biomass use for large-scale applications may provide stability in the fuel market because most of the inputs can be obtained at no cost or at low cost. Waste streams such as sewage, domestic organic waste, and municipal waste can all be used for generation of different fuel types.

Bioenergy could increase socioeconomic development because its chain supply has the potential for creating jobs, generating income, and for producing tax revenues associated with growing and harvesting of the resources. Rural empowerment could be boosted through increasing biomass penetration because the primary feedstocks are agriculture and forestry residues, which are mostly produced by rural communities.

18.9.3 Environmental Benefit

Nowadays there is great concern about environment and climatic change of which combustion of fossil fuel is considered as the major contributor or source. The use of bioenergy may reduce the toxic pollutants in the environment, increasing air quality and thereby improving public health. Biomass is perceived to be environmentally benign because it contributes nearly contributing zero net greenhouse gas emissions. A reduction in emissions could prevent many types of environmental degradation.

18.10 CHALLENGES HAMPERING EFFECTIVE APPLICATION OF BIOFUELS IN NIGERIA

In spite of the abundant biomass resources and recognition of their impact on the energy mix of the country, applications in an industrial capacity seem to be moving at a very slow rate. The following factors could be the reason.

18.10.1 Lack of Awareness of the Energy Potential and Efficiency of the Biomass

Many Nigerians are ignorant of the clean nature of renewable energy fuels and the consequences of conventional fuel to the environment. This low level of awareness deterred the demand for renewable energy commodities in the Nigerian energy market as well as the demand for inclusion of biofuels in the country energy plan. However, the low level of public awareness of biofuels' potential creates a huge distortion in the market and makes potential investors skeptical about investing in renewable energy projects.

18.10.2 Research and Development Project Funding

It is globally acknowledged that dependence on research and development activities are crucial to the sustainable socioeconomic advancement of any nation. Energy is a vital sector that can boost the nation's economy. Therefore, it is imperative that research, development, and training be given adequate attention with regard to key issues such as energy resource development and utilization. Nigeria has many centers and institutes dedicated to energy research. Most of these centers and institutes are hampered by obsolete equipment and inadequate research facilities. At present, none of the energy research centers have an existing biomass pyrolysis unit, syngas conversion plant, or standard biogas digester. Modern cutting edge research is rarely accessible in these centers of excellence. Inadequate research funds grossly affect productivity at these institutes and centers.

18.10.3 Capacity and Lack of Standard Quality Control

Applications of biomass for electricity is a complex process involving use of machines (Biomass → combustion units → processing unit of combusted gas → turbines → generator → power). Effective handling of the equipment for the generation process requires some technical expertises which seem to be in adequate in this sector. Infrastructures (components) for conversion of renewable resources to a variety of energy carriers such as electricity are scarce; this makes the country rely totally on imports of components for the maintenance of renewable energy resources used for electricity. However, lack of technical know-how regarding fabrication of simple equipment for conversion of biomass to other energy end products in most local setting in the country could be a barrier for effective utilization of biomass resources. For instance, information about the fabrication of a biogas digester, for conversion of domestic organic waste to biogas, is unknown by many rural dwellers.

Another major barrier to the development of alternative sources of energy in Nigeria is the lack of established standards and quality control systems for local and imported manufactured technologies. Quality assurance builds consumer confidence in the new and growing market of renewable energy.

18.10.4 Financial and Fiscal Incentives

Exploration, production, and conversion activities in the energy sector are characterized by huge capital demands and by advanced technology. Capital formation capabilities in the country's private sector and the level of domestic technological development are still low in relation to what is needed by the energy sector. Consequently, financial incentives from the government to encourage the establishment of bioenergy plants are inadequate or absent. Policies that support the financial incentives that could fast tract the growth of Nigerian bioenergy plants are not in place.

18.10.5 Lack of Awareness of the Importance of Energy Efficiency and Conservation

Energy utilization in Nigeria is not very efficient; this inefficiency in energy utilization could result in increases in environmental problems and in the cost of product production as well as cause wastage of the resources. In the household sector, for instance, there is considerable energy loss due to inefficient traditional three-stone stoves, which are used for cooking mainly in the rural areas. Therefore, it is imperative to promote energy conservation and efficient energy utilization in all sectors of the Nigerian economy.

18.11 STRATEGIES FOR INCREASING BIOMASS PENETRATION INTO NIGERIA'S ENERGY MIX

For an efficient and economical use of biomass beyond current applications, the adoption of some strategies is paramount. Conversely, initiation and promotion of application and market-driven biofuel research cannot be disputed from reshaping or repositioning the existing energy research institutes and centers in the country. The following are some of the adoptable strategies that could increase biomass penetration into Nigeria's energy mix.

- Increasing funding for research institutes, centers, and tertiary institutions that are undertaking research and development efforts on renewable energy resources to ensure productive research and development and the establishment of appropriate infrastructure as well as strengthening support of the shift toward increased renewable energy utilization
- Developing and promoting local capabilities in the nation's energy centers and research institutes for the design and fabrication of efficient energy devices and technologies for the utilization of renewable energy resources
- Promoting the demonstration and dissemination of renewable energy devices and technologies for their adoption and market penetration
- Encouraging result-oriented research and development in the energy sector
- Provisioning of fiscal and nonfiscal incentives for private investors that are willing to invest in the nonconventional energy sector
- Strengthening the public's knowledge of clean combustion and the efficiency of biomass for gainful acceptance

18.12 CONCLUSIONS

In recognition of energy's significance to the socioeconomic and industrial development of a nation, many reforms and policies were put in place by past and present administrations in Nigeria to improve the energy supply in that country. Although the target goal is yet to be achieved because the present energy demand outweighs its supply, this chapter showcases Nigeria's abundant energy resources. Integration of these resources into Nigeria's energy mix can improve and provide clean energy to the country's teeming population. Utilization of most of these resources, especially biomass, in large-scale applications seems to be moving at a slow rate or seems to be at a standstill. This chapter reviewed the energy scenario in Nigeria and examined the potential biomass feedstocks and the possible contributions toward solving Nigeria's energy challenge. Some barriers to advanced applications of the resources beyond the conventional were identified and strategies to be adopted were proffered. Energy projected from the improved usage of the biomass will bring an immense boost to Nigeria's socioeconomic development and open gateways for employment opportunities. It is evident that if Nigeria's dream of being among the top developed nations by the year 2020 and of competing favorably in the global scene needs to be a realized, an integrated energy matrix cannot be compromised upon. In order to be energy independent, Nigeria's biomass-based resources and other alternative energy resources (solar, wind, and hydro) have to be introduced and promoted.

ACKNOWLEDGMENTS

The author appreciated the support received from the Centre for Science and Technology of the Non-Aligned and Other Developing Countries (NAM S&T), India, for a fellowship award under RTF-DCS scheme 2012–2013, and wishes to thank the immense contributions of Dr. Thallada Bhaskar, head of the Thermocatalytic Processes Area, Bio-Fuels Division (BFD), CSIR-Indian Institute of Petroleum (IIP), Dehradun, India, for editing the manuscript and for his major inputs toward the successful composition of this chapter. Access to facilities provided by the Indian Institute of Petroleum (CSIR-IIP), Dehradun, is highly appreciated.

REFERENCES

Abel, O.A. and Afolabi, O. 2007. Estimating the quality of solid waste generation in Oyo, Nigeria. *Waste Management and Research* 25(4): 371–379.

Afolayan, S.O., Ogedengbe, K., Saleh, A., Idris, B.A., Yaduma, J.J., Shuaibu, S.M., and Muazu, Y.G. 2012. Emission stabilization using empirical evaluation of some agricultural residues as potential alternative energy and soil amendment sources. *Advanced Scientific and Technical Research* 1(2): 376–390.

Agboola, O.P. and Agboola, O.M. 2011. Nigeria's bio-ethanol: Need for capacity building strategies to prevent food crises. *Proceedings of the Word Congress on Renewable Energy*, Electronic Conference Proceedings, Linköping University Electronic Press, Linköping, Sweden, 8–13 May 2011.

Agboola, O.P., Agboola, O.M., and Egelioglu, F. 2011. Bioethanol derivation from energy crops in Nigeria: A path to food security or bio-fuel advancement. *Proceedings of the Word Congress on Engineering*, Vol. III. London, UK, July 6–8 2011.

Ajao, E.A. and Anurigwo, S. 2002. Land based sources of pollution in the Niger Delta, Nigeria. *AMBIO Journal of the Human Environment* 31(5): 371–379.

Ajayi, I.A. 2010. Physicochemical attributes of oils from seeds of different plants in Nigeria. *Bulletin of Chemical Society Ethiopia* 24(1): 145–149.

ASABE. (2011). *Terminology and Definition for Biomass Production, Harvesting and Collection, Storage, Processing, Conversion and Utilization.* ANSI/ASABE S593.1. ASABE, MI, USA, pp. 821–824.

Aziegbe, F.I. 2007. Seasonality and environmental impact status of polyethylene (cellophane) generation and disposal in Benin City, Nigeria. *Journal of Human Ecology* 22(2): 141–147.

Bourn, D., Wint, W., Blench, R., and Woollay, E. 1992. *Nigerian Livestock Survey*. A Report submitted to Federal Government of Nigeria by Resource Inventory Limited. http://fao.org/livestock/agap/Frg/FEEDback/war/t1300b/t1300bog.htm#potential%20application%20elsewhere

Chukwu, P.U., Ibrahim, I.U., Ojusu, J.O., and Iortyer, H.A. 2014. Sustainable energy future for Nigeria role of engineers. *Journal of Sustainable Developmental Studies* 6(2): 242–259.

Collins, E.R., Barker, J.C., Carr, L.E., Brodie, H.L., and Martin, J.H. (1999). *Poultry Waste Management Handbook*; Tables 1-1, 1-2, 1-5, 1-6 and 1-9, and Figure 2-1. NRAES-132. ISBN 0-935817-42-5. Ithaca, NY: Natural Resource, Agriculture, and Engineering Service (NRAES).

Dara, S.S. 2007. *A Text Book of Engineering Chemistry*. 10th revised ed. Ram Nagar, New Delhi: S. Chand and Company Ltd, pp. 73.

Demirbas, A. 2008. Biofuels sources, biofuel policy, biofuel economy and global biofuel projection. *Journal of Energy Conversion and Management* 49: 2106–2116.

Echiegu, E.A., Nwoke, O.A., Ugwuishiwu, B.O., and Opara, I.N. 2013. Calorifc value of manure from some Nigerian livestock and poultry as affected by age. *Scientific and Engineering Research* 4(11): 999–1004.

Edirin, B.A. and Nosa, A.O. 2012. A comprehensive review of biomass resources and biofuel production potential in Nigeria. *Research Journal in Engineering and Applied Science* 1(3): 149–155.

EIA. 2011. Total primary energy consumption in Nigeria. http://www.eia.gov./countries/analysisbriefs/Nigeria/images_consumptionpng

EIA. 2013. US Information and Administration. http://www.eia.gov./countries/analysisbriefs/Nigeria/nigeria.pdf

EIA. 2014. Independent Statistic Analysis by United State Energy and Information Administration. http://www.eia.gov./countries/analysisbriefs/Nigeria/nigeria.pdf

Fachagentur, N.R. 2009. Comprehensive Overview of the Biogas Situation in Germany. http://www.fnr-server.de/ftp/pdf/literatur/pdf_185-basisdaten_biogas_2009.pdf

FAO. 2010. *Statistics of Animal Production, 2009*. Rome, Italy. FAO. http://faostat.fao.org/

FAOSTAT. 2013. http://www.fao.org/docrep/x2740e/x2740e31.pdf

FAOSTAT. 2014a. FAO of the UN. http://faostat.fao.org/site/550/default.aspx#ancor

FAOSTAT. 2014b. Food and Agricultural Organization Statistic. http://Faostat.fao.org/site/567/default.aspx#anchor

FMNP. 2012. *Federal Ministry of Power 2011 Annual Report*. Federal Ministry of Power (FMN), Federal Secretariat, Abuja, pp. 1–137.

Gnansounou, E. and Dauriat, A. 2010. Techno-economic analysis of lignocellulosic ethanol a review. *Bioresource Technology* 101(13): 4980–4991.

Hassan, L.G. and Sani, N.A. 2010. Preliminary study on biofuels properties of bottle gourd seed oil. *Nigerian Journal of Renewable Energy* 11: 20–25.

Ibitoye, F.I. (2013). The millennium development goals and household energy requirement in Nigeria. *Springer Plus* 2: 529.

Iloeje, O.C. 2004. Overview of renewable energy in Nigeria, opportunities for rural development and development of renewable energy master plan. *Paper Presented at the Renewable Energy Conference "Energetic Solutions,"* Abuja, 21–26 November 2004.

Ishola, M.M., Brandberg, T., Sanni, A.S., and Taherzedeh, J.M. 2013. Biofuel in Nigeria: A critical strategic evaluation. *Journal of Renewable Energy* 55: 554–560.

Iwayemi, A. 2008. Nigeria's dual energy problems: Policy issues and challenges. *International Association of Energy Economic* 53: 17–21.

Jacobs, H. 2014. The 17 countries sitting on the most valuable energy reserve. http://www.businessinsider.in/the-17-countries-sitting-on-the-most-valuable-energy-reserves/articleshow/30355956.cms

Kaith, N.S. and Bhardwaj, J.C. 2009. Organic Manure: A Present Day Need. A Publication of GB Pant Institute of Himalayan Environment and Development. Kosi-Katarmal, Almora, ENVIS Bulletin, *Himalayan Ecology* 17: 4–8.

Luo, Z., Wang, L., and Shahbazi, A. 2014. Optimization of ethanol production from sweet sorghum (sorghum bicolor) juice using response surface methodology. *Biomass and Bioenergy* 67: 53–59.

Makwe, J.N., Akinwale, Y.O., and Atoyebi, M.K. 2012. An economic assessment of the reform of Nigerian Electricity Market. *Energy Power* 2(3): 24–32.

Mohammed, Y.S., Mustafa, M.W., Bashir, N., and Mokhtar, A.S. 2013. Renewable energy resources for distributed power generation in Nigeria: A review of the potential. *Journal of Renewable and Sustainable Energy Reviews* 22: 257–268.

Naibbi, A.I. and Healey, R.G. 2013. Northern Nigeria's Dependence on Fuelwood: Insights from Nationwide Cooking Fuel Distribution Data. *International Journal of Humanities and Social Science* 3(17): 160–173.

Nang'ayo, F., Omanya, G., Bokanga, M., Odera, M., Muchiri, N., Ali, Z., and Werehire, P. 2005. A strategy for industrialisation of cassava in Africa. *Proceedings of the Small Group Meeting*, Ibadan, Nigeria. Nairobi, Kenya: African Agricultural Technology Foundation, 14–18 November 2005.

Nebo, C. 2014. Nigeria needs 200,000MW to meet electricity demand. An interview with Nigeria's Minister of Power. *Daily Independent News Paper*, 8 August 2014, 1–24.

NNPC. 2013. Sustainable Biofuels Industries in Nigeria—A gate way to economic development. An Address by Group Manger Renewable Industries Division of Nigerian National Petroleum Corporation (NNPC) at the Second Nigerian Alternative Energy Expo. http://www.slides.net/mathesisslides/sustainable-biofuels-industries-in-nigeria

NPG. 2012. *Nigeria Power Guide*. Volume I. http://clients.squareeye.net/uploads/ipfa/Nigeria PowerGuide2012EDITION.pdf

Ohimain, E.I. 2010. Emerging bio-ethanol projects in Nigeria: Opportunities and challenges. *Energy Policy* 38(11): 7161–7168.

Oladeji, J.T. and Oyetunji, O.R. 2013. Investigation of physical and fuel characteristic of briquettes produced from cassava and yam peels. *Journal of Energy Technologies and Policy* 3(7): 1–8.

Olanyi, A. 2007. Biofuels opportunities and development of renewable energies markets in Africa: A case study of Nigeria. *A paper presented at Biofuel Market Africa conference held in Cape Town South Africa*, 5–7 November 2007. Available at http://www.cejab.org

Olaoye, J.O. 2011. *An analysis of the environmental impacts of energy crops in Nigeria towards environmental sustainability.* In: Ogunlela, A.O. (ed). Tillage for Agricultural Productivity and Environmental Sustainability. Proceedings of Nigerian Branch of International Soil Tillage Research Organization, Ilorin, Nigeria, February 21–23, 2011, pp. 204–212.

OPEC. 2014. *Annual Statistical Bulletin*. OPEC, pp. 1–112. Vienna, Austria. Available at http://www.opec.org

Oseni, M.O. 2012. Improving households' access to electricity and energy consumption pattern in Nigeria: Renewable energy alternative. *Renewable and Sustainable Energy Reviews* 16: 3967–3974.

Poonam, S.N. and Anoop, S. 2011. Review on production liquid fuels from renewable source. *Progress and Combustion Science* 37: 52–68.

Sambo, A.S. 2005. Renewable energy for rural development: The Nigerian perspective. *ISECO Science and Technology* 1: 12–22.

Sambo, A.S. 2009. Strategic development in renewable energy in Nigeria. *Proceedings of the International Association for Energy Economics*, IAEE International Conference, San Francisco, 21–24 June, pp. 15–19.

Schnurer, A. and Jarvis, A. 2010. *Microbiology Handbook for biogas Plants*. Swedish Waste Management U2009:03, Swedish gas report. Svenskt Gastekniskt Center AB, Malmö, pp. 1–134.

Simonyan, K.J. and Fasina, O. 2013. Biomass resources and bioenergy potentials in Nigeria. *African Journal of Agricultural Research* 8(40): 4975–4989.

Sokoto, M.A., Hassan, L.G., Salleh, M.A., Dangoggo, S.M., and Ahmad, H.G. 2013. Quality assessment and optimization of biodiesel from *Lagenaria Vulgaris* (calabash) seeds oil. *International Journal of Pure and Applied Sciences and Technology* 15(1): 55–66.

USDA. 2013. *United States Department of Agriculture e Foreign Agricultural Service. World Fuel Ethanol Production*. Renewable Fuels Association, Washington, DC. http://ethanolrfa.org/pages/World-Fuel-Ethanol-Production

Wang, L.J., Luo, Z.L., Xiu, S.N., and Shahbazi, A. 2011. Pretreatment and fractionation of wheat straw with acetic acid to enhance enzymatic hydrolysis and ethanol fermentation. *Energy Source* 33(13): 1230–1238.

Williams, C.M., Barker, J.C., and Sims, J.T. 1999. Management and utilization of poultry wastes. *Reviews of Environmental Contamination and Toxicology* 162: 105–157.

World Bank. 2012. World Development Indicator. http://databank.worldbank.org/ddp/home.do

Xu, C. and Bell, L. 2013. Worldwide reserves oil production post modest rise. *Oil and Gas Journal.* http://www.ogj.com/articles/print/volume-111/issue-12/special-report-worldwide-report/worldwide-reserves-oil-production-post-modest-rise.html

Yokomo, S. 2008. Asian Biomass Handbook. A publication of Ministry of Agriculture, Forestry and Fisheries. *Japan institute of Energy*. pp. 1–338.

Index

9 781138 894150